Principal Symbols (con't.)

$\mathcal{R}$ - reluctance

$\boldsymbol{S}$ - complex power

S - apparent power

S_T - total three-phase apparent power

s - slip

SF - stacking factor

T_d - electromagnetic developed torque

T_s - shaft torque

T_{FW} - friction and windage torque

$\bar{V}$ - rms phasor of $v(t)$

V - magnitude of $\bar{V}$; volume

V_a - armature terminal voltage

V_L - line voltage; load voltage

$\bar{V}_{Th}$ - Thevenin equivalent voltage

V_t - terminal voltage

V_ϕ - phase voltage

v - instantaneous voltage; speed

W_f - magnetic field energy

W_f' - magnetic field coenergy

X - reactance

X_d - direct-axis reactance

X_ℓ - leakage reactance

X_m - magnetizing reactance

X_q - quadrature-axis reactance

X_s - synchronous reactance

X_1 - primary coil leakage reactance

X_2 - secondary coil leakage reactance

X_2' - X_2 reflected to primary reference

X_ϕ - magnetizing reactance

$\boldsymbol{Y}$ - admittance

Y - magnitude of $\boldsymbol{Y}$

$\boldsymbol{Z}$ - impedance

Z - magnitude of $\boldsymbol{Z}$; no. total conductors

Z_s - synchronous impedance

$\boldsymbol{Z}_{Th}$ - Thevenin equivalent impedance ($R_{Th} + jX_{Th}$)

δ - air gap length; torque or power angle

η - efficiency

θ - phase angle; mechanical angle

λ - flux linkage

μ - permeability

μ_o - permeability of free space

μ_R - relative permeability

ρ - resistivity

σ - conductivity

Φ_m - maximum value of flux

Φ_p - flux per pole

ϕ - magnetic flux

ϕ_m - mutual flux

ω - source electrical radian frequency

ω_m - mechanical rotational speed

ω_r - rotor electrical radian frequency

ω_s - synchronous speed

ELECTRIC MACHINES:

ANALYSIS AND DESIGN
APPLYING MATLAB®

McGraw-Hill Series in Electrical and Computer Engineering

Senior Consulting Editor
Stephen W. Director, University of Michigan, Ann Arbor

Circuits and Systems
Communications and Signal Processing
Computer Engineering
Control Theory and Robotics
Electromagnetics
Electronics and VLSI Circuits
Introductory
Power
Antennas, Microwaves, and Radar

Previous Consulting Editors

Ronald N. Bracewell, Colin Cherry, James F. Gibbons, Willis W. Harman, Hubert Heffner, Edward W. Herald, John G. Linvill, Simon Ramo, Ronald A. Rohrer, Anthony E. Siegman, Charles Susskind, Frederick E. Terman, John G. Truxal, Ernst Weber, and John R. Whinnery

ELECTRIC MACHINES:

ANALYSIS AND DESIGN
APPLYING MATLAB®

Jimmie J. Cathey
Professor of Electrical Engineering
University of Kentucky

Boston   Burr Ridge, IL   Dubuque, IA   Madison, WI   New York   San Francisco   St. Louis
Bangkok   Bogotá   Caracas   Kuala Lumpur   Lisbon   London   Madrid   Mexico City
Milan   Montreal   New Delhi   Santiago   Seoul   Singapore   Sydney   Taipei   Toronto

McGraw-Hill Higher Education

A Division of The **McGraw-Hill** Companies

ELECTRIC MACHINES: ANALYSIS AND DESIGN APPLYING **MATLAB**®
Published by McGraw-Hill, an imprint of The McGraw-Hill Companies, Inc., 1221 Avenue of the
Americas, New York, NY 10020. Copyright © 2001 by The McGraw-Hill Companies, Inc. All rights
reserved. No part of this publication may be reproduced or distributed in any form or by any means, or
stored in a data base or retrieval system, without the prior written consent of The McGraw-Hill
Companies, Inc., including, but not limited to, in any network or other electronic storage or
transmission, or broadcast for distance learning.
Some ancillaries, including electronic and print components, may not be available to customers outside
the United States.

This book is printed on acid-free paper.

1 2 3 4 5 6 7 8 9 0 FGR/FGR 0 9 8 7 6 5 4 3 2 1

ISBN 0-07-242370-6

President of McGraw-Hill Science/Engineering/Math: *Kevin Kane*
Publisher: *Thomas Casson*
Sponsoring editor: *Catherine Fields Shultz*
Developmental editor: *Michelle L. Flomenhoft*
Senior marketing manager: *John T. Wannemacher*
Project manager: *Laura Ward Majersky*
Lead production supervisor: *Heather D. Burbridge*
Freelance design coordinator: *Pam Verros*
Supplement coordinator: *Nate Perry*
Senior Media technology producer developer: *Phil Meek*
Cover designer: *JoAnne Schopler*
Cover illustration: ©*Masterfile*
Compositor: *Lachina Publishing Services*
Typeface: *10/12 Times Roman*
Printer: *Quebecor World Fairfield Inc.*

Library of Congress Cataloging-in-Publication Data

Cathey, Jimmie J.
    Electric machines: analysis and design applying Matlab / Jimmie J. Cathey.
      p. cm
    Includes bibliographical references and index.
    ISBN 0-07-242370-6 (alk. paper)
    1. Electric machinery—Computer-aided design. 2. Electric machinery—Design and
construction—Computer programs. 3. MATLAB. I. Title.

TK2182.C36 2001
621.31'042'02855369—dc21

                                                         00-051753

www.mhhe.com

This work is dedicated with love to Mary Ann for being a wife that makes keeping the marriage commitment easy.

TABLE OF CONTENTS

Preface   xi

About the Author   xiv

Chapter 1

Introduction   1

1.1   What   1
1.2   When   2
1.3   Where   2
1.4   How   3

Chapter 2

Sinusoidal Steady-State Circuits   5

2.1   Introduction   5
2.2   Phasor and Impedance Principles   5
2.3   Single-Phase Network Analysis   10
2.4   Three-Phase Networks   14
    2.4.1   Connection arrangements   15
    2.4.2   Balanced conditions   16
    2.4.3   Connection transformations   21
    2.4.4   Analysis   23
2.5   Power Flow   28
    2.5.1   Single-phase circuits   28
    2.5.2   Three-phase circuits   32
    2.5.3   Power measurement   34
    2.5.4   Power factor correction   35
2.6   Multiple Frequency Circuits   38
2.7   Computer Analysis Code   39
Summary   43
Problems   44

Chapter 3

Magnetic Circuits and Energy Conversion   49

3.1   Introduction   49
3.2   Governing Laws and Rules   49
3.3   Ferromagnetism   54

3.3.1   Saturation   54
3.3.2   Hysteresis   57
3.3.3   Curie temperature and magnetostriction   57
3.4   Magnetic Circuits   57
    3.4.1   Analysis methodology   58
    3.4.2   Air gap fringing   60
    3.4.3   Leakage flux   61
    3.4.4   Series magnetic circuits   61
    3.4.5   Parallel magnetic circuits   69
3.5   Energy and Inductance   76
    3.5.1   Inductance   76
    3.5.2   Energy   80
3.6   Sinusoidal Excitation   81
3.7   Permanent Magnets   83
    3.7.1   Classification and characteristics   83
    3.7.2   Performance   85
3.8   Energy Conversion   90
    3.8.1   Model formulation   90
    3.8.2   Force and energy   92
    3.8.3   Force and coenergy   98
    3.8.4   Doubly excited systems   102
3.9   Solenoid Design   103
    3.9.1   Rough sizing   103
    3.9.2   Magnetic circuit   106
    3.9.3   Sample design   108
3.10   Computer Analysis Code   113
Summary   126
Problems   127
References   130

Chapter 4

Transformers   131

4.1   Introduction   131
4.2   Physical Construction   132
4.3   The Ideal Transformer   133
    4.3.1   Circuit diagrams   137
    4.3.2   Voltage and current relationships   137
    4.3.3   Power and impedance relationships   139
4.4   The Practical Transformer   141

4.4.1   Lossless-core transformer   141
4.4.2   Ferromagnetic core properties   143
4.4.3   Goodness of performance implications   152
4.4.4   Nameplate and coil polarity   155
4.5   Test Determination of Parameters   156
4.5.1   Preliminary tests   157
4.5.2   Short-circuit test   157
4.5.3   Open-circuit test   158
4.6   Performance Assessment   161
4.6.1   Voltage-current analysis   161
4.6.2   Approximate equivalent circuit   163
4.6.3   Efficiency   166
4.6.4   Voltage regulation   171
4.6.5   Inrush current   172
4.7   Residential Distribution Transformers   173
4.8   Autotransformers   175
4.8.1   Ideal autotransformer   176
4.8.2   Power flow   178
4.9   Three-Phase Transformers   179
4.9.1   Connection schemes   179
4.9.2   Performance analysis   182
4.10   Transformer Winding Taps   184
4.10.1   Fixed tap adjustment   184
4.10.2   Tap change under load   186
4.11   Instrument Transformers   187
4.11.1   Potential transformers   187
4.11.2   Current transformers   187
4.12   Transformer Design   188
4.12.1   Core volume sizing   188
4.12.2   Magnetic circuit analysis   190
4.12.3   Equivalent circuit parameters   192
4.12.4   Sample design   197
4.13   Computer Analysis Code   202
Summary   221
Problems   222
References   227

Chapter 5

DC Machines   229

5.1   Introduction   229
5.2   Physical Construction   229
5.3   Voltage and Torque Principles   232
5.3.1   No-load magnetic field   232
5.3.2   Armature coil induced voltage   234

5.3.3   Electromagnetic developed torque   239
5.4   Classification by Field Winding   241
5.4.1   Basic field windings   241
5.4.2   Field connection arrangements   242
5.5   Nature and Interaction of Magnetic Fields   243
5.5.1   Principal magnetic circuit   243
5.5.2   Magnetic field interaction   245
5.6   Generator Performance   251
5.6.1   Separately excited DC generator   253
5.6.2   Shunt DC generator   254
5.6.3   Series excited DC generator   256
5.6.4   Cumulative compound DC generator   256
5.6.5   Differential compound DC generator   258
5.7   Motor Performance   258
5.7.1   Shunt excited DC motor   260
5.7.2   Series excited DC motor   263
5.7.3   Cumulative compound DC motor   266
5.7.4   Differential compound DC motor   269
5.8   Motor Control   270
5.8.1   Starting control   270
5.8.2   Speed control   271
5.9   DC Motor Design   275
5.9.1   Classifications and standardizations   275
5.9.2   Volume and bore sizing   276
5.9.3   Armature design   278
5.9.4   Field pole design   285
5.9.5   Magnetic circuit analysis   287
5.9.6   Field winding design   288
5.9.7   Design refinement   289
5.9.8   Sample design   290
5.10   Computer Analysis Code   298
Summary   310
Problems   311
References   316

Chapter 6

Induction Motors   317

6.1   Introduction   317
6.2   Classification and Physical Construction   317

6.3   Stator Winding and MMF   319
    6.3.1   Stator windings   319
    6.3.2   Winding MMF   319
    6.3.3   Stator air gap traveling wave   326
    6.3.4   Synchronous speed   327
6.4   Rotor Action and Slip   328
    6.4.1   Rotor coil induced voltages   328
    6.4.2   Rotor air gap traveling wave   331
6.5   Equivalent Circuit   333
6.6   Test Determination of Parameters   335
    6.6.1   Blocked rotor test   336
    6.6.2   No-load test   337
6.7   Performance Calculations and Nature   340
    6.7.1   Induction motor power flow   340
    6.7.2   Developed torque determination   341
    6.7.3   Developed torque nature   348
    6.7.4   Frequency sensitivity of rotor
        parameters   351
    6.7.5   Machine performance evaluation   353
6.8   Reduced Voltage Starting   355
    6.8.1   Autotransformer starting   357
    6.8.2   Solid-state starter   358
6.9   Speed Control   358
    6.9.1   Rotor resistance control   359
    6.9.2   Voltage control   359
    6.9.3   Pole changing   360
    6.9.4   Frequency control   361
6.10   Single-Phase Motors   366
    6.10.1   Air gap field   367
    6.10.2   Equivalent circuit   368
    6.10.3   Performance nature   370
    6.10.4   Auxiliary starting winding   373
6.11   Induction Motor Design   375
    6.11.1   Classifications and
        standardizations   375
    6.11.2   Volume and bore sizing   377
    6.11.3   Stator design   379
    6.11.4   Rotor design   384
    6.11.5   Equivalent circuit parameters   389
    6.11.6   Design refinement   395
    6.11.7   Sample design   395
6.12   Computer Analysis Code   404
Summary   417
Problems   417
References   420

Chapter   7

Synchronous Machines   421

7.1   Introduction   421
7.2   Classification and Physical
    Construction   422
    7.2.1   Stator winding and MMF   424
    7.2.2   Rotor windings and air gap fields   426
7.3   Generated Voltages and Equivalent
    Circuit   428
    7.3.1   Coil fluxes and voltages   429
    7.3.2   Per phase equivalent circuit: round
        rotor case   432
    7.3.3   Magnetic linearization   434
7.4   Equivalent Circuit Parameters from Test
    Data   437
7.5   Generator Performance   440
    7.5.1   Synchronous generator phasor
        diagram   441
    7.5.2   Electromechanical developed
        torque   441
    7.5.3   Isolated synchronous generators   445
    7.5.4   Interconnected synchronous
        generators   448
7.6   Motor Performance   455
7.7   Salient-Pole Machine Performance   457
7.8   Self-Synchronous Motors   464
    7.8.1   Brushless DC motors   465
    7.8.2   Switched reluctance motors   470
7.9   Synchronous Machine Design   476
    7.9.1   Standards and classifications   476
    7.9.2   Volume and bore sizing   477
    7.9.3   Stator design   479
    7.9.4   Air gap sizing   483
    7.9.5   Rotor design   483
    7.9.6   Equivalent circuit parameters   491
    7.9.7   Design refinement   491
    7.9.8   Sample design   492
7.10   Computer Analysis Code   499
Summary   511
Problems   511
References   515

Appendix A

Winding Factors   517

A.1   Distribution Factor   517
A.2   Pitch Factor   520
A.3   Winding Factor   521

Appendix B

Conversion Factors   523

Appendix C

Magnetic Wire Tables   525

C.1   Round Wire with Film Insulation   525
C.2   Square Wire with Film Insulation   526

Index   527

PREFACE

A single-semester introductory course in electric machines and energy conversion is necessary to conform to the constraints of a typical electrical engineering undergraduate core curriculum. This course should be of sufficient depth to satisfy the needs of those students who pursue specialization in other than the power area. Concurrently, the course should prepare individuals in the electric power area with adequate prerequisite foundation to enter advanced courses in the electric machines and energy conversion area for the purpose of gaining a fuller breadth in the field.

This text contains sufficient material for a single-semester core course while allowing some selectivity among the topics covered by the latter sections of Chaps. 3 to 7 in order to fit the needs and emphasis of a particular program curriculum. A presently unique feature of the text is its integrated option to introduce popular interactive computer software MATLAB to handle the tedious calculations arising in electric machine analysis. As a consequence, more exact models of devices can be retained for analysis rather than the approximate models commonly introduced for the sake of computational simplicity. A computer icon appears in the margin with each introduction of MATLAB analysis.

Chapter 1 serves the purposes of motivation and orientation. The student is made aware of the widespread application of energy conversion devices. Growth trends and technology advancement directions are presented. The material can be covered either as a one-class period lecture or as a reading assignment.

Although sinusoidal steady-state circuit analysis and power flow are topics of required core circuit courses in all electrical engineering curriculums, the compression of material in these courses typically does not allow adequate drill in the sinusoidal steady-state techniques to give most students the proficiency necessary to comfortably handle energy conversion analysis. Chapter 2 provides the needed review and drill as well as familiarization with the notation peculiar to the text. Further, the chapter introduces students to the use of MATLAB for the analysis of circuits in the sinusoidal steady state, thereby establishing the basic techniques of computer analysis for use in the chapters that follow.

Chapter 3 straightway develops from Ampere's circuit law, Faraday's law, and Lorentz' force equation interpretive rules suited for use in energy conversion device understanding and analysis. The balance of the chapter is devoted to study of the nature and behavior of magnetic circuits as they appear in transformers and electric machines, serving as an intermediary stage in the energy conversion process. Computer analysis methods in this chapter introduce the student to the use of MATLAB for the purpose of generating saturation curves and flux linkage plots given the physical dimensions of a magnetic structure and the associated *B-H* curve. In addition, numerical techniques are established for determination of force developed by a variable-position magnetic structure due to change in stored magnetic energy with position $(F_e = \partial W_m / \partial x)$. Significant use of these computer analysis methods will be made in subsequent chapters.

Chapters 4 to 7 deal with the transformers, dc machines, and ac machines that are most frequently encountered in industry. Mastery of the material in these chapters should adequately prepare the reader to specify an application-appropriate energy conversion device and to predict performance after installation. In addition, Chap. 7 presents the operational principles of the brushless dc motor (recently matured technology) and the switched reluctance motor (emerging technology). Through use of the developed computer software for each energy conversion device, the reader is able to quickly calculate and plot wide-range performance data. Also, the impact of parameter sensitivity can be easily assessed.

Another unique feature of the text is coverage of basic design principles of the energy conversion equipment after study of its performance analysis in Chaps. 4 to 7. Although the design methodologies stop short of the empirical refinement necessary for commercial product development, the scope and depth are sufficient to train engineering students in the mindset of design. The reference list at the end of each chapter serves as a beginning point for the reader to pursue additional design refinement.

The text can serve as a basis either for a course in energy device principles and analysis with an optional design project, or for a capstone design course offered subsequent to an introductory course in energy device principles. Computer software developed in the first parts of Chaps. 4 to 7 is expanded and integrated into the software to support the latter section design procedures. Each of Chaps. 4 to 7 contains problems formulated to drill the reader on the individual concepts of design as each is introduced. A completed example design in each chapter serves as a coagulant for the learned individual principles.

There is no universally acceptable presentation order for the study of energy conversion devices. However, the vast majority of instructors do cover magnetic circuit analysis first followed by study of transformers. This order of presentation has been assumed in the writing of this text to justify delay of losses in ferromagnetic materials until the transformer with its sinusoidal excitation is under study. Otherwise, the material of Chaps. 5 to 7 stands reasonably independent, leaving the order of coverage to the discretion of the individual instructor. The one exception to this claim is that the air gap traveling wave of Sec. 6.3.3 must be extracted for appropriate presentation if synchronous machines are to be studied before induction machines.

Supplemental materials supporting the book are available at *www.mhhe.com/engcs/electrical/cathey*. The materials on that site include downloadable files of the MATLAB source code, a list of errata for the text, and other useful information.

The author acknowledges the many constructive suggestions offered by the manuscript reviewers. The individuals are Professors Parviz Famouri of West Virginia University, Clifford Grigg of Rose-Hulman Institute of Technology, Mo Shiva of California State University-Fullerton, Medhat Morcos of Kansas State

University, Gill Richards of University of New Orleans, Miroslav Begovic of Georgia Institute of Technology, Leon Tolbert of University of Tennessee, Maamar Taleb of University of Bahrain, and Steven Hietpas of South Dakota State University. Their implemented suggestions enhance the quality of the book.

A special thanks is offered to Ross Cutts. As a student in my electric machines course, he voluntarily spent numerous hours discussing ideas with me to make the presentation clearer from a student point of view. His contribution is valued.

Jimmie J. Cathey
Lexington, KY

ABOUT THE AUTHOR

Jimmie J. Cathey received the BSEE degree (1965) and the PhD degree (1972) both from Texas A&M University. He holds the MSEE degree (1968) from Bradley University.

He has 13 years of combined industrial experience with Caterpillar Tractor Company and Marathon LeTourneau Company working in the areas of electric machine and control design with application to off-highway vehicle propulsion. In 1980, Dr. Cathey joined the faculty of Electrical Engineering at the University of Kentucky. Over his 20 years on the faculty, he has principally taught courses dealing with electric machine analysis, electric machine design, power electronics, and electric drives. He currently holds the title of TVA Professor of Electrical Engineering.

He is a Senior Member of IEEE and a Registered Professional Engineer in the state of Texas.

1

INTRODUCTION

Energy is defined as the ability to do work. The ability to harness and control energy determines the productivity potential and subsequent lifestyle advancement of society. Electric machines have been, and will continue to be, a practical and dominant medium for achievement of productivity improvement.

The present annual worldwide electrical energy usage is approximately 15 trillion kWh, with an annual growth rate of approximately 500 billion kWh per year. Electric machines form a vital link in initially converting the bulk of this energy to electrical form for convenient distribution to the end-use point, where once again electric machines serve to convert a significant portion of the energy from electrical to mechanical form.

1.1 WHAT

Practical electric machines are bilateral energy converters that use an intermediary magnetic field link. Only machines that operate on this principle are considered in the scope of this text. Machines that process energy received in mechanical form and export energy in electrical form are known as *generators*. When a machine reverses the conversion process to absorb energy in electrical form and reformat the energy to mechanical form on a sustained basis, it is called a *motor*. A machine or device that is designed for limited mechanical motion is commonly known as an *actuator*.

The common wound-core transformer is traditionally studied along with electric machines. This transformer type functions to adjust the voltage magnitude between points of interface in an ac system. Although both input energy and output energy are in electrical form, the device does convert energy absorbed into magnetic form and then reconverts it back to electrical form. From this point of view, the wound-core transformer is also an energy converter.

1.2 WHEN

The electric machine age can be traced to 1831 with the invention by Michael Faraday of the disk machine—a true dc machine. Electric machines remained largely a laboratory and demonstration curiosity until the 1870s when Thomas Edison began commercial development of the dc generator to support electrical power distribution for the purpose of bringing electrical lighting to homes. In so doing, Edison was pioneering the concept of electric power distribution from central generation stations and thus introducing the electric grid infrastructure concept that must exist for widespread application of electric motors.

A major milestone in the history of the electric machine was the patent of the three-phase induction motor by Nikola Tesla in 1888. Tesla's concepts of alternating current were advanced by Charles Steinmetz over the next decade so that by 1900 reliable wound-core transformers were available, opening the way for long-distance power transmission. The electrification of the United States was well under way, although the process would take another 30 years to complete, with the final rural electric power distribution not being completed until the 1930s. The proliferation of electric machine applications closely tracked the expansion of the electric utility grids.

Although the concepts of the electric machines in application today originated a century ago, the refinement and improvement process has never reached stagnation. Development of better ferromagnetic materials and insulations has led to increasing power densities over an order of magnitude greater than the early electric machines. High-volume manufacturing techniques have resulted in less expensive electric machines, thereby opening the door for even more applications. Finally, the reliable, high-power-level switching devices and microprocessors resulting from the "solid-state revolution" of the recent decades have led to development of significant control improvement in electric motor drives. All this is evolvement of the ability to harness and control energy, thereby continuing to enhance lifestyle advancement.

1.3 WHERE

Electric machines are readily apparent in a manufacturing facility or in the HVAC machinery room of a building. However, many electric machines are integrated into appliances, vehicles, and service machines so that the average person is not even aware of their presence. The typical professional wakes up in the morning in his home that probably has no fewer than 30 electric machines in service. He drives to work in an automobile that will have 15 or more motors. He arrives at his office that probably has at least five motors functioning in personal office equipment alone. By the time he uses the copy machine and FAX machine, then buys a cup of coffee at the vending machine, he probably has activated another 10 or more electric machines. It is 9 A.M., and already 60 motors have played a role in his productive lifestyle.

1.4 How

The focus of this text is electric machines that process energy through an intermediary magnetic field; thus, the energy density of the magnetic core has a major impact on the size or volume of the electric machine. Energy density in a magnetic structure is given by

$$w_m = \int \mu B \, dB \qquad\qquad \textbf{[1.1]}$$

Ferromagnetic materials with their associated high permeability (μ) offer the largest energy density (w_m) of any known materials while requiring modest expenditure of energy to establish magnetic flux density (B). They are therefore the choice for the magnetic structure of electric machines. Unfortunately, the use of ferromagnetic materials introduces nonlinear relationships into the analytical study of electric machines, resulting in cumbersome solution procedures. In addition, the exact equivalent circuit to model certain electric machines requires tedious repetitive calculations to predict wide-range performance.

Either of the above cited situations can turn the analysis procedure into an exercise in mathematics distracting from the intended discovery of the electric machine performance nature. In order to circumvent this difficulty, common practice in electric machine analysis has been to make accuracy-sacrificing approximations to yield manageable calculation procedures. This text introduces MATLAB®[1] at the points where cumbersome and tedious work arises in an effort to ease the analysis intensity without a sacrifice in accuracy. Consequently, wide-range performance study and parameter variation sensitivity are not a burden. The student is encouraged to take initiative beyond the suggestions of the text and the requirements of the instructor to modify both the MATLAB code and the parameter values of the devices and thereby learn even more about the nature of the electric machines that have such a profound effect on our daily lives.

| 1. MATLAB is a registered trademark of The MathWorks, Inc.

2

SINUSOIDAL STEADY-STATE CIRCUITS

2.1 INTRODUCTION

The focus of introductory energy conversion study is steady-state performance. The dc machine and certain solenoids are analyzed in the dc steady state. The balance of energy conversion devices of interest (transformers, induction motors, synchronous machines, and ac solenoids after initial activation) operates in the sinusoidal steady state for a first-order approximation. Hence, a good proficiency in sinusoidal steady-state analysis methods is a prerequisite to their study. Although sinusoidal steady-state analysis can require some cumbersome complex number arithmetic, computer techniques will be introduced to alleviate the burden.

2.2 PHASOR AND IMPEDANCE PRINCIPLES

The steady-state response of a circuit is the resulting performance after any transient behavior has subsided. If a linear circuit is excited by a sinusoidal source, then all voltages and currents of the circuit, due to the exciting sinusoidal source, are sinusoidal quantities of the source frequency once the steady state is reached. Consequently, circuit analyses using Kirchhoff voltage law (KVL) and Kirchhoff current law (KCL) require addition and subtraction of sinusoidally varying time quantities of the same frequency.

Without loss of generality, consider the addition of the two voltages $v_1(t) = \sqrt{2}\, V_1 \cos(\omega t + \beta)$ and $v_2(t) = \sqrt{2}\, V_2 \cos(\omega t + \gamma)$.

$$v(t) = v_1(t) + v_2(t) = \sqrt{2}\, V_1 \cos(\omega t + \beta) + \sqrt{2} \cos(\omega t + \gamma) \quad \textbf{[2.1]}$$

By use of the Euler equation $e^{jx} = \cos x + j \sin x$, [2.1] can be written as

$$v(t) = \mathrm{Re}\left\{\sqrt{2}\, V_1 e^{j(\omega t + \beta)}\right\} + \mathrm{Re}\left\{\sqrt{2}\, V_2 e^{j(\omega t + \gamma)}\right\} \qquad \textbf{[2.2]}$$

$$v(t) = \mathrm{Re}\left\{\sqrt{2}\,(V_1 e^{j\beta} + V_2 e^{j\gamma})e^{j\omega t}\right\} \qquad \textbf{[2.3]}$$

The transition from [2.2] to [2.3] made use of the real part of the sum of two complex numbers being equal to the sum of the real parts and the law of exponents. Each of the quantities inside the parentheses of [2.3] is a complex number, known as a *phasor*, associated with the time functions of $v_1(t)$ and $v_2(t)$. As is apparent from this arbitrarily chosen representation in polar form, the phasor magnitudes are the rms magnitudes of $v_1(t)$ and $v_2(t)$, and the phasor angles indicate the phase shift with respect to $\cos(\omega t)$. It is noted that the phasor quantity is not equal to the time functions, but rather it is a particular complex number associated with the time function. Hence, the phasor can be thought of as a transformation of the time function.

It is obvious from [2.3] that addition of the phasors $\bar{V}_1$ and $\bar{V}_2$ allows representation of the sum $v_1(t) + v_2(t)$ as a single sinusoidal time function.

$$\bar{V} = \bar{V}_1 + \bar{V}_2 = V_1 \angle \beta + V_2 \angle \gamma = V \angle \theta \qquad \textbf{[2.4]}$$

$$v(t) = \text{Re}\left\{\sqrt{2}(V \angle \theta)\,e^{j\omega t}\right\} = \sqrt{2}\,V\cos(\omega t + \theta) \qquad \textbf{[2.5]}$$

A sketch of phasor quantities on the complex plane with the axes omitted is known as a *phasor diagram*, which serves as a valuable aid in visualizing the addition process.

Three clarifications of the introduced phasor representation are noteworthy:

1. The prefactoring of $\sqrt{2}$ in [2.3] resulted in rms-valued phasors rather than maximum value phasors. The rms-valued phasor is common practice in power system study and will be used in this text unless explicitly stated otherwise.

2. The time functions of [2.1] were arbitrarily written in terms of $\cos(\omega t)$. The development could have been carried out in terms of $\sin(\omega t)$ using the imaginary part for [2.2] and [2.3]; however, a consistency in choice of the reference time function must be maintained.

3. As one develops a comfortable feel for phasor representation, transformations to and from the time domain may become rare, with all analysis being conducted entirely within the phasor domain.

Example 2.1 | Find a single time function for $v(t) = \sqrt{2}\ 120\cos(\omega t) - 30\sin(\omega t)$ using rms phasor methods.

Rewrite $v(t)$ in terms of $\cos(\omega t)$ reference.

$$v(t) = \sqrt{2}\,120\cos(\omega t) - 30\cos(\omega t - 90°)$$

Then,

$$\bar{V} = 120\angle 0° - 30/\sqrt{2}\angle - 90°$$

$$\bar{V} = 120 - (-j21.213) = 121.86\angle 10.02°$$

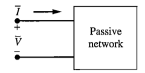

Figure 2.1
One-port passive network

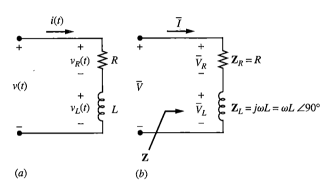

Figure 2.2
Series R-L network. (a) Time domain. (b) Phasor domain.

Transform $\bar{V}$ to time domain.

$$v(t) = \sqrt{2}\,121.86\cos(\omega t + 10.02°) = 172.31\cos(\omega t + 10.02°)$$

Impedance (**Z**) in the phasor domain is defined as the ratio of the voltage $\bar{V}$ to the current $\bar{I}$ at the port of a passive network with polarities as indicated by Fig. 2.1. The unit of measure is ohms (Ω). It is understood that the associated frequencies of $\bar{V}$ and $\bar{I}$ are identical. Thus,

$$\mathbf{Z} = \frac{\bar{V}}{\bar{I}} = \frac{V\angle\beta}{I\angle\alpha} = \frac{V}{I}\angle(\beta - \alpha) = Z\angle\theta \qquad \textbf{[2.6]}$$

It is seen that the impedance angle θ indicates the time-phase lag of the current $\bar{I}$ with respect to the voltage $\bar{V}$.

Figure 2.2a shows a series R-L circuit in the time domain. If $i(t) = \sqrt{2}\,I\sin(\omega t)$, use the impedance definition of [2.6] to show that the element impedances indicated in Fig. 2.2b are correct. Also, determine the equivalent impedance **Z** for this circuit. | **Example 2.2**

For the time domain circuit,

$$v_R(t) = Ri(t) = \sqrt{2}\,IR\sin(\omega t)$$

$$v_L(t) = L\frac{di(t)}{dt} = \omega L\sqrt{2}I\cos(\omega t)$$

Transforming $i(t)$, $v_R(t)$, and $v_L(t)$ to the phasor domain and applying [2.6] yields the individual element impedances as

$$\mathbf{Z}_R = \frac{\bar{V}_R}{\bar{I}} = \frac{IR\angle-90°}{I\angle-90°} = R$$

$$\mathbf{Z}_L = \frac{\bar{V}_L}{\bar{I}} = \frac{\omega LI\angle0°}{I\angle-90°} = \omega L\angle90° = j\omega L$$

Apply KVL to the phasor domain circuit of Fig. 2.2b to obtain

$$\bar{V} = \bar{V}_R + \bar{V}_L = \bar{I}\mathbf{Z}_R + \bar{I}\mathbf{Z}_L = \bar{I}(R + j\omega L)$$

whence

$$\mathbf{Z} = \frac{\bar{V}}{\bar{I}} = R + j\omega L = \sqrt{R^2 + (\omega L)^2}\angle\tan^{-1}(\omega L/R)$$

Admittance (**Y**) is defined as the reciprocal of impedance, or

$$\mathbf{Y} = \frac{1}{\mathbf{Z}} = \frac{\bar{I}}{\bar{V}} \qquad\qquad \textbf{[2.7]}$$

The unit of measure is siemens (S).

Example 2.3 | Figure 2.3a shows a parallel R-C circuit in the time domain. If $v(t) = \sqrt{2}V\cos(\omega t)$, show that the element impedances of Fig. 2.3b are correct and determine the equivalent impedance **Z** for this network.

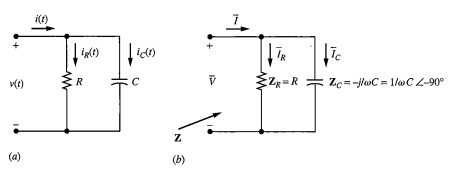

(a)   (b)

Figure 2.3
Parallel *R-C* network. (*a*) Time domain. (*b*) Phasor domain.

For the time domain circuit,

$$i_R(t) = \frac{v(t)}{R} = \frac{\sqrt{2}\,V}{R}\cos(\omega t)$$

$$i_C(t) = C\frac{dv(t)}{dt} = -\omega C\sqrt{2}\,V\sin(\omega t) = \omega C\sqrt{2}\,V\cos(\omega t + 90°)$$

Transform $i_R(t)$, $i_C(t)$, and $v(t)$ to the phasor domain, then apply [2.7] and [2.6] to give

$$\mathbf{Y}_R = \frac{\overline{I}_R}{\overline{V}} = \frac{V/R\angle 0°}{V\angle 0°} = \frac{1}{R} \qquad\qquad \mathbf{Z}_R = \frac{1}{\mathbf{Y}_R} = R$$

$$\mathbf{Y}_C = \frac{\overline{I}_C}{\overline{V}} = \frac{\omega CV\angle 90°}{V\angle 0°} = \omega C\angle 90° = j\omega C \qquad \mathbf{Z}_C = \frac{1}{\mathbf{Y}_C} = -j\frac{1}{\omega C}$$

In the time domain,

$$i(t) = i_R(t) + i_C(t) = \frac{\sqrt{2}\,V}{R}\cos(\omega t) + \omega C\sqrt{2}\,V\cos(\omega t + 90°)$$

By analogy with [2.2] and [2.3],

$$i(t) = \mathrm{Re}\left\{\sqrt{2}\left(\frac{V}{R}e^{j0} + \omega CVe^{j\pi/2}\right)e^{j\omega t}\right\} = \mathrm{Re}\left\{\sqrt{2}\,Ve^{j0}\left(\frac{1}{R} + j\omega C\right)e^{j\omega t}\right\}$$

Thus, phasor $\overline{I}$ must be

$$\overline{I} = \overline{V}\left(\frac{1}{R} + j\omega C\right)$$

and

$$\mathbf{Y} = \frac{\overline{I}}{\overline{V}} = \frac{1}{R} + j\omega C \qquad \mathbf{Z} = \frac{1}{\mathbf{Y}} = \frac{1}{1/R + j\omega C}$$

The result of Example 2.2 can be generalized to conclude that the equivalent impedance of n series-connected impedance elements is simply

$$\mathbf{Z}_{eq} = \mathbf{Z}_1 + \mathbf{Z}_2 + \cdots + \mathbf{Z}_n \qquad\qquad \textbf{[2.8]}$$

Likewise, the result of Example 2.3 leads to the general conclusion that the equivalent impedance and admittance of m parallel-connected impedance or admittance elements are given by

$$\mathbf{Y}_{eq} = \mathbf{Y}_1 + \mathbf{Y}_2 + \cdots + \mathbf{Y}_m = \frac{1}{\mathbf{Z}_1} + \frac{1}{\mathbf{Z}_2} + \cdots + \frac{1}{\mathbf{Z}_m} \qquad \textbf{[2.9]}$$

$$\mathbf{Z}_{eq} = \frac{1}{\mathbf{Y}_1 + \mathbf{Y}_2 + \cdots + \mathbf{Y}_m} = \frac{1}{1/\mathbf{Z}_1 + 1/\mathbf{Z}_2 + \cdots + 1/\mathbf{Z}_m} \qquad \textbf{[2.10]}$$

Further, the work of Examples 2.2 and 2.3 has illustrated that KVL and KCL are valid in phasor domain analysis.

2.3 SINGLE-PHASE NETWORK ANALYSIS

The equivalent impedance and admittance principles developed in the previous section, along with phasor domain application of KVL and KCL, allow use of simple systematic network reduction for analysis of sinusoidal steady-state circuits in the phasor domain. Several examples are presented to clarify the techniques.

Example 2.4

For the circuit of Fig. 2.4a, $\bar{V} = 120\angle 30°$ V. Determine $\bar{I}$, $\bar{I}_1$, and $\bar{I}_2$. Also, sketch a phasor diagram showing all currents and source voltage $\bar{V}$.

By Ohm's law,

$$\bar{I}_1 = \frac{\bar{V}}{4 + j6} = \frac{120\angle 30°}{7.21\angle 56.31°} = 16.64\angle -26.31° \text{ A}$$

$$\bar{I}_2 = \frac{\bar{V}}{-j3} = \frac{120\angle 30°}{3\angle -90°} = 40\angle 120° \text{ A}$$

Using KCL,

$$\bar{I} = \bar{I}_1 + \bar{I}_2 = (14.91 - j7.37) + (-20 + j34.64)$$

$$\bar{I} = -5.09 + j27.27 = 27.74\angle 100.6° \text{ A}$$

The phasor diagram is shown by Fig. 2.4b where the addition of $\bar{I}_1$ and $\bar{I}_2$ to yield $\bar{I}$ and corroborate KCL has been indicated.

Although not too cumbersome in the above example, the complex number arithmetic can become tedious in solution of sinusoidal steady-state circuits. The

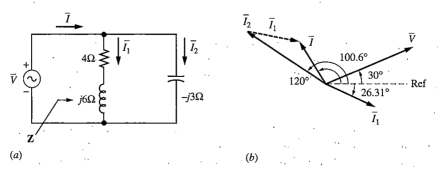

(a)   (b)

Figure 2.4
Single-phase network. (a) Phasor domain circuit. (b) Phasor diagram.

MATLAB program ⟨cratio.m⟩ has been developed to reduce a ratio of complex number products to a single numerical value. A companion MATLAB program ⟨csum.m⟩ adds a series of complex numbers. The user needs only to respond to the prompted inputs.

Use ⟨cratio.m⟩ and ⟨csum.m⟩ to determine the currents $\bar{I}_1$, $\bar{I}_2$, and $\bar{I}$ of Example 2.4. **Example 2.5**
    Currents $\bar{I}_1$ and $\bar{I}_2$ are determined by use of ⟨cratio.m⟩ by input of $\bar{V}$ as the single numerator term and the branch impedance as the single denominator term. The resulting screen displays follow:

```
   RATIO  OF  COMPLEX  NUMBER  PRODUCTS

   Form:  1-polar,  2-rectangular

How  many  numerator  numbers?  1

Form  of  1  =  1
Mag  1  =  120
Deg  1  =  30

How  many  denominator  numbers?  1

Form  of  1  =  2
Real  1  =  4
Imag  1  =  6

RATIO  =  14.92  +j  -7.376  =  16.64|_-26.31deg
```

```
   RATIO  OF  COMPLEX  NUMBER  PRODUCTS

   Form:  1-polar,  2-rectangular

How  many  numerator  numbers?  1

Form  of  1  =  1
Mag  1  =  120
Deg  1  =  30

How  many  denominator  numbers?  1

Form  of  1  =  2
Real  1  =  0
Imag  1  =  -3

RATIO  =  -20  +j  34.64  =  40|_120deg
```

Current $\bar{I}$ is determined by ⟨csum.m⟩ by input of the values of $\bar{I}_1$ and $\bar{I}_2$ found by ⟨cratio.m⟩. The screen display is

```
   SUM  OF  COMPLEX  NUMBERS

   Form:  1-polar,  2-rectangular

How  many  numbers  to  be  added?  2

Form  of  1  =  1
Mag  1  =  16.64
Deg  1  =  -26.31

Form  of  2  =  1
Mag  2  =  40
Deg  2  =  120


SUM  =  -5.084  +j  27.27  =  27.74|_100.6deg
```

Example 2.6

Use ⟨cratio.m⟩ only to determine currents $\bar{I}$ and $\bar{I}_1$ of Example 2.4.

Current $\bar{I}$ is calculated by two applications of ⟨cratio.m⟩. First, the input impedance **Z** is determined as

$$\mathbf{Z} = \frac{(4 + j6)(-j3)}{4 + j6 - j3}$$

Then the input current $\bar{I}$ is found as $\bar{I} = \bar{V}/\mathbf{Z}$. The screen displays are

```
   RATIO  OF  COMPLEX  NUMBER  PRODUCTS

   Form:  1-polar,  2-rectangular

How  many  numerator  numbers?  2

Form  of  1  =  2
Real  1  =  4
Imag  1  =  6

Form  of  2  =  2
Real  2  =  0
Imag  2  =  -3

How  many  denominator  numbers?  1
```

```
Form of 1 = 2
Real 1 = 4
Imag 1 = 3

RATIO = 1.44 +j -4.08 = 4.327|_-70.56deg
```

```
   RATIO OF COMPLEX NUMBER PRODUCTS

   Form: 1-polar, 2-rectangular

How many numerator numbers? 1
Form of 1 = 1
Mag 1 = 120
Deg 1 = 30

How many denominator numbers? 1

Form of 1 = 2
Real 1 = real (RAT)
Imag 1 = imag (RAT)

RATIO = -5.083 +j 27.27 = 27.74|_100.6deg
```

By use of current division, current $\bar{I}_1$ is given as

$$\bar{I}_1 = \frac{-j3}{4 + j6 - j3}\bar{I}$$

The resulting screen display is

```
   RATIO OF COMPLEX NUMBER PRODUCTS

   Form: 1-polar, 2-rectangular

How many numerator numbers? 2

Form of 1 = 2
Real 1 = 0
Imag 1 = -3

Form of 2 = 2
Real 2 = real (RAT)
Imag 2 = imag (RAT)

How many denominator numbers? 1
```

```
Form of 1 = 2
Real 1 = 4
Imag 1 = 3

RATIO = 14.92 +j -7.376 = 16.64|_-26.31deg
```

When executing ⟨cratio.m⟩, the answer is the variable RAT, which remains in the workspace as a complex number in rectangular form. Advantage was taken of this fact in application of ⟨cratio.m⟩ above.

Example 2.7   If the source frequency of Example 2.4 is known to be 60 Hz, determine (*a*) time domain current $i(t)$ and (*b*) the values of L and C. Assume that phasor $\bar{V}$ has a cosine reference.

(*a*) The rms-valued phasor $\bar{I} = 27.74\angle100.6°$ is known from the solution of Example 2.4; and $\omega = 2\pi f = 2\pi(60) = 120\pi$ rad/s. Hence,

$$i(t) = 27.74\sqrt{2}\cos(120\pi t + 100.6°) \text{ A}$$

(*b*) Using $\mathbf{Z}_L = jX_L = j\omega L$ and $\mathbf{Z}_C = -jX_C = -j/\omega C$,

$$L = \frac{X_L}{\omega} = \frac{6}{120\pi} = 15.91 \text{ mH}$$

$$C = \frac{1}{\omega X_C} = \frac{1}{(120\pi)(3)} = 884.2 \ \mu\text{F}$$

2.4   THREE-PHASE NETWORKS

A circuit excited by three sinusoidal sources displaying distinct time-phase relationships and connected in a common arrangement is known as a *three-phase system* or *network*. With few exceptions, electric utility grid generation and transmission systems are three-phase networks. Distribution systems serving the industrial and commercial sectors are typically three-phase systems. Power can be transmitted in three-phase form with the least conductor volume of any form resulting in minimum capital outlay for transmission lines. Three-phase power, as well as all polyphase power under balanced operation, has the property that the instantaneous power is constant in value. Consequently, electric machines operated from polyphase power exhibit less vibration and acoustical noise than single-phase machines.

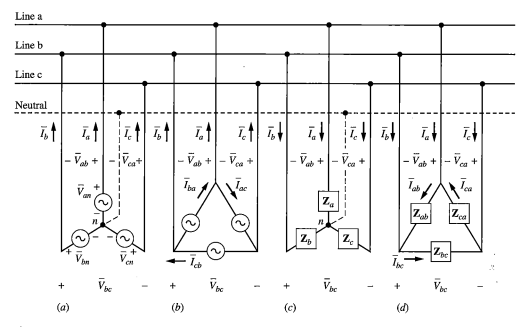

Figure 2.5
Three-phase system. (a) Wye (Y)-connected source. (b) Delta (Δ)-connected source. (c) Wye-connected load. (d) Delta-connected load.

2.4.1   CONNECTION ARRANGEMENTS

A *bus* is a circuit node that serves as a connection point for sources and loads. Busses in a three-phase system are spanned by a set of three-phase lines that practically display a distributed impedance over their length and are composed of three principal current-carrying conductors (line *a*, line *b*, line *c*). An optional fourth conductor, known as the *neutral,* may be present.

Figure 2.5 displays a three-phase system with detailed illustration of three-phase source and load connections where the two common topological arrangements (wye and delta) for source and load connections are introduced. A *phase* is one of the three branches of a three-phase source or load. For a wye connection, the collection of circuit elements between one line and the common or neutral point is known as the phase. For a delta connection, the collection of circuit elements between two lines is the phase.

The voltage across a phase and the current through a phase are known as the *phase voltage* and *phase current,* respectively. Currents flowing in the three conductors connecting a three-phase source or load to the three-phase system lines are known as *line currents* of the source or load. The voltages between these connecting conductors are called the *line voltages*. The commonly assumed directions for voltages and currents of three-phase sources and loads are indicated by Fig. 2.5, where it should be noted that currents are directed out of the positive polarity mark of phase voltage for a source and into the positive polarity mark of the phase voltage for a load. The phase currents and line currents for a wye-connected source or load are one and the same. Likewise, the phase voltages and line voltages for a delta connection are identical quantities. Each separate group of three line voltages, phase voltages, line currents, or phase currents of a three-phase source or load is known as a *three-phase set*. Any three-phase set of line voltages must sum to zero, since the three voltages form a closed loop when applying KVL. If no neutral connection is present, any three-phase set of line currents from a source or to a load must sum to zero as a consequence of KCL. In general, no further null summation constraints can be established for three-phase sets. However, for the important special case of balanced conditions to be addressed in the following section, other constraints exist, resulting in analysis simplification.

2.4.2 BALANCED CONDITIONS

If for each three-phase set of a source or load, the three voltages or currents of that set have equal magnitudes and the three phase angles are successively 120° apart, the source or load exhibits *balanced conditions*. For example, if the three-phase sets for the wye-connected source of Fig. 2.5a are

$$\bar{V}_{an} = V_\phi \angle \alpha \qquad \bar{V}_{bn} = V_\phi \angle (\alpha - 120°) \qquad \bar{V}_{cn} = V_\phi \angle (\alpha + 120°)$$

$$\bar{V}_{ab} = V_L \angle (\alpha + 30°) \qquad \bar{V}_{bc} = V_L \angle (\alpha - 90°) \qquad \bar{V}_{ca} = V_L \angle (\alpha + 150°)$$

$$\bar{I}_a = I_L \angle \beta \qquad \bar{I}_b = I_L \angle (\beta - 120°) \qquad \bar{I}_c = I_L \angle (\beta + 120°)$$

then balanced conditions exist for the source. Likewise, balanced conditions exist for the delta-connected load of Fig. 2.5d if

$$\bar{V}_{ab} = V_L \angle \alpha \qquad \bar{V}_{bc} = V_L \angle (\alpha - 120°) \qquad \bar{V}_{ca} = V_L \angle (\alpha + 120°)$$

$$\bar{I}_a = I_\phi \angle \beta \qquad \bar{I}_b = I_L \angle (\beta - 120°) \qquad \bar{I}_c = I_L \angle (\beta + 120°)$$

$$\bar{I}_{ab} = I_\phi \angle (\beta + 30°) \qquad \bar{I}_{bc} = I_\phi \angle (\beta - 90°) \qquad \bar{I}_{ca} = I_\phi \angle (\beta + 150°)$$

Two subtle points have been introduced in the above three-phase set examples. First, it can be observed that the quantity in each set with the first subscript of b lags the quantity with the first subscript of a while the quantity with first subscript of c leads the quantity with first subscript a. As a specific example, $\bar{V}_{bn}$ lags $\bar{V}_{an}$ by 120° while $\bar{V}_{cn}$ leads $\bar{V}_{an}$ by 120°. This situation is known as *a·b·c phase sequence*. The source of Fig. 2.5a would still exhibit balanced conditions if

$$V_{an} = V_\phi \angle \alpha \qquad V_{bn} = V_\phi \angle (\alpha + 120°) \qquad V_{cn} = V_\phi \angle (\alpha - 120°)$$

However, this situation is known as *a·c·b phase sequence*. The electric utility standard for transmission and distribution is *a·b·c* phase sequence, but in the end-use connection of induction motors, *a·c·b* phase sequence may be used for reversal of rotation direction.

The second subtlety of the above illustrated three-phase set is that there is a 30° phase difference between line and phase voltages associated by the first subscript for a wye-connected source or load. A similar 30° phase shift exists between line and phase currents of a delta-connected source or load. The sense of these phase shifts depends on phase sequence.

Wye Connection Voltages Application of KVL to the wye configuration of either Fig. 2.5*a* or Fig. 2.5*c* yields

$$\bar{V}_{ab} = \bar{V}_{an} - \bar{V}_{bn} \qquad \textbf{[2.11]}$$

$$\bar{V}_{bc} = \bar{V}_{bn} - \bar{V}_{cn} \qquad \textbf{[2.12]}$$

$$\bar{V}_{ca} = \bar{V}_{cn} - \bar{V}_{an} \qquad \textbf{[2.13]}$$

If the phase voltages $\bar{V}_{an}$, $\bar{V}_{bn}$, and $\bar{V}_{cn}$ are known, obviously line voltages $\bar{V}_{ab}$, $\bar{V}_{bc}$, and $\bar{V}_{ca}$ can be uniquely determined by [2.11] to [2.13] regardless of balanced or unbalanced conditions for the phase voltages. However, the converse is not true in general. Rewrite [2.11] to [2.13] in matrix format to give

$$\begin{bmatrix} 1 & -1 & 0 \\ 0 & 1 & -1 \\ -1 & 0 & 1 \end{bmatrix} \begin{bmatrix} \bar{V}_{an} \\ \bar{V}_{bn} \\ \bar{V}_{cn} \end{bmatrix} = \begin{bmatrix} \bar{V}_{ab} \\ \bar{V}_{bc} \\ \bar{V}_{ca} \end{bmatrix} \qquad \textbf{[2.14]}$$

Since the determinant of the phase voltage coefficient matrix has a zero value, knowledge of the line voltages does not allow a direct, unique calculation of the phase voltages unless a constraint exists. The general determination of phase voltages from knowledge of line voltages for unbalanced conditions can be handled by use of symmetrical components, but the procedure is considered beyond the scope of this text. However, for balanced conditions with *a·b·c* phase sequence,

$$\bar{V}_{bn} = \bar{V}_{an}e^{-j120} \qquad \bar{V}_{cn} = \bar{V}_{an}e^{j120°} \qquad \textbf{[2.15]}$$

Using [2.15] in [2.11] to [2.13] results in

$$\bar{V}_{ab} = \bar{V}_{an} - \bar{V}_{bn} = \bar{V}_{an} - \bar{V}_{an}e^{-j120°} = \bar{V}_{an}\sqrt{3}\,e^{j30°} \qquad \textbf{[2.16]}$$

$$\bar{V}_{bc} = \bar{V}_{bn} - \bar{V}_{cn} = \bar{V}_{bn} - \bar{V}_{bn}e^{-j120°} = \bar{V}_{bn}\sqrt{3}\,e^{j30°} \qquad \textbf{[2.17]}$$

$$\bar{V}_{ca} = \bar{V}_{cn} - \bar{V}_{an} = \bar{V}_{cn} - \bar{V}_{cn}e^{-j120°} = \bar{V}_{cn}\sqrt{3}\,e^{j30°} \qquad \textbf{[2.18]}$$

Hence, the line voltages for balanced conditions of *a·b·c* phase sequence are given by the phase voltage associated by the first subscript if its magnitude is multiplied by $\sqrt{3}$ and its phase angle is shifted a positive 30°. Conversely, the phase voltages are uniquely determined from known line voltages from [2.16] to [2.18] as

$$\bar{V}_{an} = \frac{\bar{V}_{ab}}{\sqrt{3}}e^{-j30°} \qquad \bar{V}_{bn} = \frac{\bar{V}_{bc}}{\sqrt{3}}e^{-j30°} \qquad \bar{V}_{cn} = \frac{\bar{V}_{ca}}{\sqrt{3}}e^{-j30°} \qquad \textbf{[2.19]}$$

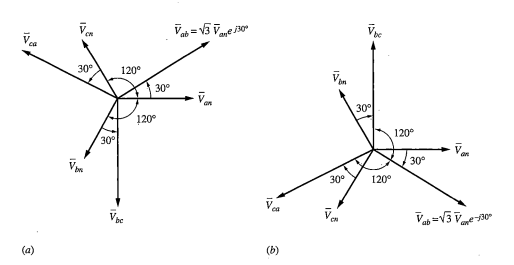

Figure 2.6
Phasor diagram of line and phase voltages for wye connection. (a) a·b·c phase sequence. (b) a·c·b phase sequence.

If the phase sequence were *a·c·b,* the signs of the exponential terms of [2.15] interchange. Following through on this change leads to

$$\bar{V}_{ab} = \bar{V}_{an}\sqrt{3}\,e^{-j30°} \qquad \bar{V}_{bc} = \bar{V}_{bn}\sqrt{3}\,e^{-j30°} \qquad \bar{V}_{ca} = \bar{V}_{cn}\sqrt{3}\,e^{-j30°}$$

[2.20]

Thus, line voltages for the *a·c·b* phase sequence are $\sqrt{3}$ greater than phase voltages, but the line voltage lags the phase voltage of the associated first subscript by 30°. The phasor diagram of Fig. 2.6 serves to clarify [2.15] to [2.20] results.

Example 2.8 | The wye-connected load of Fig. 2.5c has a phase voltage $\bar{V}_{bn} = 2400\angle -150°$ V. $\mathbf{Z} = \mathbf{Z}_a = \mathbf{Z}_b = \mathbf{Z}_c = 200\angle 30°\,\Omega$. Determine all other phase and line voltages and the line currents. Since not explicitly stated otherwise, assume *a·b·c* phase sequence. Based on [2.15],

$$\bar{V}_{an} = \bar{V}_{bn}e^{j120°} = 2400\angle -150°\,e^{j120°} = 2400\angle -30°\text{ V}$$

$$\bar{V}_{cn} = \bar{V}_{bn}e^{-j120°} = 2400\angle -150°\,e^{-j120°} = 2400\angle 90°\text{ V}$$

By [2.16] to [2.18],

$$\bar{V}_{ab} = \bar{V}_{an}\sqrt{3}\,e^{j30°} = 2400\angle -30°\,\sqrt{3}\,e^{j30°} = 4156.92\angle 0°\text{ V}$$

$$\bar{V}_{bc} = \bar{V}_{bn}\sqrt{3}\,e^{j30°} = 2400\angle -150°\,\sqrt{3}\,e^{j30°} = 4156.92\angle -120°\text{ V}$$

$$\bar{V}_{ca} = \bar{V}_{cn}\sqrt{3}\,e^{j30°} = 2400\angle 90°\,\sqrt{3}\,e^{j30°} = 4156.92\angle 120°\text{ V}$$

By use of Ohm's law,

$$\bar{I}_a = \frac{\bar{V}_{an}}{\mathbf{Z}} = \frac{2400 \angle -30°}{200 \angle 30°} = 12 \angle -60° \text{ A}$$

$$\bar{I}_b = \frac{\bar{V}_{bn}}{\mathbf{Z}} = \frac{2400 \angle -150°}{200 \angle 30°} = 12 \angle 180° \text{ A}$$

$$\bar{I}_c = \frac{\bar{V}_{cn}}{\mathbf{Z}} = \frac{2400 \angle 90°}{200 \angle 30°} = 12 \angle 60° \text{ A}$$

Delta Connection Currents Application of KCL to the delta-connected load of Fig. 2.5d gives

$$\bar{I}_a = \bar{I}_{ab} - \bar{I}_{ca} \qquad \textbf{[2.21]}$$

$$\bar{I}_b = \bar{I}_{bc} - \bar{I}_{ab} \qquad \textbf{[2.22]}$$

$$\bar{I}_c = \bar{I}_{ca} - \bar{I}_{bc} \qquad \textbf{[2.23]}$$

Placing [2.21] to [2.23] in matrix format produces

$$\begin{bmatrix} 1 & 0 & -1 \\ -1 & 1 & 0 \\ 0 & -1 & 1 \end{bmatrix} \begin{bmatrix} \bar{I}_{ab} \\ \bar{I}_{bc} \\ \bar{I}_{ca} \end{bmatrix} = \begin{bmatrix} \bar{I}_a \\ \bar{I}_b \\ \bar{I}_c \end{bmatrix} \qquad \textbf{[2.24]}$$

Similar to the case of voltages for a wye connection, knowledge of the phase currents in a delta circuit always uniquely determines the line currents through use of [2.21] to [2.23]. However, since the determinant of coefficients for the phase currents of [2.24] has a zero value, knowledge of the line currents does not allow direct, unique determination of the phase currents for the general case of unbalanced conditions. The method of symmetrical components can be used to handle this case, as mentioned earlier. If balanced conditions do exist with a·b·c phase sequence,

$$\bar{I}_{bc} = \bar{I}_{ab} e^{-j120°} \qquad \bar{I}_{ca} = \bar{I}_{ab} e^{j120°} \qquad \textbf{[2.25]}$$

By use of [2.25] in [2.21] to [2.23],

$$\bar{I}_a = \bar{I}_{ab} - \bar{I}_{ca} = \bar{I}_{ab} - \bar{I}_{ab} e^{j120°} = \bar{I}_{ab} \sqrt{3} e^{-j30°} \qquad \textbf{[2.26]}$$

$$\bar{I}_b = \bar{I}_{bc} - \bar{I}_{ab} = \bar{I}_{bc} - \bar{I}_{bc} e^{j120°} = \bar{I}_{bc} \sqrt{3} e^{-j30°} \qquad \textbf{[2.27]}$$

$$\bar{I}_c = \bar{I}_{ca} - \bar{I}_{bc} = \bar{I}_{ca} - \bar{I}_{ca} e^{j120°} = \bar{I}_{ca} \sqrt{3} e^{-j30°} \qquad \textbf{[2.28]}$$

It is apparent from [2.26] that knowledge of the line currents for a delta-connected load with balanced conditions and a·b·c phase sequence uniquely specifies the phase currents. Further, the line current lags the phase current associated by the first subscript by 30°. If the phase sequence were a·c·b, all signs of exponent terms in [2.25] to [2.28] are changed.

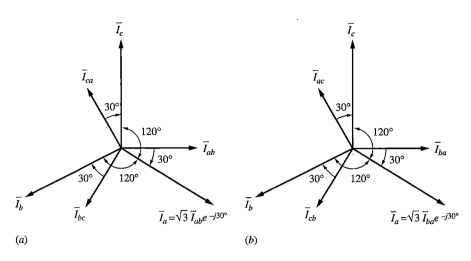

Figure 2.7
Phasor diagram for phase and line currents (a·b·c). (a) Delta-connected load. (b) Delta-connected source.

In a similar manner, the relationships for the line and phase currents for the delta-connected source of Fig. 2.5b with a·b·c phase sequence can be found as

$$\bar{I}_a = \bar{I}_{ba}\sqrt{3}\,e^{-j30°}$$ **[2.29]**

$$\bar{I}_b = \bar{I}_{cb}\sqrt{3}\,e^{-j30°}$$ **[2.30]**

$$\bar{I}_c = \bar{I}_{ac}\sqrt{3}\,e^{-j30°}$$ **[2.31]**

The direction of the phase currents was chosen so that current flows from the positive source voltage polarity marking. With this choice, it should be noted that line current is $\sqrt{3}$ times the phase current in magnitude and lags the phase current associated by the second subscript by 30°. Figure 2.7 serves to clarify the magnitude and phase relationships for the delta connections for the case of a·b·c phase sequence. Although not illustrated, all line currents would lead the associated phase current by 30° for the a·c·b phase sequence case.

Example 2.9 | If the delta-connected load of Fig. 2.5d has $\bar{I}_a = 15\angle0°$ A and $\mathbf{Z} = \mathbf{Z}_{ab} = \mathbf{Z}_{bc} = \mathbf{Z}_{ca} = 27.71\angle60°\ \Omega$, determine all voltages and currents.
For a·b·c phase sequence,

$$\bar{I}_b = 15\angle-120°\,\text{A} \qquad \bar{I}_c = 15\angle120°\,\text{A}$$

Using [2.26] to [2.28],

$$\bar{I}_{ab} = \frac{\bar{I}_a}{\sqrt{3}}e^{j30°} = \frac{15\angle0°}{\sqrt{3}}e^{j30°} = 8.66\angle30°\,\text{A}$$

$$\overline{I}_{bc} = \frac{\overline{I}_b}{\sqrt{3}} e^{j30°} = \frac{15\angle -120°}{\sqrt{3}} e^{j30°} = 8.66\angle -90° \text{ A}$$

$$\overline{I}_{ca} = \frac{\overline{I}_c}{\sqrt{3}} e^{j30°} = \frac{15\angle 120°}{\sqrt{3}} e^{j30°} = 8.66\angle 150° \text{ A}$$

By Ohm's law,

$$\overline{V}_{ab} = \overline{I}_{ab}\mathbf{Z} = (8.66\angle 30°)(27.71\angle 60°) = 239.97\angle 90° \text{ V}$$

$$\overline{V}_{bc} = 239.97\angle -30° \text{ V} \qquad \overline{V}_{ca} = 239.97\angle -150° \text{ V}$$

2.4.3 CONNECTION TRANSFORMATIONS

Analysis is simplified, especially for the case of balanced conditions, if all sources and loads can be represented as wye-connected. The transformations for the general unbalanced case of Fig. 2.8 with no neutral conductor present are as follows:

Delta-Wye

$$\mathbf{Z}_A = \frac{\mathbf{Z}_{AB}\mathbf{Z}_{CA}}{\mathbf{Z}_{AB} + \mathbf{Z}_{BC} + \mathbf{Z}_{CA}} \qquad \text{[2.32]}$$

$$\mathbf{Z}_B = \frac{\mathbf{Z}_{AB}\mathbf{Z}_{BC}}{\mathbf{Z}_{AB} + \mathbf{Z}_{BC} + \mathbf{Z}_{CA}} \qquad \text{[2.33]}$$

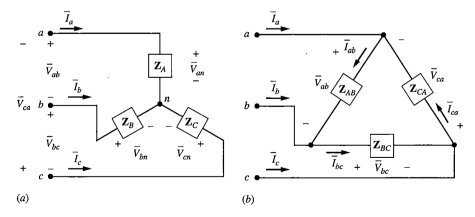

(a) (b)

Figure 2.8
General wye-delta loads. (a) Wye connection. (b) Delta connection.

$$\mathbf{Z}_C = \frac{\mathbf{Z}_{BC}\mathbf{Z}_{CA}}{\mathbf{Z}_{AB} + \mathbf{Z}_{BC} + \mathbf{Z}_{CA}} \qquad \text{[2.34]}$$

Wye-Delta

$$\mathbf{Z}_{AB} = \frac{\mathbf{Z}_A\mathbf{Z}_B + \mathbf{Z}_B\mathbf{Z}_C + \mathbf{Z}_C\mathbf{Z}_A}{\mathbf{Z}_C} \qquad \text{[2.35]}$$

$$\mathbf{Z}_{BC} = \frac{\mathbf{Z}_A\mathbf{Z}_B + \mathbf{Z}_B\mathbf{Z}_C + \mathbf{Z}_C\mathbf{Z}_A}{\mathbf{Z}_A} \qquad \text{[2.36]}$$

$$\mathbf{Z}_{CA} = \frac{\mathbf{Z}_A\mathbf{Z}_B + \mathbf{Z}_B\mathbf{Z}_C + \mathbf{Z}_C\mathbf{Z}_A}{\mathbf{Z}_B} \qquad \text{[2.37]}$$

For the special case of balanced conditions, $\mathbf{Z}_Y = \mathbf{Z}_A = \mathbf{Z}_B = \mathbf{Z}_C$ and $\mathbf{Z}_\Delta = \mathbf{Z}_{AB} = \mathbf{Z}_{BC} = \mathbf{Z}_{CA}$ for which [2.32] and [2.35] yield

$$\mathbf{Z}_Y = \frac{1}{3}\mathbf{Z}_\Delta \qquad \text{[2.38]}$$

$$\mathbf{Z}_\Delta = 3\mathbf{Z}_Y \qquad \text{[2.39]}$$

Example 2.10 A delta-connected load with balanced conditions has an impedance for each branch of 3 + j4 Ω. Determine the equivalent wye connection.
By [2.38],

$$\mathbf{Z}_Y = \frac{1}{3}\mathbf{Z}_\Delta = \frac{1}{3}(3 + j4) = \frac{1}{3}(5\angle 53.13°) = 1.667\angle 53.13° \, \Omega = 1 + j1.333 \, \Omega$$

If a delta-wye or wye-delta conversion is to be made for an unbalanced load, application of [2.32] to [2.34] or [2.35] to [2.37] generally results in cumbersome complex number arithmetic. The MATLAB program ⟨delwye.m⟩ has been developed to alleviate the tedious task.

Example 2.11 A wye-connected load has impedances $\mathbf{Z}_A = 3 + j4 = 5\angle 53.13°\,\Omega$, $\mathbf{Z}_B = 10 \angle 30°\,\Omega$, and $\mathbf{Z}_C = 3 \angle -45°\,\Omega$ using the notation of Fig. 2.8. Determine the phase impedance of the equivalent delta load using ⟨delwye.m⟩.
The screen display from execution of ⟨delwye.m⟩ is shown below, where the required delta equivalent impedance values are

$$\mathbf{Z}_{AB} = 22.153\angle 86.46° \, \Omega$$

$$\mathbf{Z}_{BC} = 13.292\angle -11.67° \, \Omega$$

$$\mathbf{Z}_{CA} = 6.646\angle 11.46° \, \Omega$$

```
Balanced Load?  (1=Yes,  2=No)  2

Conversion Direction?  (1=D->Y,  2=Y->D)  2

WYE to DELTA CONVERSION

magZa = 5
angZa = 53.13
magZb = 10
angZb = 30
magZc = 3
angZc = -45

DELTA IMPEDANCES in POLAR FORM
22.1526      86.4557
13.2915     -11.6743
 6.6458      11.4557
```

2.4.4 ANALYSIS

The circuits of Fig. 2.9 are sufficiently general to illustrate the principles of three-phase circuit analysis. In the event that a circuit with no neutral has a delta-configured load, it can be transformed to the form of Fig. 2.9*b* by use of [2.32] to [2.34].

Neutral Conductor   Application of KVL around the three loops of Fig. 2.9*a* yields in matrix format

$$\begin{bmatrix} \overline{V}_{an} \\ \overline{V}_{bn} \\ \overline{V}_{cn} \end{bmatrix} = \begin{bmatrix} \mathbf{Z}_a + \mathbf{Z}_\ell & 0 & 0 \\ 0 & \mathbf{Z}_b + \mathbf{Z}_\ell & 0 \\ 0 & 0 & \mathbf{Z}_c + \mathbf{Z}_\ell \end{bmatrix} \begin{bmatrix} \overline{I}_a \\ \overline{I}_b \\ \overline{I}_c \end{bmatrix} \qquad \textbf{[2.40]}$$

The decoupled set of algebraic equations given by [2.40] gives the currents as

$$\overline{I}_a = \frac{\overline{V}_{an}}{\mathbf{Z}_a + \mathbf{Z}_\ell} \qquad \overline{I}_b = \frac{\overline{V}_{bn}}{\mathbf{Z}_b + \mathbf{Z}_\ell} \qquad \overline{I}_c = \frac{\overline{V}_{cn}}{\mathbf{Z}_c + \mathbf{Z}_\ell} \qquad \textbf{[2.41]}$$

The neutral current follows from KCL as

$$\overline{I}_n = \overline{I}_a + \overline{I}_b + \overline{I}_c \qquad \textbf{[2.42]}$$

For the special case of balanced conditions, $\mathbf{Z} = \mathbf{Z}_a + \mathbf{Z}_b + \mathbf{Z}_c$, $\overline{V}_{bn} = \overline{V}_{an}e^{-j120°}$, and $\overline{V}_{cn} = \overline{V}_{an}e^{j120°}$, whence [2.41] and [2.42] can be expressed as

$$\overline{I}_a = \frac{\overline{V}_{an}}{\mathbf{Z} + \mathbf{Z}_\ell} \qquad \textbf{[2.43]}$$

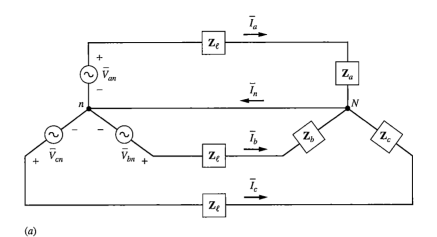

(a)

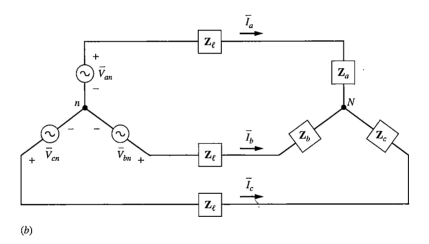

(b)

Figure 2.9
Three-phase networks. (a) With neutral conductor. (b) Without neutral conductor.

$$\bar{I}_b = \frac{\bar{V}_{an}e^{-j120°}}{\mathbf{Z} + \mathbf{Z}_\ell} = \bar{I}_a e^{-j120°} \qquad \textbf{[2.44]}$$

$$\bar{I}_c = \frac{\bar{V}_{an}e^{j120°}}{\mathbf{Z} + \mathbf{Z}_\ell} = \bar{I}_a e^{j120°} \qquad \textbf{[2.45]}$$

$$\bar{I}_n = \bar{I}_a + \bar{I}_a e^{-j120°} + \bar{I}_a e^{j120°} = 0 \qquad \textbf{[2.46]}$$

It is apparent from [2.43] to [2.46] that only a solution for $\bar{I}_a$ need be made. $\bar{I}_b$ and $\bar{I}_c$ can then be formed by $-120°$ and $120°$ phase shift, respectively, of $\bar{I}_a$. Further, since $\bar{I}_n = 0$, the neutral conductor could be removed without altering the performance. This latter observation bears out the earlier statement that the neutral conductor carries only the unbalanced current.

Example 2.12

A three-phase circuit operating at balanced conditions can be modeled by Fig. 2.9a where $\bar{V}_{bn} = 277\angle-90°$ V, line impedance $\mathbf{Z}_\ell = 0.1\angle90°\,\Omega$, and the load impedance $\mathbf{Z} = 3 + j2\,\Omega$. Determine the values of the three line currents.

Since balanced conditions exist,

$$\bar{V}_{an} = \bar{V}_{bn}e^{j120°} = 277\angle30° \text{ V}$$

Based on [2.43] to [2.45],

$$\bar{I}_a = \frac{\bar{V}_{an}}{\mathbf{Z} + \mathbf{Z}_\ell} = \frac{277\angle30°}{3 + j2 + j0.1} = \frac{2.77\angle30°}{3.662\angle34.99°} = 75.642\angle-4.99° \text{ A}$$

$$\bar{I}_b = \bar{I}_a e^{-j120°} = 75.642\angle-124.99° \text{ A}$$

$$\bar{I}_c = \bar{I}_a e^{j120°} = 75.642\angle115.01° \text{ A}$$

No Neutral Conductor For the three-phase circuit of Fig. 2.9b without a neutral conductor, applications of KVL around the two loops and KCL at node N yield in matrix format

$$[\bar{V}] = \begin{bmatrix} \bar{V}_{an} - \bar{V}_{bn} \\ \bar{V}_{bn} - \bar{V}_{cn} \\ 0 \end{bmatrix}$$

$$= \begin{bmatrix} \mathbf{Z}_a + \mathbf{Z}_\ell & -(\mathbf{Z}_b + \mathbf{Z}_\ell) & 0 \\ 0 & \mathbf{Z}_b + \mathbf{Z}_\ell & -(\mathbf{Z}_c + \mathbf{Z}_\ell) \\ 1 & 1 & 1 \end{bmatrix} \begin{bmatrix} \bar{I}_a \\ \bar{I}_b \\ \bar{I}_c \end{bmatrix} = [\mathbf{Z}][\bar{I}] \qquad \textbf{[2.47]}$$

or

$$[\bar{I}] = [\mathbf{Z}]^{-1}[\bar{V}] \qquad \textbf{[2.48]}$$

Determination of currents $\bar{I}_a$, $\bar{I}_b$, and $\bar{I}_c$ of [2.47] requires simultaneous solution of the three linear algebraic equations through use of the matrix inversion and multiplication as suggested by [2.48] or another valid approach of similar complexity. However, if balanced conditions exist, the first equation of [2.47] can be removed and manipulated to produce

$$\bar{V}_{an} - \bar{V}_{bn} = (\mathbf{Z} + \mathbf{Z}_\ell)(\bar{I}_a - \bar{I}_b)$$

$$\bar{V}_{an}\sqrt{3}\,e^{j30°} = (\mathbf{Z} + \mathbf{Z}_\ell)\bar{I}_a\sqrt{3}\,e^{j30°}$$

or

$$\bar{I}_a = \frac{\bar{V}_{an}}{\mathbf{Z} + \mathbf{Z}_\ell}$$

Then,

$$\overline{I}_b = \overline{I}_a e^{-j120°} \qquad \overline{I}_c = \overline{I}_a e^{j120°}$$

giving currents identical to [2.43] to [2.45]. The result is not surprising, since it has been concluded that the neutral conductor carries no current for balanced conditions. Thus, removal of the neutral conductor from Fig. 2.9a renders the circuit identical to Fig. 2.9b demanding that identical current result.

   The results of this section have shown that analyses of the somewhat general three-phase circuits of Fig. 2.9 are significantly simplified for the case of balanced conditions in that phase a can be isolated for independent performance computation. The performance of phases b and c can then be determined by appropriate phase shift of phase a results. No such simplified analysis exists for the unbalanced case, but rather simultaneous solution of the equation set of [2.47] must be performed. The MATLAB program ⟨Tphckt.m⟩ has been developed to handle either of the circuits of Fig. 2.9, balanced or unbalanced, without the need to carry out complex number arithmetic by a hand-held calculator. The program determines the three line currents, the neutral current, and the load phase voltages.

Example 2.13

A three-phase circuit modeled by Fig. 2.9a has a balanced source and an unbalanced load. Specific data are $\overline{V}_{an} = 240\angle30°$ V, $\mathbf{Z}_\ell = 0.2\angle90°\,\Omega$, $\mathbf{Z}_a = 4\angle30°\,\Omega$, $\mathbf{Z}_b = 5\angle24.5°\,\Omega$, and $\mathbf{Z}_c = 7\angle15°\,\Omega$. Determine the load currents, the neutral current, and the load voltages.
   Execution of ⟨Tphckt.m⟩ finds

$$\overline{I}_a = 58.484\angle-2.42°\text{ A} \qquad \overline{I}_b = 47.186\angle123.45°\text{ A}$$

$$\overline{I}_c = 34.021\angle-106.57°\text{ A} \qquad \overline{I}_n = 23.12\angle10.7°\text{ A}$$

$$\overline{V}_{aN} = 233.938\angle27.58°\text{ V} \qquad \overline{V}_{bN} = 235.933\angle147.95°\text{ V}$$

$$\overline{V}_{cN} = 238.149\angle-91.57°\text{ V}$$

where the screen display of the session is shown below.

```
THREE-PHASE CIRCUIT ANALYSIS

  Source balanced? (1=Yes, 2=No) 1
  magVan = 240
  angVan = 30

  Load balanced? (1=Yes, 2=No) 2
  magZa = 4
  angZa = 30
  magZb = 5
```

```
   angZb  =  24.5
   magZc  =  7
   angZc  =  15

 Line impedance?  (1=Yes,  2=No)  1
   magZl  =  0.2
   angZl  =  90

 Neutral connection?  (1=Yes,  2=No)  1

 LOAD  CURRENT  in  POLAR  FORM
    58.4844         -2.4190
    47.1865         123.4494
    34.0213        -106.5692

 In  =  23.12|_10.7  Load->Source

 LOAD  VOLTAGE  in  POLAR  FORM
    233.9377         27.5810
    235.9327        147.9494
    238.1489        -91.5692
```

Rework Example 2.13 if the neutral conductor is removed.   **Example 2.14**

The following screen display from execution of ⟨Tphckt.m⟩ shows the resulting currents and load voltage.

```
 THREE-PHASE  CIRCUIT  ANALYSIS

  Source  balanced?  (1=Yes,  2=No)  1
  magVan  =  240
  angVan  =  30

  Load  balanced?  (1=Yes,  2=No)  2
  magZa  =  4
  angZa  =  30
  magZb  =  5
  angZb  =  24.5
  magZc  =  7
  angZc  =  15

 Line  impedance?  (1=Yes,  2=No)  1
   magZl  =  0.2
   angZl  =  90

 Neutral  connection?  (1=Yes,  2=No)  2
```

```
LOAD CURRENT  in POLAR FORM
   48.8405        -3.8682
   50.7405       131.6615
   37.7233      -113.4311

LOAD VOLTAGE  in POLAR FORM
  195.3620        26.1318
  253.7027       156.1615
  264.0631       -98.4311
```

2.5 POWER FLOW

Flow of *instantaneous power* p(t) is of interest in any electric circuit to give direct assessment of the energy transfer nature. If the circuit is excited by a periodic source, the instantaneous power is also periodic and, thus, an *average power* can be formed to evaluate the energy transfer rate. For those circuits in the sinusoidal steady state, three additional power quantities can be formed to characterize the energy transfer process—*reactive power, apparent power,* and *complex power.* A review of these five power quantities, using the typical power system case of an equivalent *R-L* circuit in the sinusoidal steady state, is presented to refresh understanding and to clarify notation.

2.5.1   SINGLE-PHASE CIRCUITS

Consider the series *R-L* circuit of Fig. 2.2 with $v(t) = \sqrt{2}V\cos(\omega t)$. Through use of sinusoidal steady-state analysis, $i(t) = \sqrt{2}I\cos(\omega t - \theta)$ where $X = \omega L$, $Z = \sqrt{R^2 + X^2}$, $I = V/Z$, and $\theta = \tan^{-1}(X/R)$. The *instantaneous power* flowing into the port is given in units of W by

$$p(t) = v(t)i(t) = \sqrt{2}V\cos(\omega t)\sqrt{2}I\cos(\omega t - \theta) = 2VI\cos(\omega t)\cos(\omega t - \theta)$$

Application of the trigonometric identities $\cos(x)\cos(y) = \frac{1}{2}\cos(x+y) + \frac{1}{2}\cos(x-y)$ and $\cos(x-y) = \cos(x)\cos(y) + \sin(x)\sin(y)$ leads to

$$p(t) = VI\cos(2\omega t - \theta) + VI\cos\theta$$

$$= VI\cos\theta\cos(2\omega t) + VI\sin\theta\sin(2\omega t) + VI\cos\theta$$

$$p(t) = VI\cos\theta[1 + \cos(2\omega t)] + VI\sin\theta\sin(2\omega t) \qquad \textbf{[2.49]}$$

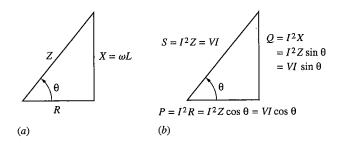

Figure 2.10
Series R-L circuit impedance and power triangles. (a) Impedance triangle. (b) Power triangle.

From the impedance triangle of Fig. 2.10a, $\cos\theta = R/Z$ and $\sin\theta = X/Z$. Substitute these relationships into [2.49] and recognize $V/Z = I$ to yield

$$p(t) = I^2R[1 + \cos(2\omega t)] + I^2X\sin(2\omega t) \qquad \textbf{[2.50]}$$

If $R = 0$, the first term of [2.50] vanishes; thus, it must describe the instantaneous power dissipated by the resistor. Instantaneous power flow to the resistor varies at twice the source frequency, but it never has a negative value. Likewise, if $L = 0$, $X = \omega L = 0$ and the second term of [2.50] vanishes, supporting the conclusion that it represents the instantaneous power flow to the inductor. Instantaneous power flows to the inductor during half of the time and from the inductor during the other half of the time at a frequency twice that of the source.

The *average power* flowing into the port is determined by the average of the instantaneous power given by [2.49] and [2.50] over a source period in units of W.

$$P = \frac{1}{2\pi}\int_0^{2\pi} p(\omega t)\, d(\omega t) = VI\cos\theta = I^2R \qquad \textbf{[2.51]}$$

The result of [2.51] emphasizes that the average power flow is only to the resistor. Average power flow to the inductor is zero; however, the peak instantaneous power flow to the inductor is known as *reactive power* and is measured in units of voltamperes reactive, or VARs.

$$Q = VI\sin\theta = I^2X \qquad \textbf{[2.52]}$$

The *power triangle* of Fig. 2.10b is a similar triangle to the impedance triangle of Fig. 2.10a formed by I^2 scaling of all sides. When fully developed, it will be found a convenient tool for use in power flow analysis. The adjacent side of the power triangle is identified as the average power to the circuit of Fig. 2.2 while the opposite side is clearly the reactive power flow. The hypotenuse is a heretofore unidentified entity that will now be known as the *apparent power* measured in units of voltamperes, or VA.

$$S = VI = I^2Z = \sqrt{P^2 + Q^2} \qquad \textbf{[2.53]}$$

The apparent power is a quantity that when multiplied by $\cos \theta$ yields the average power, or if multiplied by $\sin \theta$ gives the reactive power. Based on this observation, $\cos \theta$ and $\sin \theta$ are known as the *power factor* and *reactive factor*, respectively.

$$PF = \cos \theta \qquad RF = \sin \theta \qquad\qquad \textbf{[2.54]}$$

The angle θ is positive valued for an *R-L* circuit and negative valued for an *R-C* circuit. It can be equivalently determined by either of the two following methods:

1. Impedance triangle angle, or $\tan^{-1}(X/R)$.
2. Angle measured from $\bar{I}$ to $\bar{V}$ on the phasor diagram, or $\angle \bar{V} - \angle \bar{I}$.

Noting the quadrature relationship between P and Q for the power triangle of Fig. 2.10*b*, a quantity known as *complex power* is defined as

$$\mathbf{S} = P + jQ = VI\cos \theta + jVI \sin \theta = I^2 \mathbf{Z} = \overline{VI}^* \qquad \textbf{[2.55]}$$

Since $S = |\mathbf{S}|$, logically the unit of measure for $\mathbf{S}$ is also VA. Average power and reactive power are both scalar quantities. The sum of the average power flow to each branch of a network must give the total average power flow into the network. Also, the reactive power flow to individual branches of a network can be summed to determine the total reactive power flow into the network. Consequently, the complex power flow to each branch of a network can be summed to yield the total complex power flow into the network.

Example 2.15 | Assume that the circuit of Fig. 2.4*a* is in the sinusoidal steady state, fed by a voltage source $v(t) = 120\sqrt{2}\cos(120\pi t)$ V. (*a*) Find an expression for the instantaneous power $p(t)$ supplied by the source $v(t)$, sketch the result, and comment on the nature of the instantaneous power flow. (*b*) Determine the average power, the reactive power, apparent power, and complex power supplied by the source $v(t)$ and the power factor seen by the source.

(*a*) The results of Example 2.4 are valid if all phasors are shifted by $-30°$. Hence, $\bar{I} = 27.74\angle 70.6°$ A, and

$$i(t) = 27.74\sqrt{2}\cos(120\pi t + 70.6°) = 39.23\cos(120\pi t + 70.6°)\,\text{A}$$

whence

$$p(t) = v(t)\,i(t) = 120\sqrt{2}\cos(120\pi t)39.23\cos(120\pi t + 70.6°)$$

$$= 6657.55\cos(120\pi t)\cos(120\pi t + 70.6°)\,\text{W}$$

Using $\cos(x)\cos(y) = \frac{1}{2}\cos(x + y) + \frac{1}{2}\cos(x - y)$,

$$p(t) = 3328.77\cos(240\pi t + 70.6°) + 3328.77\cos(70.6°)\,\text{W}$$

$$p(t) = 1105.69 + 3328.77\cos(240\pi t + 70.6°)\,\text{W}$$

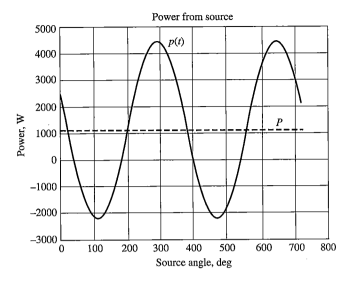

Figure 2.11
Power flow for Example 2.11

Figure 2.11 shows a plot of this instantaneous power, along with a plot of the average power where the instantaneous power is seen to vary at twice source frequency about the mean value P. During the intervals for which $p(t) < 0$, power is actually being absorbed by source $v(t)$.

(b) Based on [2.51] to [2.55],

$$P = VI \cos \theta = (120)(27.74) \cos (-70.6°) = 1105.69 \, \text{W}$$

$$Q = VI \sin \theta = (120)(27.74) \sin (-70.6°) = -3139.80 \, \text{VARs}$$

$$PF = \cos \theta = \cos (-70.6°) = 0.332 \, \text{leading}$$

$$S = VI = (120)(27.74) = 3328.80 \, \text{VA}$$

$$\mathbf{S} = \overline{VI}^* = (120\angle 0°)(27.74\angle -70.6°) = 3328.80 \angle -70.6° \, \text{VA}$$

The circuit of Fig. 2.4a is excited by the source of Example 2.15. (a) Determine the complex power flowing to each of branches 1 and 2. (b) Show that addition of these two quantities yields the complex power supplied by the source $v(t)$ as determined in Example 2.15. (c) Calculate required values and draw the power triangles for branches 1 and 2 and the composite power triangle for the source.

Example 2.16

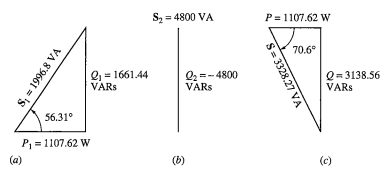

Figure 2.12
Power triangles. (a) Branch 1. (b) Branch 2. (c) Source.

(a) Using the phasor values of currents $\bar{I}_1$ and $\bar{I}_2$, after a $-30°$ phase shift to adjust for the present source being on the reference, in [2.55],

$$\mathbf{S}_1 = \overline{VI}_1^* = (120 \angle 0°)(16.64 \angle 56.31°) = 1996.8 \angle 56.31° \text{ VA}$$

$$\mathbf{S}_2 = \overline{VI}_2^* = (120 \angle 0°)(40 \angle -90°) = 4800 \angle -90° \text{ VA}$$

(b)

$$\mathbf{S} = \mathbf{S}_1 + \mathbf{S}_2 = (1107.62 + j1661.44) + (0 - j4800)$$

$$\mathbf{S} = 1107.62 - j3138.56 = 3328.27 \angle -70.6° \text{ VA}$$

(c) The power triangle for the two branches and the source are displayed by Fig. 2.12. Since $P_2 = 0$, the power triangle for branch 2 degenerates to a vertical line segment.

2.5.2 THREE-PHASE CIRCUITS

Since any delta-connected load can be converted to an equivalent wye-connected load, determination of power flow to a wye-connected load is sufficiently general for any three-phase power flow analysis. Further, instantaneous power is a scalar quantity so that addition of the power flowing to each of the three ports that share the neutral as a common point in Fig. 2.8a gives the total instantaneous power flowing to the three-phase load as

$$p_T(t) = v_{an}(t)\,i_a(t) + v_{bn}(t)\,i_b(t) + v_{cn}(t)\,i_c(t) \qquad \textbf{[2.56]}$$

It is emphasized that [2.56] is valid with or without a neutral conductor present. However, if the actual load is delta-connected, determination of the equivalent line-to-neutral voltages can be difficult. Solution of one of the problems at the end of this chapter will show that for any three-wire, three-phase load, [2.56] can be rewritten to allow calculation of three-phase instantaneous power to the load with knowledge of two of the line voltages, giving

$$p_T(t) = v_{bc}(t)\, i_b(t) - v_{ca}(t)\, i_a(t) \qquad \textbf{[2.57]}$$

The only constraint on [2.57] is that $i_a + i_b + i_c = 0$; thus, the application can be balanced or unbalanced.

Referring to Fig. 2.8a and using [2.51], the total average power flow to any three-phase load in the sinusoidal steady state is

$$P_T = V_{an}I_a \cos\theta_a + V_{bn}I_b \cos\theta_b + V_{cn}I_c \cos\theta_c \qquad \textbf{[2.58]}$$

where the voltages and currents are rms values. The angles are the impedance angles of the wye-connected loads. Based on [2.52], the total reactive power flow to any three-phase load can be written as

$$Q_T = V_{an}I_a \sin\theta_a + V_{bn}I_b \sin\theta_b + V_{cn}I_c \sin\theta_c \qquad \textbf{[2.59]}$$

For the important and common case of a balanced three-phase wye equivalent or actual load,

$$V_\phi = V_L/\sqrt{3} = V_{an} = V_{bn} = V_{cn} \qquad \textbf{[2.60]}$$

$$I_\phi = I_L = I_a = I_b = I_c \qquad \textbf{[2.61]}$$

$$\theta = \theta_a = \theta_b = \theta_c \qquad \textbf{[2.62]}$$

By use of [2.60] to [2.62], [2.58] and [2.59] yield for the balanced case

$$P_T = 3V_\phi I_\phi \cos\theta = \sqrt{3}\, V_L I_L \cos\theta \qquad \textbf{[2.63]}$$

$$Q_T = 3V_\phi I_\phi \sin\theta = \sqrt{3}\, V_L I_L \sin\theta \qquad \textbf{[2.64]}$$

It is emphasized that the angle θ is the angle of the phase impedance. A three-phase power triangle for balanced conditions is constructed in Fig. 2.13, leading to the definition of total apparent power.

$$S_T = 3V_\phi I_\phi = \sqrt{3}\, V_L I_L \qquad \textbf{[2.65]}$$

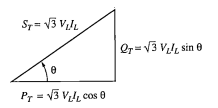

Figure 2.13
Three-phase power triangle for balanced conditions

Total complex power flow for the balanced three-phase case follows as

$$\mathbf{S}_T = \sqrt{3}\, V_L I_L \angle \theta = \sqrt{3}\, V_L I_L \cos \theta + j\sqrt{3}\, V_L I_L \sin \theta \qquad \text{[2.66]}$$

Example 2.17 Determine the total three-phase average power supplied to the unbalanced load of Example 2.13.

Based on [2.58] and the results of Example 2.13,

$$P_T = V_{aN} I_a \cos \theta_a + V_{bN} I_b \cos \theta_b + V_{cN} I_c \cos \theta_c$$

$$P_T = (233.938)(58.484)\cos(30°) + (235.933)(47.186)\cos(24.5°)$$

$$+ (238.149)(34.021)\cos(15°)$$

$$P_T = 29.805 \text{ kW}$$

Example 2.18 For the balanced load of Example 2.8, find (a) P_T, (b) Q_T, and (c) $\mathbf{S}_T$.

Using [2.63], [2.64], and [2.66],

(a) $P_T = \sqrt{3}\, V_L I_L \cos \theta = \sqrt{3}(4156.92)(12)\cos(30°) = 74.822$ kW

(b) $Q_T = \sqrt{3}\, V_L I_L \sin \theta = \sqrt{3}(4156.92)(12)\sin(30°) = 43.199$ kVARs

(c) $\mathbf{S}_T = P_T + jQ_T = 74.822 + j43.199 = 86.397 \angle 30°$ kVA

2.5.3   POWER MEASUREMENT

The instrument for measurement of average power flow is the wattmeter, denoted by WM in Fig. 2.14. The wattmeter has a voltage or potential coil (PC) and a current coil (CC) with polarity markings as indicated. When the wattmeter is connected with coil polarity markings as shown in Fig. 2.14, the meter simply reads the average value of power flowing into the single-phase network as

$$P = \frac{1}{T} \int_0^T v(t)\, i(t)\, dt \qquad \text{[2.67]}$$

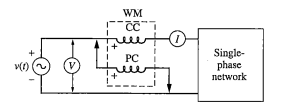

Figure 2.14
Wattmeter in single-phase application

provided the meter can fully respond to the frequencies of v and i. If both coils were reversed, the wattmeter reading would be unchanged. If either one of the coils were reversed, the meter would give the negative of the value of [2.67]. Electronic wattmeters will indicate any negative values by a minus sign in the display. Analog wattmeters will simply attempt to read down scale, and the user must reverse the potential coil connection and manually insert the negative value in recorded data.

If both v and i are void of harmonics, the apparent power flowing to the network is given by $S = VI$ where V and I are the rms values of v and i, respectively. The reactive power flow to the network is determined by $Q = [(VI)^2 - P^2]$. The power factor follows as $PF = P/S$. Electronic wattmeters may have selector switches that allow calculation and display of S, Q, and PF by the meter; thus, use of the voltmeter and ammeter shown in Fig. 2.14 is not necessary.

In concept, measurement of average power flow in three-phase circuits can be accomplished by connecting a wattmeter to read the power flowing to each phase using the connections of Fig. 2.14. The total three-phase power would then be the sum of the three wattmeter readings. However, if the network is wye-connected, the instrument connection requires that the potential coils of the three wattmeters be connected to the neutral. If the three-phase system is a three-wire connection, the neutral point is not available. On the other hand, if the three-phase system is delta-connected, each branch of the delta must be opened to insert the wattmeter current coils. This is at best difficult, and sometimes it is not feasible.

Conveniently, [2.57] suggests a simple approach to three-phase power measurement where only access to the three feed lines is required. The subscripts of v_{ca} can be reversed, allowing the equation to be rewritten as

$$p_T(t) = v_{bc}(t)i_b(t) + v_{ac}(t)i_a(t) \qquad \textbf{[2.68]}$$

The average value of [2.68] gives the total three-phase average power.

$$P_T = V_{bc}I_b\cos(\angle \overline{V}_{bc} - \angle \overline{I}_b) + V_{ac}I_a\cos(\angle \overline{V}_{ac} - \angle \overline{I}_a) \qquad \textbf{[2.69]}$$

Figure 2.15 shows a connection of two wattmeters to measure P_T as suggested by [2.69]. The total three-phase average power is the algebraic sum of the wattmeter readings. Since [2.68] is valid for any three-wire, three-phase system, whether balanced or unbalanced, this *two-wattmeter method* measures the total three-phase average power of a three-wire system even if it is unbalanced.

2.5.4 POWER FACTOR CORRECTION

The electric utility charges a customer for energy usage (Wh); however, in many industrial applications with a large percentage of motor loads, the inherent lagging power factor may decrease significantly from unity. Since the average power supplied is determined by $VI(PF)$, and since the utility grid is a near-constant voltage system, a lower value of power factor means that a larger value of current is required to supply the power that integrates over time to determine energy usage.

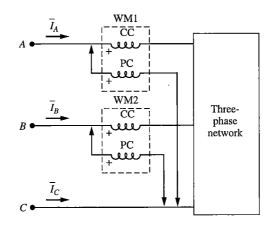

Figure 2.15
Two-wattmeter method

Consequently, to supply a specific value of energy to a low power factor load, the utility must bear the capital outlay of greater ampacity alternators, transformers, and transmission lines than would be necessary to supply the same value of energy to a high power factor load. In order to compensate for the added capital outlay resulting from low power factor loads, the utility has a pricing structure that charges a higher rate for energy supplied to a low power factor load. The break-point is typically at 0.8 PF lagging. The low power factor customer has a choice of paying the higher rate or correcting the power factor by installing shunt-connected capacitors at the substation supplying power to the plant. The power factor penalty is typically large enough that the cost of the capacitors can be amortized in less than a year of service.

Example 2.19 | The single-phase $3 + j4\ \Omega$ load of Fig. 2.16 has an input PF of 0.6 lagging. Determine the value of C that must be installed in this 60-Hz circuit to correct the input power factor to 0.9 lagging.

With the switch open,

$$\bar{I} = \frac{240\angle 0°}{3 + j4} = 48\angle -53.13°\,\text{A}$$

The complex power supplied to the load is found as

$$\mathbf{S} = \bar{V}\bar{I}^* = (240\angle 0°)(48\angle 53.13°) = 11{,}520\angle 53.13°\,\text{VA}$$

The average and reactive power are

$$P = 11{,}520\cos(53.13°) = 6912.02\,\text{W}$$

$$Q = 11{,}520\sin(53.13°) = 9215.99\,\text{VARs}$$

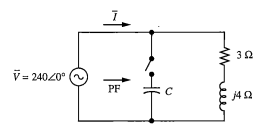

Figure 2.16
Single-phase PF correction

Addition of C will not change the average power supplied, but the leading VARs supplied by the capacitor will reduce the lagging VARs required from the source $\bar{V}$. Let Q' be the lagging VARs supplied by the source for a 0.9 PF lagging. Then,

$$\frac{Q'}{P} = \tan[\cos^{-1}(0.9)]$$

$$Q' = 6912.02\tan[25.84°] = 3347.64 \text{ VARs}$$

The reactive power that must be supplied by the capacitor is

$$Q - Q' = 9215.99 - 3347.64 = 5868.35 \text{ VARs}$$

and

$$C = \frac{Q_c}{\omega V^2} = \frac{5868.35}{120\pi(240)^2} = 270.2 \ \mu\text{F}$$

The balanced, delta-connected capacitor bank of Fig. 2.17 is installed to correct the PF of the balanced $3 + j4 \ \Omega$ delta load to unity. Determine the value of C required for this 60-Hz circuit. | **Example 2.20**

For each phase, the leading reactive VARs of the capacitor must equal the lagging reactive VARs of the inductor.

$$Q_L = Q_C$$

$$\omega C V_L^2 = \left(\frac{V_L}{Z}\right)^2 \omega L$$

or

$$C = \frac{L}{Z^2} = \frac{X_L\omega}{R^2 + X_L^2} = \frac{4/120\pi}{(3)^2 + (4)^2} = 424.4 \ \mu\text{F}$$

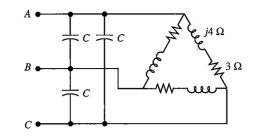

Figure 2.17
Three-phase PF correction

2.6 MULTIPLE FREQUENCY CIRCUITS

Modern electric machines are frequently fed from power electronic circuits that use high power level switching devices to fabricate output voltage waveforms. Consequently, the voltage impressed at the terminals of an energy conversion device may contain harmonic frequencies as well as the fundamental sinusoidal component. If the equivalent circuit of an energy conversion device is linear, superposition argument and phasor methods can be used to handle analysis. The procedure will be clarified by the example that follows.

Example 2.21   A series R-L circuit is fed by a voltage source $v(t) = 120\sqrt{2}\cos(120\pi t) + 15\sqrt{2}\cos(600\,\pi t + 15°)$ V. If $R = 10\ \Omega$ and $L = 40$ mH, determine the current $i(t)$.

  The time domain circuit is shown in Fig. 2.18a, where the source $v(t)$ has been shown as two sources. $v_1(t)$ models the fundamental frequency component of $v(t)$, which is 60 Hz in this case. $v_h(t) = v_5(t)$ represents the fifth-harmonic component of $v(t)$. Phasor methods can be applied only if all frequencies in a circuit are identical. Figure 2.18b is the phasor domain circuit with the 60-Hz component of $v(t)$ only applied, from which

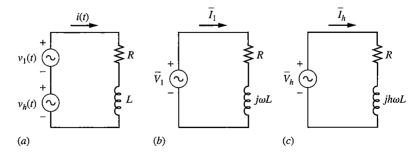

Figure 2.18
Multiple frequency source. (a) Complete circuit in time domain. (b) Phasor domain circuit for fundamental component. (c) Phasor domain circuit for harmonic component.

$$\overline{I}_1 = \frac{\overline{V}_1}{R + j\omega L} = \frac{120\angle 0°}{10 + j(120\pi)(0.040)} = 6.632\angle -56.45° \text{ A}$$

$$i_1(t) = 6.632\sqrt{2}\cos(120\pi t - 56.45°) = 9.379\cos(120\pi t - 56.45°) \text{ A}$$

From the circuit of Fig. 2.18c, the fifth-harmonic current is

$$\overline{I}_5 = \frac{\overline{V}_5}{R + jh\omega L} = \frac{15\angle 15°}{10 + j(5)(120\pi)(0.040)} = 0.197\angle -67.44° \text{ A}$$

$$i_5(t) = 0.197\sqrt{2}\cos(600\pi t - 67.44°) = 0.279\cos(600\pi t - 67.44°) \text{ A}$$

By superposition argument,

$$i(t) = i_1(t) + i_5(t)$$

$$= 9.379\cos(120\pi t - 56.43°) + 0.279\cos(600\pi t - 67.44°) \text{ A}$$

2.7 COMPUTER ANALYSIS CODE

```
%%%%%%%%%%%%%%%%%%%%%%%%%%%%%%%%%%%%%%%%%%%%%%%%%%%%%%%%%%%%%%%%%%%%%%%%%%%
%
% csum.m - forms the sum of a series of complex numbers
%          and displays the result in both polar and
%          rectangular format.
%
%%%%%%%%%%%%%%%%%%%%%%%%%%%%%%%%%%%%%%%%%%%%%%%%%%%%%%%%%%%%%%%%%%%%%%%%%%%
S=0; disp(' ');
disp(' SUM OF COMPLEX NUMBERS'); disp(' ');
disp(' Form: 1-polar, 2-rectangular'); disp(' ');
nterms=input('How many numbers to be added? '); disp(' ');
for i=1:nterms;
  F=input(['Form of ',num2str(i),' = ']);
  if (F~=1 & F~=2); disp('INVALID FORM'); end
  if F==1
    M=input(['Mag  ',num2str(i),' = ']);
    A=input(['Deg  ',num2str(i),' = '])*pi/180;
    S=S+M*exp(j*A); disp(' ');
  else
    R=input(['Real ',num2str(i),' = ']);
    I=input(['Imag ',num2str(i),' = ']);
    S=S+R+j*I; disp(' ');
  end
end
```

```
disp(' '); disp(['SUM = ' num2str(real(S)) ' +j '...
    num2str(imag(S)) ' = ' num2str(abs(S)) '|_'...
    num2str(angle(S)*180/pi) 'deg']);


%%%%%%%%%%%%%%%%%%%%%%%%%%%%%%%%%%%%%%%%%%%%%%%%%%%%%%%%%%%%%%%%%%%%%%%%
%
% cratio.m - forms the ratio of products of complex numbers
%            and displays the result in both polar and
%            rectangular format.
%
%%%%%%%%%%%%%%%%%%%%%%%%%%%%%%%%%%%%%%%%%%%%%%%%%%%%%%%%%%%%%%%%%%%%%%%%

disp(' ');
disp(' RATIO OF COMPLEX NUMBER PRODUCTS'); disp(' ');
disp(' Form: 1-polar, 2-rectangular'); disp(' ');
num=input('How many numerator numbers? '); disp(' ');
for i=1:num;
   F=input(['Form of ',num2str(i),' = ']);
   if (F~=1 & F~=2); disp('INVALID FORM'); end
   if F==1
     M=input(['Mag  ',num2str(i),' = ']);
     A=input(['Deg  ',num2str(i),' = '])*pi/180;
     N(i)=M*exp(j*A); disp(' ');
   else
     R=input(['Real ',num2str(i),' = ']);
     I=input(['Imag ',num2str(i),' = ']);
     N(i)=R+j*I; disp(' ');
   end
end
NP=N(1);  for k=2:num; NP=NP*N(k); end
den=input('How many denominator numbers? '); disp(' ');
for i=1:den;
   F=input(['Form of ',num2str(i),' = ']);
   if (F~=1 & F~=2); disp('INVALID FORM'); end
   if F==1
     M=input(['Mag  ',num2str(i),' = ']);
     A=input(['Deg  ',num2str(i),' = '])*pi/180;
     D(i)=M*exp(j*A); disp(' ');
   else
     R=input(['Real ',num2str(i),' = ']);
     I=input(['Imag ',num2str(i),' = ']);
     D(i)=R+j*I; disp(' ');
   end
end
DP=D(1);  for k=2:den; DP=DP*D(k); end; RAT=NP/DP;
disp(' '); disp(['RATIO = ' num2str(real(RAT)) ' +j '...
    num2str(imag(RAT)) ' = ' num2str(abs(RAT)) '|_'...
    num2str(angle(RAT)*180/pi) 'deg']);
```

```
%%%%%%%%%%%%%%%%%%%%%%%%%%%%%%%%%%%%%%%%%%%%%%%%%%%%%%%%%%%%%%%%%%%%%%%%%%%%
%
% delwye.m - performs delta-wye or wye-delta conversion for
%            balanced or unbalanced load.
%
%%%%%%%%%%%%%%%%%%%%%%%%%%%%%%%%%%%%%%%%%%%%%%%%%%%%%%%%%%%%%%%%%%%%%%%%%%%%

clear; disp(' ');
Bal=input('  Balanced Load? (1=Yes, 2=No) '); disp(' ');
Dir=input('  Conversion Direction? (1=D->Y, 2=Y->D) '); disp(' ');

if Dir==1
   disp('  DELTA to WYE CONVERSION'); disp(' ');
   if Bal==1
     magZab=input('  magZab = ');
     magZbc=magZab; magZca=magZab;
     angZab=input('  angZab = ')*pi/180;
     angZbc=angZab; angZca=angZab;
   elseif Bal==2
     magZab=input('  magZab = ');
     angZab=input('  angZab = ')*pi/180;
     magZbc=input('  magZbc = ');
     angZbc=input('  angZbc = ')*pi/180;
     magZca=input('  magZca = ');
     angZca=input('  angZca = ')*pi/180;
   end
   Zab=magZab*exp(j*angZab); Zbc=magZbc*exp(j*angZbc);
   Zca=magZca*exp(j*angZbc); den=Zab+Zbc+Zca;
   Za=Zab*Zca/den; Zb=Zbc*Zab/den; Zc=Zca*Zbc/den;
   disp(' '); disp(' WYE IMPEDANCES in POLAR FORM');
   disp([abs(Za) angle(Za)*180/pi; abs(Zb) angle(Zb)*180/pi; ...
     abs(Zc) angle(Zc)*180/pi]);
elseif Dir==2
   disp('  WYE to DELTA CONVERSION'); disp(' ');
   if Bal==1
     magZa=input('  magZa = ');
     magZb=magZa; magZc=magZa;
     angZa=input('  angZa = ')*pi/180;
     angZb=angZa; angZc=angZa;
   elseif Bal==2
     magZa=input('  magZa = ');
     angZa=input('  angZa = ')*pi/180;
     magZb=input('  magZb = ');
     angZb=input('  angZb = ')*pi/180;
     magZc=input('  magZc = ');
     angZc=input('  angZc = ')*pi/180;
   end
   Za=magZa*exp(j*angZa); Zb=magZb*exp(j*angZb);
```

```
   Zc=magZc*exp(j*angZc); num=Za*Zb+Zb*Zc+Zc*Za;
   Zab=num/Zc; Zbc=num/Za; Zca=num/Zb;
   disp(' '); disp(' DELTA IMPEDANCES in POLAR FORM');
   disp([abs(Zab) angle(Zab)*180/pi; ...
        abs(Zbc) angle(Zbc)*180/pi; ...
        abs(Zca) angle(Zca)*180/pi]);
end

%%%%%%%%%%%%%%%%%%%%%%%%%%%%%%%%%%%%%%%%%%%%%%%%%%%%%%%%%%%%%%%%%%%%%%%%%%%
%
% Tphckt.m - solves for currents & voltages of a 3-phase load
%            supplied by a 3-phase source. Source & load can be
%            unbalanced with or without neutral connection. Both
%            source & load are wye connected. a·b·c phase
%            sequence is assumed. Line impedance is modeled.
%
%%%%%%%%%%%%%%%%%%%%%%%%%%%%%%%%%%%%%%%%%%%%%%%%%%%%%%%%%%%%%%%%%%%%%%%%%%%

clear; disp(' ');
disp(' THREE-PHASE CIRCUIT ANALYSIS'); disp(' ');
BalV=input('  Source balanced? ( 1=Yes, 2=No )  ');
if BalV==1
   magVan=input('  magVan = ');
   angVan=input('  angVan = ')*pi/180;
   magVbn=magVan; angVbn=angVan+2*pi/3;
   magVcn=magVan; angVcn=angVan-2*pi/3;
elseif BalV==2
   magVan=input('  magVan = ');
   angVan=input('  angVan = ')*pi/180;
   magVbn=input('  magVbn = ');
   angVbn=input('  angVbn = ')*pi/180;
   magVcn=input('  magVcn = ');
   angVcn=input('  angVcn = ')*pi/180;
else; disp(' INPUT ERROR on VOLTAGE BALANCE');
end
Van=magVan*exp(j*angVan); Vbn=magVbn*exp(j*angVbn);
Vcn=magVcn*exp(j*angVcn); ;disp(' ');
BalL=input('  Load balanced? ( 1=Yes, 2=No ) ');
if BalL==1
   magZa=input('  magZa = ');
   angZa=input('  angZa = ')*pi/180;
   magZb=magZa; angZb=angZa;
   magZc=magZa; angZc=angZa;
elseif BalL==2
   magZa=input('  magZa = ');
   angZa=input('  angZa = ')*pi/180;
   magZb=input('  magZb = ');
   angZb=input('  angZb = ')*pi/180;
   magZc=input('  magZc = ');
```

```
   angZc=input(' angZc = ')*pi/180;
else; disp(' INPUT ERROR on LOAD BALANCE');
end;
disp(' '); LZ=input(' Line impedance? ( 1=Yes, 2=No ) ');
if LZ==2; magZl=0; angZl=0;
else
   magZl=input(' magZl = ');
   angZl=input(' angZl = ')*pi/180;
end
Za=magZa*exp(j*angZa); Zb=magZb*exp(j*angZb);
Zc=magZc*exp(j*angZc); Zl=magZl*exp(j*angZl); disp(' ');
N=input(' Neutral connection? ( 1=Yes, 2=No ) '); disp(' ');
if N~=1 & N~=2; disp(' INPUT ERROR on NEUTRAL'); end
if N==1
   A=[Za+Zl 0 0; 0 Zb+Zl 0; 0 0 Zc+Zl];
   I=inv(A)*[Van Vbn Vcn].';
else
   A=[Za+Zl -Zb-Zl 0; 0 Zb+Zl -Zc-Zl; 1 1 1];
   I=inv(A)*[Van-Vbn Vbn-Vcn 0].';
end
VN=[Za 0 0; 0 Zb 0; 0 0 Zc]*I;
disp(' '); disp(' LOAD CURRENT in POLAR FORM');
disp([abs(I) angle(I)*180/pi]);
if N==1
   disp(' '); In=I(1)+I(2)+I(3); angIn=angle(In)*180/pi;
   disp([' In = ', num2str(abs(In)), '|_', ...
        num2str(angIn), ' Load->Source']);
end
disp(' '); disp(' LOAD VOLTAGE in POLAR FORM');
disp([abs(VN) angle(VN)*180/pi]);
```

SUMMARY

- Numerous energy conversion devices operate in the sinusoidal steady state.
- Phasor domain methods simplify sinusoidal steady-state analysis.
- KVL and KCL are equally valid in the phasor domain as in the time domain.
- RMS-valued phasors are predominantly used in analysis of energy conversion devices.
- Reduction of phasor domain networks modeled by impedances and admittances follows the identical procedures used for time domain networks of resistances and conductances, respectively.
- MATLAB's simplicity in handling complex arithmetic makes it a valuable tool for tedious numerical manipulation of phasor domain analysis.

- Three-phase networks are the dominant circuit arrangement in electric power generation, transmission, and industrial distribution systems.

- Three-phase sources and loads may be connected in either delta or wye configuration. Conversion of a delta arrangement to an equivalent wye arrangement leads to analysis simplification.

- Three-phase loads can be balanced or unbalanced. Effort to maintain the balanced condition is made as the unbalanced case results in reduction of the advantages of three-phase systems such as minimum conductor requirement.

- For the balanced three-phase case, only one phase need be analyzed. Performance of the remaining two phases is known by appropriate $\pm 120°$ phase shift of quantities in each three-phase set.

- Five quantities or forms of power are meaningful in energy flow rate analysis—instantaneous power, average power, reactive power, apparent power, and complex power. The first of these quantities has meaning only in the time domain. Average power and reactive power lend themselves to either time domain or phasor domain analysis. The latter two quantities pertain only to phasor domain analysis.

- The power triangle, obtained as a similar triangle to the impedance triangle, is a useful tool to guide in phasor domain power flow analysis.

- Multiple frequency sources of excitation can be encountered in power electronic converter applications. Phasor domain analysis using the principle of superposition provides a convenient method of analysis if load circuits are linear.

PROBLEMS

2.1 The voltage and current for the passive network of Fig. 2.1 are $\bar{V} = 240 \angle 30°$ V and $\bar{I} = 20 \angle -60°$ A. If the source frequency is 60 Hz, determine the values of two series-connected circuit elements that form the network.

2.2 Repeat Prob. 2.1 except that the two circuit elements are connected in parallel.

2.3 Three circuit elements, $R = 10\ \Omega$, $C = 50\ \mu F$, and $L = 140\ \mu H$, are connected in series and excited by a sinusoidal source. Determine the source frequency if the source voltage and current are in phase.

2.4 $v(t) = V_m \cos(\omega t + \phi) = 10 \cos(\omega t - 30°) + 15 \sin(\omega t) - 20 \cos(\omega t + 150°)$. Use the MATLAB program $\langle$csum.m$\rangle$ as an aid in determination of the numerical values of V_m and ϕ.

2.5 The circuit of Fig. 2.19 is excited by a 240-V, 60-Hz sinusoidal source. If $C = 10\ \mu F$, determine (*a*) the magnitude of current $\bar{I}$ and (*b*) the average value of power dissipated by resistor R.

2.6 If the source of Prob. 2.5 has a phase angle of $-30°$ so that $\bar{V} = 240 \angle -30°$ V, (*a*) determine the complex power supplied by the source and (*b*) draw

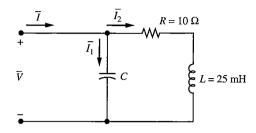

Figure 2.19

a power triangle labeling all three sides and the adjacent angle with numerical values.

2.7 The circuit of Fig. 2.19 is excited by a 120-V, 60-Hz source and $C = 10\ \mu F$, (a) Determine the power factor seen by the source. (b) To what value of capacitance must C be changed so that the power factor seen by the source is 0.9 lagging?

2.8 For the circuit of Fig. 2.19, $\bar{I}_2 = 8.75\angle -30°$ A and $C = 50\ \mu F$. If the exciting voltage source has a frequency of 60 Hz, find (a) the source voltage $\bar{V}$ and (b) the capacitor current $\bar{I}_1$. Use the MATLAB program ⟨cratio.m⟩ to solve the problem.

2.9 For the circuit of Fig. 2.20, $\bar{V}_1 = \bar{V}_2 = 120\angle 0°$ V and $\mathbf{Z}_x = \mathbf{Z}_y = \frac{1}{2}\mathbf{Z}_z = 10\angle 30°\ \Omega$. Determine (a) the phasor values of all currents and (b) the average value of power supplied to the circuit by source $\bar{V}_2$.

2.10 Repeat Prob. 2.9 if $\frac{1}{2}\mathbf{Z}_x = \mathbf{Z}_y = \frac{1}{2}\mathbf{Z}_z = 10\angle 30°\ \Omega$ and all else is unchanged.

2.11 The impedances of Fig. 2.20 form a delta arrangement. The values are given in Prob. 2.10. Use the MATLAB program ⟨delwye.m⟩ to find an equivalent wye arrangement and draw the converted circuit with impedances labeled in numerical values.

2.12 For the circuit of Fig. 2.21, determine $\mathbf{Z}_{eq}$ in symbols as a ratio of products in $\mathbf{Z}_1$, $\mathbf{Z}_2$, $\mathbf{Z}_3$, and $\mathbf{Z}_4$.

2.13 For the circuit of Fig. 2.21, $\bar{V} = 120\angle 0°$ V, $\mathbf{Z}_1 = 10\angle 15°\ \Omega$, $\mathbf{Z}_2 = 5\angle -20°$ Ω, $\mathbf{Z}_3 = 4 + j3\ \Omega$, $\mathbf{Z}_4 = 4 - j3\ \Omega$. Determine $\mathbf{Z}_{eq}$ and $\bar{I}$ by entering the

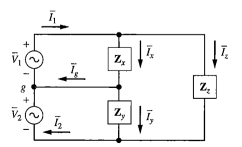

Figure 2.20

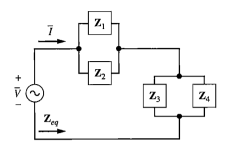

Figure 2.21

following code from the MATLAB Command Window where the semicolon
has been omitted at the end of the last three lines to echo answers:

```
k = pi/180;  Z1=10*exp(j*15*k);  Z2=5*exp(-j*20*k);
Z3 = 4+j*3;  Z4 = 4-j*3;  V = 120;
Zeq = Z1*Z2/(Z1+Z2)+Z3*Z4/(Z3+Z4)
Imag = abs(V/Zeq)
Iang = angle(V/Zeq)
```

2.14 A set of balanced three-phase voltages is impressed on a balanced, three-phase, delta-connected load. If $\overline{V}_{ab} = 7200\angle 0°$ V and $\overline{I}_a = 12\angle 0°$ A, determine (*a*) the phase current $\overline{I}_{ab}$ and (*b*) the total average power supplied to the load.

2.15 Construct a three-phase power triangle for the load of Prob. 2.14 and label all sides and the adjacent angle with numerical values.

2.16 A balanced three-phase load has a balanced set of 480-V line voltages applied. The total average power to the load is 1000 W, and the power factor is 0.8 lagging. Find (*a*) the phase impedance and (*b*) the line current magnitude if the load is delta-connected.

2.17 Work Prob. 2.16 if the load is wye-connected and all else is unchanged.

2.18 Show that for a three-wire, three-phase system, balanced or unbalanced, knowledge of two line voltages and two line currents is sufficient to determine the instantaneous power flow to the system. Specifically, verify that [2.57] is correct.

2.19 The level of current density in the conductors of an electrical transmission line determines its operating temperature for a particular set of ambient conditions, thereby limiting the ampacity rating of the line. Show that a three-phase transmission line, under balanced conditions, can transport power with one-half the total conductor volume as a single-phase transmission line if the line-to-neutral voltage of the three-phase line is equal to the voltage of the single-phase line and if both transmission lines have identical current density.

2.20 For a balanced three-phase load, start with [2.58] and show that [2.63] results.

2.21 Show that for balanced conditions delta load the angle θ in [2.63] and [2.64] can be taken as either the angle of the actual phase load or the angle of the wye equivalent load.

2.22 For a balanced three-phase load, show that the instantaneous power flow to the load is a constant with a value equal to the total average power flow to the load.

2.23 A balanced set of three-phase voltages is impressed on the circuit of Fig. 2.22, which is known as an *open delta*. Let $\mathbf{Z} = 3 + j4\,\Omega$ and $\bar{V}_{ab} = 120\angle0°\,\text{V}$. Determine (*a*) the line currents and (*b*) the total power supplied to the load.

2.24 If the open delta is completed or closed in Prob. 2.23 with an impedance of equal value to the other two phases, determine the percentage increase in total average power flow to the load over the case of the open-delta connection.

2.25 Convert the open delta of Fig. 2.22 to an equivalent wye connection. If $\mathbf{Z} = 3 + j4\,\Omega$, $\bar{V}_{ab} = 120\angle0°\,\text{V}$, and the impressed voltages form a balanced set, determine the line currents.

2.26 A delta and a wye load are connected in parallel as illustrated by Fig. 2.23. Let $\mathbf{Z}_a = 4\angle30°\,\Omega$, $\mathbf{Z}_b = 5\angle15°\,\Omega$, $\mathbf{Z}_c = 5\angle-30°\,\Omega$, $\mathbf{Z}_{ab} = 10\angle30°\,\Omega$, $\mathbf{Z}_{bc} = 20\angle0°\,\Omega$, and $\mathbf{Z}_{ca} = 10\angle20°\,\Omega$. Find the elements of a single wye-connected load to model the two given parallel loads.

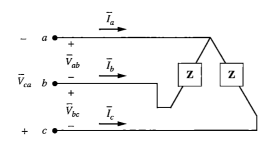

Figure 2.22

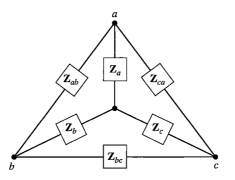

Figure 2.23

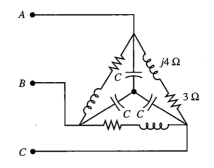

Figure 2.24

2.27 Find the elements of an equivalent delta-connected load to model the given parallel wye and delta loads of Prob. 2.26.

2.28 An unbalanced wye-connected three-phase source has $\overline{V}_{an} = 240\angle 20°$ V, $\overline{V}_{bn} = 220\angle - 100°$ V, and $\overline{V}_{cn} = 230\angle 90°$ V. Determine the phasor values of line voltages $\overline{V}_{ab}$, $\overline{V}_{bc}$, and $\overline{V}_{ca}$ using the MATLAB program $\langle\text{csum.m}\rangle$.

2.29 The unbalanced wye-connected voltage set of Prob. 2.28 feeds a balanced, three-phase, delta-connected load with a phase impedance of $20\angle 30°\,\Omega$. Use the MATLAB program $\langle\text{Tphckt.m}\rangle$ to determine the three line currents.

2.30 For the wattmeter of Fig. 2.14, let $v(t) = \sqrt{2}V_1\cos\omega t + \sqrt{2}V_3\cos 3(\omega t + \phi_3)$ and $i(t) = \sqrt{2}I_1\cos(\omega t - \phi_1) + \sqrt{2}I_3\cos 3(\omega t + \phi_3 - \theta_3)$. If the meter can fully respond to the higher frequency, find an expression for the average power indicated by the meter.

2.31 Capacitors C are added to the 60-Hz, three-phase network of Fig. 2.24 to correct the input power factor to unity. (*a*) Determine the value of C. (*b*) If $V_{AB} = 480$ V, find the total kVAR rating of the capacitor bank.

2.32 A four-wire, wye-connected load is balanced with each phase of the load consisting of a series connection of $R = 5\,\Omega$ and $L = 20$ mH. A set of phase voltages are impressed on the network with the time domain descriptions

$$v_{an}(t) = 120\sqrt{2}\cos(\omega t) + 20\sqrt{2}\cos(3\omega t)$$

$$v_{bn}(t) = 120\sqrt{2}\cos(\omega t - 120°) + 20\sqrt{2}\cos(3\omega t)$$

$$v_{cn}(t) = 120\sqrt{2}\cos(\omega t + 120°) + 20\sqrt{2}\cos(3\omega t)\ \text{V}$$

Determine the three line currents and the neutral conductor current if $\omega = 120\pi$ rad/s.

2.33 For the line-to-neutral voltage set of Prob. 2.32, determine the three line voltages $v_{ab}(t)$, $v_{bc}(t)$, and $v_{ca}(t)$.

3

MAGNETIC CIRCUITS AND ENERGY CONVERSION

3.1 INTRODUCTION

Explanation and analysis of energy conversion devices begins with a study of the cause-effect relationships and the associated bilateral energy flow between electrical circuits and their coupled magnetic fields. In the case of the transformer, this is the complete study. For all other devices with the potential for movement of a member, the interaction of two or more established magnetic fields to produce a force becomes the second dimension of study. Although final performance equations for an energy conversion device have typically eliminated any magnetic field variables through use of inductance quantities, conceptual understanding of the magnetic field variables is indispensable in explanation of the resulting device performance and associated limitation imposed by the magnetic field quantities.

3.2 GOVERNING LAWS AND RULES

Under the assumption of prior introduction through an elementary physics course, five laws and two interpretive rules are reviewed as a basis for understanding and explanation of the interfaces between electrical, magnetic, and mechanical variables for the energy conversion devices of interest in this text.

Lorentz Force Equation An *electromagnetic field* is an area of influence over which an electrical charge in motion experiences a force acting upon that charge. The force magnitude and direction are determined by the cross-product relationship

$$F = q\,u \times B \qquad\qquad \textbf{[3.1]}$$

where F = force (N)
 q = value of electric charge (C)
 u = velocity of the charge (m/s)
 B = magnetic flux density (T)

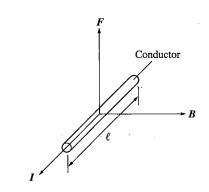

Figure 3.1
Lorentz force

The charge in motion of concern is current flow in a conductor. If a current of magnitude I flows as indicated through the conductor of Figure 3.1 over length ℓ within a uniform B-field, then $q\,\mathbf{u}$ of [3.1] can be replaced with ℓI, giving

$$F = \ell\mathbf{I} \times \mathbf{B} \qquad\qquad \textbf{[3.2]}$$

Given the shown mutually perpendicular $\mathbf{I}$ and $\mathbf{B}$, the resulting force (F) is indicated. For this special case, the magnitude of the force is given by $F = B\ell I$. If either the B-field or current density were not uniform, a force density (f) must be introduced from which the total force can be obtained by volume integration.

$$F = \int_v fdv = \int_v (\mathbf{J} \times \mathbf{B})dv \qquad\qquad \textbf{[3.3]}$$

where f = force density (N/m^3)
     $\mathbf{J}$ = current density (A/m^2)

Example 3.1   The conductor of Fig. 3.1 is in a B-field with direction shown and described by $B(t) = 5\cos\omega t$ Wb. The conductor length is 25 cm and carries a current $i(t) = 2\cos\omega t$ A with direction as indicated by I. Determine the force acting on the conductor.
     I and B are mutually perpendicular. Based on [3.2], the force magnitude is

$$F = \ell i(t)B(t)\sin(\pi/2) = (0.25)2\cos\omega t\,5\cos\omega t$$

$$F = 2.5\cos^2\omega t = 1.25 + 1.25\cos 2\omega t \text{ N}$$

The force F is directed upward.

Biot-Savart Law   The magnetic field intensity H is established by motion of charge or current flow. The direction and magnitude of H (A/m) can be determined

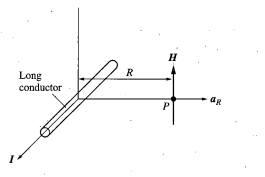

Figure 3.2
Biot-Savart law

at any point P located at a perpendicular distance R (m) from a long current-carrying conductor by

$$H = \frac{1}{2\pi R} I \times a_R \qquad \text{[3.4]}$$

where the vector I (A) of current enclosed by a circular path of radius R and unit vector a_R are clarified by Fig. 3.2. The magnetic field density B follows from H using the auxiliary equation

$$B = \mu H = \mu_R \mu_o H \qquad \text{[3.5]}$$

The factor μ (H/m), known as the *permeability*, is a property of the material lying between supporting current I and the field point P. The unitless quantity μ_R (*relative permeability*) is the ratio

$$\mu_R = \frac{\mu}{\mu_o} \qquad \text{[3.6]}$$

where $\mu_o = 4\pi \times 10^{-7}$ H/m is the permeability of free space.

The negligibly small diameter conductor of Fig. 3.2 is 10 m long and carries a constant current of 5 A. Assume that the conductor can be treated as long and describe the H- and B-fields at a distance R from the conductor. The conductor is located in air with $\mu = 4\pi \times 10^{-7}$ H/m.

Example 3.2

The field point P can be moved at a constant radius R around the conductor, giving at each point the same value of H-field. Thus, the H-field is directed counterclockwise and for a constant distance R is given by [3.4] as

$$H = \frac{I}{2\pi R} = \frac{5}{2\pi R} = \frac{0.796}{R} \text{ A/m}$$

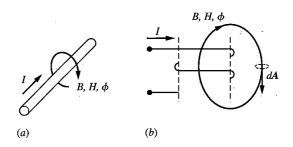

Figure 3.3
Right-hand rule illustrated. (a) Straight conductor. (b) Coil.

By [3.5],

$$B = \mu H = (4\pi \times 10^{-7})\left(\frac{0.796}{R}\right) = \frac{1 \times 10^{-6}}{R} \text{ T}$$

Right-Hand Rule Based on the Biot-Savart law, the right-hand rule gives a convenient method for determination of the magnetic field direction near a current-carrying conductor. If a conductor is grasped in the right hand with the thumb extended in the direction of current flow, then the fingers curl around the conductor in the direction of the established B- or H-field.

Magnetic flux ϕ (Wb) is given by

$$\phi = \int B \cdot dA \qquad\qquad [3.7]$$

where dA is a directed differential area through which the B-field passes. If the plane of dA is perpendicular to B, then the right hand rule also gives the flow direction of magnetic flux ϕ. Figure 3.3 clarifies magnetic field directions for a given current I.

Faraday's Law A coil of N turns linked by a changing magnetic flux ϕ has a voltage (e) induced behind the terminals of that coil with magnitude given by

$$e = N\frac{d\phi}{dt} \qquad\qquad [3.8]$$

Lenz's Law The polarity of the voltage given by [3.8] is established by Lenz's law: The voltage induced in a coil by a changing flux will be of such a polarity that if a current could flow as a result of that induced voltage, the flux established by that current would oppose the causing or original flux change. Figure 3.4 illustrates the polarity of induced voltage e for two cases. In both cases, the flux ϕ is in the same direction, but the flux is increasing in magnitude for one case while decreasing in magnitude for the other case.

$B\ell v$ Rule Consider the rectangular area of height ℓ (m) enclosed by dashed lines in Fig. 3.5. A uniform, non-time-varying B-field is directed into the page. The

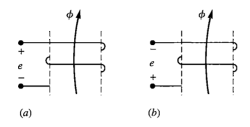

Figure 3.4
Lenz's law illustrated. (a) ϕ increasing ($d\phi/dt > 0$). (b) ϕ decreasing ($d\phi/dt < 0$).

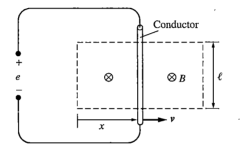

Figure 3.5
$B\ell v$ rule illustrated

conductor is moving to the right in the plane of the page through the B-field with a speed v (m/s). By [3.7], the flux directed into the page and passing through the one-turn coil formed by the conductor and its connections back to the terminals across which induced voltage e is indicated is

$$\phi = BA = B\ell v \qquad\qquad [3.9]$$

By [3.8] with $N = 1$,

$$e = \frac{d\phi}{dt} = \frac{d(B\ell x)}{dt} = B\ell\frac{dx}{dt} = B\ell v \qquad\qquad [3.10]$$

The result of [3.10] made use of $dB/dt = 0$ since the B-field is constant. Consequently, the $B\ell v$ rule is applicable only for the case of a constant B-field. The polarity indicated for induced voltage e of Fig. 3.5 can be verified as correct by Lenz's law, noting that the flux through the one-turn coil is increasing.

Ampere's Circuit Law   The cause-effect relationship between an electrical circuit and its supported magnetic field is described by

$$\oint \boldsymbol{H}\cdot d\boldsymbol{\ell} = \text{current enclosed} = \mathscr{F} \qquad\qquad [3.11]$$

where $d\ell$ is a differential length along the closed path of integration and $\mathscr{F}$ is a scalar known as *magnetomotive force* (mmf). Although evaluation of [3.11] in general can be cumbersome, if segments of the uniform H-field align with the path of

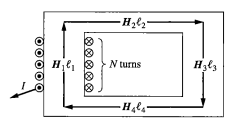

Figure 3.6
Ampere's circuital law illustrated

integration, the evaluation becomes rather straightforward. In Fig. 3.6, assume that the H-field is uniform over each of the four subscripted lengths ℓ; then application of [3.11] leads to

$$\oint H \cdot d\ell = H_1\ell_1 + H_2\ell_2 + H_3\ell_3 + H_4\ell_4 = NI = \mathscr{F} \qquad \textbf{[3.12]}$$

3.3 FERROMAGNETISM

Materials are classified as ferromagnetic or nonmagnetic depending on their magnetization or B-H characteristic. Nonmagnetic materials have a linear B-H curve while ferromagnetic materials display a nonlinear B-H characteristic. Figure 3.7 typifies the two material classifications. For operating temperatures of concern, ferromagnetism is displayed by iron, nickel, and cobalt in pure form, and by certain metal alloys.

Practical electromechanical energy conversion devices are constructed with ferromagnetic metals forming nearly the complete magnetic field paths, resulting in devices of high torque or force volume density. Although electromechanical energy conversion devices could be theoretically designed with nonmagnetic material for field paths, the value of attainable magnetic field intensity is practically limited for heating reasons by the permissible current density of the conductors supporting the magnetic field. Air-cooled conductors are limited to current densities of 1.5 to 10 MA/m² (1 to 6.5 kA/in²) depending upon accessibility of cooling air to conductors and natural convection vs. forced-air cooling methods.

3.3.1 SATURATION

The typical B-H characteristic for ferromagnetic material in Fig. 3.7 has been subdivided into two general regions—*linear region* and *saturation region*. The transition between the linear and saturation regions is known as the *knee*. In addition, the relative permeability of the ferromagnetic material has been plotted. While the relative permeability of the nonmagnetic material has a constant value $\mu_R = 1$, the

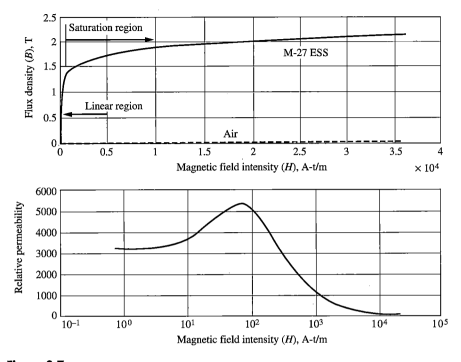

Figure 3.7
Magnetic characteristics

relative permeability of the ferromagnetic material is significantly greater than unity over the practical range of magnetic field intensity.

Electric steel sheets (ESS) specifically manufactured for use in electro-mechanical energy conversion devices undergo specific processing to enhance suitability for particular application. Ferromagnetic metals have atoms with less than full inner electron shells. Consequently, the magnetic fields established by electron spin are not completely offset at the level of the atom and thus each atom of a ferromagnetic material has a magnetic moment.[1] When a molten ferromagnetic material solidifies, a crystal formation results with layered domains of atoms. Within each domain, the magnetic moments of all atoms are aligned.

Nonoriented ESS are metallurgically processed so that domain walls form at oblique angles as illustrated by the two-dimensional concept diagram of Fig. 3.8*a*, wherein the crystal formation is magnetostatically balanced. Only a moderate *H*-field is required to reverse the magnetic moment of atoms and effectively move the domain wall within each crystal, as indicated by Fig. 3.8*b*. With the resultant magnetic moment now directed to the right, magnetic flux flow from the left to the right is supported. This domain wall movement phenomenon is associated with the linear portion of the *B-H* curve of Fig. 3.7. If the *H*-field were oriented upward, the domain wall movement would be such that the resultant magnetic moment is directed

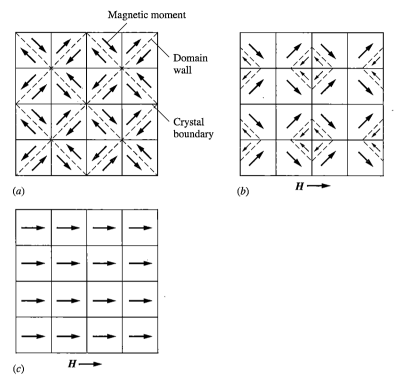

Figure 3.8
Magnetization process. (a) No external H-field. (b) Moderate H-field. (c) Large H-field.

upward, or the material can support flux flow in any direction with only a moderate external H-field applied. This potential to support flux flow in any direction makes the material suitable for use in the magnetic structure of electric motors. It is the basis for the name nonoriented ESS.

After the domain walls are totally collapsed, the magnetic moments still only have a component in the direction of the applied external magnetic field. If the external magnetic field is increased significantly, all magnetic moments can be aligned with the applied field as indicated by Fig. 3.8c. Approximately a 15 percent increase in flux flow results from the point of domain wall collapse to total alignment of the magnetic moments at the expense of significant increase in the value of applied external magnetic field. This latter phenomenon is associated with the saturation region for the B-H curve of Fig. 3.7.

Other metallurgical processing procedures to enhance elongated crystal formation and control of the direction of rolling can be implemented to yield *grain oriented ESS*. In this case, the domain walls nearly align in one direction, resulting in a ferromagnetic material that can attain high flux density levels with only the necessity of domain wall movement. Consequently, grain oriented ESS exhibits a larger B-field with a smaller applied H-field than is the case for nonoriented ESS. Grain oriented ESS finds application in transformers where the direction of the magnetic field is fixed.

3.3.2 HYSTERESIS

When an external magnetic field is applied to a specimen of ferromagnetic material and then removed, the domain walls may not return to the exact original position without application of a reverse-directed magnetic field. The underlying reason is that some of the atom magnetic moments may align with one of their several preferred axes of alignment different from the original axes. Consequently, the magnetic moments may have a nonzero resultant that supports a small B-field. Thus, the descending B-H curve does not track the ascending curve. This phenomenon, called *hysteresis*, will be addressed in more detail in Sec. 4.4.2.

3.3.3 CURIE TEMPERATURE AND MAGNETOSTRICTION

Two additional properties of ferromagnetic materials are mentioned for the sake of completeness and curiosity. A temperature known as the *Curie temperature* exists for which the magnetic moments of ferromagnetic material become sufficiently diverse in orientation that the material becomes nonmagnetic. The Curie temperature is significantly above the practical operating temperature of any electromagnetic device of interest.

Magnetostriction is the elastic deformation of ferromagnetic material in the direction of an applied external magnetic field. The dimensional change is in the order of μm/m. Although some small performance impact may result from stress introduced by magnetostriction, the primary manifestation is excitation of vibration in the audible range, giving electromagnetic devices the characteristic twice source frequency noise.

3.4 MAGNETIC CIRCUITS

Magnetic flux paths constructed of ferromagnetic materials constitute important parts of any electromechanical energy conversion device. All commercially available motors, generators, actuators, and transformers make use of one or more such paths. These magnetic flux paths along with the sources of excitation are called *magnetic circuits*. A magnetic circuit is usually designed for the purpose of establishing a predetermined flux in a definite space by the least possible excitation mmf. Physical function of electromechanical devices requires in most cases that one or more air gaps be included in the magnetic circuit. A notable exception is the transformer for which the principal flux path is ferromagnetic material over its entire length.

Once a magnetic circuit has been described, the analysis can be classified as one of two types:

> *Type 1:* Determine the excitation mmf required to establish a given value of magnetic flux.
>
> *Type 2:* Determine the magnetic flux established by a given value of excitation mmf.

If ferromagnetic material had a linear *B-H* relationship, solution of both types of problems would be straightforward; however, it will be seen that as a result of magnetic nonlinearity, the Type 2 problem always requires either a graphical or iterative solution approach, with the exception of the trivial case of a single uniform flux path. If multiple flux paths exist in the magnetic circuit, the Type 1 problem is likely to require a graphical or iterative solution.

3.4.1   ANALYSIS METHODOLOGY

Analogy between magnetic circuits with constant mmf excitation sources and dc electric circuits can be drawn. Except for the limitations resulting from magnetic nonlinearity, the analysis approach for magnetic circuits is similar to the methods utilized in electric circuit analysis.

In magnetic circuits consisting of ferromagnetic paths, the path of integration for [3.12] and the magnetic field are typically collinear. Hence, using $B = \mu H$ and $B = \phi/A$, [3.12] can be rewritten as

$$\oint \phi \frac{d\ell}{\mu A} = \mathscr{F} = NI \qquad \text{[3.13]}$$

If over the path of the closed line integration indicated by [3.13] segments of the path have a constant value of flux ϕ, [3.13] yields

$$\phi_1 \int_{\ell_1} \frac{d\ell}{\mu_1 A_1} + \cdots + \phi_n \int_{\ell_n} \frac{d\ell}{\mu_n A_n} = \mathscr{F} = NI \qquad \text{[3.14]}$$

$$\phi_1 \mathscr{R}_1 + \cdots + \phi_n \mathscr{R}_n = \mathscr{F} = NI \qquad \text{[3.15]}$$

$$\mathscr{R}_1 = \frac{\ell_1}{\mu_1 A_1} \qquad \mathscr{R}_n = \frac{\ell_n}{\mu_n A_n} \qquad \text{[3.16]}$$

The quantity $\mathscr{R}$, known as *reluctance*, can be thought of as the opposition to flux flow with direct analogy to resistance in an electric circuit. Extending the analogy to compare flux with current and an mmf source with a voltage source, [3.15] indicates that the sum of all mmf drops around any closed path must be equal to the value of the mmf sources of that path—a direct analogy with Kirchhoff's voltage law.

Consider the distributed parameter electric circuit of Fig. 3.9*a* consisting of a voltage source *V* driving current *I* through three series-connected metallic bars. The resistance of each of the metallic bars could be appropriately calculated using

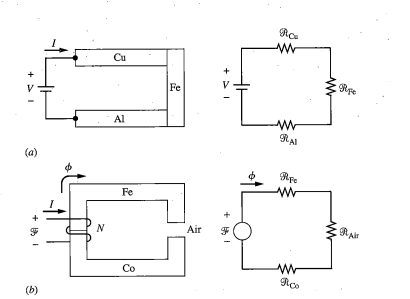

Figure 3.9
Distributed and lumped parameter circuits. (a) Electric circuit. (b) Magnetic circuit.

Table 3.1. Comparison of electrical and magnetic quantities

Electrical			*Magnetic*		
Voltage	V	(V)	mmf	$\mathscr{F}$	(A-t)
Current	I	(A)	Flux	ϕ	(Wb)
Resistance	R	(Ω)	Reluctance	$\mathscr{R}$	(H^{-1})
Electric field intensity	E	(V/m)	Magnetic field intensity	H	(A-t/m)
Current density	J	(A/m^2)	Flux density	B	(T)
Conductivity	σ	(S/m)	Permeability	μ	(H/m)

$$V = IR$$

$$J = \sigma E$$

$$R = \frac{\ell}{\sigma A} = \frac{1}{G}$$

$$\mathscr{F} = \phi \mathscr{R}$$

$$B = \mu H$$

$$\mathscr{R} = \frac{\ell}{\mu A} = \frac{1}{\mathscr{P}}$$

$\mathscr{R} = \ell/\sigma A$, allowing the lumped parameter circuit to the right to be drawn as a model of the actual distributed parameter circuit. Using the concept of reluctance introduced by [3.14] to [3.16], the magnetic circuit of Fig. 3.9b can also be modeled by the lumped parameter electric analog circuit to its right. A comparative listing of the parallels between electrical quantities and analogous magnetic quantities is presented by Table 3.1.

With the electric circuit analogy for magnetic circuits now established, the analysis of magnetic circuits would be straightforward except for the fact that permeability μ is a function of the flux density B (or the magnetic field intensity H)

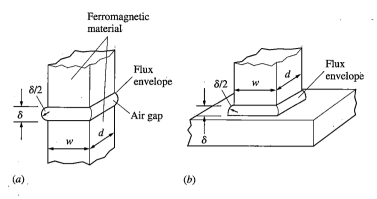

Figure 3.10
Air gap fringing. (a) Equal opposing face. (b) Plane opposing face.

in the ferromagnetic material. The resulting difficulties that arise are identical to those that would exist in analysis of an electrical circuit if the values of the resistors were functions of the current.

3.4.2 AIR GAP FRINGING

As magnetic flux crosses an air gap between two ferromagnetic surfaces, the magnetic field will spread or *fringe* over an area larger than the area of the equal opposing faces of the ferromagnetic material, as illustrated by Fig. 3.10a. For air gaps of length greater than 0.5 mm (0.020 in), the additional effective cross-sectional area utilized by the air gap magnetic field increases sufficiently to introduce significant error in any calculation of air gap magnetic field intensity based on the footprint area of the adjacent ferromagnetic material surfaces. Except for operation in the saturation region of the ferromagnetic material, the mmf drop of the air gap ($\mathcal{F}_g = \phi_g \delta / \mu_o A_g$) is dominant. Consequently, the error in air gap mmf prediction approaches the error in total magnetic circuit mmf calculation.

A common air gap arrangement is that of equal area opposing ferromagnetic material faces illustrated by Fig. 3.10a. Treating the fringing path along the edges of the air gap field as semicircular cylinders with spherical quadrants at each of the four corners, the permeance of the air gap is given by[2]

$$\mathcal{P}_g = \mu_o \left[\frac{wd}{\delta} + 0.52(w + d) + 0.308\,\delta \right] \qquad \textbf{[3.17]}$$

A reasonable computation of the air gap permeance for the plane opposing face air gap illustrated by Fig. 3.10b is

$$\mathcal{P}_g = \mu_o \left[\frac{wd}{\delta} + 1.04(w + d) + 0.616\,\delta \right] \qquad \textbf{[3.18]}$$

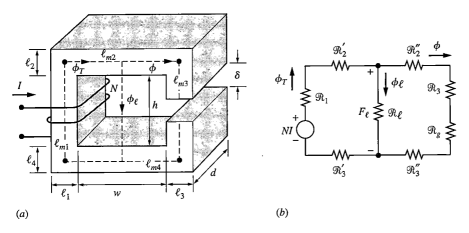

Figure 3.11
Series magnetic circuit (C-shape). (*a*) Physical. (*b*) Schematic.

3.4.3   LEAKAGE FLUX

Magnetic flux that follows a path other than the path intended by ideal design is known as *leakage flux*. Figure 3.11 illustrates the concept where leakage flux ϕ_ℓ does not cross the air gap of the magnetic structure as desired. In any practical magnetic circuit, some leakage flux flows between planes of different magnetic potential. Generally, leakage flux increases as a magnetic structure moves into saturation region operation. Leakage flux will be quantitatively addressed in examples of magnetic circuit analysis.

3.4.4   SERIES MAGNETIC CIRCUITS

A magnetic circuit with a single flux path, if leakage flux were neglected, is commonly called a *series magnetic circuit*. The analogy with a single-loop electrical circuit is obvious. Figures 3.11 and 3.9*b* illustrate the physical arrangement and electrical circuit analog of a series magnetic circuit with and without leakage flux, respectively.

For the series magnetic circuit of Fig. 3.11 with air gap δ, $\ell_1 = \ell_2 = \ell_3 = \ell_4 = d = 0.05$ m, $h = 0.10$ m, $w = 0.15$ m, and $\delta = 0.002$ m. The coil has $N = 500$ turns. The ferromagnetic core is 24-gage M-27 ESS described by Fig. 3.12. The laminated structure is assembled with a stacking factor SF = 0.95—the per unit portion of dimension d that is composed of ferromagnetic material when the lamination surface insulation and departure from flatness irregularities are considered. Assume that the flux produced by the coil is $\phi_T = 0.004$ Wb and that leakage is negligible. Determine the required coil current I.

Example 3.3

Magnetization characteristics

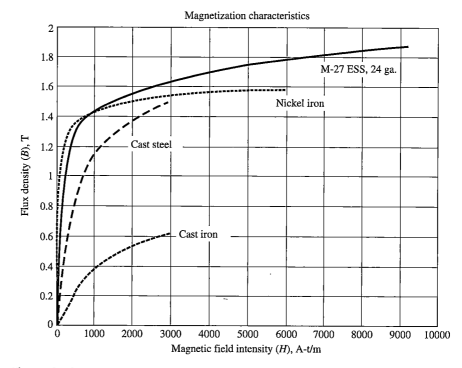

Figure 3.12
Magnetization curves for typical ferromagnetic materials

This is a Type 1 problem. Neglecting the leakage flux and noting that the cross-sectional area of the ferromagnetic core is uniform yields

$$B_c = \frac{\phi_T}{A_c} = \frac{\phi_T}{\ell_1 d SF} = \frac{0.004}{(0.05)(0.05)(0.95)} = 1.684 \text{ T}$$

Enter Fig. 3.12 with $B_c = 1.684$ T to find $H_c \cong 4050$ A-t/m. The flux mean length path within the core is

$$\ell_m = 2(w + \ell_1) + 2(h + \ell_2) - \delta$$

$$\ell_m = 2(0.15 + 0.05) + 2(0.10 + 0.05) - 0.002 = 0.698 \text{ m}$$

Based on [3.12], the core mmf requirement is

$$\mathscr{F}_c = H_c \ell_m = (4050)(0.698) = 2826.9 \text{ A-t}$$

Based on [3.17], the air gap permeance has the value

$$\mathscr{P}_g = \mu_o \left[\frac{\ell_3 d}{\delta} + 0.52(\ell_3 + d) + 0.308\delta \right]$$

$$\mathscr{P}_g = 4\pi \times 10^{-7}\left[\frac{(0.05)(0.05)}{0.002} + 0.52(0.05 + 0.05) + 0.308(0.002)\right]$$

$$= 1.637 \times 10^{-6}\ \text{H}$$

Hence, the air gap mmf follows as

$$\mathscr{F}_g = \frac{\phi_T}{\mathscr{P}_g} = \frac{0.004}{1.637 \times 10^{-6}} = 2443.5\ \text{A-t}$$

The required coil current is

$$I = \frac{\mathscr{F}_c + \mathscr{F}_g}{N} = \frac{2826.7 + 2443.5}{500} = 10.54\ \text{A}$$

The above example with uniform core area was not too tedious. However, care must be exercised in reading the value of H_c from the B-H curve. Further, if the core sections had different areas, multiple entries must be read from the B-H curve. The MATLAB program ⟨cckt1.m⟩ has been developed to analyze this particular C-shaped core with an air gap. The program is set up to handle a structure with different core section widths. The core depth d is assumed uniform, although changes to accommodate different depths for each core section would not be difficult. Dimensional values can be entered in either mks or British (fps) units with the value of the string variable *dim* controlling the conversion to mks units. The program uses a function-file ⟨hm27.m⟩ containing piecewise linear B-H arrays for 24-gage M-27 ESS. Similar function-files can be introduced for other materials. The function-file performs a linear interpolation between bounding values of B_c to determine H_c. Execution of ⟨cckt1.m⟩ with the values from Example 3.3 results in the following screen display:

```
C-SHAPED MAGNETIC CIRCUIT, No Leakage
COIL CURRENT = 10.6
```

A single performance point for a magnetic circuit does not give a comfortable understanding of the degree of saturation. The program ⟨cckt1.m⟩ also generates a wide-range plot of flux vs. coil mmf and superimposes the specified performance point on the produced plot. The resulting plot for the values of Example 3.3 is displayed by Fig. 3.13.

A series magnetic circuit has the dimensions, material, and coil of Example 3.3; however, leakage flux is not to be neglected. The flux that crosses the air gap is $\phi = 0.004$ Wb. Find: (a) the values of leakage flux ϕ_ℓ and coil flux ϕ_T; and (b) the required coil current I.

Example 3.4

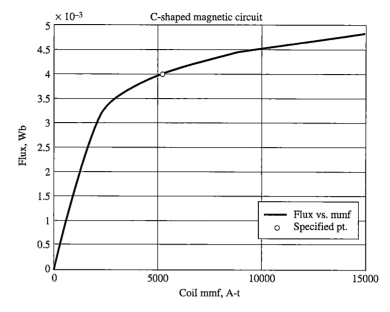

Figure 3.13
Example 3.3 data results

(a) The schematic of Fig. 3.11b is a valuable guide in solution of this problem. As long as the leakage flux is small, it is a reasonable approximation to model the leakage reluctance $\mathcal{R}_\ell$ = $1/\mathcal{P}_\ell$ as a lumped parameter acting between the midpoints of the upper and lower core sections. The leakage flux principally flows between the opposing faces of the upper and lower core members over the window opening with approximate value given by

$$\mathcal{P}_\ell = \mu_o \frac{wd}{h} = 4\pi \times 10^{-7} \frac{(0.15)(0.05)}{0.10} = 9.425 \times 10^{-8} \text{ H}$$

The air gap permeance determined in Example 3.3 is still valid.

The flux density of the right-side core paths that support flux ϕ is

$$B_{cr} = \frac{\phi}{\ell_3 dSF} = \frac{0.004}{(0.05)(0.05)(0.95)} = 1.684 \text{ T}$$

From Fig. 3.12, $H_{cr} \cong 4050$ A-t/m. With function-file ⟨hm27.m⟩ located in a MATLAB directory, from the MATLAB Command Window simply enter hm27(1.684) to find H_{cr} = 4083.5 A-t. Since this latter value is a slightly more accurate reading from the B-H curve, it will be used.

Mean-length flux paths for each of the core members are needed to continue the solution.

$$\ell_{m1} = h + \ell_2 = 0.10 + 0.05 = 0.15 \text{ m}$$

$$\ell_{m2} = \ell_{m4} = w + \ell_1 = 0.15 + 0.05 = 0.20 \text{ m}$$

$$\ell_{m3} = h + \ell_2 - \delta = 0.10 + 0.05 - 0.002 = 0.148 \text{ m}$$

The mmf drop $\mathscr{F}_\ell$ as indicated in Fig. 3.11b can now be calculated:

$$\mathscr{F}_\ell = H_{cr}\left(\frac{\ell_{m2}}{2} + \ell_{m3} + \frac{\ell_{m4}}{2}\right) + \frac{\phi}{\mathscr{P}_g}$$

$$\mathscr{F}_\ell = 4083.5\left(\frac{0.20}{2} + 0.148 + \frac{0.20}{2}\right) + \frac{0.004}{1.637 \times 10^{-6}} = 3864.5 \text{ A-t}$$

The leakage flux and coil flux follow as

$$\phi_\ell = \mathscr{F}_\ell\mathscr{P}_\ell = (3864.5)(9.425 \times 10^{-8}) = 0.364 \text{ mWb}$$

$$\phi_T = \phi + \phi_\ell = 4 + 0.364 = 4.364 \text{ mWb}$$

(b) With ϕ_T known, the flux density around the left-hand core member is

$$B_{cl} = \frac{\phi_T}{\ell_1 dSF} = \frac{0.004364}{(0.05)(0.05)(0.95)} = 1.837 \text{ T}$$

By either careful reading from Fig. 3.12 or use of ⟨hm27.m⟩, $H_{cl} \cong 8132$ A-t. The required coil mmf and current can now be determined:

$$NI = H_{cl}\left(\frac{\ell_{m2}}{2} + \ell_{m1} + \frac{\ell_{m4}}{2}\right) + \mathscr{F}_\ell$$

$$NI = 8132\left(\frac{0.20}{2} + 0.15 + \frac{0.20}{2}\right) + 3864.5 = 6710.7 \text{ A-t}$$

$$I = \frac{NI}{N} = \frac{6710.7}{500} = 13.42 \text{ A}$$

Comparing Examples 3.3 and 3.4, it is seen that the addition of the leakage flux consideration increased the computation effort to find the value of coil current. However, the resulting coil current is over 25 percent greater when the leakage flux is taken into account, leading to question of accuracy for any analysis that neglects leakage flux. Further, each graphical determination of magnetic field intensity can easily introduce propagating errors. The MATLAB program ⟨cleak1.m⟩ has been developed to accurately analyze the particular C-shaped core with leakage flux of Fig. 3.11 following the analysis procedure of Example 3.4. The screen display from execution of ⟨cleak1.m⟩ with the values of Example 3.4 is shown below.

```
C-SHAPED MAGNETIC CIRCUIT, With Leakage
   COIL CURRENT = 13.44
   AIR GAP FLUX = 0.004
   LEAKAGE FLUX = 0.0003644
```

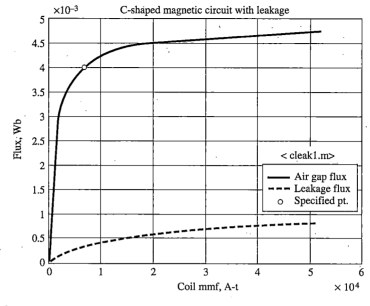

Figure 3.14
Example 3.4 data results

The wide-range plot of air gap flux and leakage flux is depicted by Fig. 3.14, from whence it can be observed that the leakage flux increases as a percentage of total or coil flux as the magnetic field of the core material increases.

The series magnetic circuits considered to this point have been Type 1 problems wherein the flux is specified in a core member and the resulting coil mmf or current to sustain that flux was determined. If a coil excitation value is given and a resulting core member flux is to be determined, the study is known as a Type 2 problem.

For the Type 2 problem of a series magnetic circuit with a uniform core cross-section area and an air gap with negligible fringing, the point of operation can be determined by a graphical technique known as the *load line method*. If the magnetic circuit is excited by an N-turn coil carrying I A, has a core mean length path ℓ_m, and has an air gap length δ, summation of mmfs yields

$$NI = H_g\delta + H_m\ell_m$$

where H_g and H_m are the magnetic field intensity of the air gap and the core, respectively. Under the assumption of negligible air gap fringing, $B_g = B_m$. Then H_g can be replaced by B_m/μ_o in the above equation and the result arranged to find

$$B_m = -\frac{\mu_o\ell_m}{\delta}H_m + \frac{\mu_o NI}{\delta} \qquad \textbf{[3.19]}$$

Equation [3.19] describes a straight line on the *B-H* characteristic of the core material with a *B*-axis intercept of $\mu_o NI/\delta$ and an *H*-axis intercept of NI/ℓ_m. If [3.19] is superimposed on the core material *B-H* characteristic, the intersection of the two curves must be the point of operation.

Let the series magnetic circuit of Fig. 3.11 be described by the data of Example 3.3. If the coil current $I = 5$ A, determine the value of core flux if leakage and air gap fringing are neglected.   **Example 3.5**

Using the values of Example 3.3, the *B* and *H* axes intercepts are

$$\frac{\mu_o NI}{\delta} = \frac{4\pi \times 10^{-7}(500)(5)}{0.002} = 1.57 \text{ T}$$

$$\frac{NI}{\ell_m} = \frac{(500)(5)}{0.698} = 3582 \text{ A-t/m}$$

Figure 3.15 displays a plot of the load line superimposed on the *B-H* curve for the M-27 core material. The intersection of the two curves gives $B_m = 1.32$ T. The core flux is then

$$\phi_m = B_m A_m = 1.32(0.05)(0.05)(0.95) = 3.13 \text{ mWb}$$

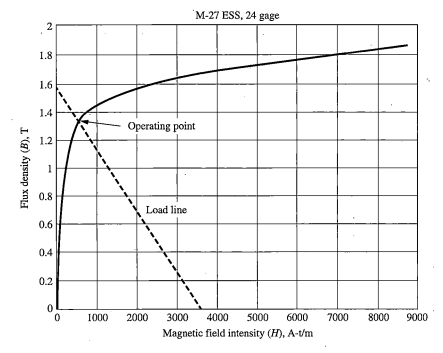

Figure 3.15
Load line solution for Example 3.5

Although the load line method has readily offered a solution to this simple problem, it does not allow consideration of air gap fringing as presented. Later in Sec. 3.7.2 dealing with permanent magnet materials, the method will be extended to allow analysis with air gap fringing. However, the load line method remains limited to series magnetic circuits with a uniform core cross section formed from a single ferromagnetic material. Analysis with leakage present is not possible. Consequently, its application is limited if accuracy is important. The Type 2 problem is best handled by generating a flux-mmf plot for the particular magnetic circuit and then entering that plot for the specified value of coil excitation to determine the resulting core member flux. Generating the necessary plot by longhand methods is a tedious task. However, the techniques of the MATLAB programs ⟨cckt1.m⟩ and ⟨cleak1.m⟩ have already been used to generate wide-range plots for the C-shaped magnetic structure of Fig. 3.11. With simple changes to these basic programs, ⟨cckt2.m⟩ and ⟨cleak2.m⟩ have been developed to accurately handle the cumbersome task of the Type 2 problem for the cases without and with leakage flux, respectively. The value of coil current is now an input. Once the wide-range arrays of flux and mmf are generated, the programs use the built-in interp1 function of MATLAB to enter the mmf array with the specified coil excitation and determine the resulting flux values.

Example 3.6 Let the series magnetic circuit of Fig. 3.11 be described by the data of Example 3.3. If the coil current $I = 15$ A, determine the value of air gap flux if leakage flux is neglected.

Solution of this Type 2 problem is handled by ⟨cckt2.m⟩. Execution of the program gives the screen display below.

```
C-SHAPED MAGNETIC CIRCUIT, No Leakage
   AIR GAP FLUX = 0.004263
   COIL CURRENT = 15
```

The flux-mmf plot of Fig. 3.16 is also generated by ⟨cckt2.m⟩, showing the resulting air gap flux to be 4.263 mWb when a coil current of 15 A flows ($NI = 500 \times 15 = 7500$ A-t).

Example 3.7 The magnetic circuit of Fig. 3.11 is described by the data of Example 3.3. Let the coil current $I = 20$ A and account for leakage flux. Determine the values of resulting air gap flux and leakage flux.

Execution of ⟨cleak2.m⟩ results in the shown screen display.

```
C-SHAPED MAGNETIC CIRCUIT, With Leakage
   AIR GAP FLUX = 0.004207
   LEAKAGE FLUX = 0.0004518
   COIL CURRENT = 20
```

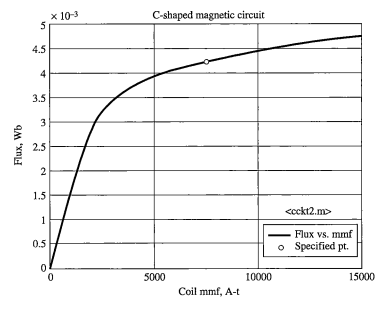

Figure 3.16
Example 3.6 result

The wide-range flux-mmf plot of Fig. 3.17 produced by ⟨cleak2.m⟩ indicates the resulting operating point with air gap flux being 4.207 mWb for a coil current of 20 A ($NI = 500 \times 20 = 10{,}000$ A-t).

3.4.5 PARALLEL MAGNETIC CIRCUITS

The broad class of magnetic circuits that have two or more flux paths, neglecting leakage flux, are classified as *parallel magnetic circuits*. Except for a few topologies with certain members treated as magnetically linear, analyses of both Type 1 and Type 2 problems require that at least a portion of the solution make use of graphical or iterative methods. The analysis techniques needed for solution of a parallel magnetic circuit, regardless of flux path complexity, can be illustrated by analysis of a magnetic circuit having principal flux paths that encompass two areas or windows of high magnetic reluctance. This general topology illustrated by Fig. 3.18 will be referred to as a *double window magnetic circuit*. Although symmetry about the horizontal centerline is used in Fig. 3.18 to reduce the number of distinct dimensions, this choice does not restrict the generality. Further, a rectangular geometry has been selected to simplify flux path length and cross-sectional area calculations; however, the analysis procedures to be introduced are equally applicable to the more geometrically complex magnetic circuit structures of electric machines.

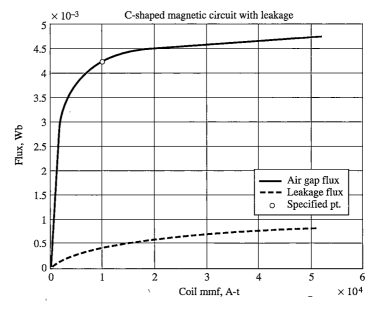

Figure 3.17
Example 3.7 result

Example 3.8

Assume that the core material for the parallel magnetic circuit of Fig. 3.18 is infinitely permeable. Let $N = 1000$ turns, $\ell_3 = 2$ in, $\ell_5 = 1$ in, $d = 3$ in, $\delta_1 = 0.100$ in, and $\delta_2 = 0.075$ in. If the flux crossing the right air gap $\phi_r = 4$ mWb, find: (a) ϕ_c and ϕ; and (b) the coil current I.

Since the ferromagnetic core material is infinitely permeable, all core reluctance values are zero. The schematic representation for this magnetic circuit is given by Fig. 3.19.

(a) Based on [3.17], with division by 39.37 to account for dimensions in inches,

$$\mathcal{P}_{g1} = \mu_o\left[\frac{\ell_3 d}{\delta_1} + 0.52(\ell_3 + d) + 0.308\delta_1\right]$$

$$\mathcal{P}_{g1} = \frac{4\pi \times 10^{-7}}{39.37}\left[\frac{(2)(3)}{0.100} + 0.52(2 + 3) + 0.308(0.100)\right] = 1.999 \times 10^{-6}\text{H}$$

$$\mathcal{P}_{g2} = \mu_o\left[\frac{\ell_5 d}{\delta_2} + 0.52(\ell_5 + d) + 0.308\delta_2\right]$$

$$\mathcal{P}_{g2} = \frac{4\pi \times 10^{-7}}{39.37}\left[\frac{(1)(3)}{0.075} + 0.52(1 + 3) + 0.308(0.075)\right] = 1.344 \times 10^{-6}\text{H}$$

$$\mathcal{F} = \frac{\phi_r}{\mathcal{P}_{g2}} = \frac{0.004}{1.344 \times 10^{-6}} = 2976.5 \text{ A-t}$$

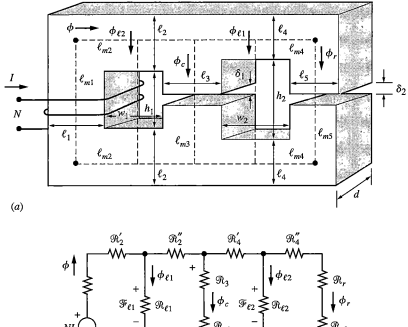

(a)

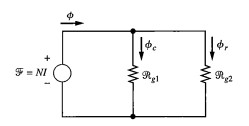

(b)

Figure 3.18
Parallel magnetic circuit (double window). (a) Physical. (b) Schematic.

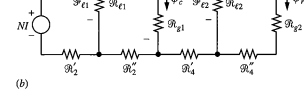

Figure 3.19
Schematic for magnetic circuit of Fig. 3.18 with $\mu = \infty$

$$\phi_c = \mathscr{F}\mathscr{P}_{g1} = (2976.5)(1.999 \times 10^{-6}) = 5.95 \text{ mWb}$$

$$\phi = \phi_c + \phi_r = 5.95 + 4.00 = 9.95 \text{ mWb}$$

(b) The coil current is

$$I = \frac{\mathscr{F}}{N} = \frac{2976.5}{1000} = 2.98 \text{ A}$$

Example 3.9

For the magnetic circuit of Example 3.8, rather than specifying ϕ_r, let coil current $I = 10$ A. Determine the coil flux ϕ and the right air gap flux ϕ_r.

This is a Type 2 problem with magnetic linearity modeled by the schematic of Fig. 3.19. Since all reluctances have constant values, the solution does not require an iterative approach. The values of air gap permeances determined in Example 3.8 are valid. The equivalent permeance seen by the coil is

$$\mathscr{P}_{eq} = \mathscr{P}_{g1} + \mathscr{P}_{g2} = 1.999 \times 10^{-6} + 1.344 \times 10^{-6} = 3.343 \times 10^{-6} \text{ H}$$

$$\phi = NI\mathscr{P}_{eq} = (1000)(10)(3.343 \times 10^{-6}) = 33.43 \text{ mWb}$$

By flux division,

$$\phi_r = \frac{\mathscr{R}_{g1}}{\mathscr{R}_{g1} + \mathscr{R}_{g2}} \phi = \frac{\mathscr{P}_{g2}}{\mathscr{P}_{g1} + \mathscr{P}_{g2}} \phi = \frac{1.344}{1.999 + 1.344}(33.43) = 13.44 \text{ mWb}$$

Example 3.10

The parallel magnetic circuit of Fig. 3.18 is built up with M-27, 24-gage core material. $N = 100$ turns. The core stacking factor is $SF = 0.95$. The dimensions in units of inches are as follows:

$\ell_1 = 3$	$d = 3$	$\delta_1 = 0.100$
$\ell_2 = 3$	$h_1 = 4$	$\delta_2 = 0.075$
$\ell_3 = 2$	$w_1 = 3.5$	
$\ell_4 = 1$	$h_2 = 3$	
$\ell_5 = 1$	$w_2 = 2.5$	

$$\ell_{m1} = h_1 + \ell_2 = 7$$

$$\ell_{m2} \cong w_1 + (\ell_1 + \ell_3)/2 = 6$$

$$\ell_{m3} = (h_1 + \ell_2 + h_2 + \ell_4)/2 - \delta_1 = 5.4$$

$$\ell_{m4} \cong w_2 + (\ell_3 + \ell_5)/2 = 4$$

$$\ell_{m5} = h_2 + \ell_4 - \delta_2 = 3.925$$

If flux $\phi_r = 3$ mWb, determine ϕ_c, ϕ, and I. Neglect any leakage.

The values of air gap permeance determined in Example 3.8 are valid. The schematic of Fig. 3.18b is applicable after removing $\mathscr{R}_{\ell 1}$ and $\mathscr{R}_{\ell 2}$. Dimensions given in inches will be converted to meters as their values are substituted into formulas:

$$B_r = \frac{\phi_r}{\ell_5 d SF} = \frac{0.003}{(0.0254)(0.0762)(0.95)} = 1.631 \text{ T}$$

From Fig. 3.12, or using hm27(1.631) from the MATLAB Command Window, H_r = 3177.2 A-t/m.

$$\mathscr{F}_c = H_r(2\ell_{m4} + \ell_{m5}) + \phi_r/\mathscr{P}_{g2}$$

$$\mathscr{F}_c = (3177.2)(2 \times 0.1016 + 0.0997) + 0.003/(1.344 \times 10^{-6}) = 3194.5 \text{ A-t}$$

At this point, a graphical solution is necessary to determine the flux ϕ_c with mmf impressed across the center core member. By choice, MATLAB was used to plot the flux-mmf characteristic of the center member shown by Fig. 3.20. Entering the curve for $\mathscr{F}_c$ = 3194.5, ϕ_c = 5.8 mWb is read:

$$\phi = \phi_c + \phi_r = 5.8 + 3 = 8.8 \text{ mWb}$$

$$B_1 = \frac{\phi}{\ell_1 dSF} = \frac{0.0088}{(0.0762)(0.0762)(0.95)} = 1.595 \text{ T}$$

From Fig. 3.12 or using hm27(1.595), H_1 = 2640 A-t/m

$$NI = H_1(\ell_{m1} + 2\ell_{m2}) + \mathscr{F}_c$$

$$NI = 2640(0.1778 + 2 \times 0.1524) + 3194.5 = 4468.6 \text{ A-t}$$

$$I = \frac{NI}{N} = \frac{4468.6}{1000} = 4.47 \text{ A}$$

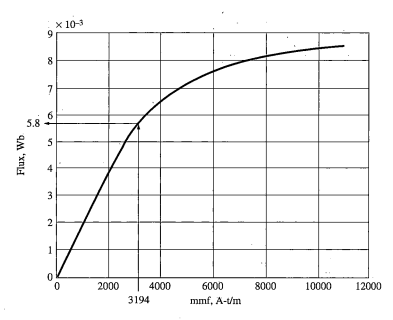

Figure 3.20
Flux-mmf plot for center member in Example 3.10

Analysis of this parallel magnetic circuit required the construction of an inter-mediate flux-mmf plot to graphically determine the flux through the center member. A similar operation is typical for solution of the parallel magnetic circuit. After the work of the above example, still only a single performance point is known without a broad-range regard to the saturation level of the magnetic circuit. MATLAB program ⟨dwckt1.m⟩ has been developed to perform a broad-range analysis of this particular parallel magnetic circuit of Fig. 3.18. The screen display upon execution of ⟨dwckt1.m⟩ using the data of Example 3.10 follows.

```
DOUBLE WINDOW MAGNETIC CIRCUIT,  No Leakage
   COIL CURRENT = 4.437
       COIL FLUX = 0.008772
AIR GAP 1 FLUX = 0.005772
AIR GAP 2 FLUX = 0.003
```

The resulting flux-mmf plot is shown by Fig. 3.21, where it is easy to assess the nature of flux in all members and the impact on coil mmf requirements if different flux values were desired.

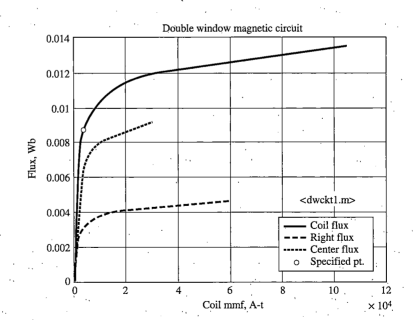

Figure 3.21
Parallel magnetic circuit, no leakage flux

Analysis of the particular parallel magnetic circuit of Fig. 3.18 by longhand methods as a Type 1 problem with leakage flux becomes unwieldy. Yet based on the observations from study of the series magnetic circuit, it is anticipated that the error due to neglect of leakage flux is significant. MATLAB program ⟨dwleak1.m⟩ has been developed to conveniently handle the analysis of this particular parallel magnetic circuit. Execution of the program using the data of Example 3.10 gives the screen display below.

```
DOUBLE WINDOW MAGNETIC CIRCUIT, With Leakage
    COIL CURRENT = 6.014
        COIL FLUX = 0.009628
AIR GAP 1 FLUX = 0.006058
AIR GAP 2 FLUX = 0.003
```

Examination of the results shows that the coil current is actually 35.5 percent greater than predicted if leakage flux were neglected. The wide-range plot of Fig. 3.22 allows evaluation of change in the operating point and assessment of the leakage flux nature.

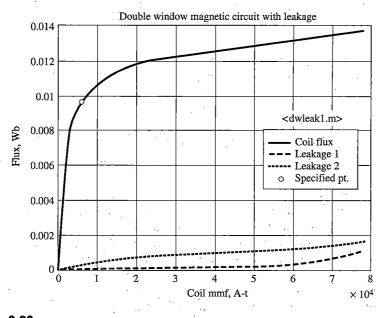

Figure 3.22
Parallel magnetic circuit with leakage flux

In order to handle the parallel magnetic circuit of Fig. 3.18 under Type 2 analysis, the MATLAB program ⟨dwleak2.m⟩ has been formed by modification of ⟨dwleak1.m⟩. The screen display results from execution of ⟨dwleak2.m⟩ using the coil and dimensional data of Example 3.10 for the case of a coil current $I = 7.5$ A are shown.

```
DOUBLE WINDOW MAGNETIC CIRCUIT, With Leakage
     COIL CURRENT = 7.5
        COIL FLUX = 0.01011
  AIR GAP 1 FLUX = 0.006375
  AIR GAP 2 FLUX = 0.003074
```

3.5   ENERGY AND INDUCTANCE

Energy can be converted from electrical form to magnetic form and vice versa. *Inductance* is an electric circuit characteristic that relates magnetic field variables to electric circuit variables; hence, its potential utility in quantitative analysis of this energy conversion process is apparent.

3.5.1   INDUCTANCE

Consider the magnetic circuit of Fig. 3.23a that is excited by the coil current i. The resulting magnetization curve is displayed by Fig. 3.23b, where the vertical axis is shown in terms of *flux linkage* ($\lambda \triangleq N\phi$), a convenience quantity adopted since the product of coil turns and the flux linking those turns appears extensively in the analysis of electromagnetic coils. Owing to the nonlinear nature of the ferromag-

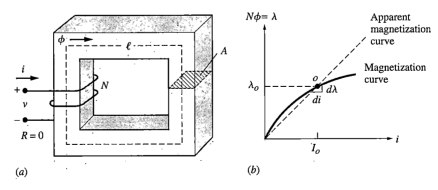

Figure 3.23
Inductance of ferromagnetic core. (a) Magnetic circuit. (b) Magnetization characteristic.

netic material, λ is a function of current; thus, $\lambda = \lambda(i)$. The voltage behind the terminals of the assumed lossless coil is determined from [3.8]:

$$v = N\frac{d\phi}{dt} = \frac{d\lambda}{dt} = \frac{\partial\lambda}{\partial i}\frac{di}{dt} \qquad \textbf{[3.20]}$$

The chain rule from calculus was applied in writing the right-hand expression of [3.20]. For the case at hand, $\partial\lambda/\partial i = d\lambda/di$ since λ is a function of only current i. However, for more general cases of magnetic circuits where either a member of the ferromagnetic core can move to alter the magnetic circuit reluctance or multiple excitation coils exist, λ may be a function of more than one variable. Thus, the partial derivative is retained. From [3.20], the definition of inductance is formed as

$$L = \frac{v}{di/dt} = \left.\frac{\partial\lambda}{\partial i}\right|_o = \left.N\frac{\partial\phi}{\partial i}\right|_o \qquad \textbf{[3.21]}$$

The value of inductance given by [3.21] is known as the *incremental inductance*. It is simply the slope of the λ-i plot at the point of operation o as indicated on Fig. 3.23b. Obviously any change in operating point requires a new evaluation of L. It is also apparent that the value of inductance decreases as the point of operation moves into the saturation region of the ferromagnetic core.

If the magnetic circuit were linear, thus a straight-line λ-i plot, evaluation of [3.21] yields a constant value regardless of the point of operation. Hence, [3.21] could be written as

$$L = \frac{\lambda}{i} = \frac{N\phi}{i} \qquad \textbf{[3.22]}$$

In electromechanical device analysis, it is common practice to treat the magnetic circuit as linear with an apparent magnetization curve passing through the point of nominal operation as illustrated by Fig. 3.23b. In such case, the inductance given by [3.22] is known as the *apparent inductance*. For devices with an air gap and only a low level of magnetic saturation, the values of the incremental and apparent inductance are approximately equal. Correction factors to account for the use of apparent inductance can be developed for cases of higher saturation. Unless explicitly stated otherwise, the common practice will be followed and any reference to inductance is understood to be apparent inductance.

The magnetic circuit of Fig. 3.23a has a core cross-sectional area $A = 25$ cm^2 and a mean length path of $\ell = 60$ cm. If the uniform core flux density $B = 1.5$ T for a current $i = 6$ A and the coil has $N = 150$ turns, determine (a) the coil inductance and (b) the core reluctance.   **Example 3.11**

(a) By [3.22],

$$L = \frac{N\phi}{i} = \frac{NBA}{i} = \frac{(150)(1.5)(25 \times 10^{-4})}{6} = 93.75 \text{ mH}$$

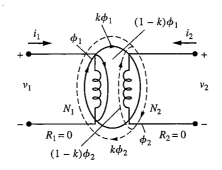

Figure 3.24
Mutual coupling flux

(b) By manipulation of [3.22],

$$L = N\frac{\phi}{i} = N^2\frac{\phi}{Ni} = \frac{N^2}{\mathcal{R}}$$

or

$$\mathcal{R} = \frac{N^2}{L} = \frac{(150)^2}{0.09375} = 2.4 \times 10^5\,\text{H}^{-1}$$

The inductance expressions of [3.21] and [3.22] were derived based on Fig. 3.23 where the flux passing through the N-turn coil is established by the current of that same coil. Such an inductance is called a *self-inductance*.

The flux ϕ of Faraday's law is simply the flux passing through the coil. Consider the pair of coils shown schematically in Fig. 3.24, where the coils are mutually coupled or share a portion of their magnetic fields. The *coefficient of coupling* (k) is introduced as the magnitude of the per unit portion of the flux established by one coil that links another coil; hence, $0 \leq k \leq 1$. In most applications of this text, good magnetic coupling of coil is desired and k has value near unity. The flux linkages for each coil of Fig. 3.24 can be formed and time differentiated to determine the voltage behind the terminals of each coil. For coil 1,

$$\lambda_1 = N_1(\phi_1 + k\phi_2) \tag{3.23}$$

$$v_1 = \frac{d\lambda_1}{dt} = N_1\frac{\partial\phi_1}{\partial i_1}\frac{di_1}{dt} + kN_1\frac{\partial\phi_2}{\partial i_2}\frac{di_2}{dt} \tag{3.24}$$

For coil 2,

$$\lambda_2 = N_2(k\phi_1 + \phi_2) \tag{3.25}$$

$$v_2 = \frac{d\lambda_2}{dt} = kN_2\frac{\partial\phi_1}{\partial i_1}\frac{di_1}{dt} + N_2\frac{\partial\phi_2}{\partial i_2}\frac{di_2}{dt} \tag{3.26}$$

The coefficient terms of di_2/dt in [3.24] and di_1/dt in [3.26] are new inductance terms that result from the mutually coupled flux and are known as *mutual inductances*. Rewrite [3.24] and [3.26] in terms of apparent inductance to yield

$$v_1 = \frac{N_1\phi_1}{i_1}\frac{di_1}{dt} + \frac{kN_1\phi_2}{i_2}\frac{di_2}{dt} = L_1\frac{di_1}{dt} + M_{12}\frac{di_2}{dt} \qquad \textbf{[3.27]}$$

$$v_2 = \frac{kN_2\phi_1}{i_1}\frac{di_1}{dt} + \frac{N_2\phi_2}{i_2}\frac{di_2}{dt} = M_{21}\frac{di_1}{dt} + L_2\frac{di_2}{dt} \qquad \textbf{[3.28]}$$

where the self-inductances of coils 1 and 2 have been recognized. The two mutual inductance terms can be manipulated to give

$$M_{12} = kN_1\frac{\phi_2}{i_2} = kN_1N_2\frac{\phi_2}{N_2 i_2} = kN_1N_2\frac{1}{\mathcal{R}_2} \qquad \textbf{[3.29]}$$

$$M_{21} = kN_2\frac{\phi_1}{i_1} = kN_1N_2\frac{\phi_1}{N_1 i_1} = kN_1N_2\frac{1}{\mathcal{R}_1} \qquad \textbf{[3.30]}$$

Since the two coils share the same flux paths, $\mathcal{R}_1 = \mathcal{R}_2$ provided the point of operation is such that $\mu_1 = \mu_2$. Thus, from [3.29] and [3.30], it is concluded that

$$M_{12} = M_{21} = M \qquad \textbf{[3.31]}$$

A mutual inductance subscript notation is not necessary for the case of two coils, as no ambiguity exists. However, for three or more coils, multiple mutual inductances exist and subscript notation is necessary. Such cases are addressed by the end of chapter problems.

Example 3.12

If the mutually coupled coil pair of Fig. 3.24 were operating with current $i_1 = I_m \sin \omega t$ A where frequency ω is known and current $i_2 = 0$, describe a laboratory test to determine the inductances L_1 and M.

With $i_2 = 0$, the phasor domain representations of [3.27] and [3.28] are

$$\bar{V}_1 = j\omega L_1 \bar{I}_1 \qquad \bar{V}_2 = j\omega M \bar{I}_1$$

Thus, if V_1 and V_2 were read with a voltmeter and I_1 were read by an ammeter, then

$$L_1 = \frac{V_1}{\omega I_1} \qquad M = \frac{V_2}{\omega I_1}$$

Example 3.13

If an open-circuit coil of 50 turns were added to the infinitely permeable magnetic circuit of Fig. 3.23a and the data of Example 3.11 remains valid, determine the value of the mutual inductance.

All flux is implicitly assumed to be contained within the ferromagnetic core; thus, $k = 1$. By [3.29],

$$M = \frac{kN_1N_2}{\mathcal{R}} = \frac{(1)(150)(50)}{2.4 \times 10^5} = 31.25 \text{ mH}$$

3.5.2   ENERGY

The electrical energy supplied to the lossless coil of Fig. 3.23a as the current increases from 0 to I_o must be equal to the energy (W_f) stored in the magnetic field of the ferromagnetic core

$$W_f = \int\limits_0^t ivdt = \int\limits_0^t i\left(\frac{d\lambda}{dt}\right)dt = \int\limits_0^{\lambda_o} id\lambda \qquad \text{[3.32]}$$

It is seen from [3.32] that this stored magnetic energy is given by the area to the left of the magnetization curve and below $\lambda = \lambda_0$. For the uniform core of Fig. 3.23a, $i = H\ell/N$ and $\lambda = NBA$, whence [3.32] can be rewritten as

$$W_f = \int\limits_0^{\lambda_o} \left(\frac{H\ell}{N}\right)NA\,dB = \ell A\int\limits_0^{B_o} H dB \qquad \text{[3.33]}$$

Since ℓA is the volume of the ferromagnetic core, a valid interpretation of [3.33] is that the area to the left of the core material B-H curve is the energy density of the core. Thus, for a fixed flux density, the volume of the ferromagnetic core is directly proportional to the energy stored in its magnetic field.

For the case of magnetic linearity, [3.22] gives $\lambda = Li$, which substituted into [3.32] leads to

$$W_f = \int\limits_0^{I_o} iL\,di = \frac{1}{2}LI_o^2 \qquad \text{[3.34]}$$

Example 3.14 | If the coil of Fig. 3.23a has a current $i = I_m\cos\omega t$ and the inductance has a value L, describe the magnetic energy stored in the core magnetic field.

From [3.34] without definite limits,

$$W_f = \int iL\,di = \frac{1}{2}Li^2 = \frac{1}{2}LI_m^2\cos^2(\omega t) = \frac{1}{4}LI_m^2(1 + \cos 2\omega t)$$

Thus, the magnetic energy is periodically stored and returned to the electrical circuit at twice source frequency and ranges from 0 to $\frac{1}{2}LI_m^2$ J in value.

3.6 SINUSOIDAL EXCITATION

Although general discussion of the flux-current characteristics for magnetic circuits has been addressed in prior sections, the sinusoidal steady-state case occurs frequently and special attention is justified. For electric machines and transformers, the excitation is a periodic waveform with zero average value. Hence, it can be represented by a Fourier series. Assuming negligible power quality problems, operation from the electric utility grid results in a fundamental component excitation voltage. However, excitation supplied from an independent power conditioner may display harmonic content.

If a magnetic circuit were linear (ϕ-i plot straight line), then the response to an excitation source would contain only the harmonics of the source. But it has been established in the prior work of this chapter that the ϕ-i plot is not linear when the magnetic circuit contains ferromagnetic material in the flux path. The MATLAB program ⟨sinex.m⟩ can be used to quantitatively determine the effects of this nonlinearity for the common case of a fundamental frequency sinusoidal voltage acting as the excitation source. The program builds a normalized, piecewise linear ϕ-i plot consisting of two straight-line segments. One line models the magnetically linear region up to the saturation breakpoint. The second line models the saturation region. The user can alter the nature of the plot by specification of the saturation breakpoint values Percentage maximum flux and Percentage maximum current. The ϕ-i plot is displayed to the screen to allow assessment of the nature of saturation specified. For a normalized impressed voltage $v(t) = \cos\omega t$, the normalized flux must be $\phi(t) = \sin\omega t$ by Faraday's law. The interp1 feature of MATLAB is used to map the current array from the flux array where the functional relationship between the two arrays is established by the ϕ-i plot. The program displays a dual plot showing the time-domain current and the Fourier spectrum of the current.

Figure 3.25 displays the result of executing ⟨sinex.m⟩ for the saturation breakpoint specified by 80 percent and 30 percent. From the Fourier spectrum, it is seen that in addition to the fundamental component, the current contains significant third- and fifth-harmonic components. Since the terminal voltage can only be the impressed fundamental component voltage, the excitation coil can be modeled as an inductance in series with current controlled voltage sources of harmonic frequency as shown by Fig. 3.26. The subscript associated with each voltage source indicates the harmonic number of the source. From the Fourier spectrum of Fig. 3.25, it is seen that the current can be written in a Fourier series as

$$i(t) \cong I_1\sin(\omega t) - I_3\sin(3\omega t) + I_5(5\omega t) \qquad \text{[3.35]}$$

where the current harmonic amplitudes are read directly from Fig. 3.25 or obtained from the MATLAB Command Window after executing ⟨sinex.m⟩ as shown below:

```
≫ f(2:2:6)
ans =
      0.7151
      0.2195
      0.0795
```

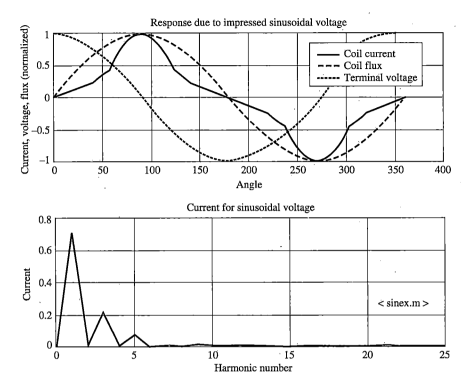

Figure 3.25
Excitation current response of a ferromagnetic core

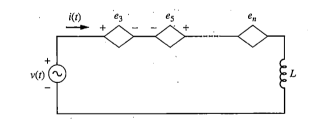

Figure 3.26
Equivalent circuit model for excitation of a ferromagnetic core

The phase angles of the harmonic currents can also be determined from the MAT-LAB Command Window by entering

```
≫ phase=angle(fftI)*180/pi;
≫ phase(2:2:6)
ans =
    -89.8242
     90.5273
    -89.1211
```

From the displayed angles, it is seen that the fundamental and fifth harmonic are in phase while the third harmonic is 180° phase shifted. Thus, the polarity markings of the controlled voltage sources in Fig. 3.26 and the negative sign of [3.35] are justified.

The harmonic component currents determined actually flow in the electric lines of the excitation voltage source. When dealing with transformers in Chap. 4, this issue of harmonic current will be discussed further in the context of an actual device. At that point, the MATLAB approach will be extended to model hysteresis phenomena.

A 120-V, 60-Hz voltage source excites a magnetic circuit. From an oscilloscope trace of the coil current, it is known that the peak value of the instantaneous coil current is 5.5 A. Determine the value of inductor L and the magnitude of the harmonic voltage sources of Fig. 3.25 if the above illustrated run of ⟨sinex.m⟩ properly reflects the saturation nature of the magnetic circuit.

Example 3.15

Since the results of ⟨sinex.m⟩ are normalized to unity, the voltage and current values must be multiplied by $\sqrt{2}\,120$ and 5.5, respectively, to yield the actual values. Superposition argument can be applied to the circuit of Fig. 3.26. Letting the subscripts denote the harmonic number, maximum value phasor analysis yields

$$L = \frac{V_1}{\omega I_1} = \frac{(1)(\sqrt{2}\,120)}{(120\pi)(0.7151 \times 5.5)} = 114.44 \text{ mH}$$

$$E_3 = 3\omega L I_3 = 3(120\pi)(0.11444)(0.2195 \times 5.5) = 156.25 \text{ V}$$

$$E_5 = 5\omega L I_5 = 5(120\pi)(0.11444)(0.0795 \times 5.5) = 94.32 \text{ V}$$

3.7 PERMANENT MAGNETS

A permanent magnet (PM) material is typically a metal alloy that, after being subjected to an H-field, retains a substantial *residual flux density* B_r. In order to reduce the flux density to zero, an H-field directed opposite in sense to the original magnetizing field must be applied. This impressed field magnitude must have a value H_c known as the *coercive force*.

3.7.1 CLASSIFICATION AND CHARACTERISTICS

The static *hysteresis loop* is the focal point in classifying PM materials. A hysteresis loop is the nonzero area enclosed by the B-H trajectory of a ferromagnetic specimen when subjected to cyclic magnetization. The point of operation traces the loop in a counterclockwise direction. Generally, the larger the area inside the hysteresis loop, the better the material performs as a PM. Information beyond this rough measure of goodness is critical to quantitative understanding. Figure 3.27

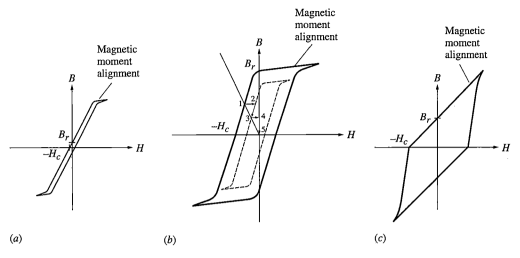

Figure 3.27
Hysteresis loops of ferromagnetic materials. (a) Low-carbon steel—poor PM. (b) Class 1—non-resilient PM. (c) Class 2—resilient PM.

depicts three distinctly different hysteresis loops. The first thin hysteresis loop is typical of low-carbon-content steel such as electric steel sheet used for ferromagnetic core material for electric machines and transformers. Even when magnetized to high levels of saturation so that all magnetic moments are unidirectional, upon removal of the external magnetic field, the residual flux density is small. Typically $B_r < 0.5$ T. More critical than the value of B_r is the fact that the coercive force has value typified by $H_c < 50$ A/m. A slight reverse field reduces the flux density to zero. This material is considered a poor PM.

The second hysteresis loop of Fig. 3.27b is characteristic of a class 1 or nonresilient (easily demagnetized) PM material such as Alnico, an alloy of aluminum, nickel, and cobalt. The material is characterized by a relatively high $B_r > 1$ T and an $H_c < 100$ kA/m. Owing to its relatively low cost, many PM device designs are built around class 1 magnet materials in spite of two less than desirable characteristics that are overcome by the class 2 PM materials:

1. By nature, the PM must operate with $-H_c < H < 0$. If the PM is to act as an mmf source for a magnetic circuit, the source mmf is determined by the product of magnet length and H. Thus, for two PM materials with the same value of B_r, the material with the larger H_c requires less volume of PM material to function as the mmf source for a particular magnetic circuit. Class 1 PMs have a relatively low value of H_c, limiting their capability to source a high-reluctance magnetic circuit. The yet to be discussed rare earth PM materials offer the smallest volume design potential, but with the penalty of greater cost.

2. The *B-H* locus followed by functional PM, due to changes in the external magnetic circuit, is known as the *recoil characteristic,* which determines its resilient nature, or lack thereof, to return to the original state of magnetization.

If a PM undergoes B-H transition while in the region of magnetic moment alignment, that is, to the right of the knee of the descending hysteresis loop, the B-H locus remains on the original major hysteresis loop. For a decreasing H transition from a point located to the left of the knee of the descending hysteresis loop, the B-H locus follows a path approximately parallel to the magnetic moment alignment region. Such a transition is indicated by the path 1-2 of Fig. 3.27b. Any increasing H transition from point 2 follows a B-H locus along an inner hysteresis loop to point 3. Points 4 and 5 indicate continued B-H transition that results in complete demagnetization of the PM material.

The demagnetization scenario for class 1 PM material can transpire for two physically likely situations. The first is assembly and disassembly of a magnetic structure. Class 1 PMs are magnetized in situ and must be remagnetized after any disassembly. Second, if the PM is part of a magnetic circuit with other mmf sources capable of decreasing the H-field of the PM material during operation, care must be taken in the design and operation to assure that the B-H trajectory during operation remains above the knee of an acceptable descending hysteresis loop.

The third hysteresis loop of Fig. 3.27c is characteristic of class 2 resilient PM material such as ceramics (barium or strontium ferrite) or rare earths (neodymium-iron-boron or samarium-cobalt alloys). The ceramics, while significantly lower in cost, are characterized by $B_r < 0.4$ T and $H_c < 300$ kA/m. The more costly rare earths have $B_r > 1$ T and $H_c > 600$ kA/m, making them a preferred material from a performance point of view. The class 2 PM materials have descending major hysteresis loops with their regions of magnetic moment alignment spanning all, or nearly all, of the second quadrant of the B-H characteristic. Consequently, design of magnetic circuits with recoil characteristic along the original major hysteresis loop inherently results. Since assembly-disassembly does not lead to demagnetization, in situ magnetization is not a necessity, although it may be a convenience.

3.7.2 PERFORMANCE

In the above discussion, it has become apparent that the performance nature of PM materials depends on the second quadrant B-H characteristic. Figure 3.28 shows these second quadrant characteristics, known as *demagnetization curves*, for selected alloy compositions among the four PM materials dominant with present technology.

Consider the series magnetic circuit of Fig. 3.29, which is similar to Fig. 3.11a except that the excitation coil has been replaced by a volume of magnetized PM material. Realizing that the PM material acts as an mmf source and neglecting any leakage flux, application of [3.11] gives

$$-H_m \ell_m = 2\ell_{Fe}H_{Fe} + \phi/\mathscr{P}_g \qquad \textbf{[3.36]}$$

Define the ratio of ferromagnetic cross-sectional area to that of the PM material as

$$k_A = A_{Fe}/A_m$$

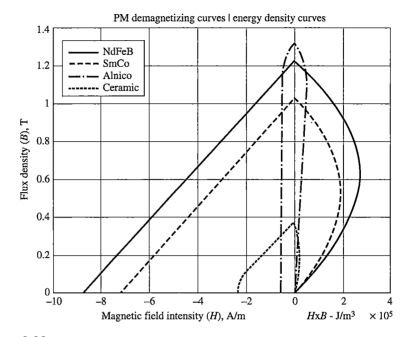

Figure 3.28
Demagnetization/energy density curves

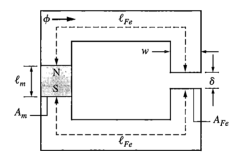

Figure 3.29
PM excited series magnetic circuit

Then using [3.37] and the relative permeability μ_R introduced in [3.6], [3.36] can be written as

$$-H_m\ell_m = \frac{2\ell_{Fe}k_AB_m}{\mu_R\mu_o} + \frac{k_AB_mA_{Fe}}{\mathcal{P}_g} \qquad \textbf{[3.38]}$$

Rearrangement of [3.38] yields the external B_m-H_m relationship of the PM material that must also simultaneously satisfy the material internal B_m-H_m relationship of the demagnetization curve.

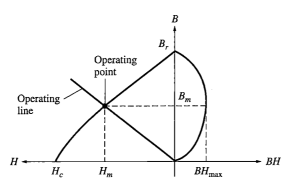

Figure 3.30
Maximum energy product PM design

$$B_m = -\left[\frac{\mu_o \mu_R \ell_m \mathscr{P}_g}{2k_A \ell_{Fe} \mathscr{P}_g + \mu_o \mu_R k_A A_{Fe}}\right] H_m \qquad \text{[3.39]}$$

This equation, or *load line*, is a straight line with negative slope passing through the origin of the PM material *B-H* curve intersecting the material demagnetization curve at the *point of operation* as illustrated by Fig. 3.30. This second-quadrant *B-H* curve load line is analogous to the earlier first-quadrant load line for a coil excited magnetic circuit discussed in Sec. 3.4.4.

Let $k_{eff} > 1$ be a multiplying factor that could be determined based on [3.17] to give the effective cross-sectional area of the air gap as $k_{eff} A_{Fe}$. A practical range is $1 < k_{eff} < 1.5$. The denominator of [3.39] can then be written as

$$\frac{2k_A \ell_{Fe} \mu_o k_{eff} A_{Fe}}{\delta} + \mu_o \mu_R k_A A_{Fe}$$

If $\mu_R \gg 2\ell_{Fe} k_{eff}/\delta$, then [3.39] reduces to

$$B_m \cong -\frac{\mu_o k_{eff} \ell_m}{k_A \delta} H_m \qquad \text{[3.40]}$$

Since the high levels of saturation are uncommon in PM circuits, μ_R is always large, allowing use of [3.40] in most applications.

PM material is typically more costly than ferromagnetic material. Hence, design economic considerations may require a minimum volume of PM material that can meet the performance specifications. The volume of PM material for the circuit of Fig. 3.29 is

$$V_m = A_m \ell_m \qquad \text{[3.41]}$$

Solve [3.40] for the magnitude of ℓ_m; use $A_m = A_g/k_{eff}k_A$ and $B_m = B_g A_g/A_m$ in [3.41].

$$V_m = A_m \ell_m = \frac{A_g}{k_{eff} k_A} \frac{B_m \delta}{\mu_o k_{eff} H_m} \frac{B_m}{B_m} = \frac{A_g \delta k_A (B_g A_g / A_m)^2}{\mu_o (k_A k_{eff})^2 B_m H_m}$$

$$V_m = \frac{A_g \delta k_A B_g^2}{\mu_o B_m H_m} = \frac{V_g k_A B_g^2}{\mu_o B_m H_m} \quad \text{[3.42]}$$

where $V_g = A_g \delta$ was recognized as the air gap volume. Eq. [3.42] indicates that for a fixed air gap volume and flux density, the smallest volume of PM material results when the energy density product $B_m H_m$ at the point of operation is maximized. The energy density curves of Fig. 3.28 are a plot of B vs. BH for each of the four common PM materials from which the value of BH_{max} is obviously the rightmost point of each curve. Figure 3.30 serves to clarify this minimum PM material volume design. Of course, design constraints may not allow the freedom to select a δ and ℓ_m to place the operating point at exactly the maximum energy density point. Further, in class 1 magnet application, maximum energy density design may place the operating point in an unfavorable position, leading to demagnetization difficulties. Class 2 magnet design typically has no such problems.

Example 3.16

A series magnetic circuit with the geometric shape of Fig. 3.29 has $A_m = A_{Fe} \doteq 25$ cm^2 and $\delta = 20$ mm. The core cross section is square with $\ell_{Fe} = 10$ cm. If the minimum volume ceramic permanent magnet with cross-sectional area equal to that of the iron core is used, determine the length of the magnet. The core material is M-27, 24-gage ESS with $SF = 0.96$.

From Fig. 3.28 at BH_{max}, $B_m \cong 0.2$ T and $H_m \cong 1.33 \times 10^5$ A-t/m. .

$$B_{Fe} = \frac{B_m}{SF} = \frac{0.2}{0.96} = 0.208 \, \text{T}$$

From either Fig. 3.12 for $B_{Fe} = 0.208$ or entering hm27(0.208) from the MATLAB Command Window, $H_{Fe} \cong 30$ A-t/m. Entering Fig. 3.7 for $H_{Fe} = 30$ A-t/m, $\mu_R \cong 4800$.

Since the core is square,

$$w = d = \sqrt{A_{Fe}} = \sqrt{25} = 5 \, \text{cm}$$

By [3.17],

$$\mathscr{P}_g = \mu_o \left[\frac{wd}{\delta} + 0.52(w + d) + 0.308\delta \right]$$

$$\mathscr{P}_g = 4\pi \times 10^{-7} \left[\frac{0.0025}{0.02} + 0.52(0.05 + 0.05) + 0.308(0.02) \right]$$

$$= 2.302 \times 10^{-7} \, \text{H}$$

Based on [3.39], the magnet length is

$$\ell_m = \frac{[2k_A \ell_{Fe} \mathscr{P}_g + \mu_o \mu_R k_A A_{Fe}] B_m}{\mu_o \mu_R \mathscr{P}_g H_m}$$

$$\ell_m = \frac{[2(1)(0.1)(2.302 \times 10^{-7}) + 4\pi \times 10^{-7}(4800)(1)(0.0025)](0.2)}{4\pi \times 10^{-7}(4800)(2.302 \times 10^{-7})(1.33 \times 10^5)}$$

$$= 16.38 \text{ cm}$$

Rework Example 3.16 using [3.42].   **Example 3.17**

   In [3.42] derivation, B_g is the actual air gap flux density with fringing while V_g is the air gap volume including fringing. Based on [3.17], the air gap multiplying factor to account for fringing is

$$k_{eff} = \frac{\mathscr{P}_g}{\mu_o w d/\delta} = \frac{2.302 \times 10^{-7}}{4\pi \times 10^{-7}(0.0025)/0.02} = 1.465$$

Hence,

$$B_g = \frac{B_{Fe}}{k_{eff}} = \frac{0.2}{1.465} = 0.1365 \text{ T}$$

From [3.41] and [3.42],

$$\ell_m = \frac{V_m}{A_m} = \frac{V_g k_A B_g^2}{\mu_o B_m H_m A_m} = \frac{\delta k_{eff} A_{Fe} k_A B_g^2}{\mu_o B_m H_m A_m}$$

$$\ell_m = \frac{(0.02)(1.465)(0.0025)(1)(0.1365)^2}{4\pi \times 10^{-7}(0.2)(1.33 \times 10^5)(0.0025)} = 16.33 \text{ cm}$$

Although [3.42] neglects the core mmf, the resulting value of ℓ_m compares nearly identically with the result of Example 3.16, indicating that the core mmf drop was truly negligible compared to the air gap mmf drop.

For the series magnetic circuit of Fig. 3.29, let $2A_m = A_{Fe} = 25$ cm^2, $\ell_{Fe} = 15$ cm, and $\delta =$ **Example 3.18**
0. If the core flux density is $B_{Fe} = 0.5$ T, determine the necessary length of a SmCo magnet.

$$B_m = k_A B_{Fe} = 2(0.5) = 1 \text{ T}$$

From Fig. 3.28 with $B_m = 1$ T,

$$H_m \cong 5778 \text{ A-t/m}$$

From Fig. 3.12 with $B_{Fe} = 0.5$ T,

$$H_{Fe} \cong 75 \text{ A-t/m}$$

$$\mu_o \mu_R = \frac{B_{Fe}}{H_{Fe}} = \frac{0.5}{75} = 0.00667 \text{ H}^{-1}$$

For $\delta = 0$, $\mathscr{P}_g \to \infty$. From [3.39] with $\mathscr{P}_g$ infinitely large,

$$B_m = \frac{\mu_o \mu_R \ell_m}{2k_A \ell_{Fe}} H_m$$

or

$$\ell_m = \frac{2k_A \ell_{Fe} B_m}{\mu_o \mu_R H_m} = \frac{2(2)(0.15)(1)}{(0.00667)(5778)} = 1.557 \, \text{cm}$$

3.8 ENERGY CONVERSION

Electromechanical energy conversion is the process of converting energy from electrical form to mechanical form or vice versa. The conversion is not direct but rather requires an intermediate conversion of energy into magnetic form. The process is reversible except for energy dissipated as heat loss. The intent of this section is to develop the concepts of electromechanical energy conversion from first principles. The results will be general and readily applicable to the electric machines of later study; however, these results, being based on first principles, will be expressed in terms of the dependent variables of the intermediary magnetic circuit. The later steady-state electric machine study will have as a goal analysis of the devices by means of an equivalent circuit wherein the magnetic circuit variables have been accounted for by means of electric circuit variables. Although steady-state electric machine analysis can be conducted largely from an electrical circuit, transient analysis must return to the concepts and procedures of this section.

3.8.1 MODEL FORMULATION

The two electromechanical systems of Fig. 3.31 are sufficiently general to develop the energy based analysis techniques. Both the translational and rotational systems are singly excited, giving one electrical degree of freedom. Also, the mechanical variable is limited to one degree of freedom in each configuration. Had systems with additional degrees of freedom been chosen, the principles to be developed would be no different in end result.

Conservation of energy must hold. By arbitrary choice, yet consistent with tradition, it will be assumed that energy flows into the electrical terminal pair and is absorbed by the mechanical load, or motoring action. Consequently, the results must be interpreted in this context. For example, a positive value of torque indicates that energy is actually being converted from electrical to mechanical form. A negative value of torque indicates that energy flow is reversed, or the so-called generating mode exists.

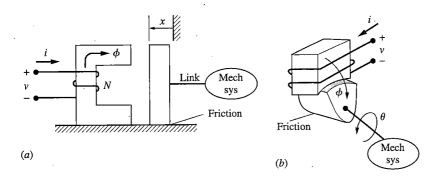

Figure 3.31
Electromechanical system. (a) Translational. (b) Rotational.

An energy audit that spans all history for either of the systems of Fig. 3.31 yields in equation form

$$
\begin{Bmatrix} \text{Electrical} \\ \text{energy} \\ \text{input} \end{Bmatrix} - \begin{Bmatrix} \text{Coil} \\ \text{ohmic} \\ \text{losses} \end{Bmatrix} - \begin{Bmatrix} \text{Magnetic} \\ \text{core} \\ \text{losses} \end{Bmatrix}
$$

$$
= \begin{Bmatrix} \text{Energy to} \\ \text{mechanical} \\ \text{system} \end{Bmatrix} + \begin{Bmatrix} \text{Friction} \\ \text{losses} \end{Bmatrix} + \begin{Bmatrix} \text{Stored} \\ \text{magnetic} \\ \text{energy} \end{Bmatrix} \qquad \textbf{[3.43]}
$$

Of the six terms in [3.43], three describe the losses that are irreversible heat dissipation in the three subsystems—electrical circuit, magnetic circuit, and interactive mechanical structure. It will be assumed that these loss mechanisms can be separated from the energy conversion and storage processes without significant sacrifice in accuracy. The loss accounting will then be placed external to a conceptual area wherein now conservative, bilateral energy conversion can occur. The coil ohmic losses can be modeled simply as a series resistor. Coulomb and viscous friction losses can be modeled as constant and speed-dependent forces or torques, respectively, at the point of restraining action by the coupled external mechanical system. Accurate modeling of the magnetic core losses is difficult even without attempting to separate them from the energy conversion and storage process. Reasonably accurate prediction of these hysteresis and eddy current losses in the sinusoidal steady state is possible. For other waveforms, no simple and accurate prediction methods exist. In the following chapter, it is shown that the parallel resistor model of Fig. 4.18 suffices for the sinusoidal steady-state case. This will be adopted as a first-order approximation for the more general excitation case at hand. After separating the loss accounting from the conservative conceptual energy area, the electromechanical systems of Fig. 3.31 can be visualized as shown in Fig. 3.32.

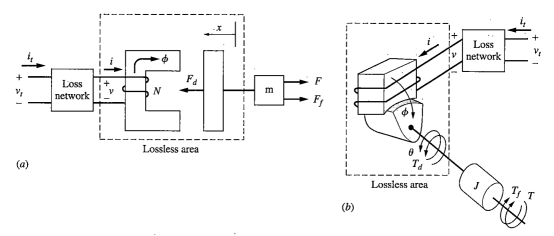

Figure 3.32
Conservative energy converter model. (a) Translational. (b) Rotational.

The mass and inertia of the movable members are also modeled external to the conservative area; hence, there is no mechanical energy storage inside the area. The force (torque) terms are clarified:

$F_d(T_d)$—mechanical force (torque) developed by energy conversion

$F_f(T_f)$—friction force (torque)

$F(T)$—restraining force (torque) of coupled mechanical system

$F_d(T_d)$ acts to increase $x(\theta)$. $F_f(T_f)$ and $F(T)$ act opposite to $x(\theta)$.

3.8.2 FORCE AND ENERGY

For the conservative or lossless energy conversion area of Fig. 3.32, a conservation of energy incremental audit over any arbitrary period of time Δt results in

$$\left\{\begin{array}{c}\text{Change in}\\\text{electrical}\\\text{energy input}\end{array}\right\} = \left\{\begin{array}{c}\text{Change in}\\\text{magnetic}\\\text{energy stored}\end{array}\right\} + \left\{\begin{array}{c}\text{Change in}\\\text{energy to}\\\text{mechanical system}\end{array}\right\} \qquad \textbf{[3.44]}$$

or

$$\Delta W_e = \Delta W_f + \Delta W_m \qquad \textbf{[3.45]}$$

If the arbitrary period of time is taken quite small, [3.45] can be expressed in differential form:

$$dW_e = dW_f + dW_m \qquad \textbf{[3.46]}$$

At this point, the translational system is singled out for specific study. After the derivation is complete for the translational system, extension to the rotational system will be made. If there is no change in x, then $dW_m = 0$; however, for a differential movement

$$dW_m = F_d dx \qquad\qquad \textbf{[3.47]}$$

Since electrical power is the time rate of change in electrical energy, then

$$dW_e = vidt = N\frac{d\phi}{dt}idt = \frac{d\lambda}{dt}idt = id\lambda \qquad\qquad \textbf{[3.48]}$$

where Faraday's law was used. Substitute [3.47] and [3.48] into [3.46] and rearrange to find

$$F_d dx = -dW_f + id\lambda$$

Following Example 3.3, the ϕ-i plot was generated for a series magnetic circuit with a fixed air gap by use of the MATLAB program ⟨cckt1.m⟩. The program was modified to plot flux linkage vs. coil current and multiple runs made to give the λ-i plots for δ ranging from 2 to 10 mm shown as Fig. 3.33a. Next, the core mean length paths were set to zero so that the core mmf drop was zero. In effect, this is identical to the case of infinitely permeable core material. Figure 3.33b displays the result. From these two plots, it is concluded that the flux linkage of the coil in Fig. 3.32a must be a function of both coil current and position x. Then by application of the chain rule of calculus,

$$d\lambda = \frac{\partial \lambda}{\partial i}di + \frac{\partial \lambda}{\partial x}dx \qquad\qquad \textbf{[3.50]}$$

In Sec. 3.5.2, it has been concluded that the area to the left of the λ-i plot is energy stored in the magnetic field. Referring again to Fig. 3.33, δ could be held at a constant value and the stored energy changed by changing coil current. Further, if current were maintained at a constant value (vertical line) and the value of δ changed, the stored magnetic energy also changes. Thus, it is concluded that stored magnetic energy in Fig. 3.32a depends on the two independent variables x and i. Again by use of the chain rule,

$$dW_f = \frac{\partial W_f}{\partial i}di + \frac{\partial W_f}{\partial x}dx \qquad\qquad \textbf{[3.51]}$$

Substitute [3.50] and [3.51] into [3.49] and rearrange the result to give

$$F_d + \frac{\partial W_f}{\partial x} - i\frac{\partial \lambda}{\partial x} = \left(-\frac{\partial W_f}{\partial i} + i\frac{\partial \lambda}{\partial i}\right)\frac{di}{dx} \qquad\qquad \textbf{[3.52]}$$

Having already established that i and x are independent variables, $di/dx = 0$ and [3.52] yields an expression for the developed mechanical force as

$$F_d = -\frac{\partial W_f(i, x)}{\partial x} + i\frac{\partial \lambda(i, x)}{\partial x} \qquad\qquad \textbf{[3.53]}$$

The above work arbitrarily treated i and x as the independent variables and λ as a dependent variable. Since for each value of i there is a corresponding value of

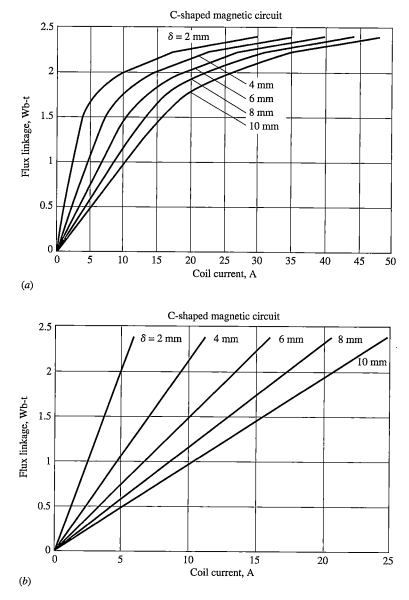

Figure 3.33
Flux linkage-current plots with variable air gap. (a) Nonzero core mmf. (b) Zero core mmf.

λ, a derivation can be made treating λ and x as independent variables and i as the dependent variable. In such case,

$$dW_f = \frac{\partial W_f}{\partial \lambda} d\lambda + \frac{\partial W_f}{\partial x} dx \qquad \textbf{[3.54]}$$

Substitute [3.54] into [3.49] and rearrange the result.

$$F_d + \frac{\partial W_f}{\partial x} = \left(-\frac{\partial W_f}{\partial x} + i \right) \frac{d\lambda}{dx} \qquad \textbf{[3.55]}$$

Since λ and x are independent, $d\lambda/dx = 0$. Thus,

$$F_d = -\frac{\partial W_f(\lambda, x)}{\partial x} \qquad \textbf{[3.56]}$$

Both [3.53] and [3.56] yield the same result. The choice depends upon the formulation of the problem.

If the rotational case had been considered rather than the translational case, the differential of mechanical energy is

$$dW_m = F_d r d\theta = T_d d\theta \qquad \textbf{[3.57]}$$

Comparing [3.57] with [3.47], it is seen that θ replaces x for the rotational case and that torque T_d replaces force F_d. The equations analogous to [3.47] and [3.57] can be written by symmetry argument as

$$T_d = -\frac{\partial W_f(i,\theta)}{\partial \theta} + i \frac{\partial \lambda(i,\theta)}{\partial \theta} \qquad \textbf{[3.58]}$$

$$T_d = -\frac{\partial W_f(\lambda,\theta)}{\partial \theta} \qquad \textbf{[3.59]}$$

The developed force or developed torque determined by [3.53], [3.56], [3.58], or [3.59] is always in the direction of the coordinate (x or θ) as a direct result of [3.47]. However, the following rules, given without justification, can be used for a valid determination for the direction of F_d or T_d. The force or torque acts in such a direction to

1. Decrease stored energy for constant flux linkages or flux.

2. Increase stored energy for constant coil current or mmf.

3. Decrease reluctance or increase permeance.

4. Increase inductance.

5. Align B-fields (multiple excitation coils).

Example 3.19

For the magnetic circuit of Fig. 3.34, $d = 5$ cm and $w = 10$ cm. The slug is originally positioned at $\ell = 5$ mm. The 50-turn coil conducts a constant current $i_t = 10$ A. Neglect magnetic losses and determine (a) the coil flux linkage, (b) the coil inductance, and (c) the developed force on the slug as a function of position x.

(a) Since the core material is infinitely permeable, the coil mmf must equal the air gap mmf drop. In order to neglect magnetic losses, $R_m = \infty$ and $i_t = i$. Using [3.17] for $\mathcal{P}_g$,

$$\lambda = N\phi = N(Ni\mathcal{P}_g) = \mu_o N^2 i \left[\frac{wd}{\ell - x} + 0.52(w + d) + 0.308(\ell - x) \right] \qquad \textbf{[1]}$$

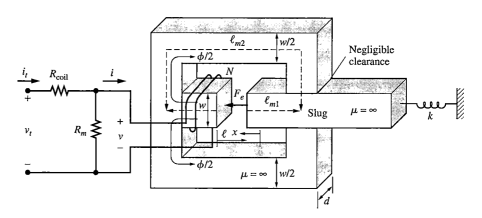

Figure 3.34
Magnetic circuit for Example 3.19

$$\lambda = 4\pi \times 10^{-7}(50)^2(10)\left[\frac{(0.1)(0.05)}{0.005 - x} + 0.52(0.1 + 0.05) + 0.308(0.005 - x)\right]$$

$$\lambda(x) = 0.001\left[\frac{0.1571}{0.005 - x} + 2.5 - 9.68x\right]\text{Wb-t} \qquad \textbf{[2]}$$

(b) For this magnetically linear circuit, [3.22] yields

$$L = \frac{\lambda}{i} = \frac{\lambda}{10} = 0.0001\left[\frac{0.1571}{0.005 - x} + 2.5 - 9.68x\right]\text{H} \qquad \textbf{[3]}$$

(c) The family of flux linkage-current plots for this linear magnetic circuit is similar to Fig. 3.33b. If current i and position x are treated as the independent variables, the force can be determined by [3.53]:

$$i\frac{\partial\lambda}{\partial x} = i\frac{\partial(N^2 i \mathcal{P}_g)}{\partial x} = N^2 i^2 \frac{\partial\mathcal{P}_g}{\partial x} \qquad \textbf{[4]}$$

The stored magnetic energy is the area to the left of the applicable λ-i plot, or

$$W_f(i, x) = \frac{1}{2} i\lambda = \frac{1}{2} i(N^2 i \mathcal{P}_g) = \frac{1}{2} N^2 i^2 \mathcal{P}_g \qquad \textbf{[5]}$$

Substitution of [4] and [5] into [3.53] gives the desired force:

$$F_d = -\frac{\partial(\frac{1}{2}N^2 i^2 \mathcal{P}_g)}{\partial x} + N^2 i^2 \frac{\partial\mathcal{P}_g}{\partial x} = \frac{1}{2} N^2 i^2 \frac{\partial\mathcal{P}_g}{\partial x}$$

$$F_d = \frac{1}{2} N^2 i^2 \mu_o\left[\frac{wd}{(\ell - x)^2 - 0.308}\right]$$

$$F_d = \frac{1}{2}(50)^2(10)^2(4\pi \times 10^{-7})\left[\frac{(0.1)(0.05)}{(0.005 - x)^2} - 0.308\right]$$

$$= \frac{0.001}{(0.005 - x)^2} - 0.062 \text{ N}$$

Determine the developed force for the magnetic circuit of Example 3.19 using [3.56].   **Example 3.20**
The independent variables are now λ and x. From $\lambda = N^2 i \mathscr{P}_g$,

$$i = \frac{\lambda}{N^2 \mathscr{P}_g}$$

The stored magnetic energy is

$$W_f(\lambda, x) = \int i\, d\lambda = \int \frac{\lambda\, d\lambda}{N^2 \mathscr{P}_g} = \frac{1}{2} \frac{\lambda^2}{N^2 \mathscr{P}_g}$$

By [3.56],

$$F_d = -\frac{\partial W_f}{\partial x} = -\frac{1}{2}\frac{\lambda^2}{N^2}\left(-\frac{1}{\mathscr{P}_g^2}\frac{\partial \mathscr{P}_g}{\partial x}\right) = \frac{1}{2}\frac{\lambda^2}{N^2 \mathscr{P}_g^2}\frac{\partial \mathscr{P}_g}{\partial x}$$

Substituting for λ leads to

$$F_d = \frac{1}{2}\frac{(N^2 i \mathscr{P}_g)^2}{N^2 \mathscr{P}_g^2}\frac{\partial \mathscr{P}_g}{\partial x} = \frac{1}{2} N^2 i^2 \frac{\partial \mathscr{P}_g}{\partial x}$$

Since this is identically the same expression for F_d determined in Example 3.19, performing the differentiation and substituting values gives

$$F_d = \frac{0.001}{(0.005 - x)^2} - 0.062 \text{ N}$$

For the series magnetic circuit of Fig. 3.11, assume that the core material is infinitely permeable and find an expression for the attractive force between the two air gap faces for a constant value of current I.   **Example 3.21**

The approach used is to set up the problem as if one air gap face could move toward the other and then evaluate for zero movement. Arbitrarily choose to treat I and x as independent variables; thus, F_d is found by use of [3.53]. The solution proceeds exactly as in Example 3.19, giving

$$F_d = \frac{1}{2} N^2 i^2 \frac{\partial \mathscr{P}_g}{\partial x}$$

Based on [3.17] air gap permeance is

$$\mathscr{P}_g = \mu_o\left[\frac{\ell_s d}{\delta - x} + 0.52(\ell_s + d) + 0.308(\delta - x)\right]$$

Then,

$$F_d = \frac{1}{2} N^2 i^2 \mu_o \left[\frac{\ell_5 d}{(\delta - x)^2} - 0.308 \right]$$

Evaluate for $x = 0$ to give the desired force.

$$F_d = \frac{1}{2} \mu_o N^2 i^2 \left(\frac{\ell_5 d}{\delta^2} - 0.308 \right)$$

3.8.3   FORCE AND COENERGY

A different approach is to be taken for derivation of the developed force equations. The end result, if the magnetic circuit is linear, allows calculation of forces using the coil inductance. The area to the left of the λ-i plot was shown to equal the stored magnetic energy by [3.32]. Rewriting that equation to clearly indicate the independent variables gives

$$W_f(i, x) = \int i \, d\lambda(i, x) \qquad\qquad \textbf{[3.60]}$$

If [3.60] is integrated by parts with x treated as constant,

$$W_f(i, x) = i\lambda - \int \lambda \, di \qquad\qquad \textbf{[3.61]}$$

The second term on the right side of [3.61] is the area below the λ-i plot as indicated by Fig. 3.35. It is defined as the *coenergy*, or

$$W'_f(i, x) = \int \lambda \, di \qquad\qquad \textbf{[3.62]}$$

With this definition, [3.61] can be written as

$$i\lambda(i, x) = W_f(i, x) + W'_f(i, x) \qquad\qquad \textbf{[3.63]}$$

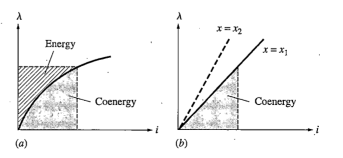

Figure 3.35
Magnetic coenergy. (a) Nonlinear. (b) Linear.

Form the total differential of [3.63] and then rearrange the result.

$$id\lambda + \lambda di = dW_f + dW'_f$$

$$id\lambda - dW_f = dW'_f - \lambda di \qquad \textbf{[3.64]}$$

From [3.49], the left side of [3.64] is $F_d dx$. Make this substitution and form the total differential of W'_f.

$$F_d dx = \frac{\partial W'_f}{\partial x} dx + \frac{\partial W'_f}{\partial i} di - \lambda di \qquad \textbf{[3.65]}$$

Rearrange [3.65] and realize that since i and x are independent, then $di/dx = 0$.

$$F_d = \frac{\partial W'_f}{\partial x} + \left(\frac{\partial W'_f}{\partial i} - \lambda\right)\frac{di}{dx}$$

$$F_d = \frac{\partial W'_f(i, x)}{\partial x} \qquad \textbf{[3.66]}$$

Although [3.66] expresses the developed force in terms of magnetic coenergy, its application for the general case offers no special advantage. However, consider the magnetically linear case illustrated by Fig. 3.35b, from which

$$W'_f(i, x) = \frac{1}{2} i\lambda(i, x) \qquad \textbf{[3.67]}$$

Since [3.22] is applicable for this magnetically linear case,

$$W'_f(i, x) = \frac{1}{2} iL(x)i = \frac{1}{2} i^2 L(x) \qquad \textbf{[3.68]}$$

If the x coordinate were to change from $x = x_1$ to $x = x_2$ in Fig. 3.35b, the value of L changes. Hence, the dependence of inductance L on position x has been explicitly indicated in [3.68]. Use of [3.68] in [3.66] leads to

$$F_d = \frac{1}{2} i^2 \frac{\partial L(x)}{\partial x} = \frac{1}{2} i^2 \frac{dL(x)}{dx} \qquad \textbf{[3.69]}$$

By the same symmetry argument used to form [3.58] and [3.59], the analogous equations to [3.66] and [3.69] for the rotational case are

$$T_d = \frac{\partial W'_f(i, \theta)}{\partial \theta} \qquad \textbf{[3.70]}$$

$$T_d = \frac{1}{2} i^2 \frac{dL(\theta)}{d\theta} \qquad \textbf{[3.71]}$$

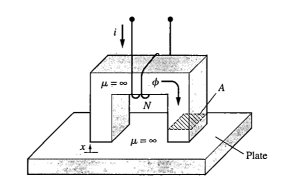

Figure 3.36
Lifting magnet

Example 3.22

The lifting magnet of Fig. 3.36 is being lowered to pick up the steel plate. The magnet is energized with current i flowing in the N-turn coil. As the magnet approaches the plate, both air gaps are of identical length. Find an expression for the lifting force exerted on the plate.

The permeance of each air gap is described by [3.18]. Since the air gaps are in series, the total permeance is

$$\mathcal{P}_{gT} = \frac{1}{2}\,\mathcal{P}_g = \frac{1}{2}\,\mu_o\left[\frac{A}{-x} + 1.04(w + d) - 0.616x\right]$$

where the negative sign for x is used, since the advancement of the magnet is opposite to the direction of the coordinate. The inductance of the coil is

$$L(x) = N^2\mathcal{P}_{gT} = \frac{1}{2}\,\mu_o N^2\left[\frac{A}{-x} + 1.04(w + d) - 0.616x\right]$$

Application of [3.69] gives the force.

$$F_d = \frac{1}{2}\,i^2\,\frac{dL(x)}{dx} = \frac{1}{4}\,\mu_o N^2\left[\frac{A}{x^2} - 0.616\right]\text{N}$$

Example 3.23

The rotatable slug of Fig. 3.37 is pinned to the lower core so that it can move through the angle θ. The clearance between the slug and the fixed core is such that the coil inductance is given by $L(\theta) = L_1 + L_2\cos\theta = 5 + 10\cos\theta$ mH over the limited range of travel. Determine the developed torque exerted on the slug when $i = 2\sin\omega t$ A and the angle $\theta = 45°$.

By [3.71],

$$T_d = \frac{1}{2}\,i^2\,\frac{dL}{d\theta} = -\frac{1}{2}\,i^2 L_2\sin\theta\;\text{N}\cdot\text{m}$$

$$T_d = -\frac{1}{2}(2\sin\omega t)^2\,0.01\sin(45°) = -7.07(1 - \cos 2\omega t)\,\text{mN}\cdot\text{m}$$

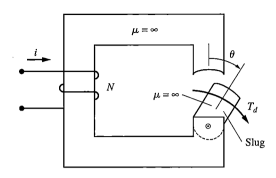

Figure 3.37
Rotatable slug

Nonlinear Analysis Analyses of energy conversion illustrated to this point have considered the core material as infinitely permeable, rendering λ-i plots linear and allowing formulation of simple analytical expressions for permeance. When finite permeability of the core material is introduced, the λ-i plots become non-linear, requiring either a graphical or numerical solution for developed forces.

Since the magnetic coenergy is the area below the λ-i plot, an approximation to [3.66] for a particular position x and current i is given by

$$F_d \cong \frac{\Delta W_f(i, x)}{\Delta x} = \frac{\int_0^i \lambda \, di \big|_{x - \Delta x/2} - \int_0^i \lambda \, di \big|_{x + \Delta x/2}}{\Delta x} \qquad \textbf{[3.72]}$$

The MATLAB program ⟨coenFe.m⟩ has been developed to perform this numerical evaluation of force F_d for the magnetic circuit of Fig. 3.34 with cast-steel core material. Leakage flux is neglected, although the refinement could be added. The dimension ℓ_{m1} is the length of the center ferromagnetic material flux path if the air gap is closed. Values of i and x are specified. The program forms λ-i arrays for values of x that are ±1 percent from the nominal value. Simple Simpson's rule integration is implemented to evaluate the indicated integration of [3.72] and allows approximate determination of force F_d. The screen display from execution of ⟨coenFe.m⟩ for $d = w = 0.05$ m, $\ell_{m1} = 0.2$ m, $\ell_{m2} = 0.4$ m, and $N = 100$ turns is shown.

```
COENERGY FORCE ANALYSIS
   Air gap = 0.002
   Current = 10
       Fd = 265.2
```

Extension of ⟨coenFe.m⟩ to handle other magnetic circuits requires only change of the magnetic circuit description in ⟨magckt1.m⟩.

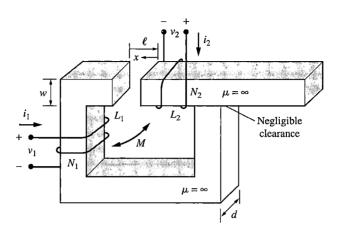

Figure 3.38
Doubly excited magnetic circuit

3.8.4 DOUBLY EXCITED SYSTEMS

Consider the series linear magnetic circuit of Fig. 3.38 with two lossless excitation coils. By holding position x constant, the differential of total electric energy supplied to the two coils must be exactly equal to the total stored magnetic energy. Further, for this linear circuit, coenergy and energy are equal. Thus,

$$dW_e = dW'_f(i_1,i_2,x) \qquad \textbf{[3.73]}$$

The coil voltages are given by [3.27] and [3.28]. So,

$$dW_e = (v_1i_1 + v_2i_2)dt = L_1i_1di_1 + Mi_1di_2 + Mi_2di_1 + L_2i_2di_2$$

$$dW_e = L_1i_1di_1 + Md(i_1i_2) + L_2i_2di_2 \qquad \textbf{[3.74]}$$

Integrating [3.73] where dW_e is given by [3.74] yields the expression for magnetic coenergy.

$$W'_f(i_1,i_2,x) = \int dW_e = \frac{1}{2}L_1i_1^2 + Mi_1i_2 + \frac{1}{2}L_2i_2^2 \qquad \textbf{[3.75]}$$

where the dependency of L_1, L_2, and M on x is understood.

   The derivation of [3.69] is general and can be extended directly to the case of multiple currents. Application to [3.75] yields the developed force as

$$F_d = \frac{\partial W'_f(i_1,i_2,x)}{\partial x} = \frac{1}{2}i_1^2\frac{dL_1(x)}{dx} + i_1i_2\frac{dM(x)}{dx} + \frac{1}{2}i_2^2\frac{dL_2(x)}{dx} \qquad \textbf{[3.76]}$$

By symmetry, the torque for a rotational case is found by replacing x by θ in [3.76].

Assume no leakage flux and neglect air gap fringing for the magnetic circuit of Fig. 3.38. If the air gap face has an area A, find an expression for the force acting on the movable slug.

Example 3.24

The reluctance of the linear magnetic circuit is

$$\mathcal{R} = \frac{1}{\mu_o} \frac{\ell - x}{A}$$

Based on the result of Example 3.11 and [3.29],

$$L_1 = \frac{N_1^2}{\mathcal{R}} = \frac{\mu_o N_1^2 A}{\ell - x} \qquad L_2 = \frac{N_2^2}{\mathcal{R}} = \frac{\mu_o N_2^2 A}{\ell - x}$$

$$M = \frac{kN_1 N_2}{\mathcal{R}} = \frac{\mu_o N_1 N_2 A}{\ell - x}$$

By [3.76],

$$F_d = \frac{1}{2} i_1^2 \frac{\mu_o N_1^2 A}{(\ell - x)^2} + i_1 i_2 \frac{\mu_o N_1 N_2 A}{(\ell - x)^2} + \frac{1}{2} i_2^2 \frac{\mu_o N_2^2 A}{(\ell - x)^2}$$

$$F_d = \frac{\mu_o A (N_1^2 i_1^2 + 2N_1 N_2 i_1 i_2 + N_2^2 i_2^2)}{2(\ell - x)^2}$$

3.9 Solenoid Design

The solenoid is a magnetic circuit typically used as a fast-acting linear actuator. The produced performance is a large force with a short travel distance. As would be expected based on the result of Example 3.19, the force ideally varies inversely with the square of air gap distance. Depending upon the application, the duty requirement may be either continuous or intermittent.

Design procedures for solenoids have never evolved to the level of formality attained for electric machines and transformers. Design work seems to fall into two categories. Designs with totally new geometry are cut-and-try after an initial guess based on magnetic linearity. If a similar geometry design is available, then extrapolation is done based on force varying with coil mmf squared and flux density varying directly with coil mmf. The analysis methods developed in this chapter are sufficient background to approach solenoid design.

3.91 Rough Sizing

Typically a solenoid is designed to have a specific holding force at air gap closure and a pickup force that must be developed at the maximum air gap extension. By the nature of the operation, the holding force requirement occurs at conditions of ferromagnetic material saturation and thus is independent of coil current beyond the level to sustain the saturation level mmf of the ferromagnetic core material. On the other hand, the pickup force requirement must be met for negligible level of saturation wherein the permeance of the magnetic circuit is essentially the permeance of the air gap. The holding force requirement to a large degree dictates base cross-sectional area of the ferromagnetic core components. The pickup force determines the minimum level of excitation coil mmf.

Base Area A_b Consider the λ-i plot of a magnetic circuit shown by Fig. 3.39 where the saturation of the ferromagnetic material is treated as abrupt and severe. Two different air gap positions (x) are shown where $x_2 > x_1$. For a base area (A_b) to be used in the core design, the saturation flux linkage is given by

$$\lambda_s = NB_s A_b \qquad \textbf{[3.77]}$$

where N is the number of exciting coil turns and B_s is the saturation flux density. Since the flux linkage is indirectly specified by the ferromagnetic material, it is an independent variable. The force developed by the solenoid can be approximated from [3.56] as

$$F_d \cong -\frac{W_{f1} - W_{f2}}{x_1 - x_2} \qquad \textbf{[3.78]}$$

Let $x_1 = 0$ to model the condition of air gap closure. As $x_1 \to 0$, the area to the left of the λ-i plot of Fig. 3.39 collapses and $W_{f1} \to 0$. Hence, [3.78] becomes

$$F_{dc} \cong \frac{W_{f2}}{x_2} = \frac{\frac{1}{2}\lambda_s i_2}{x_2} \qquad \textbf{[3.79]}$$

where the subscript c denotes closure condition and x_2 is taken arbitrarily small to determine the force produced by any attempt to move the solenoid from the closure position.

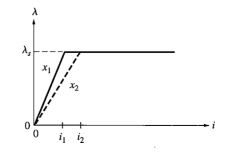

Figure 3.39
Flux linkage for abrupt saturation

Current i_2 required to produce the air gap mmf can be determined from

$$Ni_2 = \frac{\phi_s}{\mathcal{P}_g} = \frac{B_s A_p}{\mu_o A_p/x_2} = \frac{B_s x_1}{\mu_o}$$ [3.80]

If i_2 determined by [3.80] and λ_s of [3.77] are substituted into [3.79], the result can be rearranged to give the base area.

$$A_b \cong \frac{2\mu_o F_{dc}}{B_s^2}$$ [3.81]

The solenoid design will size all cross-sectional paths for flux flow to approximately match A_b given by [3.81].

Base or Pickup Current I_b For the case of operation at hand, the magnetic circuit is linear and the current is being specified independently of the coil flux linkage. Based on the result of Example 3.19 with air gap fringing neglected,

$$F_{dp} = \frac{1}{2} N^2 i_p^2 \mu_o \frac{A_b}{\delta_p^2}$$ [3.82]

where the subscript p denotes pickup condition and the air gap area is assumed to be equal to the base area. Solution of [3.82] for coil pickup mmf yields

$$Ni_p = \left[\frac{2\delta_p^2 F_{dp}}{\mu_o A_b} \right]^{1/2}$$ [3.83]

The mmf requirement of [3.83] can be utilized in more than one way. Obviously, if a number of turns were selected, the value of pickup current follows directly. However, if a conductor size (A_c) and current density (J_c) were selected, the equation determines the number of turns as

$$N = \left[\frac{2\delta_p^2 F_{dp}}{\mu_o (J_c A_c)^2 A_b} \right]^{1/2}$$ [3.84]

Yet another utility of [3.83] is to select a conductor area, current density, and a coil fill factor (CF) to estimate the coil cross-section area as

$$A_{coil} = \frac{NA_c}{CF} = \left[\frac{2\delta_p^2 F_{dp}}{\mu_o (J_c CF)^2 A_b} \right]^{1/2}$$ [3.85]

By use of [3.81] and [3.83] or one of the variations given by [3.84] and [3.85] as guidelines, the solenoid design process can be carried out.

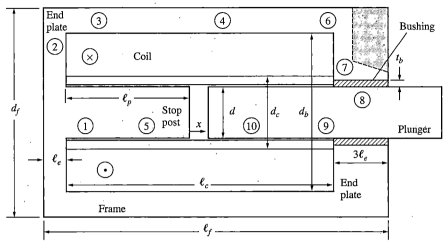

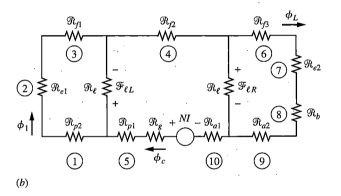

(b)

Figure 3.40
Cylindrical solenoid. (a) Physical. (b) Schematic.

3.9.2   MAGNETIC CIRCUIT

Solenoid geometry can vary significantly, ranging from the so-called iron-clad structure or cylindrical solenoid in which the coil is encased in a ferromagnetic structure to open-frame designs with configuration similar to a single-coil version of Fig. 3.34. Since the cylindrical solenoid structure offers a geometry not covered to this point, it is chosen for specific illustration of design techniques.

The cylindrical solenoid is shown by Fig. 3.40a. In operation, the frame is fixed and the plunger connects to the mechanical load to be displaced. A thin, non-magnetic bushing offers a low-friction guide for the plunger. The bushing wall thickness is maintained as small as possible, since it acts as an air gap requiring mmf but not contributing to the developed force. The flux path passes through the plunger-air gap-stop post and then moves radially outward across one end plate, along the frame, and returns radially inward through the opposite end plate.

The base area A_b is determined by application of [3.81] using the specified holding force and saturation flux density of the chosen ferromagnetic material. Insofar as practical, all cross-section areas are sized to match this value, giving uniform flux density throughout the structure. The solenoid structure is fashioned around the required coil area, satisfying any overall length and diameter specifications.

Ten specific sections are indicated on both the device drawing of Fig. 3.40a and the analog electric schematic of Fig. 3.40b. Application of [3.11] from a point at half the coil length with the air gap collapsed leads to the conclusion that leakage flux is directed radially outward across the coil area between the center member (stop post or plunger) and the frame and in the opposite sense over the other half coil. The permeance of this leakage flux path is

$$\mathcal{P}_\ell = \mu_o \frac{\pi \left(\dfrac{d_b + d}{2} \right) \dfrac{\ell_p}{2}}{(d_b - d)/2} = \frac{\mu_o \pi}{2} \left(\frac{d_b + d}{d_b - d} \right) \qquad \textbf{[3.86]}$$

The air gap permeance is placed between the axial midpoint of the stop post and a point radially outward on the frame for the left side of the coil and the same distance to the left of the right end plate.

The air gap permeance follows from [3.17] where the area wd is replaced by $\pi d^2/4$, the periphery length $w + d$ is replaced by πd, and the quarter sphere term diminishes to zero.

$$\mathcal{P}_g = \mu_o \left[\frac{\pi d^2}{4x} + 0.52\pi d \right] \qquad \textbf{[3.87]}$$

Areas and lengths as seen by flux passing through each of the 10 sections are given by the formulas of Table 3.2.

Any attempt to solve this magnetic circuit by longhand methods is a tedious challenge. When numerical analysis is later carried out, the approach is as follows:

1. The far left branch of Fig. 3.40b is treated as a Type 1 problem, allowing calculation of a $\mathcal{F}_{\ell L}$-ϕ_1 array. For each value of $\mathcal{F}_{\ell L}$ the leakage is calculated as $\mathcal{F}_{\ell L}\mathcal{P}_\ell$ and added to ϕ_1 to yield the $\mathcal{F}_{\ell L}$-ϕ_L array that describes the combined leakage path and the far left branch.

2. A similar analysis of the far right branch and its parallel leakage permeance yields a $\mathcal{F}_{\ell R}$-ϕ_R array.

3. With the arrays from steps 1 and 2 available, a Type 1 analysis is made for the center loop to determine the source mmf NI. Since $\phi_C = \phi_L = \phi_R$, the $\mathcal{F}_{\ell L}$-ϕ_L and $\mathcal{F}_{\ell R}$-ϕ_R arrays can be entered at each value of ϕ_C to determine the appropriate value of $\mathcal{F}_{\ell L}$ and $\mathcal{F}_{\ell R}$. For each value of ϕ_C, the source mmf is

$$NI = \phi_C(\mathcal{R}_g + \mathcal{R}_{p1} + \mathcal{R}_{f2} + \mathcal{R}_{a1}) + \mathcal{F}_{\ell L} + \mathcal{F}_{\ell R}$$

Table 3.2   Area and length formulas

Section	Length (ℓ)	Area (A)
1	$\frac{1}{2}(\ell_p + \ell_e)$	$\frac{\pi}{4}d^2$
2	$\frac{1}{4}(d_b + d_f)$	$\frac{\pi}{4}(d_b + d_f)\ell_e$
3	$\frac{1}{2}(\ell_p + \ell_e)$	$\frac{\pi}{4}(d_f^2 + d_b^2)$
4	$\frac{1}{2}\ell_p$	$\frac{\pi}{4}(d_f^2 - d_b^2)$
5	$\frac{1}{2}\ell_p$	$\frac{\pi}{4}d^2$
6	$\frac{1}{2}(\ell_p + 3\ell_e)$	$\frac{\pi}{4}(d_f^2 - d_b^2)$
7	$\frac{1}{4}(d_b + d_f) - t_b$	$\frac{\pi}{4}(d_b + d_f)3\ell_e$
8	t_b	$\pi(d + t_b)3\ell_e$
9	$\frac{1}{2}(\ell_p + 3\ell_e)$	$\frac{\pi}{4}d^2$
10	$\frac{1}{2}\ell_s - x$	$\frac{\pi}{4}d^2$

3.9.3   SAMPLE DESIGN

The cylindrical solenoid to be designed is specified as follows:

Holding force, $F_{dc} = 240$ lb

Pickup force, $F_{dp} = 40$ lb at 0.25 in

Total length of plunger travel $= 1$ in

Maximum diameter, $df \leq 3$ in

Intermittent duty, 20 percent duty cycle

Operate from a 50-V dc source

Rough Sizing   Cast steel is chosen as the ferromagnetic core material for all sections. Extension of the *B-H* curve of Fig. 3.12 shows the saturation flux density of cast steel to be $B_s \cong 1.9$ T. The base area is calculated from [3.81].

$$A_b \cong \frac{2\mu_o F_{dc}}{B_s^2} = \frac{2(4\pi \times 10^{-7})(240/0.2248)}{(1.9)^2} = 7.433 \times 10^{-4} \text{ m}^2$$

$$A_b = 1.15 \text{ in}^2$$

The plunger and stop post diameter are sized to match the area A_b with their final dimensions rounded up to a standard dimension value.

$$d = \sqrt{\tfrac{4}{\pi}A_b} = \sqrt{\tfrac{4}{\pi}(1.15)} = 1.21 \text{ in}$$

The 20 percent duty cycle results in an rms current with value of energized current value divided by $\sqrt{20} = 4.47$. Hence, the current density can be selected for a value of 4.47 times the value of current density that can be cooled adequately if it existed on a continuous basis. This totally enclosed structure has no internal air flow and only convection air flow externally. It is estimated that a current density of 800 to 900 A/in² can be cooled on a continuous basis. Thus, the conservative design goal current density is $4.47(800) = 3576$ A/in² (5.5 MA/m²). A coil fill factor of 65 percent is realizable if the coil is carefully wound. By [3.85],

$$A_{coil} = \left[\frac{2\delta_p^2 F_{dp}}{\mu_o(J_c CF)^2 A_b} \right]^{1/2}$$

$$= \left[\frac{2\left(\dfrac{0.25}{39.37}\right)^2\left(\dfrac{40}{0.2248}\right)}{(4\pi \times 10^{-7})(5.5 \times 10^6 \times 0.65)^2(7.433 \times 10^{-4})} \right]^{1/2}$$

$$A_{coil} = 10.96 \times 10^{-4} \text{m}^2 = 1.7 \text{ in}^2$$

After some trial-and-error work, it is decided to use a 2000-turn coil wound from 22-gage round copper wire. The diameter (d_{cond}) of 22-gage round wire over enamel insulation is approximately 0.026 in. For a perfectly layered arrangement,

$$N = \frac{A_{coil}}{d_{cond}^2} = \frac{1.7}{(0.026)^2} = 2514 \text{ turns}$$

Although theoretically the coil area allows for 25 percent more turns than selected, the extra area is needed for ground insulation and to allow for irregularities in coil layers. The excitation current to meet the pickup force based on [3.83] is

$$i_p = \left[\frac{2\delta_p^2 F_{dp}}{\mu_o N^2 A_b} \right]^{1/2} = \left[\frac{2\left(\dfrac{0.25}{39.37}\right)^2\left(\dfrac{40}{0.2248}\right)}{(4\pi \times 10^{-7})(2000)^2(7.433 \times 10^{-4})} \right]^{1/2} = 1.96 \text{ A}$$

A current of 2 A will be used for the study.

The solenoid frame must now be fashioned around this required coil area. A 0.125-in spacing is allowed between the plunger and the coil inside diameter. Part of this distance is to allow clearance for the plunger movement, and the balance forms the bobbin on which the coil is wound. Consequently, the coil inside diameter is

$$d_c = d + 2 \times 0.125 = 1.50 \text{ in}$$

Since the design specification requires that the maximum value of d_f be no more than 3 in, a choice of coil cross section is chosen to be 0.57 $\times$ 3 in. A bushing thickness of 0.030 in is deemed adequate. The balance of a set of dimensions that satisfy the constraints are chosen for a trial design.

$$d_b = d_c + 2 \times 0.57 = 2.64 \, \text{in}$$

$$\ell_c = 3.00 \, \text{in} \qquad \ell_p = \frac{1}{2} \, \ell_c = 1.50 \, \text{in}$$

$$\frac{\pi}{4}(d_f^2 - d_b^2) = A_b$$

or

$$d_f = \left[\frac{4}{\pi} A_b + d_b^2 \right]^{1/2} = \left[\frac{4}{\pi}(1.15) + (2.64)^2 \right]^{1/2} = 2.78 \, \text{in}$$

The left end plate thickness is chosen so that the mean cross-section area seen by the flux equals A_b, or

$$\ell_e = \frac{A_b}{\frac{\pi}{4} (d_f + d_b)} = \frac{1.15}{\frac{\pi}{4}(2.78 + 2.64)} = 0.27 \, \text{in}$$

The right end plate is arbitrarily chosen as three times this value of ℓ_e to provide adequate bearing surface for the plunger. The frame outside diameter is sized so that the axial traveling flux sees a cross-section area equal to A_b.

The MATLAB program ⟨soldes.m⟩ has been developed to predict the performance of cylindrical solenoids of this present design. The function m-file ⟨magckt2.m⟩ calculates a λ-i array for a specified value of air gap. Dimensional data is entered through the m-file ⟨soldata.m⟩. The function m-file ⟨Hcs.m⟩ contains a *B-H* array for cast steel that when entered with a value of flux density returns the corresponding value of magnetic field intensity. The program ⟨soldes.m⟩ performs four major analysis procedures:

1. *Flux Density Checks.* Neglecting leakage, flux density is calculated for each of the 10 sections. The result for the present design is displayed by Fig. 3.41. The lower flux densities indicated for sections 7 and 8 are the result of choosing the right end plate to have the dimension $3\ell_e$ for full height. The shaded area indicated in Fig. 3.40a can be removed, saving material while still providing adequate bearing surface for the plunger.

2. *Weight, Size, and Coil Characteristics.* The screen display of these calculated quantities is shown below, allowing a quick assessment of the solenoid weight and dimensions. The coil resistance of 23.14 Ω allows the necessary current for pickup using the specified 50-V dc source.

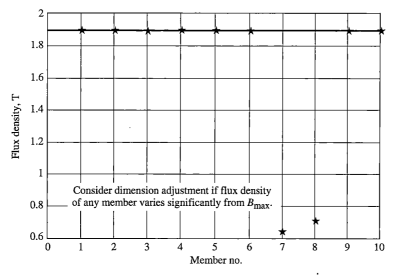

Figure 3.41
Flux densities from solenoid design

```
Current for Force Study = 2
WEIGHT, SIZE & COIL CHARACTERISTIC ESTIMATES
        Core Weight(lbs) = 4.505
        Coil Weight(lbs) = 2.868
            Frame OD(in) = 2.984
        Frame length(in) = 4.135
   Conductor area(sq in) = 0.000585
   Current Density(Apsi) = 3419
   Resistance(Ohm,150 C) = 23.12
      Inductance(H,1/8") = 1.294
```

3. *Flux-Current Plots.* For a set of air gap positions specified by vector x in ⟨soldes.m⟩, flux-current plots are generated to allow assessment of the saturation nature of the solenoid. Figure 3.42 shows this set of plots for the sample design ranging from closed air gap to a 1-in air gap.

4. *Force vs. Displacement.* The program uses the technique of ⟨coenFe.m⟩ to calculate the developed force vs. air gap position, allowing assessment of performance to compare with specification. The initial run using the values of $d = 1.21$ in and $d_b = 2.64$ in met the pickup specification of 40 lb at 0.5 in air gap. However, the holding force produced was 225 lb. Adjustment was made to $d = 1.27$ in and $d_b = 2.70$ in. Figure 3.43 shows the resulting force-displacement plot for these new dimensions, where it is seen that the specifications for holding force and pickup force are clearly met.

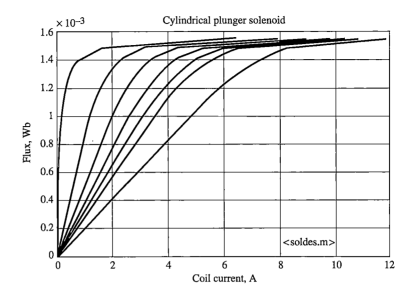

Figure 3.42
Flux-current for solenoid design

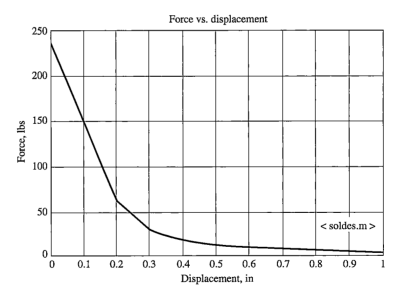

Figure 3.43
Force-displacement for solenoid design

3.10   COMPUTER ANALYSIS CODE

```
%%%%%%%%%%%%%%%%%%%%%%%%%%%%%%%%%%%%%%%%%%%%%%%%%%%%%%%%%%%%%%%%%%%%%%%%%%%%%%%%
%
% cckt1.m - Analysis of C-shaped magnetic circuit.
%            Calculates coil current for specified
%            value of flux & plots Flux vs. mmf.
%            (Type 1 problem)
%
%%%%%%%%%%%%%%%%%%%%%%%%%%%%%%%%%%%%%%%%%%%%%%%%%%%%%%%%%%%%%%%%%%%%%%%%%%%%%%%%
clear;
% Units of input dimensions - 'MKS' or 'FPS' (British)
dim='MKS';
if dim=='MKS'; k=1; elseif dim=='FPS'; k=39.37; else; end
w=0.15; h=0.10;        % Window width & height
d=0.05; del=0.002;     % Core depth & air gap length
l1=0.05; l2=0.05; l3=0.05; l4=0.05; % Core widths
phi=0.0040;            % Flux for analysis point - Wb
N=500; SF=0.95;        % Coil turns, stacking factor
w=w/k; h=h/k; d=d/k; del=del/k;
l=[ l1 l2 l3 l4]/k;   % All dimensions now in MKS units
lm=[h+(l(2)+l(4))/2 w+(l(1)+l(3))/2 h+(l(2)+l(4))/2-del ...
      w+(l(1)+l(3))/2];
Pg=4e-7*pi*(l(3)*d/del+0.52*(l(3)+d)+0.308*del); F1=0;
% Specified point analysis
for i=1:4
  F1=F1+hm27(phi/l(i)/d/SF)*lm(i);
end
F1=F1+phi/Pg; I1=F1/N;
disp([blanks(4) 'C-SHAPED MAGNETIC CIRCUIT, No Leakage']);
disp(' '); disp([blanks(8) 'COIL CURRENT = ', num2str(I1)]);
% Flux vs. mmf plot
phimax=2.0*d*min(l)*SF;
phip=linspace(0,phimax,100); npts=length(phip);
for j=1:npts
  F(j)=0;
  for i=1:4;
    F(j)=F(j)+hm27(phip(j)/l(i)/d/SF)*lm(i);
  end
  F(j)=F(j)+phip(j)/Pg;
end
plot(F,phip,F1,phi,'o'); grid
title('C-shaped magnetic circuit');
xlabel('Coil mmf, A-t'); ylabel('Flux, Wb');
legend('Flux vs mmf', 'Specified pt.', 4);
```

```
%%%%%%%%%%%%%%%%%%%%%%%%%%%%%%%%%%%%%%%%%%%%%%%%%%%%%%%%%%%%%%%%%%%%%%%%%%%%%
%
% Hm27.m - B-H interpolation routine
%
%%%%%%%%%%%%%%%%%%%%%%%%%%%%%%%%%%%%%%%%%%%%%%%%%%%%%%%%%%%%%%%%%%%%%%%%%%%%%
function y = Hx(Bx)
% B-H values that follow are valid for M-27, 24 ga. ESS
B=[0 3 7 20 30 40 50 60 70 80 86 90 95 100 108 120 130 140 ...
150 160 170 180 190 ]*0.0155; % T
H=[0 0.25 0.5 1.3 1.75 2.38 3.25 4.65 6.7 10.4 14 19 32 ...
50 97 222 460 950 2750 4550 6033 8100 10000]*39.37; % A-t/m
% Activate to plot B-H curve
% m=15; plot(H(1:m),B(1:m)); grid; pause; % Linear plot
% m=23; semilogx(H(2:m),B(2:m));grid; pause; % Semilog plot
n=length(B); k=0;
if Bx==0; k=-1; y=0; end
if Bx<0; k=-1; y=0; disp('WARNING - Bx < 0, Hx = returned'); end
if Bx>B(n); y=H(n); k=-1; disp('CAUTION - Beyond B-H curve'); end
for i=1:n
if k==0 & (Bx-B(i))<=0; k=i; break; end
end
if k>0;
y=H(k-1)+(Bx-B(k-1))/(B(k)-B(k-1))*(H(k)-H(k-1));
else;
end


%%%%%%%%%%%%%%%%%%%%%%%%%%%%%%%%%%%%%%%%%%%%%%%%%%%%%%%%%%%%%%%%%%%%%%%%%%%%%
%
% cleak1.m - Analysis of C-shaped magnetic circuit with
%            leakage flux.Calculates coil current for
%            value of air gap flux & plots Flux vs. mmf.
%            (Type 1 problem)
%
%%%%%%%%%%%%%%%%%%%%%%%%%%%%%%%%%%%%%%%%%%%%%%%%%%%%%%%%%%%%%%%%%%%%%%%%%%%%%
clear;
% Units of input dimensions - 'MKS' or 'FPS' (British)
dim='MKS';
if dim=='MKS'; k=1; elseif dim=='FPS'; k=39.37; else; end
w=0.15; h=0.10;        % Window width & height
d=0.05; del=0.002;    % Core depth & air gap length
l1=0.05; l2=0.05; l3=0.05; l4=0.05; % Core widths
phig=0.0040;          % Air gap flux for analysis point - Wb
N=500; SF=0.95;       % Coil turns, stacking factor
w=w/k; h=h/k; d=d/k; del=del/k;
l=[ l1 l2 l3 l4]/k;  % All dimensions now in MKS units
lm=[h+(l(2)+l(4))/2 w+(l(1)+l(3))/2 h+(l(2)+l(4))/2-del ...
      w+(l(1)+l(3))/2];
Pg=4e-7*pi*(l(3)*d/del+0.52*(l(3)+d)+0.308*del);
Pl=4e-7*pi*w*d/h;
% Specified point analysis
```

```
F1=phig/Pg;
F1=F1+hm27(phig/l(3)/d/SF)*lm(3)+hm27(phig/l(2)/d/SF) ...
   *lm(2)/2+hm27(phig/l(4)/d/SF)*lm(4)/2;
phil=F1*Pl; phigl=phig+phil;    % Leakage & coil fluxes
F1=F1+hm27(phigl/l(1)/d/SF)*lm(1)+hm27(phigl/l(2)/d/SF) ...
   *lm(2)/2+hm27(phigl/l(4)/d/SF)*lm(4)/2;
I1=F1/N;
disp([blanks(4) 'C-SHAPED MAGNETIC CIRCUIT, With Leakage']);
disp(' '); disp([blanks(8) 'COIL CURRENT = ', num2str(I1)]);
disp(' '); disp([blanks(8) ' AIR GAP FLUX = ', num2str(phig)]);
disp(' '); disp([blanks(8) 'LEAKAGE FLUX = ', num2str(phil)]);
% Flux vs. mmf plot
phimax=2.0*d*min(l)*SF;
phip=linspace(0,phimax,100); npts=length(phip);
for j=1:npts
  F(j)=phip(j)/Pg;
  F(j)=F(j)+hm27(phip(j)/l(3)/d/SF)*lm(3)+hm27(phip(j)/ ...
    l(2)/d/SF)*lm(2)/2+hm27(phip(j)/l(4)/d/SF)*lm(4)/2;
  phil(j)=F(j)*Pl; phigl=phip(j)+phil(j);
  F(j)=F(j)+hm27(phigl/l(1)/d/SF)*lm(1)+hm27(phigl/l(2)/d/SF) ...
    *lm(2)/2+hm27(phigl/l(4)/d/SF)*lm(4)/2;

end
plot(F,phip,F,phil,'--',F1,phig,'o'); grid
title('C-shaped magnetic circuit with leakage');
xlabel('Coil mmf, A-t'); ylabel('Flux, Wb');
legend('Air gap flux', 'Leakage flux', 'Specified pt.', 0);

%%%%%%%%%%%%%%%%%%%%%%%%%%%%%%%%%%%%%%%%%%%%%%%%%%%%%%%%%%%%%%%%%%%%%%%%%
%
% cckt2.m - Analysis of C-shaped magnetic circuit.
%          Calculates air gap flux for specified
%          value of coil current. Plots Flux vs. mmf.
%          (Type 2 problem)
%
%%%%%%%%%%%%%%%%%%%%%%%%%%%%%%%%%%%%%%%%%%%%%%%%%%%%%%%%%%%%%%%%%%%%%%%%%
clear;
% Units of input dimensions - 'MKS' or 'FPS' (British)
dim='MKS';
if dim=='MKS'; k=1; elseif dim=='FPS'; k=39.37; else; end
w=0.15; h=0.10;        % Window width & height
d=0.05; del=0.002;     % Core depth & air gap length
l1=0.05; l2=0.05; l3=0.05; l4=0.05; % Core widths
I=15.0;                % Current for analysis point - A
N=500; SF=0.95;        % Coil turns, stacking factor
w=w/k; h=h/k; d=d/k; del=del/k;
l=[ l1 l2 l3 l4]/k;   % All dimensions now in MKS units
lm=[h+(l(2)+l(4))/2 w+(l(1)+l(3))/2 h+(l(2)+l(4))/2-del ...
      w+(l(1)+l(3))/2];
Pg=4e-7*pi*(l(3)*d/del+0.52*(l(3)+d)+0.308*del); F1=0;
```

```
% Flux vs. mmf values
phimax=2.0*d*min(l)*SF;
phip=linspace(0,phimax,100); npts=length(phip);
for j=1:npts
  F(j)=0;
  for i=1:4;
    F(j)=F(j)+hm27(phip(j)/l(i)/d/SF)*lm(i);
  end
  F(j)=F(j)+phip(j)/Pg;
end
% Specified point
phig=interp1(F, phip, N*I);
disp([blanks(4) 'C-SHAPED MAGNETIC CIRCUIT, No Leakage']);
disp(' '); disp([blanks(8) ' AIR GAP FLUX = ', num2str(phig)]);
disp(' '); disp([blanks(8) 'COIL CURRENT = ', num2str(I)]);
plot(F,phip,N*I,phig,'o'); grid
title('C-shaped magnetic circuit');
xlabel('Coil mmf, A-t'); ylabel('Flux, Wb');
legend('Flux vs mmf', 'Specified pt.', 4);


%%%%%%%%%%%%%%%%%%%%%%%%%%%%%%%%%%%%%%%%%%%%%%%%%%%%%%%%%%%%%%%%%%%%%%%%%
%
% cleak2.m - Analysis of C-shaped magnetic circuit with
%            leakage flux.Calculates air gap flux for
%            specified value of coil current. Plots Flux
%            vs. mmf. (Type 2 problem)
%
%%%%%%%%%%%%%%%%%%%%%%%%%%%%%%%%%%%%%%%%%%%%%%%%%%%%%%%%%%%%%%%%%%%%%%%%%
clear;
% Units of input dimensions - 'MKS' or 'FPS' (British)
dim='MKS';
if dim=='MKS'; k=1; elseif dim=='FPS'; k=39.37; else; end
w=0.15; h=0.10;          % Window width & height
d=0.05; del=0.002;       % Core depth & air gap length
l1=0.05; l2=0.05; l3=0.05; l4=0.05; % Core widths
I=20.0;                  % Coil current for analysis point - A
N=500; SF=0.95;          % Coil turns, stacking factor
w=w/k; h=h/k; d=d/k; del=del/k;
l=[ l1 l2 l3 l4]/k;   % All dimensions now in MKS units
lm=[h+(l(2)+l(4))/2 w+(l(1)+l(3))/2 h+(l(2)+l(4))/2-del ...
      w+(l(1)+l(3))/2];
Pg=4e-7*pi*(l(3)*d/del+0.52*(l(3)+d)+0.308*del);
Pl=4e-7*pi*w*d/h;
% Flux vs. mmf curve
phimax=2.0*d*min(l)*SF;
phip=linspace(0,phimax,100); npts=length(phip);
for j=1:npts
  F(j)=phip(j)/Pg;
```

```
   F(j)=F(j)+hm27(phip(j)/l(3)/d/SF)*lm(3)+hm27(phip(j)/ ...
      l(2)/d/SF)*lm(2)/2+hm27(phip(j)/l(4)/d/SF)*lm(4)/2;
   phil(j)=F(j)*P1; phigl=phip(j)+phil(j);
   F(j)=F(j)+hm27(phigl/l(1)/d/SF)*lm(1)+hm27(phigl/l(2)/d/SF) ...
      *lm(2)/2+hm27(phigl/l(4)/d/SF)*lm(4)/2;
end
% Specified point analysis
phig=interp1(F, phip, N*I);
phill=interp1(F, phil, N*I);
disp([blanks(4) 'C-SHAPED MAGNETIC CIRCUIT, With Leakage']);
disp(' '); disp([blanks(8) ' AIR GAP FLUX = ', num2str(phig)]);
disp(' '); disp([blanks(8) 'LEAKAGE FLUX = ', num2str(phill)]);
disp(' '); disp([blanks(8) 'COIL CURRENT = ', num2str(I)]);
plot(F,phip,F,phil,'--',N*I,phig,'o'); grid
title('C-shaped magnetic circuit with leakage');
xlabel('Coil mmf, A-t'); ylabel('Flux, Wb');
legend('Air gap flux', 'Leakage flux', 'Specified pt.', 0);


%%%%%%%%%%%%%%%%%%%%%%%%%%%%%%%%%%%%%%%%%%%%%%%%%%%%%%%%%%%%%%%%%%%%%%%%%%%
%
% dwckt1.m - Analysis of double window magnetic circuit.
%           Calculates coil current for specified value
%           of air gap 2 flux & plots Flux vs. mmf.
%           (Type 1 problem)
%
%%%%%%%%%%%%%%%%%%%%%%%%%%%%%%%%%%%%%%%%%%%%%%%%%%%%%%%%%%%%%%%%%%%%%%%%%%%
clear;
% Units of input dimensions - 'MKS' or 'FPS' (British)
dim='FPS';
if dim=='MKS'; k=1; elseif dim=='FPS'; k=39.37; else; end
w1=3.50; h1=4.00;                % Window 1 width & height
w2=2.50; h2=3.00;                % Window 2 width & height
d=3.00; del1=0.100; del2=0.075;  % Core depth & air gap length
l1=3.00; l2=3.00; l3=2.00; l4=1.00; l5=1.00;   % Core widths
phig2=0.0030;                    % Air gap 2 flux (analysis point) - Wb
N=1000; SF=0.95;                 % Coil turns, stacking factor
lm1=h1+l2; lm2=w1+(l1+l3)/2;     % Mean length paths
lm3=(h1+l2+h2+l4)/2-del1; lm4=w2+(l3+l5)/2; lm5=h2+l4-del2;
w1=w1/k; w2=w2/k; h1=h1/k; h2=h2/k; d=d/k;
l=[ l1 l2 l3 l4 l5]/k; del1=del1/k; del2=del2/k;
lm=[lm1 lm2 lm3 lm4 lm5]/k;    % All dimensions now in MKS units
Pg1=4e-7*pi*(l(3)*d/del1+0.52*(l(3)+d)+0.308*del1);
Pg2=4e-7*pi*(l(5)*d/del2+0.52*(l(5)+d)+0.308*del2);
% Generate flux vs. Fc & Fr arrays (center & right legs)
phimax=2.5*d*min([l(4) l(5)])*SF; npts=250;
if phimax<phig2; disp('SPECIFIED FLUX TOO HIGH');end
phir=linspace(0,phimax,npts);
for i=1:npts
```

```
   Fr(i)=phir(i)/Pg2+hm27(phir(i)/l(5)/d/SF)*lm(5)+ ...
      2*hm27(phir(i)/l(4)/d/SF)*lm(4);
end
phimax=2.5*d*l(3)*SF; phic=linspace(0,phimax,npts);
for i=1:npts
   Fc(i)=phic(i)/Pg1+hm27(phic(i)/l(3)/d/SF)*lm(3);
end
Fp=interp1(phir, Fr, phig2);
if Fp>max(Fc); disp('SPECIFIED POINT NOT POSSIBLE');end
% Generate coil flux(phi) vs. mmf(FT) array
Fmax=min([max(Fc) max(Fr)]); F=linspace(0,Fmax,npts);
for i=1:npts
   phi(i)=interp1(Fc, phic, F(i))+interp1(Fr, phir, F(i));
   FT(i)=F(i)+2*hm27(phi(i)/l(2)/d/SF)*lm(2)+ ...
      hm27(phi(i)/l(1)/d/SF)*lm(1);
end
% Specified point analysis
phig1=interp1(Fc, phic, Fp); phip=phig1+phig2;
Fp=interp1(phi, FT, phip);I=Fp/N;
disp([blanks(4) 'DOUBLE WINDOW MAGNETIC CIRCUIT, No Leakage']);
disp(' '); disp([blanks(8) '   COIL CURRENT = ', num2str(I)]);
disp(' '); disp([blanks(8) ' COIL FLUX = ', num2str(phip)]);
disp(' '); disp([blanks(8) 'AIR GAP 1 FLUX = ', num2str(phig1)]);
disp(' '); disp([blanks(8) 'AIR GAP 2 FLUX = ', num2str(phig2)]);
plot(FT,phi,Fr,phir,'--',Fc,phic,'-.',Fp,phip,'o'); grid
title('Double window magnetic circuit');
xlabel('Coil mmf, A-t'); ylabel('Flux, Wb');
legend('Coil flux','Right flux','Center flux', 'Specified pt.', 4);

%%%%%%%%%%%%%%%%%%%%%%%%%%%%%%%%%%%%%%%%%%%%%%%%%%%%%%%%%%%%%%%%%%%%%%%%%
%
% dwleak1.m - Analysis of double window magnetic circuit
%            with leakage flux. Calculates coil current
%            for specified value of air gap 2 flux & plots
%            Flux vs. mmf. (Type 1 problem)
%
%%%%%%%%%%%%%%%%%%%%%%%%%%%%%%%%%%%%%%%%%%%%%%%%%%%%%%%%%%%%%%%%%%%%%%%%%
clear;
% Units of input dimensions - 'MKS' or 'FPS' (British)
dim='FPS';
if dim=='MKS'; k=1; elseif dim=='FPS'; k=39.37; else; end
w1=3.50; h1=4.00;                  % Window 1 width & height
w2=2.50; h2=3.00;                  % Window 2 width & height
d=3.00; del1=0.100; del2=0.075;    % Core depth & air gap length
l1=3.00; l2=3.00; l3=2.00; l4=1.00; l5=1.00; % Core widths
phig2=0.0030;                 % Air gap 2 flux (analysis point) - Wb
N=1000; SF=0.95;              % Coil turns, stacking factor
```

```
lm1=h1+l2; lm2=w1+(l1+l3)/2;        % Mean length paths
lm3=(h1+l2+h2+l4)/2-del1; lm4=w2+(l3+l5)/2; lm5=h2+l4-del2;
w1=w1/k; w2=w2/k; h1=h1/k; h2=h2/k; d=d/k;
l=[ l1 l2 l3 l4 l5]/k; del1=del1/k; del2=del2/k;
lm=[lm1 lm2 lm3 lm4 lm5]/k; % All dimensions now in MKS units
Pg1=4e-7*pi*(l(3)*d/del1+0.52*(l(3)+d)+0.308*del1);
Pg2=4e-7*pi*(l(5)*d/del2+0.52*(l(5)+d)+0.308*del2);
Pl1=4e-7*pi*w1*d/h1; Pl2=4e-7*pi*w2*d/h2;
% Generate flux vs. Fc & Fr arrays (center & right legs)
phimax=2.2*d*min([l(4) l(5)])*SF; npts=250;
if phimax < phig2; disp('SPECIFIED FLUX TOO HIGH');end
phir=linspace(0,phimax,npts);
for i=1:npts
   Fr(i)=phir(i)/Pg2+hm27(phir(i)/l(5)/d/SF)*lm(5)+ ...
      hm27(phir(i)/l(4)/d/SF)*lm(4);
   phil1(i)=Fr(i)*Pl1; phirt(i)=phir(i)+phil1(i);
   Fr(i)=Fr(i)+hm27(phirt(i)/l(4)/d/SF)*lm(4);
end
phimax=2.2*d*l(3)*SF; phic=linspace(0,phimax,npts);
for i=1:npts
   Fc(i)=phic(i)/Pg1+hm27(phic(i)/l(3)/d/SF)*lm(3);
end
Fp=phig2/Pg2+hm27(phig2/l(5)/d/SF)*lm(5)+ ...
   hm27(phig2/l(4)/d/SF)*lm(4); phig2l=Fp*Pl1;
Fp=Fp+hm27((phig2l+phig2)/l(4)/d/SF)*lm(4);
if Fp > max(Fc); disp('SPECIFIED POINT NOT POSSIBLE');end
% Generate coil flux(phit) vs. mmf(FT) array
Fmax=min([max(Fc) max(Fr)]); F=linspace(0,Fmax,npts);
for i=1:npts
   phi(i)=interp1(Fc, phic, F(i))+interp1(Fr, phirt, F(i));
   FT(i)=F(i)+hm27(phi(i)/l(2)/d/SF)*lm(2);
   phil2(i)=FT(i)*Pl2; phit(i)=phi(i)+phil2(i);
   FT(i)=FT(i)+hm27(phit(i)/l(2)/d/SF)*lm(2)+ ...
      hm27(phit(i)/l(1)/d/SF)*lm(1);
end
phig1=interp1(Fc, phic, Fp); % Specified point analysis
phig1l=(Fp+hm27((phig2+phig2l+phig1)/l(2)/d/SF)*lm(2))*Pl2;
phip=phig2+phig2l+phig1+phig1l; Fp=interp1(phit,FT,phip);I=Fp/N;
disp([blanks(4) 'DOUBLE WINDOW MAGNETIC CIRCUIT, With Leakage']);
disp(' '); disp([blanks(8) ' COIL CURRENT = ', num2str(I)]);
disp(' '); disp([blanks(8) '   COIL FLUX = ', num2str(phip)]);
disp(' '); disp([blanks(8) 'AIR GAP 1 FLUX = ', num2str(phig1)]);
disp(' '); disp([blanks(8) 'AIR GAP 2 FLUX = ', num2str(phig2)]);
plot(FT,phit,FT,phil1,'--',FT,phil2,'-.',Fp,phip,'o'); grid
title('Double window magnetic circuit with leakage');
xlabel('Coil mmf, A-t'); ylabel('Flux, Wb');
legend('Coil flux','Leakage 1','Leakage 2', 'Specified pt.', 0);
```

```
%%%%%%%%%%%%%%%%%%%%%%%%%%%%%%%%%%%%%%%%%%%%%%%%%%%%%%%%%%%%%%%%%%%%%%%%%%%%%%%%%%%%%%%%%%
%
% dwleak2.m - Analysis of double window magnetic circuit
%             with leakage flux. Calculates air gap fluxes
%             for specified value of coil current & plots
%             Flux vs. mmf. (Type 2 problem)
%
%%%%%%%%%%%%%%%%%%%%%%%%%%%%%%%%%%%%%%%%%%%%%%%%%%%%%%%%%%%%%%%%%%%%%%%%%%%%%%%%%%%%%%%%%%
clear;
% Units of input dimensions - 'MKS' or 'FPS' (British)
dim='FPS';
if dim=='MKS'; k=1; elseif dim=='FPS'; k=39.37; else; end
w1=3.50; h1=4.00;                 % Window 1 width & height
w2=2.50; h2=3.00;                 % Window 2 width & height
d=3.00; del1=0.100; del2=0.075;  % Core depth & air gap length
l1=3.00; l2=3.00; l3=2.00; l4=1.00; l5=1.00;   % Core widths
I=7.5;               % Coil current (analysis point) - A
N=1000; SF=0.95;     % Coil turns, stacking factor
lm1=h1+l2; lm2=w1+(l1+l3)/2;        % Mean length paths
lm3=(h1+l2+h2+l4)/2-del1; lm4=w2+(l3+l5)/2; lm5=h2+l4-del2;
w1=w1/k; w2=w2/k; h1=h1/k; h2=h2/k; d=d/k;
l=[ l1 l2 l3 l4 l5]/k; del1=del1/k; del2=del2/k;
lm=[lm1 lm2 lm3 lm4 lm5]/k; % All dimensions now in MKS units
Pg1=4e-7*pi*(l(3)*d/del1+0.52*(l(3)+d)+0.308*del1);
Pg2=4e-7*pi*(l(5)*d/del2+0.52*(l(5)+d)+0.308*del2);
Pl1=4e-7*pi*w1*d/h1; Pl2=4e-7*pi*w2*d/h2;
% Generate flux vs. Fc & Fr arrays (center & right legs)
phimax=2.2*d*min([l(4) l(5)])*SF; npts=250;
phir=linspace(0,phimax,npts);
for i=1:npts
  Fr(i)=phir(i)/Pg2+hm27(phir(i)/l(5)/d/SF)*lm(5)+ ...
    hm27(phir(i)/l(4)/d/SF)*lm(4);
  phil1(i)=Fr(i)*Pl1; phirt(i)=phir(i)+phil1(i);
  Fr(i)=Fr(i)+hm27(phirt(i)/l(4)/d/SF)*lm(4);
end
phimax=2.2*d*l(3)*SF; phic=linspace(0,phimax,npts);
for i=1:npts
 Fc(i)=phic(i)/Pg1+hm27(phic(i)/l(3)/d/SF)*lm(3);
end
% Generate coil flux(phit) vs. mmf(FT) array
Fmax=min([max(Fc) max(Fr)]); F=linspace(0,Fmax,npts);
for i=1:npts
  phi(i)=interp1(Fc,phic,F(i))+interp1(Fr,phirt,F(i));
  FT(i)=F(i)+hm27(phi(i)/l(2)/d/SF)*lm(2);
  phil2(i)=FT(i)*Pl2; phit(i)=phi(i)+phil2(i);
  FT(i)=FT(i)+hm27(phit(i)/l(2)/d/SF)*lm(2)+ ...
    hm27(phit(i)/l(1)/d/SF)*lm(1);
end
```

```
if N*I > max(FT); disp('CURRENT TOO HIGH'); end
% Specified point analysis
flxp=interp1(FT, phit, N*I);
F=N*I-hm27(flxp/l(2)/d/SF)*lm(2)-hm27(flxp/l(1)/d/SF)*lm(1);
flxl1=F*Pl1; F=F-hm27((flxp-flxl1)/l(2)/d/SF)*lm(2);
flxg1=interp1(Fc, phic, F); phiR=flxp-flxl1-flxg1;
F=F-hm27(phiR/l(4)/d/SF)*lm(4); flxl2=F*Pl2; flxg2=phiR-flxl2;
disp([blanks(4) 'DOUBLE WINDOW MAGNETIC CIRCUIT, With Leakage']);
disp(' '); disp([blanks(8) ' COIL CURRENT = ', num2str(I)]);
disp(' '); disp([blanks(8) '   COIL FLUX = ', num2str(flxp)]);
disp(' '); disp([blanks(8) 'AIR GAP 1 FLUX = ', num2str(flxg1)]);
disp(' '); disp([blanks(8) 'AIR GAP 2 FLUX = ', num2str(flxg2)]);
plot(FT,phit,FT,phil1,'--',FT,phil2,'-.',N*I,flxp,'o'); grid
title('Double window magnetic circuit with leakage');
xlabel('Coil mmf, A-t'); ylabel('Flux, Wb');
legend('Coil flux','Leakage 1','Leakage 2', 'Specified pt.', 0);


%%%%%%%%%%%%%%%%%%%%%%%%%%%%%%%%%%%%%%%%%%%%%%%%%%%%%%%%%%%%%%%%%%%%%%%%%%%%%
%
% sinex.m - Plots voltage-flux-current relationships for a
%           magnetic circuit that exhibits saturation. The
%           flux-current plot is formed by piecewise linear
%           approximation. Normalized flux and current
%           responses formed for sinusoidal voltage input.
%
%%%%%%%%%%%%%%%%%%%%%%%%%%%%%%%%%%%%%%%%%%%%%%%%%%%%%%%%%%%%%%%%%%%%%%%%%%%%%
clear;
% Set saturation break point for magnetization curve
satphi=input('Percentage maximum flux = ')/100;
sati=input('Percentage maximum current = ')/100;
% Build piecewise linear flux-current arrays
npts=200; n1=fix(npts/2*satphi); n2=npts-n1;
y1=[[linspace(0,satphi,n1)] [linspace(satphi,1,n2+1)]]; y1(n1)=[];
x1=[[linspace(0,sati,n1)] [linspace(sati,1,n2+1)]]; x1(n1)=[];
nz1=fix(0.2*n1); nz2=ceil(0.2*n2);        % Chamfer corner
ya=y1(n1-nz1); yb=y1(n1+nz2); xa=x1(n1-nz1); xb=x1(n1+nz2);
m=(ya-yb)/(xa-xb); b=(xa*yb-xb*ya)/(xa-xb);
for i=n1-nz1:n1+nz2; y1(i)=m*x1(i)+b; end
curr=[-fliplr(x1) x1]; curr(npts)=[];   % Completed arrays
phi=[-fliplr(y1) y1]; phi(npts)=[];
plot(curr,phi); grid
title('Magnetization curve');
xlabel('Current');ylabel('Flux');pause
% Build response for the case of an impressed sinsuiodal voltage.
np=1024; ang=linspace(0,2*pi,1024)'; deg=180/pi;
V=max(phi)*cos(ang);         % Voltage time array
Flx=max(phi)*sin(ang);       % Flux time array
I=interp1(phi, curr, Flx);   % Current time array
```

```
figure(2);
subplot(2,1,1); plot(ang*deg,I,ang*deg,Flx,'--',ang*deg,V,'-.');
grid; title('Response due to impressed sinusoidal voltage');
ylabel('Current, voltage, flux (normalized)'); xlabel('Angle');
legend('Coil current','Coil flux','Terminal voltage', 0);
fftI=fft(I); f=abs(fftI); f(1)=f(1)/2; f=f/(np/2);    % FFT of current
subplot(2,1,2); plot([0:25],f(1:26)); grid
title('Current for sinusoidal voltage');
ylabel('Current'); xlabel('Harmonic number');

%%%%%%%%%%%%%%%%%%%%%%%%%%%%%%%%%%%%%%%%%%%%%%%%%%%%%%%%%%%%%%%%%%%%%%%%%%%%%
%
% coenFe.m - evaluates developed force Fd for a magnetic
%            circuit for a specified current and air gap
%            value. Calls magckt1.m for flux linkage -
%            current array.
%
%%%%%%%%%%%%%%%%%%%%%%%%%%%%%%%%%%%%%%%%%%%%%%%%%%%%%%%%%%%%%%%%%%%%%%%%%%%%%
clear;
x=input(' Air gap length = ');
if x==0; x=1e-06; else; end; del=0.01*x;
I=input('Current for Fd = ');
% Generate pair of flux linkage - current plots
LamI=magckt1(x-del); m=length(LamI);
Lam1(1:m)=LamI(1,1:m); I1(1:m)=LamI(2,1:m);
LamI=magckt1(x+del);
Lam2(1:m)=LamI(1,1:m); I2(1:m)=LamI(2,1:m);
if (I>max(I1)) | (I>max(I2)); disp(' ');disp('INVALID CURRENT'); end
Wf1=0; Wf2=0;
for i=2:m; % Calculate force
   if (I1(i)>I) | (I2(i)>I)
     Wf1=Wf1+(interp1(I1,Lam1,I)+Lam1(i-1))/2*(I-I1(i-1));
     Wf2=Wf2+(interp1(I2,Lam2,I)+Lam2(i-1))/2*(I-I2(i-1));
     break
   else
     Wf1=Wf1+(Lam1(i)+Lam1(i-1))/2*(I1(i)-I1(i-1));
     Wf2=Wf2+(Lam2(i)+Lam2(i-1))/2*(I2(i)-I2(i-1));
   end
end
   Fd=(Wf1-Wf2)/2/del;
disp(' '); disp([blanks(8) 'COENERGY FORCE ANALYSIS']);
disp(' '); disp([blanks(12) [' Air gap = ',num2str(x)]]);
disp(' '); disp([blanks(12) ['Current = ',num2str(I)]]);
disp(' '); disp([blanks(12) ['   Fd = ',num2str(Fd)]]);
% Activate to check lambda-i plots
plot(I1,Lam1,I2,Lam2); grid;
xlabel('Coil current, A'); ylabel('Flux linkage, Wb-t');
```

```
%%%%%%%%%%%%%%%%%%%%%%%%%%%%%%%%%%%%%%%%%%%%%%%%%%%%%%%%%%%%%%%%%%%%%%%%%%%%
%
% magckt1.m - Analysis of Fig 3.34 magnetic circuit.
%             Calculates flux linkage & current arrays
%             for specified value of air gap length x.
%             Used by coenFe.m.
%
%%%%%%%%%%%%%%%%%%%%%%%%%%%%%%%%%%%%%%%%%%%%%%%%%%%%%%%%%%%%%%%%%%%%%%%%%%%%
function [LamI] = magckt1(x);
% Assumes that input dimensions are MKS units.
w=0.05; d=0.05;        % Core width & depth
lm1=0.20; lm2=0.40;    % Mean length paths
N=100;                 % Coil turns
npts=50; B=linspace(0,1.6,npts); LamI=zeros(2,npts);
Pg=4e-7*pi*(w*d/x+0.52*(w+d)+0.308*x);
for i=1:npts
   LamI(1,i)=N*B(i)*w*d;
   LamI(2,i)=(B(i)*w*d/Pg+(lm1-x)*Hcs(B(i))+lm2*Hcs(B(i)))/N;
end


%%%%%%%%%%%%%%%%%%%%%%%%%%%%%%%%%%%%%%%%%%%%%%%%%%%%%%%%%%%%%%%%%%%%%%%%%%%%
%
% soldes.m - solenoid design program that uses the data
%            provided by soldata.m & calls magckt2.m for
%            calculation of flux linkages-current at each
%            specified value of air gap position.
%
%%%%%%%%%%%%%%%%%%%%%%%%%%%%%%%%%%%%%%%%%%%%%%%%%%%%%%%%%%%%%%%%%%%%%%%%%%%%
clear;
soldata,
I=input('   Current for Force Study = ');
% Approximate flux density checks
flx=Bmax*pi*d^2/4;   % Plunger for specified Bmax
A=[A1 A2 A3 A4 A5 A6 A7 A8 A9 A10]; mA=length(A);
figure(1);
for i=1:mA; Bck(i)=flx/A(i); end
plot([1:mA],Bck,'p',[0 mA],[Bmax Bmax]); grid;
ylabel('Flux density, T'); xlabel('Member no.');
X1=(Bmax+min(Bck))/2; X2=0.9*X1;
text(1,X1,'Consider dimension adjustment if flux density');
text(1,X2,'of any member varies significantly from Bmax.');
% Weight, size & current density estimates
J=N*I/((db-dc)/2*lc)/39.37^2/0.65; % 65% copper fill
WTFe=pi*df^2*le+pi*(df^2-db^2)/4*lc+pi*d^2/4*1.1*lc;
WTFe=WTFe*0.283*39.37^3;
WTCu=0.75*pi*(db^2-dc^2)/4*lc*0.322*39.37^3;
Ac=0.65*(db-dc)/2*lc/N*39.37^2;
```

```
Rc=1.025e-06*N*pi*(db+dc)*39.37/2/Ac;
Lc=4e-07*pi*N^2*pi/4*d^2/(0.125/39.37);
disp(' '); disp( '   WEIGHT, SIZE & COIL CHARACTERISTIC ESTIMATES');
disp(' '); disp(['        Core weight(lbs) = ' num2str(WTFe)]);
disp(' '); disp(['        Coil weight(lbs) = ' num2str(WTCu)]);
disp(' '); disp(['          Frame OD(in) = ' num2str(df*39.37)]);
disp(' '); disp(['        Frame length(in) = ' num2str(lf*39.37)]);
disp(' '); disp(['   Conductor area(sq in) = ' num2str(Ac)]);
disp(' '); disp(['   Current density(Apsi) = ' num2str(J)]);
disp(' '); disp(['    Resistance(Ohm,150 C) = ' num2str(Rc)]);
disp(' '); disp(['      Inductance(H,1/8") = ' num2str(Lc)]);
disp(' ');
% Generate a set of flux-current plots for air gap lengths
% specified by x. Enter dimensions in inches.
x=[1e-12 0.1 0.2 0.3 0.4 0.5 1.0]/39.37; m=length(x);
figure(2);
for i=1:m
  LamI=magckt2(x(i)); n=length(LamI);x  x
  plot(LamI(2,1:n),LamI(1,1:n)/N); grid; hold on
  xlabel('Coil current, A'); ylabel('Flux, Wb');
  title('Cylindrical plunger solenoid');
end
% Force study for specified current I
for i=1:m
  % Generate pair of flux linkage - current plots
  del=x(i)/100; LamI=magckt2(x(i)-del); n=length(LamI);
  Lam1(1:n)=LamI(1,1:n); I1(1:n)=LamI(2,1:n);
  LamI=magckt2(x(i)+del);
  Lam2(1:n)=LamI(1,1:n); I2(1:n)=LamI(2,1:n);
  if (I>max(I1)) | (I>max(I2));
    disp(' ');disp('INVALID CURRENT');
  end
  Wf1=0; Wf2=0;
  for j=2:n; % Calculate force
    if (I1(j)>I) | (I2(j)>I)
      Wf1=Wf1+(interp1(I1,Lam1,I)+Lam1(j-1))/2*(I-I1(j-1));
      Wf2=Wf2+(interp1(I2,Lam2,I)+Lam2(j-1))/2*(I-I2(j-1));
      break
    else
      Wf1=Wf1+(Lam1(j)+Lam1(j-1))/2*(I1(j)-I1(j-1));
      Wf2=Wf2+(Lam2(j)+Lam2(j-1))/2*(I2(j)-I2(j-1));
    end
  end
  Fd(i)=(Wf1-Wf2)/2/del;
end
figure(3); plot(x*39.37,Fd*0.2248); grid
title('Force vs. displacement');
xlabel('Displacement, in'); ylabel('Force, lbs');
```

```
%%%%%%%%%%%%%%%%%%%%%%%%%%%%%%%%%%%%%%%%%%%%%%%%%%%%%%%%%%%%%%%%%%%%%%%%%%%%
%
% magckt2.m - Analysis of Fig 3.40 magnetic circuit.
%             Calculates flux linkage & current arrays
%             for specified value of air gap length x.
%             Leakage flux is considered in calculations.
%             Used by soldes.m.
%
%%%%%%%%%%%%%%%%%%%%%%%%%%%%%%%%%%%%%%%%%%%%%%%%%%%%%%%%%%%%%%%%%%%%%%%%%%%%
function [LamI] = magckt1(x);
soldata,
if x==0; x=1e-12; else; end; del=0.01*x;
% Air gap & leakage permeances
Pg=4e-07*pi*(pi*d^2/4/x+0.52*pi*d);
Pl=4e-07*pi^2*(d+db)*lp/(db-d)/2;
% Generate left & right flux-mmf curves
npts=25; phimax=Bmax*pi*d^2/4; phi=linspace(0,phimax,npts);
for i=1:npts
  FL(i)=l1*Hcs(phi(i)/A1)+l2*Hcs(phi(i)/A2)+l3*Hcs(phi(i)/A3);
  phiL(i)=phi(i)+FL(i)*Pl;
  FR(i)=l6*Hcs(phi(i)/A6)+l7*Hcs(phi(i)/A7)+l8*Hcs(phi(i)/A8)+ ...
    l9*Hcs(phi(i)/A9);
  phiR(i)=phi(i)+FR(i)*Pl;
end
% Solve center loop looking up FL & FR at same values of flux
for i=1:npts
  mmfR=interp1(phiR, FR, phi(i));
  mmfL=interp1(phiL, FL, phi(i));
  F(i)=l4*Hcs(phi(i)/A4)+l5*Hcs(phi(i)/A5)+(lp/2-x)* ...
    Hcs(phi(i)/A10)+phi(i)/Pg+mmfR+mmfR;
  LamI(1,i)=N*phi(i); LamI(2,i)=F(i)/N;
  philR(i)=mmfR*Pl; philL(i)=mmfL*Pl;
end
% Activate to plot leakage & plunger fluxes
% Can handle from command window: magckt1(x)
% plot(F/N,phi,F/N,philL,F/N,philR); grid;
% ylabel('Flux, Wb'); xlabel('Coil current, A');


%%%%%%%%%%%%%%%%%%%%%%%%%%%%%%%%%%%%%%%%%%%%%%%%%%%%%%%%%%%%%%%%%%%%%%%%%%%%
%
% soldata.m - Input data for soldes.m, Also called by magckt2.m
%
%%%%%%%%%%%%%%%%%%%%%%%%%%%%%%%%%%%%%%%%%%%%%%%%%%%%%%%%%%%%%%%%%%%%%%%%%%%%
% Units of input dimensions - 'MKS' or 'FPS' (British)
dim='FPS';
if dim=='MKS'; k=1; elseif dim=='FPS'; k=39.37; else; end
d=1.27; dc=1.50; db=2.70;        % Plunger dia, coil OD&ID
tb=0.030; lc=3.00;               % Bushing thick, cavity depth
```

```
N=2000; Bmax=1.90;                    % Coil turns, plunger max B
d=d/k;dc=dc/k;db=db/k;tb=tb/k;lc=lc/k;   % Dimensions now MKS
df=sqrt(d^2+db^2); le=d^2/(db+df); lf=lc+4*le; lp=lc/2;
% Areas & length of flux paths
l1=(lp+le)/2; A1=pi*d^2/4; l2=(db+df)/4; A2=pi*(db+df)/4*le;
l3=(lp+le)/2; A3=pi*(df^2-db^2)/4; l4=lc/2; A4=A3;
l5=lp/2; A5=pi*d^2/4; l6=(lp+3*le)/2; A6=A3;
l7=(db+df)/4-tb; A7=pi*(db+df)/4*3*le; l8=tb; A8=pi*(d+tb)*3*le;
l9=l6; A9=A1; A10=A1;


%%%%%%%%%%%%%%%%%%%%%%%%%%%%%%%%%%%%%%%%%%%%%%%%%%%%%%%%%%%%%%%%%%%%%%%%%%
%
% Hcs.m - B-H interpolation routine
%
%%%%%%%%%%%%%%%%%%%%%%%%%%%%%%%%%%%%%%%%%%%%%%%%%%%%%%%%%%%%%%%%%%%%%%%%%%%
function y = Hx(Bx)
% B-H values that follow are valid for Cast Steel
Bcs=[0 0.04 0.1 0.2 0.25 0.3 0.5 0.73 1.15 1.38 1.51 1.63 1.8];
Hcs=[0 50 90 140 170 182 275 400 1000 2000 3000 6000 20000];
B=linspace(0,max(Bcs),200); H=spline(Bcs,Hcs,B);
% Extending beyond range that spline yields monotonic curve
B=[B 1.85 1.9]; H=[H 40000 150000]; m=length(B);
% Activate to plot B-H curve
% plot(H(1:m),B(1:m)); grid; pause; % Linear plot
% semilogx(H(2:m),B(2:m));grid; pause; % Semilog plot
n=length(B); k=0;
if Bx==0; k=-1; y=0; end
if Bx<0; k=-1; y=0; disp('WARNING - Bx < 0, Hx = 0 returned'); end
if Bx>B(n);
   y=H(n)+(Bx-B(n))/(pi*4e-07); % Use after complete saturation
   k=-1; disp('CAUTION - Beyond B-H curve');
end
for i=1:n
if k==0 & (Bx-B(i))<=0; k=i; break; end
end
if k>0;
y=H(k-1)+(Bx-B(k-1))/(B(k)-B(k-1))*(H(k)-H(k-1));
else;
end
```

SUMMARY

- Electromechanical energy conversion devices by nature change energy into magnetic form as an intermediate step.

- The first principles of electromagnetic energy conversion can always be explained by elementary laws of physics.

- Use of ferromagnetic material to form the principal flux paths of electro-mechanical energy converters leads to high power density machines.

- Ferromagnetic materials exhibit a nonlinear *B-H* characteristic that significantly increases analytical difficulty in the study of electromechanical energy conversion devices.

- Electrical circuit analogs to magnetic circuits enhance understanding of magnetic circuits and serve to guide the analysis process.

- There is no magnetic insulator of the same quality available for electrical circuits. Consequently, leakage flux flows across undesirable paths and frequently reaches levels that cannot be ignored without sacrifice of analytical accuracy.

- MATLAB has features for array manipulation that make it a valuable tool for use in analysis of nonlinear magnetic circuits.

- The nonlinearity of magnetic circuits can lead to low-order odd harmonics for the case of sinusoidal excitation.

- Permanent magnets can be used for the excitation of magnetic circuits in place of current-carrying coils offering the advantage of improved efficiency due to the absence of the ohmic losses. However, the levels of flux density attainable by permanent magnet excitation are less than possible with current-carrying coils.

- Useful methods can be developed to predict the force or torque produced by an electromechanical energy conversion device based on the change of either stored energy or coenergy in the intermediary magnetic fields.

- The principles and procedures of the chapter are put to practical use in design of solenoids to meet performance specifications.

PROBLEMS

3.1 Rework Example 3.1 if the *B*-field is directed at an angle of 30° from the conductor centerline.

3.2 The conductor of Fig. 3.2 has a diameter *D*. Find an expression to describe the *H*-field within the conductor.

3.3 Assume that the *B*-field of Fig. 3.5 is uniform but not constant being described by $B(t) = B_1 + B_2 \sin \omega t$. Determine the voltage *e* as indicated.

3.4 Rework Example 3.3 if $\ell_2 = \ell_4 = 0.10$ m and all else is unchanged. Use the MATLAB program ⟨cckt1.m⟩ for convenience.

3.5 Rework Example 3.3 assuming no fringing of air gap flux. What is the percentage error introduced by this approximation?

3.6 Rework Example 3.4 with the air gap increased to $\delta = 0.003$ m. Determine the percentage increase in leakage flux over the value found in Example 3.4.

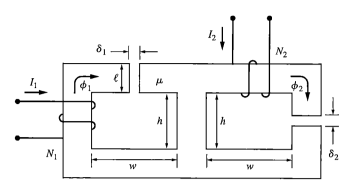

Figure 3.44

3.7 Let $w = d = 50$ mm and $\delta = 2$ mm. Determine the percentage difference in air gap permeance for the two air gap configurations of Fig. 3.10.

3.8 Use [3.17] to find an effective area for an air gap with fringing present. With this expression for effective air gap area known, derive an expression similar to [3.19] that allows extension of the load line method to include air gap fringing.

3.9 Rework Example 3.5 using the load line method with air gap fringing developed in Prob. 3.8.

3.10 Draw an analog electric schematic for the parallel magnetic circuit of Fig. 3.44 for both the cases of $\mu = \infty$ and $\mu \neq \infty$.

3.11 For the parallel magnetic circuit of Fig. 3.44, assume that the core material is infinitely permeable. Let $\delta_1 = 3$ mm and $\delta_2 = 2$ mm. The thickness of all core members is $\ell = 50$ mm. The core has a uniform depth into the page of 75 mm. $N_1 = 2N_2 = 100$ turns. Neglect air gap fringing.

(a) If $I_2 = 0$ and $\phi_1 = 15$ mWb, find the value of I_1.

(b) If $I_1 = 10$ A and $I_2 = 20$ A, determine ϕ_1 and ϕ_2.

3.12 For the parallel magnetic circuit of Fig. 3.44, the core material is M-27, 24-gage ESS. Let $\delta_1 = 3$ mm, $\delta_2 = 2$ mm, $w = 125$ mm, and $h = 150$ mm. As in Prob. 3.11, the thickness of all core members is 50 mm and the core depth into the page is a uniform $d = 75$ mm. $N_1 = 2N_2 = 100$ turns. If $I_2 = 0$ and $\phi_2 = 1$ mWb, find flux ϕ_1 and current I_1. Neglect leakage, but consider air gap fringing. Assume SF = 0.95.

3.13 Develop a MATLAB program to solve for the flux ϕ_2 and the current I_1 for the parallel magnetic circuit of Fig. 3.44 if $I_2 = 0$, ϕ_1 is known, and the core material is M-27, 24-gage ESS. Assume that all core members are of uniform thickness ℓ and depth d where ℓ and d are known values. Neglect leakage flux.

3.14 For the mutually coupled coil pair of Fig. 3.24, the left coil has a self-inductance L_1, the right coil has a self-inductance L_2, and the two coils have a mutual inductance M. If the terminals of the right coil are shorted, find an equivalent inductance (L_{eq}) looking into the terminals of the left coil in terms of L_1, L_2, and M.

3.15 Show that for two mutually coupled coils having self-inductances L_1 and L_2 and a coefficient of coupling k, the mutual inductance is given by $M = k\sqrt{L_1 L_2}$. Clearly state the assumption underlying this expression for M.

3.16 Assume that three coils are mutually coupled. Three coefficients of coupling given by k_{12}, k_{13}, and k_{23} are necessary to describe the coupling fluxes. Draw a schematic analogous to Fig. 3.24 for this condition and write the three voltage equations analogous to [3.27] and [3.28]. Clearly define three mutual inductances $M_{12} = M_{21}$, $M_{13} = M_{31}$, and $M_{23} = M_{32}$.

3.17 Complete the table below by determining the correct number of mutual inductances required for the given number of mutually coupled coils.

No. coils	2	3	4	5	6
No. mutual inductances	1	3			

3.18 Use the mean length path ℓ and cross-sectional area A of Fig. 3.23 to express the inductance as given by [3.22] in terms of the magnetic core geometry. Determine three physical changes that would each result in an increase of the coil self-inductance if the core flux density is unchanged.

3.19 For the ferromagnetic core of Fig. 3.23a, let $A = 25$ cm^2 and $\ell = 60$ cm. If the core material is M-27, 24-gage ESS, develop a MATLAB program that uses Simpson's rule integration to predict the magnetic energy stored for $B = 1.8$ T based on [3.33].

3.20 Execute ⟨sinex.m⟩ for both Percentage maximum flux and Percentage maximum current equal to 99.9 percent. This gives as near a linear flux-current plot as possible. Comment on the nature of the current harmonic content.

3.21 Using [3.17], find an expression for k_{eff} of [3.40].

3.22 Use [3.40] to determine the relative permeability of a specific PM material at the point of operation in a magnetic circuit.

3.23 Rework Example 3.16 except use a SmCo magnet and $\delta = 30$ mm.

3.24 Rework Example 3.16 treating the ferromagnetic material as infinitely permeable and determine the error in the resulting magnet length.

3.25 Assume that the ceramic magnet specified in Example 3.16 is installed in the series magnetic circuit of Fig. 3.29. Add a 200-turn coil around the magnetic core structure and inject a current I into the coil in such a direction that the resulting coil mmf opposes the flux established by the PM. Determine the value of I to render the air gap flux zero.

3.26 For the magnetic circuit of Fig. 3.34 with infinitely permeable core material, find an expression relating the coil flux linkage (λ) to the coil current (i) and the slug position (x). Neglect air gap fringing. Explain how this expression justifies Fig. 3.33b.

3.27 If the magnetic circuit for a singly excited, translational device is linear, show that [3.56] can be written as $F_d = \frac{1}{2} i \frac{d\lambda(x)}{dx}$.

3.28 The rectangular-shaped core for the magnetic circuit of Fig. 3.38 has a uniform depth d into the page. Assume that $i_2 = 0$ and find expressions for the force acting on the slug.

3.29 The core material for the cylindrical solenoid of Fig. 3.40 can be considered infinitely permeable. The guide bushing has a thickness $t_b = 1$mm, the end plate thickness $\ell_e = 3$ cm, and the plunger diameter is $d = 5$ cm. The coil has 500 turns and conducts a 5 A current. The air gap permeance is described by [3.87]. Find an expression for the developed force acting on the plunger as a function of air gap length x.

3.30 Find an expression for the error in force that would result if Example 3.24 were solved by superposition. By superposition, it is meant that the force is determined as $F_{eT} = F_{e1}(i_1 = 0) + F_{e2}(i_2 = 0)$.

3.31 Add an N-turn coil with flexible leads around the rotatable slug in Fig. 3.37. Neglect any leakage flux. If the inductance of the coil on the stationary member is given by $L_1 = L_a + L_b \cos\theta$, find an expression for the torque acting on the slug.

3.32 Use ⟨soldes.m⟩ to design a cylindrical solenoid that has a pickup force of at least 30 lb when the air gap is 0.5 in and has a holding force of at least 350 lb. The core material is cast steel and the coil current density must not exceed 2000 A/in².

REFERENCES

1. G. R. Slemon, *Magnetoelectric Devices: Transducers, Transformers, and Machines,* John Wiley & Sons, New York, 1966.
2. H. C. Roters, *Electromagnetic Devices*, John Wiley & Sons, New York, 1941.

TRANSFORMERS

4.1 INTRODUCTION

The transformer is the analog of the transmission in mechanical systems. When well designed, both devices process incoming power from input to output with only small internal losses. Also, both devices implement an inverse adjustment in variable values between input and output and are capable of bilateral energy flow. For the transmission of Fig. 4.1*a*, neglect of losses gives the torque-speed ratios $T_1/\omega_2 = T_2/\omega_1$. The analogous relationship for the transformer of Fig. 4.1*b* is

$$\frac{V_1}{I_2} = \frac{V_2}{I_1} \qquad \text{or} \qquad V_1 I_1 = V_2 I_2 \qquad\qquad [4.1]$$

It is the constant power voltage-current relationship of [4.1] that establishes the transformer as an invaluable asset in electrical power transmission and distribution systems. A transformer is used to raise voltage while lowering current for transmission of power over long distances. If voltage is raised by a factor of 10 while current is reduced by a factor of 10, the cross-sectional area (thus, volume) of the transmission-line conductor is reduced by a factor of 10 for a particular level of current density. Significant savings in capital outlay for transmission systems result. Further, if the resistance per conductor is increased an order of magnitude as a result of conductor area reduction while the current is reduced an order of magnitude, the ohmic losses

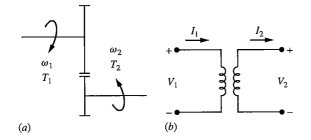

(a)   (b)

Figure 4.1
Transformer compared with mechanical transmission. (a) Transmission. (b) Transformer.

(I^2R) are reduced by a factor of 10. Hence, significant savings in operating costs are also realized. Of course, the complementing ability of the transformer to reduce voltages to levels that are consistent with both personnel safety and practical energy conversion equipment design at the point of end use is of equal importance with the voltage increase capability in realization of a high-voltage transmission system.

4.2 PHYSICAL CONSTRUCTION

Technically, any arrangement of two or more coils that share a time-varying mutual flux could be considered a transformer. However, conventional usage of the name *transformer* has been limited to sets of mutually coupled coils designed for through-put of power with performance characterized by low losses (high efficiency) and a small normalized voltage drop. It will be later understood that the latter performance characteristic can only be satisfied if a high percentage of the magnetic flux established by one coil links all other coils. In power frequency transformer design, this near perfect magnetic coupling is accomplished by use of a ferromagnetic (iron) core. The core material must be laminated to reduce eddy current losses produced by the cyclic magnetic field so as to not compromise the former performance characteristic.

Two basic ferromagnetic material constructions for transformers have evolved. The first is the *core-type* structure as illustrated by Fig. 4.2*a* wherein the mutual

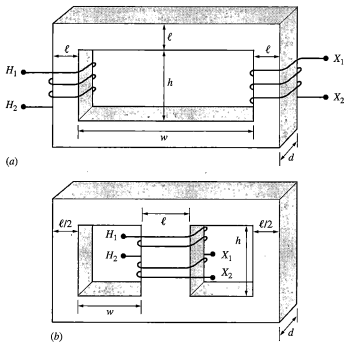

(a)

(b)

Figure 4.2
Transformer construction types. (a) Core type. (b) Shell type.

flux established by one coil links a second coil by a single central or core path. The second is the *shell-type* structure of Fig. 4.2b for which the mutual flux generated by one coil links a second coil by two distinct paths, forming a shell that partially encases the linked coils. Neither transformer type is dominant in low-voltage devices. The shell-type design generally has a shorter flux path than the core-type transformer; however, the mean length turn of the windings for the shell-type tends to be generally longer than for the core-type transformer. Since more coil space can be made available for the core-type transformer, it is the common choice for high-voltage transformer design.

The coil of a transformer that acts as an input port for electrical energy flow is called the *primary winding*. The coil that forms an output port for energy flow is known as a *secondary winding*. Consequently, the primary and secondary windings cannot be declared until the application is defined. Regardless of the ferromagnetic construction, the primary and secondary coils are subdivided and assembled in an interleaved pattern that results in reduced leakage flux. Spacers can be installed between coil layers to provide passageways for cooling fluids. The primary-secondary interleaved winding arrangement of Fig. 4.3a and c is known as a *concentric-layered winding*. This winding arrangement is suitable for cooling by natural-convection air or liquid flow. Figure 4.4 illustrates a ventilated or natural-convection, air-cooled transformer. Figure 4.5 displays a three-phase transformer with concentric-layered windings using natural-convection oil cooling. The *vertical-* (or *disk-*) *layered winding* arrangement illustrated by Fig. 4.3b and d with appropriate baffling installed is suitable for cooling by natural-convection cooling liquid flow. Special transformer oils with high dielectric strength are used as the cooling liquids. Pole-mount distribution transformers typically transfer heat removed by the cooling liquid to the tank, where it conducts through the walls and is radiated to the atmosphere. Figure 4.6 shows a pole-mount transformer and the associated shell-type transformer that fits in the tank. High power level substation transformers typically have side-mounted radiator cores for passage of the naturally convected cooling liquid that are more effective in heat removal than is the case for radiation from the mounting tank. Figure 4.7 displays transformer designs that utilize the radiator cooling technique.

4.3 THE IDEAL TRANSFORMER

The ideal transformer forms a logical point to begin study of the performance characteristics of a practical transformer. After all, the equivalent circuit of a practical transformer is developed around the ideal transformer nucleus. The magnetically and electrically perfect or ideal transformer is characterized by three performance postulates:

1. Negligible losses exist.
2. Perfect magnetic coupling is displayed between winding pairs.
3. Negligible current is required to establish mutual flux.

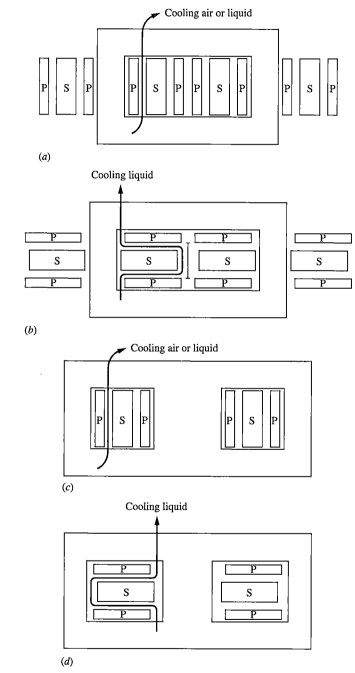

Figure 4.3

Transformer coil arrangements. (a) Core type with concentric-layered winding. (b) Core type with vertical-layered winding. (c) Shell type with concentric-layered winding. (d) Shell type with vertical-layered winding.

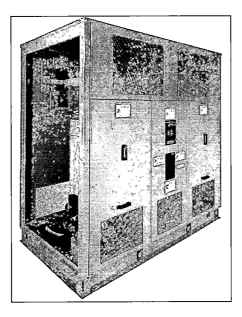

Figure 4.4
Pad-mount, concentric-layered wound, natural-convection air-cooled, core-type transformer.
(*Courtesy of Square D/Schneider Electric.*)

Figure 4.5
Concentric-layered wound, natural-convection oil-cooled, shell-type three-phase transformer.
(*Courtesy of Kuhlman Electric Corporation.*)

(a)   (b)

Figure 4.6
Oil-cooled transformer using tank radiation. (*Courtesy of Kuhlman Electric Corporation.*)
(a) Complete pole-mount distribution transformer. (b) Oil submersible transformer assembly.

Figure 4.7
Oil-cooled substation transformers using radiators with forced-air ventilation. (a) Step-up transmission transformer. (*Courtesy of Kentucky Utilities Company.*) (b) Step-down substation transformer. (*Courtesy of Waukesha Electric Systems.*)

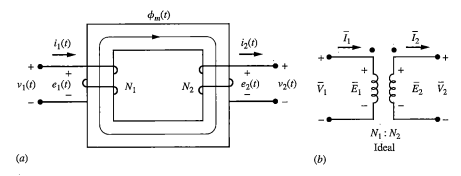

Figure 4.8
Ideal transformer. (a) Physical arrangement. (b) Schematic representation.

4.3.1 CIRCUIT DIAGRAMS

The single-phase, two-winding ideal transformer is shown with both physical circuit arrangement and schematic circuit representation in Fig. 4.8. The physical arrangement of Fig. 4.8a has all variables labeled in the time domain while the schematic of Fig. 4.8b has phasor domain variables indicated. The variables could just as well have been interchanged, as both figures convey identical information. Voltages $v_1(t)$ and $v_2(t)$ are the voltages across the coil terminals. The time-varying mutual flux $\phi_m(t)$ perfectly couples coils 1 and 2, and it induces voltages $e_1(t)$ and $e_2(t)$ behind the coils. For the ideal transformer with lossless coils and no leakage flux, $v_1(t) = e_1(t)$ and $v_2(t) = e_2(t)$. If the mutual flux $\phi_m(t)$ of Fig. 4.8a is allowed to increase, it can be verified by Lenz's law that $e_1(t)$ and $e_2(t)$ are simultaneously positive-valued with the polarity indicated on the circuit diagram of Fig. 4.8a. Conversely, the voltage polarities are negative for decreasing $\phi_m(t)$. Thus, it is concluded that voltages $e_1(t)$ and $e_2(t)$ are in phase.

4.3.2 VOLTAGE AND CURRENT RELATIONSHIPS

The coil voltages of Fig. 4.8a can be determined by Faraday's law using the mutual flux $\phi_m(t)$ that perfectly couples both coils.

$$v_1(t) = e_1(t) = N_1 \frac{d\phi_m(t)}{dt} \qquad\qquad \textbf{[4.2]}$$

$$v_2(t) = e_2(t) = N_2 \frac{d\phi_m(t)}{dt} \qquad\qquad \textbf{[4.3]}$$

Forming the ratio of [4.2] to [4.3] yields

$$\frac{v_1}{v_2} = \frac{e_1}{e_2} = \frac{N_1}{N_2} = a \qquad\qquad \textbf{[4.4]}$$

Or it can be stated that the terminal voltages of the ideal transformer vary directly with the coil turns ratio regardless of the nature of the time-varying flux $\phi_m(t)$.

Example 4.1

The ideal transformer of Fig. 4.8a with $N_1 = 100$ turns and $N_2 = 200$ turns has a mutual coupling flux described by the parabolic function $\phi_m(t) = -0.05(t^2 - 2t)$ Wb. Determine the coil terminal voltages.

By Faraday's law,

$$v_1 = N_1 \frac{d\phi_m}{dt} = 100 \frac{d}{dt}[-0.05(t^2 - 2t)] = -10(t - 1) \text{ V}$$

Applying [4.4],

$$v_2 = \frac{N_2}{N_1} v_1 = \frac{200}{100}[-10(t - 1)] = -20(t - 1) \text{ V}$$

For the case of sinusoidal steady-state operation, the mutual coupling flux of Fig. 4.8a can be expressed as $\phi_m(t) = \Phi_m \sin\omega t$. The resulting coil voltages follow from [4.2] and [4.3] as

$$v_1(t) = N_1 \frac{d}{dt}(\Phi_m \sin\omega t) = \omega N_1 \Phi_m \cos\omega t \qquad \textbf{[4.5]}$$

$$v_2(t) = N_2 \frac{d}{dt}(\Phi_m \sin\omega t) = \omega N_2 \Phi_m \cos\omega t \qquad \textbf{[4.6]}$$

Transforming [4.5] and [4.6] to rms valued phasors,

$$\overline{V}_1 = \overline{E}_1 = \frac{\omega N_1 \Phi_m}{\sqrt{2}} \angle 0° = 4.44 N_1 \Phi_m f \angle 0° \qquad \textbf{[4.7]}$$

$$\overline{V}_2 = \overline{E}_2 = \frac{\omega N_2 \Phi_m}{\sqrt{2}} \angle 0° = 4.44 N_2 \Phi_m f \angle 0° \qquad \textbf{[4.8]}$$

where use has been made of $\omega = 2\pi f$. It should be noted that although [4.7] and [4.8] yield rms values for voltage, Φ_m is the maximum value of mutual flux. The ratio of [4.7] to [4.8] shows that the phasor domain terminal voltages of Fig. 4.8b also vary as the direct turns ratio.

$$\frac{\overline{V}_1}{\overline{V}_2} = \frac{\overline{E}_1}{\overline{E}_2} = \frac{N_1}{N_2} \qquad \textbf{[4.9]}$$

The physical magnetic circuit of Fig. 4.8a can be modeled by the electrical circuit analog of Fig. 4.9. The third ideal transformer performance postulate can hold only if the H-field within the core is zero. Hence, the core material is infinitely permeable ($\mu = B/H$) and the core reluctance must be zero ($\mathcal{R}_c = \ell_c/\mu A_c$). Summation of the mmf's of Fig. 4.9 gives

$$N_1 i_1 - N_2 i_2 = \phi_m \mathcal{R}_c = 0 \qquad \textbf{[4.10]}$$

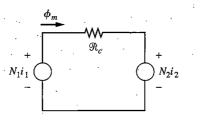

Figure 4.9
Transformer magnetic circuit model

From [4.10], it is observed that for the ideal transformer, the input-output currents vary with the indirect turns ratio.

$$\frac{i_1}{i_2} = \frac{N_2}{N_1} = \frac{1}{a} \qquad \textbf{[4.11]}$$

Since the ratio of current time functions in [4.11] is equal to a positive constant value, it can be concluded that i_1 and i_2 are identical time functions with identical phase relationships except for magnitude. It follows that [4.11] holds in the phasor domain, yielding

$$\frac{\overline{I}_1}{\overline{I}_2} = \frac{N_2}{N_1} = \frac{1}{a} \qquad \textbf{[4.12]}$$

The observations on voltages and currents allow the following rules to be established pertaining to the dotted terminals of Fig. 4.8b:

1. Voltages from dotted to undotted terminals are in phase.
2. Current into a dotted terminal is in phase with the current out of a dotted terminal.
3. Both dots can be moved to opposite terminal without invalidating the above rules.

4.3.3 POWER AND IMPEDANCE RELATIONSHIPS

The reciprocal of [4.11] can be equated to [4.4] to conclude that the instantaneous input and output powers are identical for the ideal transformer, thus corroborating postulate 1 for the ideal transformer.

$$v_1 i_1 = v_2 i_2 \qquad \textbf{[4.13]}$$

The implications of [4.13] are manyfold and worthy of explicit recognition. If the voltages and currents of [4.13] are periodic time functions, then the average values of input and output power are equal. Since the analogous phasor domain ratios of [4.9] and [4.12] could have been used to form a frequency domain equivalent of [4.13] as

$$\overline{V}_1 \overline{I}_1 = \overline{V}_2 \overline{I}_2 \qquad \textbf{[4.14]}$$

it is apparent that the complex powers flowing into and out of the ideal transformer are identical. Since average powers and complex powers are equal, reactive powers flowing into and out of the ideal transformer must be equal.

Example 4.2

For the ideal transformer of Fig. 4.8a, let $v_1(t) = \sqrt{2}\,V_s\cos(\omega t)$, $i_1(t) = \sqrt{2}\,I_s\cos(\omega t - 30°)$, and $a = 2$. Find an expression for $p_2(t)$, the instantaneous power flow out of the ideal transformer.

Based on [4.4] and [4.11],

$$v_2 = v_1/a = \frac{\sqrt{2}\,V_s}{2}\cos(\omega t) \qquad i_2 = ai_1 = 2\sqrt{2}\,I_s\cos(\omega t - 30°)$$

Hence,

$$p_2(t) = v_2 i_2 = 2V_s I_s\cos(\omega t)\cos(\omega t - 30°)$$

$$= V_s I_s\cos(30°) + V_s I_s\cos(2\omega t - 30°)$$

Example 4.3

Determine the average power (P) and reactive power (Q) flowing into the ideal transformer of Example 4.2 if $V_s = 125$ (V) and $I_s = 10$ (A).

Since input and output average power flows are identical, the expression for $p_2(t)$ from Example 4.2 could be averaged for a period T to yield

$$P = \frac{1}{T}\int_0^T p_2(t)dt = V_s I_s\cos(30°) = (125)(10)\left(\frac{\sqrt{3}}{2}\right) = 1082.53 \text{ W}$$

As an alternate computation method, a phasor domain approach could be taken.

$$\overline{V}_s = 125\angle 0° \text{ V} \qquad \overline{I}_s = 10\angle{-30}° \text{ A}$$

$$P = V_s I_s\cos\theta = (125)(10)\cos[0° - (-30°)] = 1082.53 \text{ W}$$

The resulting reactive power is

$$Q = V_s I_s\sin\theta = (125)(10)\sin[0° - (-30°)] = 625 \text{ VARs}$$

Having established that voltages and currents undergo a magnitude transformation by the transformer, it is logical to explore the impedance transformation nature of the ideal transformer under sinusoidal steady-state operation. Let an impedance Z_2 be connected across the output terminals of the ideal transformer of Fig. 4.8b. Then, by use of [4.9] and [4.12], the impedance as seen from the input terminals is

$$Z_1 = \frac{\overline{V}_1}{\overline{I}_1} = \frac{a\overline{V}_2}{\overline{I}_2/a} = a^2\frac{\overline{V}_2}{\overline{I}_2} = a^2 Z_2 \qquad\qquad \textbf{[4.15]}$$

Thus, an impedance on one side of an ideal transformer appears to have a value that is adjusted by the direct turns ratio squared on the opposite side of the transformer. Equation [4.15] is quite useful when reflecting loads and line impedance to a common side of the transformer for analysis purposes. Although it has no direct application in power systems work, the impedance reflection characteristic of [4.15] is commonly used in the electronics area to match a load impedance with amplifier internal impedance for conditions of maximum power transfer. In such an application, the amplification performance benefit outweighs the resulting 50 percent efficiency.

4.4 THE PRACTICAL TRANSFORMER

Infinitely permeable, lossless ferromagnetic material does not exist. Practical magnet wire with negligible resistance for coil winding is not a reality. Although the ideal transformer may be approached, it cannot be attained. The small departures from perfection, in all but a few applications, have sufficient impact on performance that the effects must be considered for accurate analysis.

4.4.1 LOSSLESS-CORE TRANSFORMER

With proper parameter specification, the pair of coupled coils in Fig. 4.10a have all the properties of a practical transformer except that the mutual flux coupling path has no magnetic losses. The coil pair is modeled by the circuit of Fig. 4.10b using controlled sources to represent the voltage induced behind one coil by a change in current of the other coil. Each coil has a resistance (R), a self-inductance (L), and an associated mutual inductance (M). The p-operator indicates time differentiation.

Apply KVL around the left and right loops, respectively, in Fig. 4.10b. If an arbitrarily chosen quantity is added and then subtracted to the left-hand side of each equation, the equality is maintained. The resulting expressions are

$$v_1 = R_1 i_1 + L_1 p i_1 - M p i_2 + a M p i_1 - a M p i_1 \qquad \textbf{[4.16]}$$

$$v_2 = -R_2 i_2 - L_2 p i_2 + M p i_1 + \frac{1}{a} M p i_2 = \frac{1}{a} M p i_2 \qquad \textbf{[4.17]}$$

Rearrange [4.16] and [4.17] and define e_1 and e_2 to give

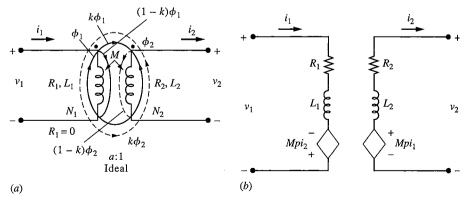

(a)

(b)

Figure 4.10
Transformer with lossless core. (a) Schematic. (b) Equivalent circuit.

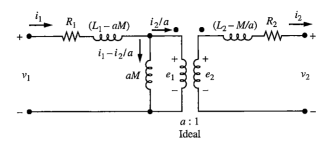

Figure 4.11
Time domain equivalent circuit of lossless-core transformer

$$v_1 = R_1 i_1 + (L_1 - aM)pi_1 + e_1 \qquad \textbf{[4.18]}$$

where

$$e_1 = aMp(i_1 - i_2/a) \qquad \textbf{[4.19]}$$

and

$$v_2 = -R_2 i_2 - (L_2 - M/a)pi_2 + e_2 \qquad \textbf{[4.20]}$$

where

$$e_2 = e_1/a = Mp(i_1 - i_2/a) \qquad \textbf{[4.21]}$$

If the arbitrary constant a is selected as the turns ratio ($a = N_1/N_2$) of an ideal transformer, then the equivalent circuit of Fig. 4.11 simultaneously satisfies [4.18] to [4.21].

Definitions of self- and mutual inductances given by [3.22] and [3.29] can be used to offer an interpretation of the inductances of the equivalent circuit of Fig. 4.11. Let $i_2 = 0$ and $i_1 \neq 0$; then

$$L_{\ell 1} = L_1 - aM = \frac{N_1 \phi_1}{i_1} - \frac{N_1}{N_2}\frac{N_2 k\phi_1}{i_1} = (1 - k)\frac{N_1 \phi_1}{i_1} = (1 - k)L_1$$

$$\textbf{[4.22]}$$

Now if coil 2 were perfectly coupled by the flux established by coil 1, the coefficient of coupling $k = 1$ and $L_1 - aM = 0$. Obviously, $L_1 - aM$ exists only when leakage flux is present; thus, it is known as the leakage inductance of coil 1.

Similarly, let $i_1 = 0$ and $i_2 \neq 0$ to find

$$L_{\ell 2} = L_2 - \frac{1}{a}M = \frac{N_2 \phi_2}{i_2} - \frac{N_2}{N_1}\frac{N_1 k\phi_2}{i_2} = (1 - k)\frac{N_2 \phi_2}{i_2} = (1 - k)L_2 \quad \textbf{[4.23]}$$

whence it is apparent that $L_2 - M/a$ exists only in the presence of leakage flux and is rightfully called the leakage inductance of coil 2.

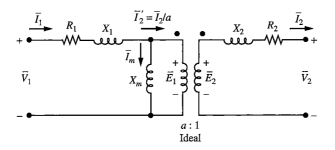

Figure 4.12
Phasor domain equivalent circuit of lossless-core transformer

Consider the shunt inductance aM if $i_2 = 0$ and $i_1 \neq 0$.

$$L_m = aM = \frac{N_1}{N_2} \frac{N_2 k\phi_1}{i_1} = k\frac{N_1\phi_1}{i_1} = kL_1 \qquad \textbf{[4.24]}$$

The inductance L_m is an inductance associated with the portion of ϕ_1 that traverses the complete path of the lossless core linking coil 2 and establishes a coupling magnetic field within the core. It is known as a *magnetizing inductance*.

The equivalent circuit of Fig. 4.11 is valid for any impressed voltage $v_1(t)$. The dominant use of the transformer is in applications where sinusoidal steady-state analysis is pertinent. The reactances associated with the leakage inductances and magnetizing inductances are formed as

$$X_1 = \omega L_{\ell 1} = \omega(L_1 - aM) \qquad \textbf{[4.25]}$$

$$X_2 = \omega L_{\ell 2} = \omega(L_2 - M/a) \qquad \textbf{[4.26]}$$

$$X_m = \omega L_m = \omega aM \qquad \textbf{[4.27]}$$

Along with these reactances, voltages and currents are expressed as phasors to change the time domain equivalent circuit of Fig. 4.11 to the sinusoidal steady-state equivalent circuit of Fig. 4.12.

4.4.2 FERROMAGNETIC CORE PROPERTIES

Introduction of a ferromagnetic core into the transformer flux path increases manyfold the energy that is converted from electrical to magnetic form per ampere of coil current. With this advantage comes the disadvantage that both losses and harmonic current distortion occur which are not accounted for in the circuit model of Fig. 4.12. After gaining some understanding of these changes, logical modification of the circuit will be made to account for the new losses. Assessment of the harmonic impact will be carried out to justify their neglect under most operating conditions.

Ferromagnetic material saturation phenomena were discussed in Sec. 3.3.1. It was pointed out that alignment of atom magnetic moments within the crystalline

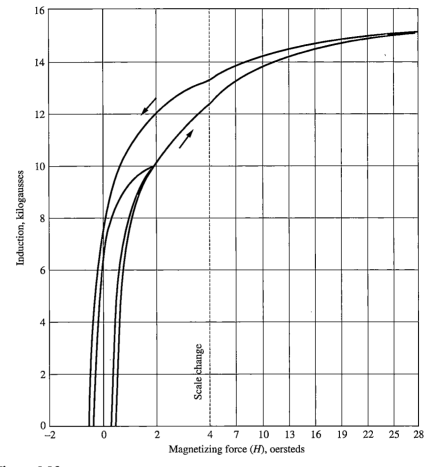

Figure 4.13
Hysteresis loops (upper halves)

structure reinforces any magnetic field impressed along the ferromagnetic material.[1] Not all energy expended in this magnetic moment alignment is recoverable, but rather a portion is dissipated as heat in the process. As a direct consequence, not all atoms return to the exact orientation that existed before application of the external magnetic field upon removal of this field. Under cyclic application of a symmetric forward and reverse magnetic field to a specimen of ferromagnetic material, its *B-H* plot traces out a *hysteresis loop.*

The graph of Fig. 4.13 displays the characteristic shape for the upper half of the hysteresis loop. The complete loop displays odd symmetry with a trace that depends upon the maximum level of flux density attained. A loop for a lesser maximum level of maximum flux density will be embedded inside a loop for a greater level of maximum flux density. Dynamically, the hysteresis loop is traversed in a counterclockwise direction, as indicated by the arrows of Fig. 4.13. As discussed in Sec. 3.5.2, the area to the left of the *B-H* plot is energy density. Since the area to the left of the ascending

Hysteresis loops

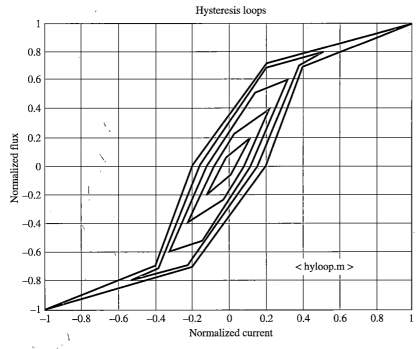

Figure 4.14
Piecewise linear hysteresis loops

portion of the hysteresis loop is greater than the area to the left of the descending portion, it can be concluded that the area inside the hysteresis loop is equal to the energy density expended as heat loss per cycle of the magnetization process.

The MATLAB program ⟨hyloop.m⟩ generates a family of piecewise linear, normalized hysteresis loops in units of flux vs. current, clearly indicating the embedded nature of loops. The number of loops is specified by the value of *nloops*. Loop widths are determined by choice of the maximum coercive current *Icmax*. The knee of the hysteresis loops is determined by specification of the saturation flux density (*puFsat*) and the associated saturation excitation current (*puIsat*). Figure 4.14 displays a typical result where ⟨hyloop.m⟩ was executed with *Ic*max = 0.2, *puFsat* = 0.7, *puIsat* = 0.3, and *nloops* = 5. In addition, the program calculates the per unit energy loss W_{mpu} of each loop and displays the values to the screen. The actual value of energy loss per cycle W_m is given if this per unit value were multiplied by the product of base flux (Wb), base current (A), and coil turns (N).

A 100-turn (N) coil is wound around a ferromagnetic core with base or saturation flux density (B_{sat}) of 1.5 T occurring with a coil excitation (I_{sat}) of 5 A (base current). The saturation curve knee is located at 70 percent base flux and 30 percent base current. Cross-section area of the ferromagnetic core is 40 cm². Maximum coercive current is 1 A. Use ⟨hyloop.m⟩ to predict the average value of hysteresis power loss (P_h) when the coil is excited by a 60-Hz source resulting in a cyclic peak flux density of 1.2 T.

Example 4.4

The values of *Icmax* (0.2), *puFsat* (0.7), *puIsat* (0.3) are correct in ⟨hyloop.m⟩ without change. The peak per unit flux density is 1.2/1.5 = 0.8; thus, with the existing value of *nloops* = 5, the fourth value of per unit energy below as displayed to the screen upon executing ⟨hyloop.m⟩ is applicable.

```
   Per Unit Energy Loss of Hysteresis Loop (min -> max)
ans =
    0.0199     0.0901     0.2145     0.3520     0.4800
```

$$P_h = W_m f = N B_{sat} A I_{sat} W_{mpu} f$$

$$P_h = (100)(1.5)(40 \times 10^{-4})(5)(0.3520)(60) = 63.36 \text{ W}$$

In addition to the heating loss associated with the hysteresis phenomenon, the nonlinear ϕ-i relationship results in distortion of either the coil voltage or the coil current, depending on whether the excitation is considered a current source or voltage source. The MATLAB program ⟨hyst.m⟩ uses the principles of ⟨hyloop.m⟩ for constructing a single piecewise linear hysteresis loop for an interactive input Percentage maximum flux. After displaying a plot of the generated hysteresis loop for inspection as shown by Fig. 4.15, the program then forms a normalized sinusoidal current and uses the look-up table feature of MATLAB to generate the corresponding array of flux as the hysteresis loop is traversed over a cycle of current. The flux array is then time differentiated to form the coil-induced voltage in accordance with Faraday's law. The normalized current, flux, and voltage are then plotted for inspection. To specifically quantize the voltage distortion, its Fourier spectrum is then plotted. Figure 4.16 displays the resulting normalized flux and voltage and the associated voltage spectrum for the case of an impressed sinusoidal excitation current with value to yield 100 percent saturation.

In power transformers, the realistic excitation is an impressed fundamental frequency sinusoidal voltage which can only support a fundamental frequency sinusoidal flux. The latter part of the analysis made by ⟨hyst.m⟩ deals with this case. A normalized sinusoidal flux is formed with peak value equal to the user-specified maximum flux amplitude for the hysteresis loop. Through use of the interpolation feature, the current array resulting from traversing the hysteresis loop over a cycle is generated. The flux array is time differentiated to form the necessary excitation voltage. The normalized current, flux, and voltage are plotted for inspection followed by a Fourier spectrum of the current to assess the introduced distortion. In addition, the magnitude and phase angle of the fundamental component of the coil current is determined from the Fourier spectrum. It is then converted back to the time domain and added to the plot of total coil current, flux, and voltage. Results

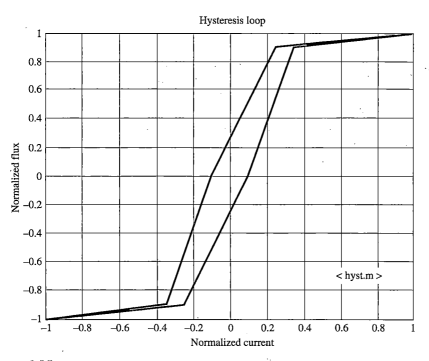

Figure 4.15
Normalized hysteresis loop

of the response due to sinusoidal voltage excitation for the case of an impressed voltage of magnitude to yield 100 percent saturation are depicted by Fig. 4.17.

Since average power flow can only take place for voltage and current of the same frequency, it can be concluded that the hysteresis power loss for the core modeled by Fig. 4.17 can be calculated by knowledge of the coil voltage and the fundamental component of coil current. Since the phase shift of the fundamental component of coil current is approximately $81° < 90°$ for the specific example, a small, but nonzero, average power flow is indicated. If the fundamental component of current is resolved into a power producing component (i_{hc}) that is in phase with the coil voltage (v_c) and a component with $90°$ lagging phase shift from v_c, then the model of Fig. 4.18 is justified where resistor R_h carries the average power-producing component of coil current. Current i_m flowing through L_m is the total coil current less i_{hc}.

An additional power loss mechanism exists in a ferromagnetic core with cyclic magnetization. In accordance with Faraday's law, distributed voltages are induced along effective short-circuit loops. As a consequence, *eddy currents* flow around these loops within the core material along planes perpendicular to the flux paths. These paths are shortened to reduce the eddy current losses by laminating the ferromagnetic

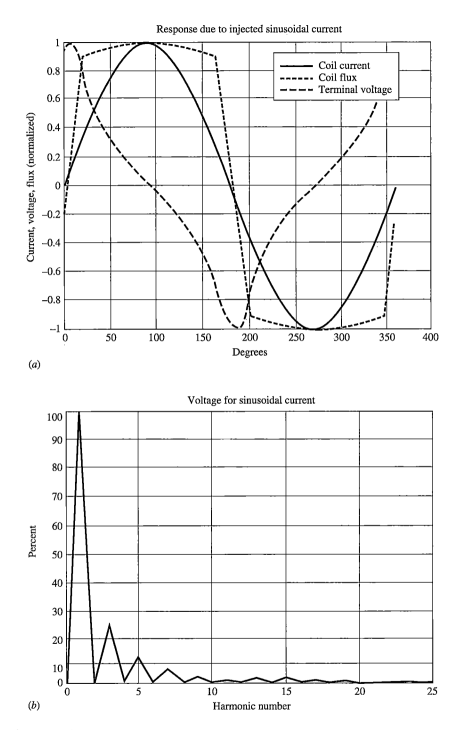

Figure 4.16
Hysteretic response of ferromagnetic core with sinusoidal current excitation. (a) Time domain response. (b) Fourier spectrum of induced voltage.

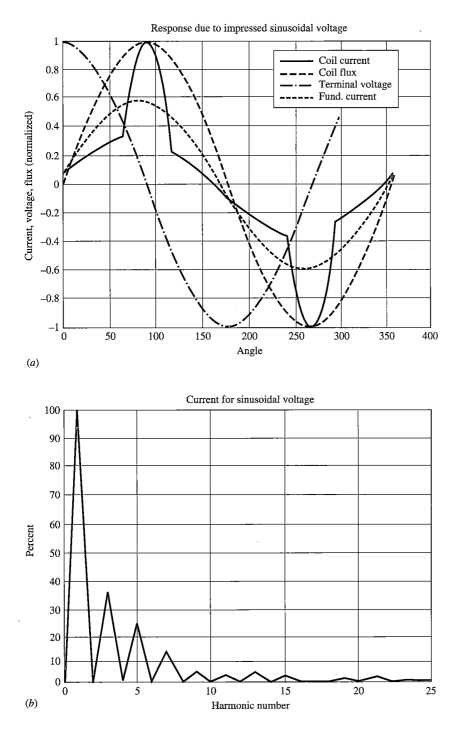

Figure 4.17
Hysteretic response of ferromagnetic core with sinusoidal voltage excitation. (a) Time domain response. (b) Fourier spectrum of current.

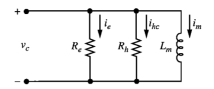

Figure 4.18
Model of ferromagnetic core with losses

material along the direction of flux flow. Eddy current volume power density for the case of a sinusoidal magnetic field is described by the well-known relationship[1]

$$p_e = k_e B_m^2 f^2 \qquad \qquad [4.28]$$

where k_e is a material-dependent constant, B_m is the peak flux density of the sinusoidal varying B-field, and f is the frequency. Since B_m is directly proportional to V_c/f where V_c is the rms value of v_c, [4.28] can be written as

$$p_e = k_e' V_c^2 = V_c^2/R_e \qquad \qquad [4.29]$$

Equation [4.29] suggests that the eddy current losses can also be modeled by a resistor as indicated in Fig. 4.18.

For the moment, neglect the harmonic currents flowing through L_m in Fig. 4.18. Under this assumption, L_m would be identically the magnetizing inductance associated with the magnetizing reactance (X_m) of Fig. 4.12. If the parallel combination of eddy current resistor R_e and hysteresis resistor R_h is denoted by core loss resistor R_c, then the exact equivalent circuit model for the transformer shown in Fig. 4.19 results. Based on the development of core loss resistance R_c, a particular value would be dependent on both flux density and frequency.

The sinusoidal steady-state, practical transformer equivalent circuit of Fig. 4.19 has been completely justified except for the question of error introduced by neglect of the harmonic currents flowing through the magnetizing inductance. The MATLAB program ⟨imthd.m⟩ is introduced to handle the somewhat tedious task of quantitatively assessing the approximation error. Three reasonable approximations are used in the computation results of ⟨imthd.m⟩: voltage drop across the primary resistance and leakage inductance can be neglected; full-load core losses are equal to 2 percent of the load power; and the current required to establish the fundamental component of mutual flux is 3 percent of the rated full-load current. Since the load power factor can slightly alter results, it is considered in calculations. The program uses the harmonic profile of magnetizing current (I_m) previously calculated by ⟨hyst.m⟩ as displayed by Fig. 4.17a, for all evaluations. Hence, ⟨hyst.m⟩ should be executed with the desired hysteresis loop characteristics and the desired level of saturation prior to use of ⟨imthd.m⟩ so that the appropriate magnetizing current profile is stored.

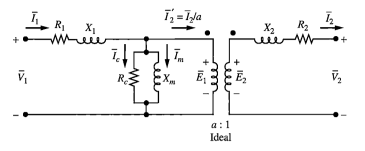

Figure 4.19
Exact equivalent circuit for transformer in the sinusoidal steady state

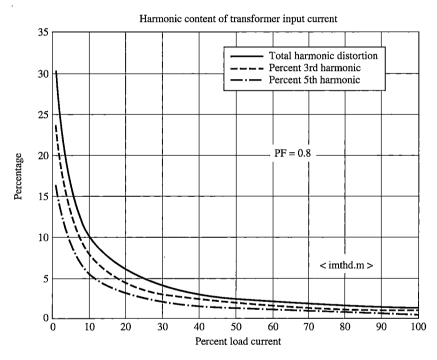

Figure 4.20
Harmonic content of transformer input current

Harmonic assessment of the transformer input current is presented in two forms by ⟨imthd.m⟩. First, the percentage magnitude of the load current for the two most significant harmonics (third and fifth) is determined. Second, the total harmonic distortion considering all harmonics through the thirty-first is computed. Figure 4.20 shows the harmonic assessments as a function of load current for the typical transformer load characteristic of 0.8 PF lagging. Obviously, a lightly loaded transformer has input current with significant harmonic distortion if the core is operated such that magnetic saturation exists over a portion of the field excursion. However, power level transformers are seldom operated at near no-load conditions. For load current greater than 25 percent of rated, the total harmonic distortion is less than 5 percent. The typical loading range of 50 to 100 percent results in total harmonic distortion of less than 2.5 percent, a value that is acceptable in power system operation. Considering the relatively small amplitudes of the harmonic currents compared to the fundamental currents, neglect of the harmonic currents in transformer analysis has negligible impact on the accuracy of results. Hence, the fundamental frequency, sinusoidal steady-state equivalent circuit of Fig. 4.19 is considered a model within engineering accuracy for all analysis work wherein the terminal voltage does not exceed rated value.

4.4.3   GOODNESS OF PERFORMANCE IMPLICATIONS

Although the equivalent circuit of Fig. 4.19 has been established as a reasonable model for the ferromagnetic core transformer, specific values of the parameters for that circuit have not been addressed. Test work to determine appropriate values will be discussed in Sec. 4.5; however, accuracy justification for that test work will depend upon a general relative size relationship declared to be typical for transformers of good design. The relative size relationships follow from performance criteria, as the succeeding examination will show.

Since the primary and secondary windings of a transformer are interleaved, the lengths of the average turn (*mean length turn*) ℓ_m of both the primary and secondary windings are nearly equal. The total lengths of the primary and secondary winding are then

$$\ell_1 = N_1 \ell_m \qquad \ell_2 = N_2 \ell_m \qquad\qquad \textbf{[4.30]}$$

For reasons of cooling and minimum losses, the winding current density $(J = I/A_c)$ should be a low value. On the other hand, reduced conductor material to minimize cost suggests increased current density. It is reasonable to assume that the compromise value of current density for a particular transformer is applied to both coils, or equating current densities and using [4.11] gives

$$1 = \frac{J_1}{J_2} = \frac{I_1/A_{c1}}{I_2/A_{c2}} = \frac{I_1 A_{c2}}{I_2 A_{c1}} = \frac{N_2 A_{c2}}{N_1 A_{c1}} \qquad\qquad \textbf{[4.31]}$$

The ratio of coil resistances is formed under the assumption of identical resistivities (identical conductor materials and temperatures), and [4.30] and [4.31] are used to yield

$$\frac{R_1}{R_2} = \frac{\rho \ell_1/A_{c1}}{\rho \ell_2/A_{c2}} = \frac{\ell_1}{\ell_2}\frac{A_{c2}}{A_{c1}} = \frac{N_1 \ell_m}{N_2 \ell_m}\frac{N_1}{N_2} = \left(\frac{N_1}{N_2}\right)^2 = a^2 \qquad \textbf{[4.32]}$$

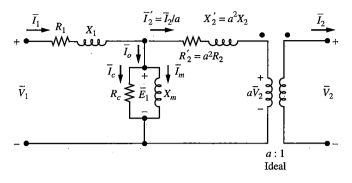

Figure 4.21
Exact equivalent circuit of transformer in primary frame of reference

Since the flux leakage paths for the two coils are identical, the reluctances are equal; hence, the ratio of leakage reactances is formed as

$$\frac{X_1}{X_2} = \frac{\omega L_{\ell 1}}{\omega L_{\ell 2}} = \frac{N_1^2/\Re}{N_2^2/\Re} = \left(\frac{N_1}{N_2}\right)^2 = a^2 \qquad \textbf{[4.33]}$$

Based on the impedance reflection property of an ideal transformer as determined by [4.15], equations [4.32] and [4.33] indicate that

$$R_1 = a^2 R_2 = R_2' \qquad X_1 = a^2 X_2 = X_2' \qquad \textbf{[4.34]}$$

Therefore, the secondary resistance and leakage reactance can be reflected across the ideal transformer in the equivalent circuit of Fig. 4.19 to give the transformer equivalent circuit of Fig. 4.21, where all resistances and reactances are now expressed in the primary frame of reference.

A good transformer design should have the two following performance attributes:

1. High efficiency—a typical value being 98 percent.
2. Small percentage voltage drop from input to output—a typical value being 3 percent.

Normalized or per unit values are introduced as below:

$$\begin{aligned} \textit{Base voltage } V_B &= \text{ rated voltage} \\ \textit{Base current } I_B &= \text{ rated current} \\ \textit{Base power } S_B &= V_B I_B \\ \textit{Base impedance } Z_B &= V_B / I_B \end{aligned}$$

Referring to Fig. 4.21, a reasonable interpretation of the performance criteria is

1. Power loss by each of R_1 and R_2' no more than 0.5 percent of full-load power
2. Power loss by R_c no more than 1 percent of full-load power
3. Voltage drop across each of $R_1 + jX_1$ and $R_2' + jX_2'$ no more than 1.5 percent of rated voltage
4. Magnetizing current I_m no more than 3 percent of rated current

Working across the transformer equivalent circuit of Fig. 4.21 while using the above interpretation of the performance criteria for $I_1 = I_B$ and $V_1 = V_B$ leads to the relative size conclusions for the equivalent circuit parameters.

1. $$I_1^2 R_1 = I_B^2 R_1 \leq 0.005 S_B = 0.005 V_B I_B$$

   or

   $$R_1 \leq 0.005 \frac{V_B}{I_B} = 0.005 Z_B$$

2. $$I_B |R_1 + jX_1| = I_B Z_1 \leq 0.015 V_B$$

   or

   $$Z_1 \leq 0.015 \frac{V_B}{I_B} = 0.015 Z_B$$

3. If condition 2 is satisfied, $E_1 \cong V_1 = V_B$ and the core losses are

   $$P_c = \frac{E_1^2}{R_c} \cong \frac{V_B^2}{R_c} \leq 0.01 S_B$$

   or

   $$R_c \geq 100 \frac{V_B^2}{S_B} = 100 \frac{V_B}{I_B} = 100 Z_B$$

   whence

   $$I_c \leq \frac{V_B}{100 Z_B} = 0.01 I_B$$

4. For $I_m \leq 0.03 I_B$,

   $$X_m \doteq \frac{E_1}{I_m} \cong \frac{V_B}{I_m} \geq \frac{V_B}{0.03 I_B} = 33.3 Z_B$$

5. It was established earlier in this section that $R_2' \cong R_1$ and $X_2' \cong X_1$.

It is now apparent that R_c and X_m are two orders of magnitude larger than R_1, R_2', X_1, and X_2'. With relative sizes of the equivalent circuit parameters established,

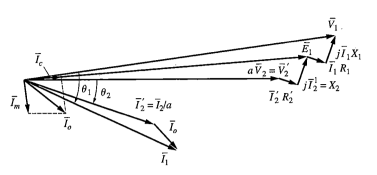

Figure 4.22
Phasor diagram for transformer modeled by the equivalent circuit of Fig. 4.21

the complete phasor diagram for the equivalent circuit of Fig. 4.21 can be drawn as displayed by Fig. 4.22.

Example 4.5

For a transformer with primary rated voltage and current of 240 V and 20.83 A, calculate a set of typical values for the primary reference frame equivalent circuit of Fig. 4.21 based on the above interpretation of performance criteria.

Base quantities are

$$V_B = 240 \text{ V} \qquad\qquad I_B = 20.83 \text{ A}$$

$$Z_B = \frac{V_B}{I_B} = \frac{240}{20.83} = 11.52 \ \Omega \qquad S_B = V_B I_B = (240)(20.83) = 5000 \text{ W}$$

The equivalent circuit values can now be calculated.

$$R_1 = R_2' \leq 0.005 Z_B = 0.005(11.52) = 0.0576 \ \Omega$$

$$Z_1 = Z_2' \leq 0.015 Z_B = 0.015(11.52) = 0.1728 \ \Omega$$

$$X_1 = X_2' \leq \sqrt{Z_1^2 - R_1^2} = \sqrt{(0.1728)^2 - (0.0576)^2} = 0.1629 \ \Omega$$

$$R_c \geq 100 Z_B = 100(11.52) = 1152 \ \Omega$$

$$X_m \geq 33.3 Z_B = 33.3(11.52) = 383.62 \ \Omega$$

4.4.4 NAMEPLATE AND COIL POLARITY

The manufacturer nameplate gives as minimum information voltage ratings, frequency rating, and apparent power rating for a transformer. Larger transformer nameplates frequently have information on temperature rise, percent leakage reactance, connection diagrams, serial number, weight, insulation class, and coolant information if liquid-cooled. American National Standards Institute (ANSI) markings

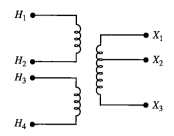

Figure 4.23
ANSI markings

of leads are used for larger transformers. $(H_1 \cdots H_n)$ and $(X_1 \cdots X_n)$ markings indicate high-voltage and low-voltage coil leads, respectively. A sample set of transformer lead markings are shown in Fig. 4.23, where the high-voltage H-coils can be connected in either parallel or series; thus, this transformer would have a dual voltage rating for the high-voltage winding. The low-voltage winding has a tapped winding in this case. For purposes of polarity markings, a dot can be consistently placed by the winding label of each coil with either the smallest or largest subscript.

In case a transformer does not have ANSI terminal markings, the dotted terminals can be established by a simple test as illustrated by Fig. 4.24. The sinusoidal source V_s can have a magnitude much less than rated value, if desired for safety reasons. With the transformer energized by $\bar{V}_s$, voltage readings V_1, V_2, and V_3 are recorded. Voltmeter reading $V_3 = |\bar{V}_1 - \bar{V}_2|$. If $\bar{V}_1$ and $\bar{V}_2$ are in phase, $V_3 = V_1 - V_2$, and the second dot is placed at the top terminal of the second coil. Otherwise, $\bar{V}_1$ and $\bar{V}_2$ are 180° out of phase, $V_3 = V_1 + V_2$, and the second dot is placed at the bottom terminal of the second coil.

4.5 TEST DETERMINATION OF PARAMETERS

Study to this point has clearly established the equivalent circuit topology and the physical meaning of its elements. A procedure to determine typical values for equivalent circuit elements based on performance criteria has been developed in

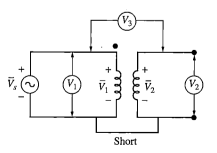

Figure 4.24
Polarity test

Sec. 4.4.3. Although these values are reasonable approximations that are valid for relative size comparisons and approximate performance prediction, more exact values are required for accurate performance analysis of a specific transformer. Reduction of data from a series of laboratory tests can yield the actual equivalent circuit values for a particular transformer.

4.5.1   PRELIMINARY TESTS

If the primary and secondary voltage ratings are stated on the nameplate, then the ratio of these voltages can be used for the turns ratio in accordance with [4.4]. In the absence of nameplate rated voltages, any convenient, safe, good-quality sinusoidal voltage below saturation level can be impressed on one of the windings to be tested. The ratio of the impressed voltage magnitude to the magnitude of the open-circuit voltage recorded at the terminals of the second coil to be tested is the direct turns ratio from the source winding to the open circuit winding. Normally this test would be performed at rated frequency, but that is not an essential condition.

In addition to the turns ratio determination, the dc resistance of the primary and secondary winding should be accurately measured. Since available wire size choices may have led to small differences in current density of the primary and secondary winding, the dc values can be used to apportion the later determined equivalent resistance between R_1 and R_2.

4.5.2   SHORT-CIRCUIT TEST

The terminals of one winding are connected short-circuit while a reduced voltage is applied across the terminals of the other winding under test. The frequency of the impressed voltage should be the frequency at which the transformer equivalent circuit is to be applied. Although reactances can be easily adjusted for frequency differences, a slight skin effect and nonuniform current distribution within a conductor act to increase the resistance of a coil a few percent above the dc value. This resistance variation is a nonlinear function of frequency that cannot be simply ratioed to determine the value for another frequency. Common practice is to perform the short-circuit test at approximately the value of rated current. If the current is not maintained at rated value sufficiently long for the windings to heat to rated temperature, then the synthesized values of R_1 and R_2 must be adjusted to rated temperature.

The choice of winding to energize, although arbitrary, is commonly the high-voltage winding, since its rated current value is less than that of the low-voltage winding by a factor of $1/a$. Consequently, an excitation source of lesser current capacity is adequate if the high-voltage winding is energized. After application of excitation sufficiently long to attain normal winding temperatures, the data recorded at the energized winding are power (P_{SC}), voltage (V_{SC}), and current (I_{SC}).

From the equivalent circuit of Fig. 4.21 with the secondary winding terminals shorted $(V_2 = aV_2 = 0)$, it is apparent that the impedance seen by a source connected to energize the primary is given by

$$Z_{SC} = R_1 + jX_1 + R_c\|jX_m\| + a^2(R_2 + jX_2) \qquad \textbf{[4.35]}$$

In Sec. 4.4.3, it was concluded that for good design R_c and X_m are two orders of magnitude greater than a^2R_2 and a^2X_2. Thus, with only small error, [4.35] can be written as

$$Z_{SC} \cong (R_1 + a^2R_2) + j(X_1 + a^2X_2) = R_{eq} + jX_{eq} \qquad \textbf{[4.36]}$$

Hence, the data recorded allows the synthesis of the series R_{eq} and X_{eq}.

$$R_{eq} = \frac{P_{SC}}{I_{SC}^2} \qquad \textbf{[4.37]}$$

$$Z_{SC} = \frac{V_{SC}}{I_{SC}} \qquad \textbf{[4.38]}$$

$$X_{eq} = \sqrt{Z_{SC}^2 - R_{eq}^2} \qquad \textbf{[4.39]}$$

Based on [4.34],

$$X_1 = X_2' = \frac{1}{2}X_{eq} \qquad \textbf{[4.40]}$$

Without knowledge of the coil dc resistance values, R_{eq} could be halved to determine R_1 and R_2'. However, with the coil dc resistance values known, R_1 and R_2' can be apportioned to account for any current density difference between the primary and secondary winding by the following procedure:

$$R_1 = \frac{R_{1dc}}{R_{1dc} + a^2R_{2dc}}R_{eq} \qquad \textbf{[4.41]}$$

$$R_2' = R_{eq} - R_1 \qquad \textbf{[4.42]}$$

4.5.3   OPEN-CIRCUIT TEST

The terminals of one winding remain open-circuit while a voltage is impressed across the terminals of the opposite coil. The impressed voltage should be of rated magnitude and of rated frequency for the excited coil. R_c and X_m are determined from this test data. X_m is sensitive to any core saturation, and core loss resistance R_c is sensitive to both core flux density and frequency. Further, the eddy current portion of core losses is temperature-sensitive through the resistivity of the ferromagnetic core material. Good test practice is to maintain the excitation until the core temperature has elevated to near operating temperature for accurate determination of R_c.

The choice of winding to energize is arbitrary; however, the low-voltage winding is a common choice, since its selection allows the safety of recording data while dealing with lesser values of voltage than if the high-voltage winding were chosen for excitation. Equally important in consideration is the fact that sensitivity of secondary amperes per primary volt is lower when the high-voltage winding is excited. Consequently, the burden of control of the small applied voltage is eased.

After attaining the normal core temperature, the data recorded for the winding under test are power (P_{OC}), voltage (V_{OC}), and current (I_{OC}).

From the equivalent circuit of Fig. 4.21 with $I_2 = I'_2 = 0$, it is concluded that the impedance seen by the exciting source is

$$Z_{OC} = R_1 + jX_1 + R_c\|jX_m \qquad \textbf{[4.43]}$$

Since it was concluded in Sec. 4.4.3 that if good design prevails, $(R_c, X_m) \gg (R_1, X_1)$, then with little error [4.43] can be written as

$$Z_{OC} \cong R_c\|jX_m \qquad \textbf{[4.44]}$$

The approximation of [4.44] implies that $V_{oc} \cong E_1$. The recorded data then allow synthesis of the parallel connected R_c and X_m.

$$R_c = \frac{V_{OC}^2}{P_{OC}} \qquad \textbf{[4.45]}$$

From Fig. 4.21, it is seen that $\overline{I}_c$ and $\overline{I}_m$ must be orthogonal; thus, $I_o^2 = I_c^2 + I_m^2$. X_m follows as

$$X_m = \frac{V_{OC}}{I_m} = \frac{V_{OC}}{\sqrt{I_{OC}^2 - (V_{OC}/R_c)^2}} \qquad \textbf{[4.46]}$$

The following 60-Hz data were recorded for a 480:240-V, 25-kVA, 60-Hz transformer with windings and core at approximately normal operating temperatures: | **Example 4.6**

Short circuit (high side):

$$V_{SC} = 37.2\,\text{V} \qquad I_{SC} = 51.9\,\text{A} \qquad P_{SC} = 750\,\text{W}$$

Open circuit (low side):

$$V_{OC} = 240\,\text{V} \qquad I_{OC} = 9.7\,\text{A} \qquad P_{OC} = 720\,\text{W}$$

In addition, the dc resistances of the two windings were measured and the values found to be $R_{1dc} = 0.110\,\Omega$ and $R_{2dc} = 0.029\,\Omega$. Determine the values of the equivalent circuit elements for the circuit of Fig. 4.19.

By use of [4.37] to [4.42],

$$R_{eq} = \frac{P_{SC}}{I_{SC}^2} = \frac{750}{(51.9)^2} = 0.2784\,\Omega$$

$$Z_{SC} = \frac{V_{SC}}{I_{SC}} = \frac{37.2}{51.9} = 0.7168$$

$$X_{eq} = \sqrt{Z_{SC}^2 - R_{eq}^2} = \sqrt{(0.7168)^2 - (0.2784)^2} = 0.6605$$

$$X_1 = X'_2 = \frac{1}{2}X_{eq} = \frac{1}{2}(0.6605) = 0.3303\,\Omega$$

$$a = \frac{V_1}{V_2} = \frac{480}{240} = 2$$

$$X_2 = \frac{X_2'}{a^2} = \frac{0.3303}{2^2} = 0.0826 \ \Omega$$

$$R_1 = \frac{R_{1dc}}{R_{1dc} + a^2 R_{2dc}} R_{eq} = \frac{0.110}{0.110 + (2)^2(0.029)}(0.2784) = 0.1355 \ \Omega$$

$$R_2' = (R_{eq} - R_1) = 0.2784 - 0.1355 = 0.1429$$

$$R_2 = \frac{R_2'}{a^2} = \frac{0.1429}{(2)^2} = 0.03573$$

Since the open-circuit test was performed on the low-voltage winding, the values of voltage and current used in [4.45] and [4.46] must be reflected to the high-voltage side.

$$R_c = \frac{(aV_{OC})^2}{P_{OC}} = \frac{(2 \times 240)^2}{720} = 320 \ \Omega$$

$$X_m = \frac{aV_{OC}}{\sqrt{(I_{OC}/a)^2 - (aV_{OC}/R_c)^2}} = \frac{a^2 V_{OC}}{\sqrt{I_{OC}^2 - a^4(V_{OC}/R_c)^2}}$$

$$X_m = \frac{(2)^2(240)}{\sqrt{(9.7)^2 - (2)^4(240/320)^2}} = 104.1 \ \Omega$$

Reduction of the data to calculate the equivalent circuit element values has introduced some approximations that are logical to question. The open-circuit data reduction ignored any voltage drop across $R_1 + jX_1$, treating $V_{OC} = E_1$, and any power dissipated by R_1 was considered dissipated by R_c. The short-circuit data reduction ignored any current flowing through the excitation branch and considered any power dissipated by R_c as power dissipated by R_{eq}. The MATLAB program ⟨traneqckt.m⟩ performs the classical data reduction outlined by [4.36] to [4.42] and [4.44] to [4.46]. When completed, the program then refines the equivalent circuit parameter realization by accounting for the voltage drop across $R_1 + jX_1$ and the power dissipated by R_1 for the open-circuit data reduction. The short-circuit data reduction accounts for the small current flow through the excitation branch. The sets of equivalent circuit parameters for both the classical data reduction and the refined data reduction are displayed to the screen. The analysis results for the open-circuit and short-circuit data of Example 4.6 are shown below.

```
      TRANSFORMER  EQUIVALENT  CIRCUIT  PARAMETERS
               Classical  Data  Reduction
   R1(ohm)       R2(ohm)       X1(ohm)       X2(ohm)
   0.1355        0.03573       0.3302        0.08256

   Rc(ohm)       Xm
   320           104.1
```

```
TRANSFORMER  EQUIVALENT  CIRCUIT  PARAMETERS
           Refined  Data  Reduction
R1(ohm)      R2(ohm)      X1(ohm)      X2(ohm)
0.1353       0.03568      0.3296       0.0824

Rc(ohm)      Xm
319.1        103.7
```

Examination of the refined data reduction shows that the results are within a fraction of a percent of the values obtained from the classical data reduction. As long as good transformer design prevails so that $(R_c, X_m) \gg (R_{eq}, X_{eq})$, the refined data reduction will not show significant improvement in parameter value accuracy. However, in the case of a transformer with significant saturation or high leakage reactance, the refined data reduction can offer accuracy improvement.

4.6 PERFORMANCE ASSESSMENT

The power transformer functions as a two-port device favorably interfacing two sections of a larger network. Engineering analysis of the larger network will require that the voltages and currents for the equivalent circuit of Fig. 4.21 be calculated from one of two constraining conditions.

Condition 1. Load voltage V_2, load current I_2, and load PF are specified, and the resulting value of input voltage V_1 and current I_1 must be calculated.

Condition 2. Input voltage V_1 is specified, with the load treated as a constant impedance requiring that the input current I_1, load current I_2, and load voltage V_2 be calculated.

Preferred performance is that the practical transformer approach the ideal transformer in characteristics. Quantitative assessments of the departure from ideal are made by calculation of the efficiency and voltage regulation.

4.6.1 VOLTAGE-CURRENT ANALYSIS

A common transformer analysis problem is that of Condition 1 above, where it is assumed that the voltage of a fixed load is maintained constant. Since both V_2 and I_2 of Fig. 4.21 are specified, analysis begins at the output port and progresses to the input port to determine V_1 and I_1 through use of KVL, KCL, and the voltage-current reflection properties of the ideal transformer.

Given a 240:120-V, 5-kVA, 60-Hz transformer with the following equivalent circuit parameter values: | **Example 4.7**

$$R_1 = 0.06\,\Omega \qquad R_2 = 0.015\,\Omega \qquad R_c = 1200\,\Omega$$

$$X_1 = 0.18\,\Omega \qquad X_2 = 0.045\,\Omega \qquad X_m = 400\,\Omega$$

The 120-V secondary is supplying a 5-kVA load with 0.8 PF lagging at rated voltage. Determine the input voltage, current, and PF.

Assume the output voltage on the reference, or $V_2 = 120\angle 0°$. The output current for the equivalent circuit of Fig. 4.21 is

$$I_2 = \frac{S}{V_2} = \frac{5000}{120} = 41.67 \text{ A}$$

$$\theta_2 = -\cos^{-1}(PF_2) = -\cos^{-1}(0.8) = -36.87°$$

$$\bar{I}_2 = 41.67\angle -36.87°$$

The turns ratio is $a = V_1/V_2 = 240/120 = 2$. KVL leads to

$$\overline{E}_1 = [a^2R_2 + ja^2X_2]\bar{I}_2/a + a\overline{V}_2$$

$$\overline{E}_1 = [(2)^2(0.015) + j(2)^2(0.045)](41.67/2\angle -36.89°) + (2)120\angle 0°$$

$$\overline{E}_1 = [0.1897\angle 71.56°](20.83\angle -36.89°) + 240\angle 0° = 243.3\angle 0.53° \text{ V}$$

Current $\bar{I}_1$ follows from application of KCL.

$$\bar{I}_1 = \bar{I}_2/a + \bar{I}_c + \bar{I}_m = \bar{I}_2/a + \frac{\overline{E}_1}{R_c} + \frac{\overline{E}_1}{jX_m}$$

$$\bar{I}_1 = \frac{41.67\angle -36.87°}{2} + \frac{243.3\angle 0.53°}{1200} + \frac{243.3\angle 0.53°}{400\angle 90°}$$

$$\bar{I}_1 = 20.83\angle -36.87° + 0.203\angle 0.53° + 0.608\angle -89.47° = 21.37\angle -37.84° \text{ A}$$

KVL applied around the input loop yields

$$\overline{V}_1 = (R_1 + jX_1)\bar{I}_1 + \overline{E}_1 = (0.06 + j0.18)(21.37\angle -37.84°) + 243.3\angle 0.53°$$

$$\overline{V}_1 = 246.7\angle 1.05° \text{ V}$$

The input PF follows as

$$PF_{in} = \cos(\angle \overline{V}_1 - \angle \bar{I}_1) = \cos(1.05° + 37.84°) = 0.778 \text{ lagging}$$

The MATLAB program ⟨tranperf1.m⟩ has been developed to perform the analysis of the exact equivalent circuit of Fig. 4.21 for the case of constant-load voltage with fixed load. The program is configured so that the code must be edited to set the values of the transformer equivalent circuit, winding voltage ratings, and apparent power. The program is interactive with the user to set the percentage load and power factor. Thus, once a transformer has been characterized, numerous load conditions can be quickly analyzed. All voltages and currents are displayed to the screen so that the user can easily highlight and use the print selection option to obtain a hard copy. The screen display for the solution of Example 4.7 is shown below.

```
    RATED  VOLTAGE  LOAD  CHARACTERISTIC
 Percent  Load(  0->100  )  =  100
 Load  Power  Factor(  0->1  )  =  0.8
 PF  Sense(  leading,lagging  )  =  lagging
```

```
              TRANSFORMER  PERFORMANCE .
  V1 mag.        V1 ang         V2 mag         V2 ang
  246.7          1.046           ·120            · 0

  E1 mag         E1 ang
  243.3           0.53

  I1 mag         I1 ang          I2 mag         I2 ang
  21.37         -37.84         ⁄ 41.67         -36.87

  Im mag         Im ang         Ic mag         Ic ang
  0.6082        -89.47          0.2027          0.53

  PFin           PFS1           PFout          PFS2
  0.7785        lagging          0.8           lagging

  Pin            Pout            eff            %Reg
  4103           4000           97.5           2.725
```

4.6.2   APPROXIMATE EQUIVALENT CIRCUIT

The calculations of Example 4.7 above are cumbersome at best. It can be observed
from these calculations that $\overline{V}_1 \cong \overline{E}_1$ and that since $(\overline{I}_m, \overline{I}_c) \ll \overline{I}_2/a, \overline{I}_1 \cong \overline{I}_2/$
$a = \overline{I}'_2$. As long as good design prevails so that $(R_c, X_m) \gg (R_1, R'_2, X_1, X'_2)$,
these observed approximations will hold, and the *approximate equivalent circuit* of
Fig. 4.25 will produce reasonably accurate predictions of the transformer voltages
and currents. Calculations made by hand-held calculator for this approximate
equivalent circuit are less cumbersome than for the exact equivalent circuit. In
power system load flow study where efficiency of the transformer is not an analysis objective, the excitation branch of the transformer is frequently discarded,
reducing the approximate equivalent circuit of Fig. 4.25 to simply $R_{eq} + jX_{eq}$.

Rework Example 4.7 except use the approximate equivalent circuit of Fig. 4.25 rather than the exact equivalent circuit. | **Example 4.8**

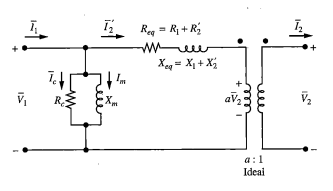

Figure 4.25
Transformer approximate equivalent circuit

Based on Example 4.7,

$$a\bar{V}_2 = (2)120\angle 0° = 240\angle 0° \text{ V} \qquad \bar{I}_2' = \bar{I}_2/a = 20.83\angle -36.87° \text{ A}$$

Now,

$$R_{eq} = R_1 + R_2' = 0.06 + (2)^2(0.015) = 0.12 \text{ } \Omega$$

$$X_{eq} = X_1 + X_2' = 0.18 + (2)^2(0.045) = 0.36 \text{ } \Omega$$

By KVL,

$$\bar{V}_1 = (R_{eq} + jX_{eq})\bar{I}_2' + a\bar{V}_2 = (0.12 + j0.36)(20.83\angle -36.87°) + 240\angle 0°$$

$$= 246.54\angle 1.05° \text{ V}$$

By KCL,

$$\bar{I}_1 = \bar{I}_2' + \bar{I}_c + \bar{I}_m = \bar{I}_2'\frac{\bar{V}_1}{R_c} + \frac{\bar{V}_1}{jX_m}$$

$$\bar{I}_1 = 20.83\angle -36.87° + \frac{246.54\angle 1.05°}{1200} + \frac{246.54\angle 1.05°}{400\angle 90°}$$

$$= 21.36\angle -37.82° \text{ A}$$

$$PF_{in} = \cos(\angle\bar{V}_1 - \angle\bar{I}_1) = \cos(1.05° + 37.82°) = 0.778 \text{ lagging}$$

It is clearly seen from comparison of the results of Examples 4.7 and 4.8 that the use of the approximate equivalent circuit has introduced negligible error in analysis of this transformer. The approximate equivalent circuit becomes the choice for any hand-held calculator analysis due to the reduction in calculation labor. Even with the approximate equivalent circuit, the computation becomes mundane and subject to careless error.

The transformer analysis problem described by Condition 2 is encountered when a study of the impact of change in power system voltages is made. The secondary load is modeled as a passive impedance with values determined for rated-load voltage condition. The analysis then moves to the input terminals, where the voltage available from the power system is impressed. An equivalent impedance seen looking into the transformer is formed, and the input current is calculated. By use of KVL, KCL, and the voltage-current reflection properties of the transformer, the analysis progresses from the input to the load terminals to predict the actual load voltage and current.

Example 4.9 | For the transformer of Example 4.7, assume that the 0.8 PF lagging load can be modeled by a passive impedance, but the transformer primary is connected to a 240-V bus. Determine the resulting load voltage and current using the approximate equivalent circuit of Fig. 4.25.

The load impedance is determined by the rated voltage and current of Example 4.7.

$$\mathbf{Z}_L = \frac{\bar{V}_2}{\bar{I}_2} = \frac{120\angle 0°}{41.67\angle -36.87°} = 2.88\angle 36.87° \text{ } \Omega$$

The impedance seen to the right of the excitation branch in Fig. 4.25 is

$$\mathbf{Z}_2' = R_{eq} + jX_{eq} + a^2 Z_L = 0.12 + j0.36 + (2)^2 2.88 \angle 36.87°$$

$$\mathbf{Z}_2' = 11.83 \angle 37.91°$$

Taking $\overline{V}_1$ on the reference, the current $\overline{I}_2'$ is

$$\overline{I}_2' = \frac{\overline{V}_1}{\mathbf{Z}_2'} = \frac{240 \angle 0°}{11.83 \angle 37.91°} = 20.29 \angle -37.91° \text{ A}$$

$$\overline{I}_2 = a\overline{I}_2' = (2)20.29 \angle -37.91° = 40.58 \angle -37.91°$$

By voltage division,

$$a\overline{V}_2 = \frac{a^2 Z_L}{\mathbf{Z}_2'} \overline{V}_1 = \frac{(2)^2 2.88 \angle 36.87°}{11.83 \angle 37.91°}(240 \angle 0°) = 233.71 \angle -1.04°$$

$$\overline{V}_2 = \frac{a\overline{V}_2}{a} = \frac{233.71 \angle -1.04°}{2} = 116.85 \angle 1.04° \text{ V}$$

The MATLAB program ⟨tranperf2.m⟩ has been developed to handle the Condition 2 analysis of Example 4.9. However, it performs the analysis using the exact equivalent circuit of Fig. 4.19. Hence, values of $\overline{E}_1$, $\overline{I}_1$, and the excitation branch currents are also determined. The program is set up similar to ⟨tranperf1.m⟩ in that it is interactive with the user for values of percent load and PF to determine the fixed-load impedance. Also, the input voltage is supplied through the keyboard. Consequently, numerous conditions can be examined with ease. A printout of the screen display for the solution of Example 4.9 is shown below.

```
DESCRIPTION OF LOAD @ RATED VOLTAGE
Percent Load( 0->100 ) = 100
Load Power Factor( 0->1 ) = 0.8
PF Sense( leading,lagging ) = lagging
Input Voltage( V ) = 240
                TRANSFORMER PERFORMANCE
V1 mag       V1 ang        V2 mag       V2 ang
240              0          116.8        -1.046

E1 mag       E1 ang
236.7        -0.5156

I1 mag       I1 ang        I2 mag       I2 ang
20.79        -38.88        40.54        -37.92

Im mag       Im ang        Ic mag       Ic ang
0.5917       -90.52        0.1972       -0.5156

PFin         PFS1          PFout        PFS2
0.7785       lagging        0.8         lagging
```

4.6.3 EFFICIENCY

In an electric utility application, energy losses in a transformer are lost revenue. Typically, energy is processed through four or more transformers prior to reaching the end-use point. Hence, transformer efficiency is of utmost concern.

Losses in a transformer occur in two places—ohmic losses of the windings (R_1, R_2) and losses in the ferromagnetic core (R_c). The efficiency in percent for any transformer model is given by

$$\eta = \frac{100 P_{out}}{P_{in}} = \frac{100 V_2 I_2 PF_2}{V_1 I_1 PF_1} \qquad \textbf{[4.47]}$$

For the equivalent circuit models of Figs. 4.18 and 4.22, the efficiency in percent can be written as, respectively,

$$\eta = \frac{100 P_{out}}{P_{out} + \text{losses}} = \frac{100 V_2 I_2 PF_2}{V_2 I_2 PF_2 + \dfrac{E_1^2}{R_c} + I_1^2 R_1 + (I_2')^2 R_2'} \qquad \textbf{[4.48]}$$

$$\eta = \frac{100 P_{out}}{P_{out} + \text{losses}} = \frac{100 V_2 I_2 PF_2}{V_2 I_2 PF_2 + \dfrac{V_1^2}{R_c} + (I_2')^2 R_{eq}} \qquad \textbf{[4.49]}$$

Example 4.10 | Determine the efficiency for the transformer of Example 4.7 when modeled by both the exact and approximate equivalent circuit.

Using the results of Example 4.7 in [4.48],

$$\eta = \frac{(100)(120)(41.67)\cos(-36.87°)}{(120)(41.67)\cos(-36.87°) + \dfrac{(243.3)^2}{1200} + (21.37)^2(0.06) + \left(\dfrac{41.67}{2}\right)^2 (2)^2 (0.015)}$$

$$\eta = 97.49\%$$

Results for the approximate equivalent circuit analysis of Example 4.8 can be used in [4.49].

$$\eta = \frac{100(120)(41.67)\cos(-36.87°)}{(120)(41.67)\cos(-36.87°) + \dfrac{(246.54)^2}{1200} + (20.83)^2(0.12)}$$

$$\eta = 97.5\%$$

It is observed that the efficiency values determined by the exact and the approximate equivalent circuit are nearly equal.

The efficiency of a transformer can vary depending on apparent power processed and the power factor of the coupled load. An approximate prediction of transformer efficiency variation with load can be made based on the open-circuit and short-circuit loss test data. P_{OC} is usually recorded for rated voltage, and since the excitation voltage E_1 varies only a small percentage from rated value with load change, the transformer core losses remain close in value to P_{OC} as load changes. Thus, the core losses can be considered a fixed loss. Owing to the near-constant voltage operation of a transformer, the winding currents vary almost directly with the apparent power supplied to the load. Thus, the winding ohmic losses $(I_1^2 R_1, I_2^2 R_2)$ can be thought of as variable losses that vary almost directly with the square root of load apparent power (S_2). If the short circuit test were performed for the typical condition of rated current, then the winding losses for any load condition are approximated by $\sqrt{S_2/S_{2R}} P_{SC}$, where the subscript R denotes rated condition.

For the 25-kVA transformer of Example 4.6, the short-circuit loss data has been measured sufficiently close to the rated value of current for the high-voltage winding that the above approximate efficiency prediction can be applied. Estimate (*a*) the full-load efficiency and (*b*) the half-load efficiency of this transformer if the load power factor in both cases is 0.8 lagging. | **Example 4.11**

(*a*) For the full-load condition,

$$P_{out} = S_{2R}(PF_2) = 25{,}000(0.8) = 20{,}000 \text{ W}$$

$$\text{Losses} \cong P_{OC} + \left[\frac{S_2}{S_{2R}}\right]^{1/2} P_{SC} = 720 + \left[\frac{25{,}000}{25{,}000}\right]^{1/2}(750) = 1470 \text{ W}$$

From application of [4.48],

$$\eta = \frac{100(20{,}000)}{20{,}000 + 1470} = 93.15\%$$

(*b*) The half-load condition gives

$$P_{out} = \frac{1}{2} S_{2R}(PF_2) = 0.5(25{,}000)(0.8) = 10{,}000 \text{ W}$$

$$\text{Losses} \cong P_{OC} + \left[\frac{S_2}{S_{2R}}\right]^{1/2} P_{SC} = 720 + \left[\frac{12{,}500}{2500}\right]^{1/2}(750) = 907.5 \text{ W}$$

$$\eta = \frac{100(10{,}000)}{10{,}000 + 907.5} = 91.68\%$$

Accurate study of efficiency for multipoint operation for a transformer using hand-held calculator computation can be a tedious task. The MATLAB program ⟨traneff.m⟩ has been formed to alleviate this task. Given the transformer equivalent circuit values and rating of the transformer, the program calculates the efficiency vs. apparent power (load current) generating a family of curves with PF as a parameter. The program also determines the point of maximum efficiency for each PF value. Figure 4.26 displays a set of results from ⟨traneff.m⟩. Three general conclusions can be reached concerning the nature of transformer efficiency from inspection of Fig. 4.26.

1. Efficiency decreases significantly for small load values. In [4.49], it is seen that all terms decrease with a decrease in current except the core loss term

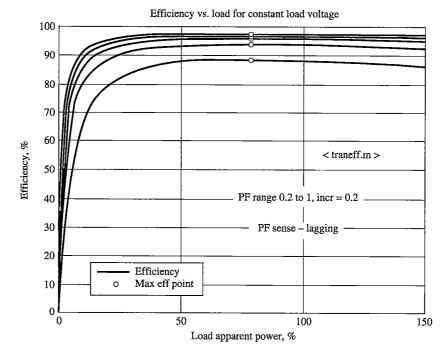

Figure 4.26
Variation of transformer efficiency with load and PF

V_1^2/R_c. Thus, as current approaches zero, the efficiency must approach zero.

2. Efficiency decreases for high load values. Again referring to [4.49], it is seen that the ohmic loss term $(I_2')^2 R_{eq}'$ increases as current squared while the core loss term remains essentially constant. All other terms increase linearly with current. Mathematically, efficiency approaches zero as current becomes infinitely large.

3. Efficiency decreases with decrease in PF for a particular value of load current. Looking at [4.49] for a particular value of current and near constant input voltage, it is apparent that the numerator decreases directly with decrease in $PF^2 = \cos\theta_2$ while the denominator decreases at a less than direct rate.

While Fig. 4.26 gives a broad insight into the efficiency vs. load nature of a transformer, it does not readily provide a means for economic assessment of the operating cost of a transformer. The point of maximum efficiency for the transformer of Fig. 4.26 occurs at approximately 79 percent of full load. By increasing the cross-sectional area of the winding conductors, this maximum efficiency point can be moved to the neighborhood of full power. The increase in material would result in an increase in the initial capital cost of the transformer. However, if this were a residential distribution transformer that operates a large percentage of the hours in each service day at less than full load, the choice of a new transformer with maximum efficiency at near full load could be a doubly poor economic choice—higher capital outlay with higher operating or variable cost.

In electric utility application, the operating cost of a transformer is quantitatively assessed by a method known as *all-day efficiency*. Based on customer energy use patterns, a load profile for a 24-h period of service can be established. Then given a particular transformer, the energy loss for the 24-h period can be calculated by Simpson's rule integration of power loss with respect to time. The product of this energy loss (kWh) and market value of energy ($/kWh) gives the variable cost of having that transformer in service. Two transformer designs can then be compared to determine the desirable transformer as the one with lower variable cost.

Example 4.12

A small manufacturing plant works three shifts per day for 5 days per week. Machining operations and heat-treating ovens are active on all three 8-h shifts and are characterized by a nearly constant 500-kVA load at 0.85 PF lagging. During the 8-h day shift, a larger workforce is present to conduct assembly and finishing operations so that the nearly constant total load for the day shift increases to 1000 kVA at 0.85 PF lagging. The plant plans to replace their present substation transformer, which is overloaded. There are no projections for expansion in the foreseeable future, so the 1000-kVA service for the day shift determines the rating of the new transformer. Two 1000-kVA transformers have been bid by suppliers with the following efficiency characteristics at 0.85 PF lagging:
Transformer A:

$$\eta = 95.1\% \text{ at } 500 \text{ kVA}$$

$$\eta = 97.2\% \text{ at } 1000 \text{ kVA}$$

Transformer B:

$$\eta = 96.9\% \text{ at } 500 \text{ kVA}$$

$$\eta = 96.4\% \text{ at } 1000 \text{ kVA}$$

The cost of energy is 0.08$/kWh. Transformer B has been quoted $250 higher in price than Transformer A. However, it will have lower operating cost due to the efficiency advantage at half load for this particular plant. Determine the annual savings in operating cost if Transformer B is installed rather than Transformer A.

Assume plant shutdown for two weeks per year for vacation and exclude the near no-load energy losses during the vacation period and over weekends. The total first shift hours per year are $50 \times 8 \times 5 = 2000$ hours. The combined second and third shift hours are $50 \times 16 \times 5 = 4000$ hours. From [4.48] with η in per unit,

$$\text{Losses} = \frac{1 - \eta}{\eta} P_o$$

The annual operating or variable cost for Transformer A is given by

$$\text{Cost A} = \frac{1 - 0.972}{0.972}(1000)(0.85)(2000)(0.08)$$

$$+ \left[\frac{1 - 0.951}{0.951}\right](500)(0.85)(4000)(0.08)$$

$$\text{Cost A} = \$10,925.06 \text{ per year}$$

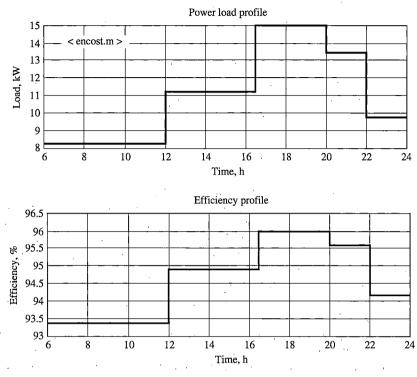

Figure 4.27
Load and resulting efficiency profiles for all-day efficiency study

Transformer B has an annual operating cost of

$$\text{Cost B} = \left[\frac{1 - 0.964}{0.964}\right](1000)(0.85)(2000)(0.08)$$

$$+ \left[\frac{1 - 0.969}{0.969}\right](500)(0.85)(4000)(0.08)$$

$$\text{Cost B} = \$9429.72 \text{ per year}$$

The annual savings for choice of Transformer B is $1495.34.

The MATLAB program ⟨encost.m⟩ has been developed to handle the task of calculating variable cost and all-day efficiency with ease. Equivalent circuit parameters and rated values for the particular transformer under study must be supplied. The apparent power (LOAD) and power factor (PF) arrays describe the transformer service load, with the duration of each load point specified in the time array (HRS). The program calculates the efficiency at each load point and the cost of loss energy over the 24-hour load profile. The all-day efficiency is simply a time-weighted average of the individual load point efficiencies. Figure 4.27 shows the average power load profile and the corresponding efficiency profile for the transformer and profiles of the program listing. The daily energy cost and all-day efficiency results are displayed to the screen as the selection printed below:

```
   ALL  DAY  EFFICIENCY  STUDY
  24-Hour  Energy  cost  —    $1.45
  All-Day  Efficiency(%)  —  94.64
```

4.6.4 VOLTAGE REGULATION

Voltage regulation of a transformer is simply a measure of the percentage increase in load voltage magnitude as the load changes from full-load condition to no-load condition while the input voltage remains constant. In equation form,

$$Reg = \frac{100(V_{NL} - V_{FL})}{V_{FL}} \qquad [4.50]$$

To clarify the significance of voltage regulation by example, suppose the transformer supplying a residence had $Reg = 15$ percent, and the voltage at a wall plug were 125 V when the power demands were peak. Then at some point when the service demands drop to near zero, the wall plug voltage will have increased to a value of $(1.15)(125) \cong 144$ V. Although a connected reading lamp may be bright, the bulb life will be short.

For evaluation of the regulation, the values of V_{FL} and V_{NL} in [4.50] need to be related to equivalent circuit calculation. If a full-load analysis is performed using the approximate equivalent circuit of [4.22], then when the load is reduced to zero $(I_2 = I'_2 = 0)$ with V_1 unchanged from the full-load condition value, $V_2 = V_1/a = V_{NL}$. If the analysis were performed modeling the transformer by the exact equivalent circuit of Fig. 4.21, at no-load condition, I_1 is quite small compared to the full-load value and $V_2 = E_1/a \cong V_1/a$. Hence, using the values of voltages determined from a full-load analysis, [4.50] can be rewritten as

$$Reg = \frac{100(V_1/a - V_2)}{V_2} \qquad [4.51]$$

Example 4.13

Calculate the regulation for the transformer of Example 4.7.
    Using the values calculated in Example 4.7, [4.51] gives

$$Reg = \frac{100(246.7/2 - 120)}{120} = 2.7\%$$

The value of regulation depends on the coil resistance and leakage reactance values. Generally, an increase in the value of either quantity leads to an increase in regulation for a particular load. Regulation is also dependent on load power factor.

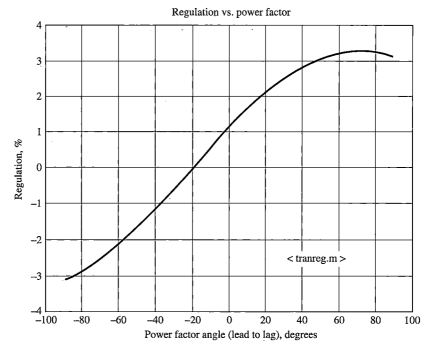

Figure 4.28
Voltage regulation vs. power factor for a transformer

The MATLAB program ⟨tranreg.m⟩ performs the regulation calculation for the complete range of load PF from 0 PF leading to 0 PF lagging given the equivalent circuit parameters of a particular transformer and its rated values. The program uses the exact equivalent circuit and actually calculates the exact no-load voltage. Figure 4.28 is a plot of the results produced by ⟨tranreg.m⟩ for the listed transformer. Inspection of the plot shows that regulation can actually have negative values for a sufficiently leading PF load. A negative regulation value means that the output transformer voltage decreases as the load is reduced.

4.6.5 INRUSH CURRENT

Although study of the transformer performance is largely an analysis of sinusoidal steady-state operation, there is one case of transient phenomena that deserves attention. Upon energizing an unloaded transformer, the initial transient current flow, called *inrush current,* can attain peak levels with values several times the rated value of current for the device. Sizing of protective breakers and explanation of system disturbance require knowledge of the inrush current nature.

The problem is one of energizing an *R-L* circuit with a sinusoidal source. The steady-state component of the current is superimposed on exponentially decaying transient response necessary to satisfy the initial condition of zero current. In a linear circuit, this value of current can reach no greater value than twice the steady-state value. However, in a ferromagnetic device where the inductance is a function of the current, the inductance decreases to smaller values as the current increases. In particular, the solution is that of a first-order differential equation with a coefficient that is a function of the dependent variable. Conclusions drawn from knowledge of linear circuit theory can be significantly in error.

The describing nonlinear differential equation formed by KVL applied to the primary winding of an unloaded transformer is

$$V_m\sin(\omega t - \zeta) = R_1 i_1 + \frac{d\lambda}{dt} = R_1 i_1 + \frac{\partial\lambda}{\partial i_1}\frac{di_1}{dt} \qquad \textbf{[4.52]}$$

where λ is the flux linkages of the primary coil and $\partial\lambda/\partial i_1$ is the inductance of the primary coil. Rearranging [4.52] leads to

$$\frac{di_1}{dt} = \frac{-R_1}{\partial\lambda/\partial i_1}i_1 + \frac{V_m}{\partial\lambda/\partial i_1}\sin(\omega t - \zeta) \qquad \textbf{[4.53]}$$

If a plot of input voltage vs. input current with the secondary winding open circuit is available for the transformer, the plot can be converted to a flux linkage vs. input current plot by dividing the voltage values by $4.44f$ based on [4.7]. Neglecting the core loss component of the no-load current, the slope of this curve at a particular value of current is used during the solution of [4.53].

The MATLAB program ⟨inrush.m⟩ starts with arrays of open-circuit voltage and current, forms the flux linkage array, and numerically solves [4.53] using a fixed-increment, fourth-order Runge-Kutta routine. The inductance $(\partial\lambda/\partial i_1)$ is numerically evaluated during each solution increment using the table1 built-in function of MATLAB. Figure 4.29 displays the resulting current for the worst case of $\zeta = 0°$. If ζ were 90°, the inrush current would have no transient component and would immediately settle down to the sinusoidal steady state with maximum value of approximately 0.9 A.

4.7 RESIDENTIAL DISTRIBUTION TRANSFORMERS

Figure 4.30 schematically shows a center-tapped residential distribution transformer. The center tap is grounded and forms the common connection (white lead) for the 120-V service wiring of a residence. Ideally, half of the 120-V load appears as load L_1 and the other half forms load L_2; however, equal load distribution rarely exists. The 240-V load (L_3) spans the full secondary. The safety ground (green) lead for the 240-V service connects to the transformer center tap to assure breaker trip if any ground insulation failure were to occur in the 240-V load.

Example 4.14

The residential distribution transformer of Fig. 4.30 is rated as 15 kVA, 2400:240/120 V, 60 Hz. The upper 120-V load L_1 is 1.5 kW at 0.8 PF lagging. The 240-V load L_3 is 5 kW at 0.8 PF lagging. Treat the transformer as ideal and determine the maximum unity PF 120-V load L_2 that can be installed without any winding current exceeding rated value.

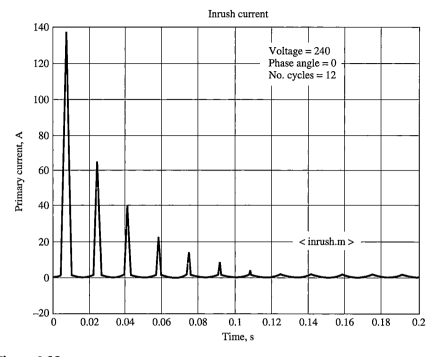

Figure 4.29
Transformer inrush current

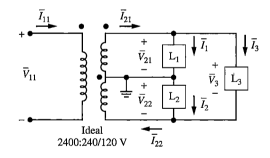

Figure 4.30
Residential distribution transformer

Assume $\bar{V}_1$ on the reference; then

$$\bar{I}_1 = \frac{1500}{120(0.8)}\angle -\cos^{-1}(0.8) = 15.625\angle -36.87°$$

$$\bar{I}_3 = \frac{5000}{240(0.8)}\angle -\cos^{-1}(0.8) = 26.042\angle -36.87° = 20.83 - j15.62\ \text{A}$$

Rated current for each of the secondary windings is

$$I_{21R} = I_{22R} = \frac{15,000}{240} = 62.50 \text{ A}$$

Since $\overline{V}_1$ is on the reference, so are $\overline{V}_{21}$ and $\overline{V}_{22}$. Thus, $\overline{I}_1$ is also on the reference. By KCL,

$$\overline{I}_{21R} = \overline{I}_2 + \overline{I}_3$$

By the Pythagorean theorem,

$$I_{21R}^2 = (I_2 + \text{Re}\{\overline{I}_3\})^2 + (\text{Im}\{\overline{I}_3\})^2$$

$$(62.50)^2 = (I_2 + 20.83)^2 + (15.62)^2$$

$$I_2^2 + 41.66 I_2 - 3228.06 = 0$$

Discarding the extraneous root from the quadratic formula solution,

$$I_2 = 39.68 \text{ A}$$

Load L_2 is

$$V_{22} I_2 = 120(39.68) = 4.762 \text{ kW}$$

4.8 AUTOTRANSFORMERS

The transformer as discussed up to this point in the chapter has been formed by two (or more) windings that are magnetically coupled but conductively isolated. An *autotransformer* is a transformer configuration that has part of its winding common to both the input and output ports. As a direct consequence, a path is introduced whereby part or all of the input current, depending on whether voltage step-up or step-down configuration, flows directly to the output after passing through a winding segment. Thus, the autotransformer is not conductively isolated, and the constraint of [4.12] no longer holds for the input and output currents. However, [4.12] is still a valid relationship for the currents of individual winding segments.

The autotransformer is principally found in two particular applications:

1. The autotransformer is used to implement a small voltage step-up at the end of a long transmission line or distribution feeder to compensate for voltage drop due to line reactance.

2. The autotransformer can provide an economical (high power density) variable voltage supply. For such case, a continuous coil is connected across the constant input voltage. One output lead is connected common with the input. The second output lead is provided with a sliding contact for adjustment of the output voltage.

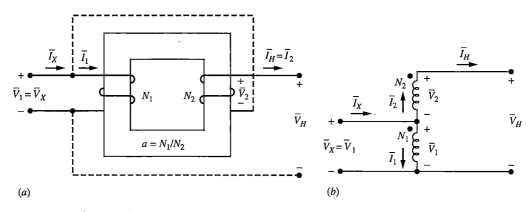

Figure 4.31
Step-up autotransformer connection. (a) Reconnection of two-winding transformer.
(b) Schematic.

4.8.1   IDEAL AUTOTRANSFORMER

Figure 4.31a illustrates by dashed lines reconnection of a two-winding transformer to form an autotransformer with the schematic of Fig. 4.31b. The autotransformer is commonly analyzed as ideal. Later work will show that in typical application configuration, its efficiency is even higher than the already high efficiency two-winding transformer; thus, ideal modeling is justified.

Addition of the connection between the primary coil N_1 and the secondary coil N_2 in Fig. 4.31a does not in any way alter the two-winding transformer magnetic circuit flux–mmf relationships. Hence, relationships between sets $(\overline{V}_1, \overline{V}_2)$ and $(\overline{I}_1, \overline{I}_2)$ given by [4.9] and [4.12], respectively, as shown on the schematic remain valid. However, these sets are no longer input-output sets. The ratio of the autotransformer input and output voltages of Fig. 4.31b must be

$$\frac{\overline{V}_H}{\overline{V}_X} = \frac{\overline{V}_1 + \overline{V}_2}{\overline{V}_1} = \frac{\dfrac{N_1}{N_2}\overline{V}_2 + \overline{V}_2}{\dfrac{N_1}{N_2}\overline{V}_2} = \frac{N_1 + N_2}{N_1} = \frac{a + 1}{a} \qquad \textbf{[4.54]}$$

The corresponding current ratio is

$$\frac{\overline{I}_H}{\overline{I}_X} = \frac{\overline{I}_2}{\overline{I}_1 + \overline{I}_2} = \frac{\overline{I}_2}{\dfrac{N_2}{N_1}\overline{I}_2 + \overline{I}_2} = \frac{N_1}{N_1 + N_2} = \frac{a}{a + 1} \qquad \textbf{[4.55]}$$

It is apparent from [4.54] and [4.55] that the input-output voltage and current ratios have a reciprocal relationship similar to those of the two-winding transformer given by [4.9] and [4.12]; however, the magnitudes of the ratios have different values. It is also pointed out that [4.54] and [4.55] hold only for the additive coil con-

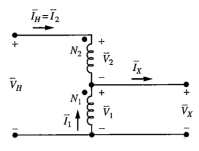

Figure 4.32
Step-down autotransformer

nection. If either coil N_1 or coil N_2 were inverted so that the coils were subtractive, a different set of voltage and current ratio results. A problem at the end of the chapter addresses this case.

The schematic of Fig. 4.32 is a step-down autotransformer $(V_X < V_H)$. Comparing this schematic with Fig. 4.31b, it is seen that the step-down autotransformer is the mirror image of the step-up autotransformer except that the directions of all current arrows are reversed. Consequently, the relationships of [4.54] and [4.55] are valid.

Example 4.15

Let the original two-winding transformer of Fig. 4.31a be rated as 2400:240 V, 50 kVA. Determine the input-output voltage and current ratios for the autotransformer if the voltages and currents of coils N_1 and N_2 are maintained.

The turns ratio is $a = 2400/240 = 10$. The rated current values for the two-winding transformer are

$$I_1 = \frac{S}{V_1} = \frac{50{,}000}{2400} = 20.83 \text{ A} \qquad I_2 = \frac{S}{V_2} = \frac{50{,}000}{240} = 208.33 \text{ A}$$

The autotransformer voltages and currents follow by scalar addition since $\overline{V}_1$ and $\overline{V}_2$ are in phase as well as $\overline{I}_1$ and $\overline{I}_2$.

$$V_X = V_1 = 2400 \qquad V_H = V_1 + V_2 = 2400 + 240 = 2640 \text{ V}$$

$$I_X = I_1 + I_2 = 20.83 + 208.33 = 229.16 \qquad I_H = I_2 = 208.33 \text{ A}$$

whence

$$\frac{V_X}{V_H} = \frac{2400}{2640} = 0.909 \qquad \frac{I_X}{I_H} = \frac{229.16}{208.33} = 1.1$$

These ratios could have also been determined by use of [4.54] and [4.55], giving

$$\frac{V_X}{V_H} = \frac{a}{a+1} = \frac{10}{10+1} = 0.909 \qquad \frac{I_X}{I_H} = \frac{a+1}{a} = \frac{10+1}{10} = 1.1$$

4.8.2   POWER FLOW

The complex powers flowing into and out of the autotransformer of Fig. 4.31*b* are given by

$$S_X = \overline{V}_X \overline{I}_X^* \qquad\qquad\text{[4.56]}$$

$$S_H = \overline{V}_H \overline{I}_H^* = \left(\frac{a+1}{a}\right)\overline{V}_X\left(\frac{a}{a+1}\right)(\overline{I}_X^*) = \overline{V}_X \overline{I}_X^* \qquad\text{[4.57]}$$

where [4.54] and [4.55] were used. It is concluded that the input and output apparent powers, average powers, and reactive powers must be individually equal, or

$$S_X = S_H \qquad P_X = P_H \qquad Q_X = Q_H \qquad\text{[4.58]}$$

An apparent power rating of the autotransformer in terms of the apparent power rating of the original two-winding transformer can be formed.

$$V_X I_X = V_1(I_1 + I_2) = V_1(I_1 + aI_1) = (1 + a)V_1 I_1 \qquad\text{[4.59]}$$

Example 4.16 | Determine the apparent power rating of the autotransformer of Example 4.15 and estimate the full-load, unity-PF efficiency of the autotransformer if the original two-winding transformer had a full-load efficiency of 98 percent for a unity PF.

Based on [4.59], the original 50-kVA transformer now has an apparent power rating of

$$V_X I_X = (1 + 10)50\,\text{kVA} = 550\,\text{kVA}$$

The autotransformer windings carry identical currents as for the case of the two-winding transformer, and the core flux density and frequency are unchanged. Thus, the autotransformer has the same losses as the original two-winding transformer, or

$$\text{Losses} = \frac{P_o}{\eta_{pu}} - P_o = \frac{50}{0.98} - 50 = 1.02\,\text{kW}$$

The efficiency of the autotransformer is then

$$\eta = \frac{100\,V_X I_X}{\text{losses} + V_X I_X} = \frac{100(550)}{1.02 + 550} = 99.81\%$$

Example 4.16 has vividly introduced the power and efficiency advantages of the autotransformer when configured for a small step-up or step-down voltage. Earlier discussion pointed out the loss of conductive isolation for the autotransformer that would generally be considered a disadvantage safetywise. However, in an application to boost voltage at the end of a long transmission line, the safety issue vanishes and the high power density of the autotransformer makes it an economical choice.

4.9 THREE-PHASE TRANSFORMERS

Electric utility companies generate and transmit power in three-phase form. Within the distribution networks, some loads are three-phase and the balance are single-phase loads with connection distributed among the phases to form a quasi-balanced three-phase load. The need for three-phase transformers to match different voltage levels within the system divisions from generation to distribution is apparent.

Three-phase power may be transformed by a properly connected bank of single-phase transformers or by a single-unit three-phase transformer. The configuration makes no difference from a performance point of view; however, the latter can take advantage of the time-phase differences in the three fluxes to better utilize the ferromagnetic core material. As a result, the single-unit three-phase transformer requires less volume of core material than three single-phase transformers translating to a lower-cost device. Nonetheless, a bank of three single-phase transformers may still be used in certain distribution applications for one of the following reasons:

1. A heavy single-phase load must be serviced. In such case, one large and two smaller single-phase transformers may be more economical than a single-unit three-phase transformer whose rating is determined by the heavy single-phase load.

2. When a three-phase transformer bank is formed by connecting the transformers delta-delta (clarified later), one transformer can be removed for repair with reduction in service capacity, but without interruption of service.

4.9.1 CONNECTION SCHEMES

The most common three-phase transformer winding connections are shown by Fig. 4.33. Other definite purpose connections exist[2] but will not be addressed in the scope of this text. The labeling of input lines (A,B,C) and output lines (A',B',C') is set up so that for the Y-Δ and Δ-Y connections, $\overline{V}_{A'B'}$ lags $\overline{V}_{AB}$ by 30°. The standard U.S. practice is to have the smaller magnitude line voltage lag the larger magnitude line voltage by 30°. Consequently, if the transformer turns ratio (a) were such that the secondary line voltage were larger than the primary line voltage, the Y-Δ and Δ-Y connections of Fig. 4.33 are different. An end of the chapter problem addresses the change.

The voltage-current relationships of [4.9] and [4.12] between paired primary and secondary coils are unaltered when a three-phase connection is introduced. With this recognition and application of the principles of balanced three-phase circuits from Chap. 2, the voltage and current magnitude relationships of Fig. 4.33 are easily verified.

It is reasonable to question the need to introduce a delta connection in a three-phase transformer bank. In Sec. 4.4.2, it was concluded that either the magnetizing current or its associated flux must be distorted due to the nonlinear ϕ-i characteristic of the transformer ferromagnetic core material. The principal distortion was determined to be third-harmonic. There must be either a third-harmonic component

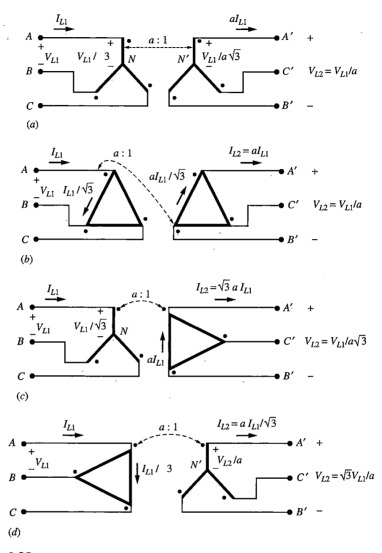

Figure 4.33
Three-phase transformer connections. (a) Wye-wye (Y-Y). (b) Delta-delta (Δ-Δ). (c) Wye-delta (Y-Δ). (d) Delta-wye (Δ-Y).

of magnetizing current or a third-harmonic component of flux, thus voltage. The more desirable choice is to allow a third-harmonic component of magnetizing current so that no third-harmonic distortion of voltages exists. By nature, the third-harmonic currents in all phases have identical time-phase relationships. The delta connection provides a path for the third-harmonic currents to flow, leaving the line current free of third-harmonic current. As an alternate solution, a wye-wye connection can be used provided the neutral point of one winding is grounded to create a return path for the third-harmonic current to flow. In a grounded neutral connection, the line currents do contain the third-harmonic component of magnetizing current, but its amplitude is usually significantly less than the total fundamental component.

In addition to provision for a third-harmonic current path, a delta connection in a three-phase transformer acts to reduce the degree of unbalance in primary currents for the case of an unbalanced secondary load. Improvement is typically observed in both angle and magnitude imbalance. The underlying reason for this phenomenon is the averaging that results from the line current flowing into a node of the delta connection being a linear combination of two phase currents.

Determine rated voltages and currents for a 10,000-kVA, 230-kV:4160 V, 60-Hz, (nameplate data) three-phase transformer if it were designed as (a) a Y-Δ transformer or (b) a Δ-Δ transformer. | **Example 4.17**

(a) In three-phase transformer rating, it is understood that the nameplate data gives total three-phase apparent power (S_T) and line-to-line voltages (V_L) in rms values. Referring to Fig. 4.33c and using the nameplate values, [2.19], and [4.9],

$$V_{L1} = 230 \text{ kV}$$

$$V_{\phi 1} = \frac{V_{L1}}{\sqrt{3}} = \frac{230}{\sqrt{3}}\text{kV} = 132.8 \text{ kV}$$

$$V_{L2} = 4160 \text{ V}$$

$$a = \frac{V_{\phi 1}}{V_{L2}} = \frac{132,800}{4160} = 31.92$$

Based on [2.65], [4.12], and [2.26],

$$I_{L1} = \frac{S_T}{\sqrt{3}\,V_{L1}} = \frac{10 \times 10^6}{\sqrt{3}(230 \times 10^3)} = 25.1 \text{ A}$$

$$I_{\phi 2} = aI_{L1} = (31.92)(25.1) = 801.3 \text{ A}$$

$$I_{L2} = \sqrt{3}I_{\phi 2} = \sqrt{3}(801.3) = 1387.8 \text{ A}$$

As an alternate solution,

$$I_{L2} = \frac{S_T}{\sqrt{3}\,V_{L2}} = \frac{10 \times 10^6}{\sqrt{3}(4160)} = 1387.9 \text{ A}$$

(b) With Fig. 4.33b and the nameplate values, the Δ-Δ voltages and currents are as follows:

$$V_{L1} = 230 \text{ kV}$$

$$V_{L2} = 4160 \text{ V}$$

Based on [4.9] and [2.65],

$$a = \frac{V_{L1}}{V_{L2}} = \frac{230{,}000}{4160} = 55.29$$

$$I_{L1} = \frac{S_T}{\sqrt{3}\,V_{L1}} = \frac{10 \times 10^6}{\sqrt{3}(230 \times 10^3)} = 25.1 \text{ A}$$

Application of [2.26] and [4.12] leads to

$$I_{\phi 1} = \frac{I_{L1}}{\sqrt{3}} = \frac{25.1}{\sqrt{3}} = 14.49 \text{ A}$$

$$I_{\phi 2} = aI_{\phi 1} = (55.29)(14.49) = 801.1 \text{ A}$$

$$I_{L2} = \sqrt{3}I_{\phi 2} = \sqrt{3}(801.1) = 1387.5 \text{ A}$$

or

$$I_{L2} = \frac{S_T}{\sqrt{3}\,V_{L2}} = \frac{10 \times 10^6}{\sqrt{3}(4160)} = 1387.9 \text{ A}$$

4.9.2   PERFORMANCE ANALYSIS

Each phase of a three-phase transformer can be modeled by a single-phase transformer equivalent circuit as in Fig. 4.25 if regulation or efficiency calculations need to be carried out. Phase voltages and currents must be appropriately used in the equivalent circuit model. Unless efficiency is a direct concern of the analysis, the magnetizing branch can be dropped from the model leaving only the per phase $R_{eq} + jX_{eq}$. Initial study of the three-phase transformer typically is an analysis of the transformer as ideal. Only the ideal three-phase transformer is to be considered in this text.

Example 4.18 | Determine all voltages and currents for the Δ-Y transformer bank of Fig. 4.34 with balanced three-phase conditions and $a \cdot b \cdot c$ phase sequence if $\bar{I}_{A'} = 36\angle{-30°}\,\text{A}$ and $\bar{V}_{A'B'} = 240\angle 30°\,\text{V}$.

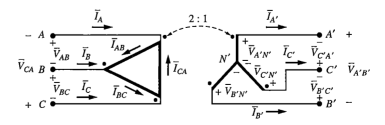

Figure 4.34
Delta-wye transformer bank

Since $a \cdot b \cdot c$ phase sequence and balanced conditions exist, the remainder of the output voltage and current three-phase sets can be immediately written based on the given quantities.

$$\bar{V}_{B'C'} = 240\angle -90° \text{ V} \qquad \bar{V}_{C'A'} = 240\angle 150° \text{ V}$$

$$\bar{I}_{B'} = 36\angle -150° \text{ A} \qquad \bar{I}_{C'} = 36\angle 90° \text{ A}$$

Recall that for $a \cdot b \cdot c$ phase sequence, line voltage leads the associated phase voltage by 30°. Thus,

$$\bar{V}_{A'N'} = \frac{\bar{V}_{A'B'}}{\sqrt{3}}\angle -30° = 138.57\angle 0° \text{ V}$$

$$\bar{V}_{B'N'} = \frac{\bar{V}_{B'C'}}{\sqrt{3}}\angle -30° = 138.57\angle -120° \text{ V}$$

$$\bar{V}_{C'N'} = \frac{\bar{V}_{C'A'}}{\sqrt{3}}\angle -30° = 138.57\angle 120° \text{ V}$$

Application of the voltage transformation ratio to primary and secondary coil pairs yields

$$\bar{V}_{AB} = -2\bar{V}_{B'N'} = (-2)(138.57)\angle -120° = 277.14\angle 60° \text{ V}$$

$$\bar{V}_{BC} = -2\bar{V}_{C'N'} = (-2)138.57\angle 120° = 277.14\angle -60° \text{ V}$$

$$\bar{V}_{CA} = -2\bar{V}_{A'N'} = (-2)138.57\angle 0° = 277.14\angle 180° \text{ V}$$

Comparing $\bar{V}_{A'B'}$ and $\bar{V}_{AB}$, it is noted that $V_{AB} > V_{A'B'}$. Consistent with the standard U.S. practice, the larger voltage $(\bar{V}_{AB})$ leads the smaller voltage $(\bar{V}_{A'B'})$ by 30°. From use of the current transformation ratio,

$$\bar{I}_{AB} = -\frac{1}{2}\bar{I}_{B'} = \left(-\frac{1}{2}\right)36\angle -150° = 18\angle 30° \text{ A}$$

$$\bar{I}_{BC} = -\frac{1}{2}\bar{I}_{C'} = \left(-\frac{1}{2}\right)36\angle 90° = 18\angle -90° \text{ A}$$

$$\bar{I}_{CA} = -\frac{1}{2}\bar{I}_{A'} = \left(-\frac{1}{2}\right)36\angle -30° = 18\angle 150° \text{ A}$$

The input line currents follow by use of KCL at each delta node point.

$$\bar{I}_A = \bar{I}_{AB} - \bar{I}_{CA} = 18\angle 30° - 18\angle 150° = 18\sqrt{3}\angle 0° = 31.18\angle 0° \text{ A}$$

$$\bar{I}_B = \bar{I}_{BC} - \bar{I}_{AB} = 18\angle -90° - 18\angle 30° = 31.18\angle -120° \text{ A}$$

$$\bar{I}_C = \bar{I}_{CA} - \bar{I}_{BC} = 18\angle 150° - 18\angle -90° = 31.18\angle 120° \text{ A}$$

Example 4.19

A 345-kV transmission line feeds a distribution substation that in turn has radial feeds to a 4160/2400-V, four-wire distribution network. Determine the voltage and current ratings and the turns ratio of the Δ-Y connected, 100-MVA transformer in this substation.

For the Δ-Y transformer, let L and ϕ denote line and phase quantities, respectively. Then,

$$I_{L1} = \frac{S_T}{\sqrt{3}V_{L1}} = \frac{100 \times 10^6}{\sqrt{3}(345 \times 10^3)} = 167.35 \text{ A}$$

$$I_{\phi 1} = \frac{I_{L1}}{\sqrt{3}} = \frac{167.35}{\sqrt{3}} = 96.62 \text{ A}$$

$$V_{\phi 2} = \frac{V_{L2}}{\sqrt{3}} = \frac{4160}{\sqrt{3}} = 2401.8 \cong 2400 \text{ V}$$

$$I_{L2} = I_{\phi 2} = \frac{S_T}{\sqrt{3}V_{L2}} = \frac{100 \times 10^6}{\sqrt{3}(4160)} = 13,879 \text{ A}$$

$$a = \frac{V_{L1}}{V_{\phi 2}} = \frac{345 \times 10^3}{2401.8} = 143.64$$

4.10   TRANSFORMER WINDING TAPS

The transformer presentation to this point has only recognized a fixed ratio $a = N_1/N_2$. Two cases are identified in application where there is need to alter the transformer ratio. The first case arises at the service end of a feeder system where in-line voltage drop reduces the available primary voltage of the service transformer to the point that the available secondary voltage supplied to connected load has a lower value than acceptable. The corrective action for this case is accomplished by *fixed tap adjustment*. The second case typically occurs at the power generation station where the turboalternator has a constant regulated output voltage. If the secondary voltage of the generator transformer that serves to step up the voltage from the alternator to the transmission grid can be varied, the reactive power flow from the alternator can be controlled nearly independent of average power flow. Such an adjustment may be desired several times during the course of a day and must be accomplished without interruption of generator service. The generator transformer is designed with extra hardware to allow *tap change under load*.

4.10.1 FIXED TAP ADJUSTMENT

Typically, the primary winding of the transformer simply has connection access to points in the winding as indicated by Fig. 4.35. If the range of fixed tap adjustment is ± 5 percent, and if the transformer design is such that 1000 turns of active primary winding yields the nominal turns ratio, the actual primary winding would have 1050 turns. The highest tap position connects to the 1050 turns point of the primary winding while the lowest tap position connects to the 950 turns point, and

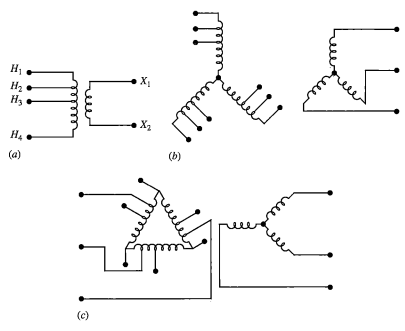

H_1

H_2

H_3

H_4

X_1

X_2

(a)

(b)

(c)

Figure 4.35
Fixed tap transformers. (a) Single-phase. (b) Wye-wye. (c) Delta-wye.

the center tap position connects to the 1000 turns point. The design could well have other intermediate tap positions to allow finer adjustment. Any tap connection changes are made with the transformer de-energized; thus, interruption of service occurs.

A small manufacturing company with single-phase electrical service was the first to establish operation in an industrial park under development. The 2400:240-V pad-mounted transformer to supply the company was set up with the transformer on nominal tap, and the 240-V service to the company loads was always in the range of 237 to 242 V. Over the last 5 years, the number of occupants of the industrial park has increased. The added loads result in increased feeder line voltage drop. The point has been reached where the original company observes that some of their processing ovens are not holding desired temperature. A voltage check determines that the 240-V bus of the original company now consistently shows voltage ranging from 230 to 235 V. If the pad-mounted transformer has five equally spaced taps that range ±5 percent, determine the best connection choice.

Example 4.20

It must be assumed that the 242 V that was the original high value was acceptable; thus, the desired percentage increase in secondary voltage is

$$\frac{242 - 235}{235}(100\%) = 2.98\%$$

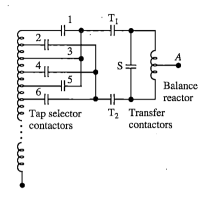

Figure 4.36
Tap change under load

The closest tap change is 2.5 percent. The tap should be moved from the nominal number 3 position to the number 4 position, giving a new turns ratio of

$$a = \frac{0.975(2400)}{240} = 9.75$$

The 240-V bus will now range from 235.7 to 240.9 V over the course of a day.

4.10.2 TAP CHANGE UNDER LOAD

Numerous schemes are available to allow tap change without interruption or significant disturbance of load voltage. The arrangement of Fig. 4.36 typifies the principles common to all tap change transformers. Tap selector contactors 1 to 6 are never opened or closed to active current flow. The transfer contactors (T_1, T_2, S) must have current-interrupting capabilities. If the transformer is operating on tap 1 position, only contactors 1, T_1, and S are closed. With contactor S closed, the balance reactor forms two mutually coupled inductors connected anti-parallel; thus, the inductance is near zero in value resulting in negligible voltage drop. If the tap position is to be decreased, selector contactor 2 is closed. Then S is opened and T_2 closed in a rapid sequential action. With this contactor configuration, the balance reactor forms an autotransformer so that the voltage at line A is midpoint between taps 1 and 2. If additional decrease in voltage is desired, T_1 is opened and S is closed in rapid sequence. Tap selector switch 1 can now be opened. The balance inductor once again forms a pair of antiparallel mutually coupled inductors presenting negligible voltage drop.

By continuing the above described switching pattern, the voltage of line A in Fig. 4.36 can be placed at 11 distinct values. A typical generator transformer is designed to allow ± 10 percent voltage adjustment from nominal value with more than 30 steps.

4.11   INSTRUMENT TRANSFORMERS

Instrument transformers serve the dual purpose of reducing the sensed voltage or current to a level compatible with input values for available instruments while providing the safety of conductive isolation from high-voltage lines. For ammeters a common value is 5 A. Voltmeters are readily available with ranges to 600 V. The operating principle of instrument transformers is no different from that of power transformers; however, the design is conservative in flux density and current density to render the final product close to an ideal transformer. Consequently, the error in application is minimized.

4.11.1 POTENTIAL TRANSFORMERS

Wattmeters, voltmeters, and any other instruments that must sense the potential difference between two lines commonly use a *potential transformer* for interface between the lines and the instrument. For example, if the voltage of a 12.5-kV line is to be sensed, a 100:1 potential transformer can be installed to yield a voltmeter input of 125 V.

Two sources of error are possible in the potential transformer application. Any voltage drop across the coil resistances and leakage reactances leads to an error in the sensed voltage. In a wattmeter application, any phase shift between the primary voltage and the secondary voltage results in an error in sensed value of power. The conservative design serves to reduce such errors to a minimal level. If additional accuracy is necessary, the potential transformer manufacturer supplies correction factors for magnitude and phase angle deviation based on volt-ampere loading of the potential transformer secondary.

4.11.2 CURRENT TRANSFORMERS

Wattmeters, ammeters, and other instruments that must sense line current are interfaced with the line by use of a *current transformer* when the value of line current exceeds instrument ratings. The primary of the current transformer is designed for a small number of turns to exhibit a low intrusive impedance in the line. The current transformer is commonly a toroidal design so that the line for which current is to be sensed is simply passed through the center of the core to form the primary winding. Consequently, the secondary of the current transformer requires a large number of turns to realize the needed current magnitude reduction. For example, if a line current of 10,000 A is to be sensed by an ammeter rated for 5 A full scale and the line is passed through the core center to form a one-turn primary winding, the secondary must have 2000 turns.

Unlike the normal power transformer application where a voltage is impressed across the primary winding, the primary of the current transformer can be considered connected to an ac current source. For normal operation, the secondary of the current transformer is connected across the low-impedance input terminals of an ammeter or wattmeter current coil, forming a practical short circuit. Hence, the secondary winding mmf is near in value to the primary winding mmf. The transformer core flux is small and, thus, the induced voltage behind the secondary winding is small in magnitude. However, if for any reason the secondary of a current transformer becomes open circuit, there is no offsetting secondary mmf. The primary mmf then cyclically drives the core flux to saturation level. The secondary voltage can become large in value, presenting danger to personnel and possibly resulting in dielectric failure of the secondary windings. Current transformers are usually designed with a secondary shorting bar that should always be closed prior to removal of an instrument to guard against the possibility of a damaging overvoltage occurrence.

4.12   TRANSFORMER DESIGN

The intended objective of this section is to present one basic approach for single-phase two-winding transformer design. Detailed design procedure for physical sizing, magnetic circuit analysis, and equivalent circuit parameter determination will be examined. The issues of thermal design and insulation design will only be addressed indirectly. British units will be used in the design equations simply because such is the common U.S. industrial practice. References are available to give the reader details of certain derivations and guidance for more refined design.[3-6]

4.12.1   CORE VOLUME SIZING

The two transformer physical core constructions to be considered are the core-type and the shell-type shown in Fig. 4.2. The dimensions ℓ, w, h, and d are sufficient to completely specify the core design. Both transformer types are commonly used in commercial practice. The choice of type is somewhat a designer discretion. Although an acceptable single-phase design can be developed with either type, the core-type design usually results in lower leakage inductance and improved cooling capability than the shell-type. Both of these advantages are directly attributable to a larger proportion of the core-type coils being outside of the ferromagnetic core material. On the other hand, the shell-type design tends to serve as a physical constraint to distortion of the coils due to separating forces on the coils during fault conditions. Also, the single-coil structure of the shell design can offer reduced winding and assembly costs.

The core volume sizing to be developed is not an unconstrained approach. It assumes that a full-load efficiency and particular voltage must be satisfied. Four dimensions (ℓ, w, h, d) are necessary to uniquely define the core; however, a coil window height-to-width ratio ($r_w = h/w$) is specified to reduce the decision to three dimensions. Further, the number of primary coil turns is treated as known a priori. If the resulting core size is unsatisfactory, r_w and the number of primary turns can be changed for another trial.

Shell-Type Transformer   The ferromagnetic core losses (P_m) can be written as

$$P_m = \gamma_m p_m vol_m = 2\gamma_m p_m \ell dSF[\ell + (r_w + 1)w] \qquad \textbf{[4.60]}$$

where γ_m = core material density (lb/in³)
p_m = core loss power density at rated flux density (W/lb)
SF = lamination stacking factor

The coil ohmic losses (P_c) are given by

$$P_c = k_\ell P_m = 2I_1^2 R_1 = 2\rho_c J_c^2 k_c r_w w^2(\ell + d + w) \qquad \textbf{[4.61]}$$

where k_ℓ = ratio of ohmic to core losses

ρ_c = conductor resistivity at operating temperature ($\Omega \cdot$ in)

J_c = rated current conductor current density (A/in²)

k_c = coil window conductor fill factor

Using $\Phi_m = B_m ASF$, [4.7] can be rewritten as

$$V_1 = 4.44 N_1 f B_m \ell dSF/(39.37)^2 \qquad \textbf{[4.62]}$$

where ℓ and d are in inches and B_m is in T.

Core-Type Transformer   The analogous equations for the core-type transformer that differ only in the P_m expression are as follows:

$$P_m = 2\gamma_m p_m \ell dSF[2\ell + (r_w + 1)w] \qquad \textbf{[4.63]}$$

$$P_c = k_\ell P_m = 2\rho_c J_c^2 k_c r_w w^2(d + \ell + w) \qquad \textbf{[4.64]}$$

$$V_1 = 4.44 N_1 f B_m \ell dSF/(39.37)^2 \qquad \textbf{[4.65]}$$

The appropriate set of the above nonlinear algebraic sizing equations can be solved numerically for a set of values for d, w, and ℓ. MATLAB program ⟨spcore.m⟩ has been developed to handle the solution of [4.60] to [4.62] for the single-phase core-type transformer core dimensions. The program uses two supporting m-files:

1. ⟨data1cor.m⟩ serves as an input file for the transformer specification and constants. The file also computes the core loss based on design goal efficiency and apportionment of core and conductor ohmic losses. The window conductor fill factor is evaluated from an empirical formula that can be user-adjusted if a better relationship is known.

2. ⟨core.m⟩ evaluates the sizing equations and their Jacobian as required by ⟨spcore.m⟩ as it executes a Newton-Raphson solution of the sizing equations.

In addition to solution for the core dimension, ⟨spcore.m⟩ calculates an estimate of the magnetizing current for the core with no lamination joints and estimates the weight of the ferromagnetic core material and the copper conductor weight. The screen display results for core sizing of a 2400-V, 15-kVA, 0.8 PF lagging, core-type transformer with a 97.5 percent efficiency are shown below:

```
CORE SIZING for CORE-TYPE TRANSFORMER
1 (in)      d (in)     w (in)    h (in)    N (turns)
1.0e+003 *
0.0027      0.0036     0.0027    0.0054    1.0000

Im (A)      puIm
0.0283      0.0045

Wfe (lb)    Wc (lb)
70.8383     32.0602
```

MATLAB program ⟨spshell.m⟩ solves the sizing equation set [4.63] to [4.65] for the shell-type transformer for the core dimensions with an estimate of the magnetizing current, core material weight, and copper conductor weight. Analogous to ⟨spcore.m⟩, ⟨spshell.m⟩ uses ⟨data1sh.m⟩ for input data and ⟨shell.m⟩ in a Newton-Raphson solution for the core dimension. The resulting screen display for sizing of a 2400-V, 15-kVA, 0.8 PF lagging transformer with a 97.5 percent efficiency is shown below:

```
CORE SIZING for SHELL-TYPE TRANSFORMER
1 (in)     d (in)     w (in)    h (in)    N (turns)
3.2785     3.9853     1.9442    4.8605    750.0000

Im (A)     puIm
0.0382     0.0061

Wfe (lb)   Wc (lb)
70.8383    24.6983
```

It should be pointed out that neither of these two sizing results is necessarily intended to reflect a final design choice. They simply illustrate a first-pass use of the sizing programs.

4.12.2 MAGNETIC CIRCUIT ANALYSIS

The flux path of a transformer ferromagnetic core does not have an air gap by design; however, the core punchings are two-piece construction to allow insertion of the wound coils. For a core-type transformer, the punchings are a *U-I* pair while an *E-I* pair is used for the shell-type transformer. Consequently, there are two joints that must be traversed by the mutual flux linking primary and secondary coils. Since the joints are staggered, the magnetic circuit in the region of a joint is as illustrated by Fig. 4.37*a*, where an air gap due to a lamination joint is in parallel

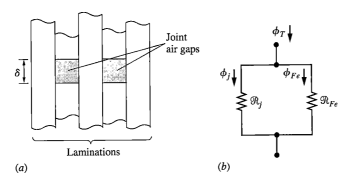

Figure 4.37
Core joints. (a) Physical arrangement. (b) Magnetic circuit model.

with a shunt path of ferromagnetic material. The magnetic circuit of Fig. 4.37b models the flux flow over two adjacent laminations in the region of a joint under the assumption that a plane of equal mmf exists immediately above and below the joint. If the level of flux density is sufficiently low, flux will bridge the joint air gap by flowing through the shunt ferromagnetic path. As the level of flux density increases, the shunt path begins to saturate and flux will also flow across the joint air gap path. If a value of uniform magnetic field intensity H_{Fe} is assumed to exist along the shunt lamination path, the joint mmf drop is given by

$$F_T = H_{Fe}\delta \qquad\qquad \textbf{[4.66]}$$

Since this identical mmf must also exist across the air path of the joint, the effective flux density across the joint-shunt lamination pair is

$$B_T = \frac{1}{2}(B_{Fe} + \mu_o H_{Fe}) \qquad\qquad \textbf{[4.67]}$$

The MATLAB program ⟨joint.m⟩ has been developed to evaluate the joint mmf drop. A value of flux density (B_{Fe}) in the shunt lamination path is assumed. The corresponding H_{Fe} is determined from the B-H curve of the core material through function-file ⟨Hm7.m⟩. Pairs of joint mmf (F_T) and effective flux density (B_T) can be calculated over a wide range of core flux. The plotted result of Fig. 4.38 provides an assessment of the addition coil MMF required due to presence of lamination joints as a function of the uniform core flux density.

Although the transformer operates at near constant flux density, an assessment of the magnetization curve over a range of values that extends above rated flux density is of value in clear assessment of the degree of saturation, inrush current nature, and the overvoltage characteristics. MATLAB program ⟨tranmag.m⟩ determines and plots the magnetization curve (terminal voltage vs. current). Input data and transformer type are supplied by the m-file ⟨sptdata.m⟩. Whether the transformer is shell-type or core-type, ⟨tranmag.m⟩ branches to appropriate magnetic circuit description. The program assumes uniform core flux density and determines the magnetic field intensity from linear interpolation between stored B-H data

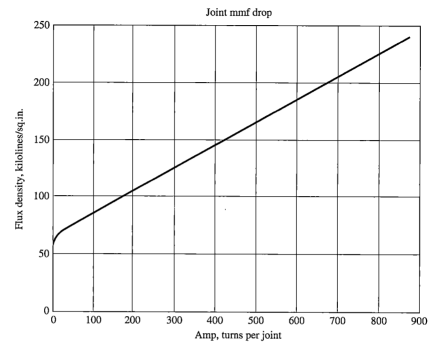

Figure 4.38
Lamination joint mmf drop

points through use of the function-files ⟨Hm7.m⟩ or ⟨Hm7cg.m⟩, depending upon whether the point of analysis is in the with-grain or cross-grain section of the core punching. Figure 4.39 shows the magnetization curve displayed by execution of ⟨tranmag.m⟩. The program also calculates the core losses at the rated voltage point from the core loss values available from the function-file ⟨Pcm7.m⟩. Using core losses and magnetizing current at rated voltage, core loss resistance R_c, and magnetizing reactance X_m for use in the equivalent circuit are determined and displayed to the screen as shown below:

```
MAGNETIZING  REACTANCE  &  CORE  LOSS  RESISTANCE
        Xm  (ohms)          Rc  (ohms)
        1.246e+004          6.702e+004
```

4.12.3 EQUIVALENT CIRCUIT PARAMETERS

Values for the magnetizing reactance X_m and the core loss resistance R_c are available from execution of ⟨tranmag.m⟩ as already discussed. However, theoretical justification of these values is in order.

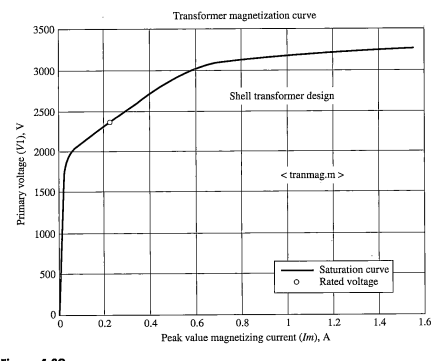

Figure 4.39
Transformer magnetization curve

Magnetizing Reactance The value of magnetizing reactance is determined by the rated value of excitation branch voltage divided by the associated value of magnetizing current.

$$X_m = \frac{E_1}{I_m} \cong \frac{0.99\,V_1}{I_m} \qquad\qquad \textbf{[4.68]}$$

When referring to the saturation curve of Fig. 4.39, it should be noted that V_1 is an rms value while I_m is peak value. Hence, a value of I_m from Fig. 4.39 must be divided by $\sqrt{2}$ prior to use in [4.68]. If ⟨tranmag.m⟩ were not available, the rms value of magnetizing current can be determined by $I_m = F_T/(\sqrt{2}N_1)$ where F_T is the peak value of the total core mmf requirement at the rated value of mutual flux density. F_T must include joint mmf drops and account for any increase in magnetic field intensity in core areas of cross-grain flux flow.

Core Loss Resistance Again, if ⟨tranmag.m⟩ were not available, the core loss power density (p_c) can be determined from the electric sheet steel manufacturer's published Epstein loss curves at the rated level of core flux density. The total core loss (P_c) is found as the product of loss power density and core volume (vol_c). Due

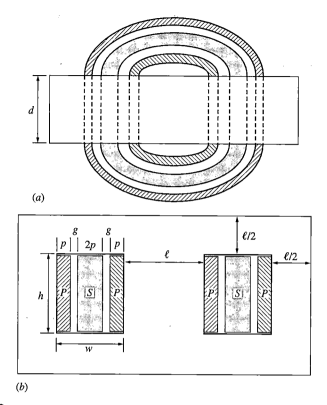

(a)

(b)

Figure 4.40
Shell-type transformer with concentric layered winding. (a) Top view. (b) Front view.

to clamping force and punching burrs, the Epstein loss value should be increased by a factor of 2 or more. The core-loss resistance follows as

$$R_c = \frac{E_1^2}{P_c} \cong \frac{(0.9V_1)^2}{2p_c vol_c}.$$
                                                      [4.69]

Leakage Reactances . Calculation of leakage reactance values requires knowledge of the coil configuration as illustrated by Fig. 4.3. A shell-type transformer with a concentric layer winding has been redrawn with dimensions in Fig. 4.40. Similar redrawn figures for the other arrangements of Fig. 4.3 could be done. A certain symmetry of each layered coil structure is assumed. The outer and inner layers are primary winding conductors plus insulation of unit thickness p. The number of layers between the inner and outer layers must be an odd number, and each has a combined conductor and insulation thickness of 2 unit layers $(2p)$. These inside layers alternate between primary and secondary conductors, always beginning with secondary conductors. A spacing of thickness (g) between each coil is provided for cooling air or liquid flow.

The described layer pattern exists for both concentric-layered and vertical-layered coils of both core-type and shell-type transformers. The number of layers per coil structure will always be an odd number. Since a shell-type transformer has only one coil structure, the total number of layers or coil sections n will always be an odd integer greater than or equal to 3 ($n \geq 3$, odd). The core-type transformer has two coil structures; hence, the total number of layers or coil sections must be an even integer given by $n = 2(k + 2)$ for k odd. The number of turns per unit layer of the primary and secondary are, respectively,

$$n_1 = N_1/(n - 1) \qquad n_2 = N_2/(n - 1) \qquad \textbf{[4.70]}$$

where N_1 and N_2 are the total primary and secondary turns.

The unit layer thickness (p), the primary coil leakage inductance $(L_{\ell 1})$, and the secondary coil leakage inductance $(L_{\ell 2})$ for each of the arrangements of Fig. 4.3 with dimensions in inches are given as follows:

Shell-type, concentric-layer:

$$p = \frac{w - (n - 1)g}{2(n - 1)} \qquad \textbf{[4.71]}$$

$$L_{\ell 1} = 6.384 \times 10^{-8}(n - 1)n_1^2(p/3 + g/2)\left[\frac{d + \pi(\ell + w)}{w}\right] \text{(H)} \quad \textbf{[4.72]}$$

$$L_{\ell 2} = 6.384 \times 10^{-8}(n - 1)n_2^2(p/3 + g/2)\left[\frac{d + \pi(\ell + w)}{w}\right] \text{(H)} \quad \textbf{[4.73]}$$

Shell-type, vertical-layer:

$$p = \frac{h - (n - 1)g}{2(n - 1)} \qquad \textbf{[4.74]}$$

$$L_{\ell 1} = 6.384 \times 10^{-8}(n - 1)n_1^2(p/3 + g/2)\left[\frac{d + \pi(\ell + w)}{h}\right] \text{(H)} \quad \textbf{[4.75]}$$

$$L_{\ell 2} = 6.384 \times 10^{-8}(n - 1)n_2^2(p/3 + g/2)\left[\frac{d + \pi(\ell + w)}{h}\right] \text{(H)} \quad \textbf{[4.76]}$$

Core-type, concentric-layer:

$$p = \frac{w - (n - 1)g}{2(n - 2)} \qquad \textbf{[4.77]}$$

$$L_{\ell 1} = 6.384 \times 10^{-8}(n - 2)n_1^2(p/3 + g/2)\left[\frac{2d + \pi(\ell + w/2)}{w}\right] \text{(H)} \quad \textbf{[4.78]}$$

$$L_{\ell 2} = 6.384 \times 10^{-8}(n - 2)n_2^2(p/3 + g/2)\left[\frac{2d + \pi(\ell + w/2)}{w}\right] \text{(H)} \quad \textbf{[4.79]}$$

Core-type, vertical-layer:

$$p = \frac{h - (n/2 - 1)g}{n - 2} \tag{4.80}$$

$$L_{\ell 1} = 6.384 \times 10^{-8}(n - 2)n_1^2(p/3 + g/2)\left[\frac{2d + \pi(\ell + w/2)}{h}\right] \text{(H)} \tag{4.81}$$

$$L_{\ell 2} = 6.384 \times 10^{-8}(n - 2)n_2^2(p/3 + g/2)\left[\frac{2d + \pi(\ell + w/2)}{h}\right] \text{(H)} \tag{4.82}$$

The leakage reactance values are determined by multiplying each leakage inductance by $\omega = 2\pi f$.

Coil Resistances   Let *puc* denote the per unit portion of each layer that is a conductor, whereas the balance of the area is insulation. Then the conductor cross-sectional area (A_c) and the mean length turn of a conductor (MLT_c) can be determined for each of the arrangements of Fig. 4.3.

Shell-type, concentric-layer:

$$A_c = \frac{pucp(h - g)}{n_1} \text{ (in}^2) \tag{4.83}$$

$$MLT_c = 2(\ell + d + w) \tag{4.84}$$

Shell-type, vertical-layer:

$$A_c = \frac{pucp(w - g)}{n_1} \tag{4.85}$$

$$MLT_c = 2(\ell + d + w) \tag{4.86}$$

Core-type, concentric-layer:

$$A_c = \frac{pucp(h - g)}{n_1} \tag{4.87}$$

$$MLT_c = 2(\ell + d + w) \tag{4.88}$$

Core-type, vertical-layer:

$$A_c = \frac{pucp(w - g)}{n_1} \tag{4.89}$$

$$MLT_c = 2(\ell + d + w) \tag{4.90}$$

Assuming that the ac value of resistance is 5 percent greater than the dc value and using the conductor resistivity (ρ_c) value for operating temperature in units of $\Omega \cdot$in, the coil resistance values are given by

$$R_1 = 1.05\rho_c MLT_c/A_c \qquad\qquad \textbf{[4.91]}$$

$$R_2 = \left(\frac{N_2}{N_1}\right)^2 R_1 \qquad\qquad \textbf{[4.92]}$$

The MATLAB program ⟨leakind.m⟩ has been developed to compute the values of leakage inductance, leakage reactance, and coil resistance. The program also calculates the conductor current density and cross-sectional area for the primary and secondary conductors. Based on the conductor current density, the user can assess adequacy of the window area with the insulation areas selected indirectly by specification of *puc*. Also, the particular primary and secondary conductors can be chosen with knowledge of conductor cross-sectional areas. A screen display from execution of ⟨leakind.m⟩ is shown below:

```
RESISTANCE  &  LEAKAGE  REACTANCE  CALCULATIONS
      Style            Wndg  sects         Wndg  type
   core__type               6             conc_layer
   R1  (ohms)          R2  (ohms)         Jc  (apsi)
      5.266              0.05266            835.1
  Ac1  (sq  in)       Ac2  (sq  in)
    0.003742            0.03742
   L1  (H)      L2  (H)      X1  (ohms)    X2  (ohms)
   0.02327    0.0002327      8.772          0.08772
```

4.12.4 SAMPLE DESIGN

The two-winding, single-phase transformer to be designed is specified as follows:

480:240 V   15 kVA   60 Hz

Core-type   Concentric-layered winding

97% full-load efficiency at 0.8 PF lagging

Air cooled   Class H insulation

The first procedure of the design process is to select a core size using ⟨spcore.m⟩ to solve the nonlinear equations [4.63] to [4.65] for a set of core dimensions. Values for ⟨data1cor.m⟩ not explicitly dictated by the design specification must be selected. It is decided to use 0.014 in thick, M-7 electric sheet steel with a rated point flux density of 1.5 T. Hence, the values of $GBm = 1.5$ T, $GHm = 0.92$ Oe, and $Gpm = 0.7$ W/lb with a 2.5 multiplier are appropriate. The stacking factor of 0.95 and core material density of 0.283 lb/in³ are somewhat standard choices. The primary turns should be in the range of 2 to 3 V/turn. A selection of

GN = 220 turns gives 2.18 V/turn. Selection of a smaller value of turns increases the turn insulation requirement and increases either the flux density or the core size, both of which tend to increase the core losses. The design specification only gave an efficiency specification for full-load. However, this low-voltage transformer will find end-use application with inherent load variation. By choosing $Gk1$ = 1.25, the transformer will exhibit maximum efficiency for approximately 100/1.25% = 80% load.

Since the transformer has Class H (180°C) insulation, it can operate at compatible temperatures. A choice of $Gdelc$ = 1250 A/in^2 with convection air flow is consistent with average coil temperatures in the 150°C range allowing for 180°C hot spot. The associated resistivity is $Grhoc$ = 1.02 × 10^{-6} Ω·in. Finally, a somewhat common window height to width ratio of Grw = 2 is chosen. Execution of ⟨spcore.m⟩ using the selected values yields

```
CORE  SIZING  for  CORE-TYPE  TRANSFORMER
l (in)      d (in)      w (in)      h (in)      N (turns)
2.9233      3.0473      3.1410      6.2820      220.0000

Im (A)      puIm
0.1476      0.0047

Wfe (lb)    Wc (lb)
73.1429     43.4699
```

The results of ⟨spcore.m⟩ give a guideline for core dimensions. Typically, a set of values not significantly different should be used. Based on the above values, choose $\ell = d = w$ = 3.00 in and h = 6.00 in.

The second step of the design process is to use ⟨tranmag.m⟩ to check for any undue saturation at the operating point and to compute the magnetizing reactance and core loss resistance. Execution with N = 220 turns, $\ell = d = w$ = 3.00, and h = 6.00 produces

```
MAGNETIZING  REACTANCE  &  CORE  LOSS  RESISTANCE
Xm (ohms)      Rc (ohms)
   392.4          1807
```

The magnetization curve shows a typical magnetizing current that is approximately 3.8 percent of rated current; hence, no saturation problem is apparent.

The third design procedure is to utilize ⟨leakind.m⟩ to determine the leakage reactance and coil resistance values and to calculate the conductor area for conductor sizing. The minimum number of winding sections (n = 6 for core winding)

should be used unless the leakage reactance is large enough to result in an unacceptable regulation value. A practical value of $g = 0.125$ in is selected for intersectional coil spacing which can be implemented by insertion of vertical oriented 0.125-in square bars of insulation material during the coil winding process. This choice of spacers occupies $(n - 1)g/w = 5 \times 0.125/3 = 0.21$ or 21 percent of the window area. Use of square or rectangular conductors that give a good fill factor is anticipated. Estimating that 20 percent of the allotted winding area after spacers are selected is required for ground and turn insulation, $puc = 0.8$ is selected. With the prior determined values of $\ell, d, w, h, g, N_1, N_2(= N_1/a), rhoc$ and puc placed in $\langle$leakind.m$\rangle$, the results are

```
RESISTANCE  &  LEAKAGE  REACTANCE  CALCULATIONS
      Style          Wndg  sects      Wndg  type
   core__type             6           conc_layer
   R1  (ohms)       R2  (ohms)       Jc  (apsi)
     0.1672           0.04179           1232
 Ac1  (sq in)     Ac2  (sq in)
     0.02537          0.05074
    L1   (H)          L2  (H)        X1  (ohms)      X2  (ohms)
   0.0008372        0.0002093         0.3156          0.0789
```

The fourth design procedure is to assemble the equivalent circuit parameter values now known and enter them into $\langle$tranperf1.m$\rangle$ to assess the success in meeting the required efficiency performance. The screen display from the resulting run is shown below:

```
          TRANSFORMER  PERFORMANCE
  V1 mag      V1  ang      V2  mag      V2  ang
  500.7        1.075         240           0

  E1 mag      E1  ang
  490.1        0.556

  I1 mag      I1  ang      I2  mag      I2  ang
  32.24       -38.34        62.5        -36.87
  Im mag      Im  ang      Ic  mag      Ic  ang
  1.249       -89.44       0.2712        0.556

  PFin         PFS1         PFout        PFS2
  0.7726      lagging        0.8        lagging

  Pin          Pout          eff         %Reg
  12.47         12          96.23        4.224
```

Inspection of the above results shows that the required efficiency of 97 percent has not been met. Hence, design refinement is necessary. The core losses, the winding losses, or a combination of both must be reduced. Any loss reduction has the penalty of added material.

As a quick check, the core losses are $P_c \cong V_1^2/R_c = (480)^2/1807 = 127.5$ W. Since this represents approximately 1 percent of the total losses of the present design, an attempt to reduce the core losses is not practical. Attention will be directed to reduction of the winding losses.

The approach will be to increase the winding window size to allow increased conductor cross-sectional area to reduce the winding resistance values. The process becomes an exercise in trial and error. Meeting the efficiency goal means that winding losses will have to be reduced to approximately two-thirds of the present value. Choose:

$$\ell = 3 \qquad d = 3 \qquad w = 3.5 \qquad h = 6.5 \qquad \text{in}$$

This is an increase in window area of approximately 25 percent. Since 80 percent of the window is actual conductor, this choice allows significant reduction in conductor cross-sectional area and thus winding resistance. Further, the 25 percent reduction in winding resistance yields reduction in the winding temperature and an associated 25 percent reduction in the value of conductor resistivity *rhoc*. Using the newly selected values of w and h along with $rhoc = 1.02 \times 10^{-6}/1.25 \cong 0.81 \times 10^6 \, \Omega \cdot \text{in}$, the screen displays from ⟨tranmag.m⟩, ⟨leakind.m⟩, and ⟨tranperf1.m⟩, respectively, are displayed below:

```
MAGNETIZING  REACTANCE  &  CORE  LOSS  RESISTANCE
Xm  (ohms)    Rc  (ohms)
386.7         1694
```

```
RESISTANCE  &  LEAKAGE  REACTANCE  CALCULATIONS
      Style            Wndg  sects        Wndg  type
    core__type             6             conc_layer

   R1  (ohms)           R2  (ohms)         Jc  (apsi)
     0.1067               0.02667           937.8

  Ac1  (sq  in)        Ac2  (sq  in)
     0.03332              0.06665

   L1  (H)       L2  (H)       X1  (ohms)     X2  (ohms)
 0.0008418     0.0002104        0.3173         0.07933
```

TRANSFORMER PERFORMANCE			
V1 mag	V1 ang	V2 mag	V2 ang
497.8	1.361	240	0
E1 mag	E1 ang		
488.7	0.6956		
I1 mag	I1 ang	I2 mag	I2 ang
32.26	-38.34	62.5	-36.87
Im mag	Im ang	Ic mag	Ic ang
1.264	-89.3	0.2885	0.6956
PFin	PFS1	PFout	PFS2
0.7694	lagging	0.8	lagging
Pin	Pout	eff	%Reg
12.36	12	97.12	3.618

It is seen from the ⟨tranperf1.m⟩ output that the efficiency goal of 97 percent has now been satisfied. Figure 4.41 displays results from ⟨traneff.m⟩ for the efficiency

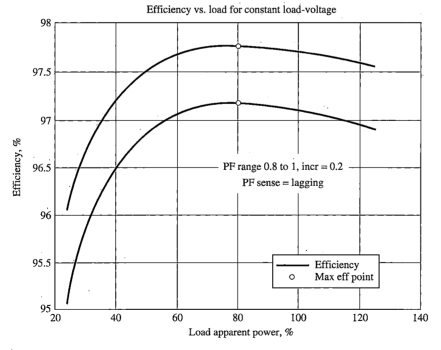

Figure 4.41
Efficiency vs. load for sample design

load characteristic for both 0.8 PF lagging and unity PF. The primary and secondary conductors can now be selected to finish the design. For this small transformer, it is desirable to use a standard square wire size if possible. The closest standard wire sizes to satisfying A_{c1} and A_{c2} are 5 AWG square and 2 AWG square, respectively. Each of these choices has approximately 6 percent less than the desired cross-sectional area; however, the efficiency goal has been exceeded by 0.12 percent, allowing a slight reduction in the values of A_{c1} and A_{c2} listed above to accommodate this reduced conductor cross section and still satisfy the efficiency requirement of 97 percent.

4.13   COMPUTER ANALYSIS CODE

```
%%%%%%%%%%%%%%%%%%%%%%%%%%%%%%%%%%%%%%%%%%%%%%%%%%%%%%%%%%%%%%%%%%%%%%%%%%%%
%
% hyloop.m - Builds embedded hysteresis loops & calculates areas.
%            Techniques of this program used by hyst.m.
%
%%%%%%%%%%%%%%%%%%%%%%%%%%%%%%%%%%%%%%%%%%%%%%%%%%%%%%%%%%%%%%%%%%%%%%%%%%%%
clear; clf;
Icmax=0.2; % Maximum coercive current. Change to alter loop widths.
puFsat=0.7; % Per unit flux at saturation curve knee.
puIsat=0.3; % Per unit current at saturation curve knee.
nloops=5; pctsat=linspace(0,1,nloops+1); % No. loops & peak fluxes
for k=1:nloops+1
   y1=[-1 -puFsat 0 puFsat 1]'; x1=[-1 -puIsat 0 puIsat 1]';
   Ic=pctsat(k)*Icmax;
   if pctsat(k) > y1(4); x1max=interp1(y1,x1,pctsat(k));
      if x1max <= x1(4)+Ic/2; x1max=1.01*(x1(4)+Ic/2); else; end
   x1(1)=-x1max; x1(5)=x1max; y1(1)=-pctsat(k); y1(5)=pctsat(k);
   xu=x1+[0 Ic/2 Ic Ic/2 0]'; xd=x1-[0 Ic/2 Ic Ic/2 0]'; yu=y1; yd=y1;
   wu=0; wd=0;
      for i=1:4   % Calculate hysteresis loop area
      wu=wu+(xu(i)+xu(i+1))/2*(yu(i+1)-yu(i));
      wd=wd+(xd(i)+xd(i+1))/2*(yd(i+1)-yd(i));
      end
   W(k)=wu-wd;
   else
   xu=x1+[0 Ic/2 Ic Ic/2 0]'; xd=x1-[0 Ic/2 Ic Ic/2 0]'; yu=y1; yd=y1;
   xmaxu=interp1(yu, xu, pctsat(k));
   ydwn=pctsat(k)^2/y1(4);
   if abs(ydwn-pctsat(k))<1e-4; ydwn=0.85*pctsat(k); end
   xdwn=interp1(yd, xd, ydwn);
```

```
   xu=[-xmaxu -xdwn Ic xmaxu]'; xd=-flipud(xu);
   yu=[-pctsat(k) -ydwn 0 pctsat(k)]'; yd=-flipud(yu);
   wu=0; wd=0;
   for i=1:3      % Calculate hysteresis loop area
      wu=wu+(xu(i)+xu(i+1))/2*(yu(i+1)-yu(i));
      wd=wd+(xd(i)+xd(i+1))/2*(yd(i+1)-yd(i));
      end
   W(k)=wu-wd;
   end
% Plot hysteresis loop
plot(xu,yu,xd,yd); hold on;
end
grid; title('Hysteresis loops');
xlabel('Normalized Current'); ylabel('Normalized flux');
hold off
disp(' '); disp(' ');
disp(' Per Unit Energy Loss of Hysteresis Loops (min -> max)');
Energy=[W(2:nloops+1)]; disp(' '); disp(Energy);


%%%%%%%%%%%%%%%%%%%%%%%%%%%%%%%%%%%%%%%%%%%%%%%%%%%%%%%%%%%%%%%%%%%%%%%%%
%
% hyst.m — Plots voltage-flux-current relationships for a coil
%          wound on a ferromagnetic structure that exhibits
%          hysteresis. The hysteresis loop is formed by piece-
%          wise linear approximation.
%
%%%%%%%%%%%%%%%%%%%%%%%%%%%%%%%%%%%%%%%%%%%%%%%%%%%%%%%%%%%%%%%%%%%%%%%%%
clear;
y1=[-1 -0.9 0 0.9 1]'; x1=[-1 -0.3 0 0.3 1]'; % Base curve for
Icmax=0.1;  %Maximum coercive current  % hysteresis buildup
pctsat=input('Percentage maximum flux = ')/100; Ic=pctsat*Icmax;
if pctsat > 1; pctsat=1; else; end
if pctsat > y1(4); x1max=interp1(y1,x1,pctsat);
   if x1max <= x1(4)+Ic/2; x1max=1.01*(x1(4)+Ic/2); else; end
   x1(1)=-x1max; x1(5)=x1max; y1(1)=-pctsat; y1(5)=pctsat;
   xu=x1+[0 Ic/2 Ic Ic/2 0]'; xd=x1-[0 Ic/2 Ic Ic/2 0]'; yu=y1; yd=y1;
else
   xu=x1+[0 Ic/2 Ic Ic/2 0]'; xd=x1-[0 Ic/2 Ic Ic/2 0]'; yu=y1; yd=y1;
   xmaxu=interp1(yu,xu,pctsat);
   ydwn=pctsat^2/y1(4);
   if abs(ydwn-pctsat)<1e-4; ydwn=0.85*pctsat; end
   xdwn=interp1(yd,xd,ydwn);
   xu=[-xmaxu -xdwn Ic xmaxu]'; xd=-xu;
   yu=[-pctsat -ydwn 0 pctsat]'; yd=-yu;
end
figure(1);         % Plot hysteresis loop
plot(xu,yu,xd,yd); grid; title('Hysteresis loop');
```

```
xlabel('Normalized current'); ylabel('Normalized flux'); pause; clf;
% Build response for the case of an injected sinusoidal current.
ang=linspace(0,2*pi,1024)'; x=max(xu)*sin(ang); npts=length(x);
y=zeros(npts,1); y(1)=interp1(xu,yu,x(1));
for i=2:npts
  if x(i) > x(i-1)
    y(i)=interp1(xu,yu,x(i));
  else y(i)=interp1(xd,yd,x(i));
  end;
end
% Frequency domain differentiation of flux to yield voltage
dy=fft(y); deg=180/pi;
for i=1:1024; dy(i)=-j*i*dy(i); end;
dy=ifft(dy); dy=real(dy*max(y)/max(abs(dy)));
plot(ang*deg,x,ang*deg,y,'--',ang*deg,dy,'-.'); grid;
title('Response due to injected sinusoidal current');
ylabel('Current, voltage, flux (normalized)'); xlabel('Degrees');
legend('Coil current','Coil flux','Terminal voltage'); pause
legend off; f=abs(fft(dy)); f(1)=f(1)/2; f=f/max(f)*100;
plot([0:25],f(1:26)); grid
title('Voltage for sinusoidal current');
ylabel('Percent'); xlabel('Harmonic number');pause;
% Build response for the case of an impressed sinusoidal voltage.
ang=linspace(0,2*pi,1024)'; y=max(yu)*sin(ang); npts=length(y);
x=zeros(npts,1); x(1)=interp1(yu,xu,y(1));
for i=2:npts
  if y(i) > y(i-1)
    x(i)=interp1(yu,xu,y(i));
  else x(i)=interp1(yd,xd,y(i));
  end
end
% Frequency domain differentiation of flux to yield voltage
dy=fft(y); deg=180/pi;
for i=1:1024; dy(i)=-j*i*dy(i); end;
dy=ifft(dy); dy=real(dy*max(y)/max(abs(dy)));
% Determine fundamental component of current
x1=fft(x); x1=abs(x1(2))*cos(ang+angle(x1(2)))/512;
plot(ang*deg,x,ang*deg,y,'--',ang*deg,dy,'-.',ang*deg,x1,':');
grid; title('Response due to impressed sinusoidal voltage');
ylabel('Current, voltage, flux (normalized)'); xlabel('Angle');
legend('Coil current','Coil flux','Terminal voltage', ...
   ' Fund. current'); pause;
legend off; f=abs(fft(x)); f(1)=f(1)/2; f=f/max(f)*100;
plot([0:25],f(1:26)); grid
title('Current for sinusoidal voltage');
ylabel('Percent'); xlabel('Harmonic number');
legend off;
Im=abs(fft(x)); save Imag Im % Im saved for use with imthd.m
```

```
%%%%%%%%%%%%%%%%%%%%%%%%%%%%%%%%%%%%%%%%%%%%%%%%%%%%%%%%%%%%%%%%%%%%%%%%%
%
% imthd.m — Calculates percent total harmonic distortion
%            and percent 3rd & 5th harmonic content of the
%            input current to a single-phase transformer
%            based on harmonic profile determined by hyst.m.
%            Assumes that magnetizing current is 3% of full-
%            load current and that a 2% of full-load core loss
%            fundamental component of current exists in phase
%            with terminal voltage. Must have executed hyst.m
%            in past so that Imag.mat exists with stored Im
%            harmonic profile for desired percentage flux.
%
%%%%%%%%%%%%%%%%%%%%%%%%%%%%%%%%%%%%%%%%%%%%%%%%%%%%%%%%%%%%%%%%%%%%%%%%%
clear; load Imag.mat Im; PF=0.80; % Neg PF indicates leading
npts=50; puload=linspace(0.01,1,npts); % Per unit load
% Add core loss component & fundamental of Im to·load current to
% Form total fundamental per unit input current.
for i=1:npts
  puI(i)=abs(0.02+puload(i)*exp(-j*PF/abs(PF)*acos(abs(PF)))- ...
    j*Im(2)/max(Im)*0.03);
end
for i=1:npts
  z=0;
  for j=4:2:32 ·     % Odds thru 31st harmonic
    z=z+(Im(j)/max(Im)*0.03/puI(i))^2;
  end
  thd(i)=sqrt(z)*100;       % Total harmonic distortion
  hd3(i)=Im(4)/max(Im)*0.03/puI(i)*100; % Percentage 3rd harmonic
  hd5(i)=Im(6)/max(Im)*0.03/puI(i)*100; % Percentage 5th harmonic
end
load=puload*100;
plot(load,thd,load,hd3,'--',load,hd5,'-.'); grid
title('Harmonic content of transformer input current');
ylabel('Percentage'); xlabel('Percent load current');
legend('Total harmonic distortion','Percent 3rd harmonic', ...
  'Percent 5th harmonic',1);
text(65,0.7*max(thd),['PF = ',num2str(PF)]);



%%%%%%%%%%%%%%%%%%%%%%%%%%%%%%%%%%%%%%%%%%%%%%%%%%%%%%%%%%%%%%%%%%%%%%%%%
%
% traneqckt.m - Determines transformer equivalent circuit
%               parameters from open-circuit & short-circuit
%               test data. Assumes high voltage winding is
%               primary and assigns subscript 1.
%
%%%%%%%%%%%%%%%%%%%%%%%%%%%%%%%%%%%%%%%%%%%%%%%%%%%%%%%%%%%%%%%%%%%%%%%%%
```

```
clear;
V1=480; V2=240; n=V1/V2;   % Rated voltage values
Vsc=37.2; Isc=51.9; Psc=750; scside='hgh';      % Test data
Voc=240; Ioc=9.7; Poc=720; ocside='low';
% dc resistance of windings — if not known, set R1dc=R2dc=1
R1dc=0.110; R2dc=0.029;
if scside == 'low'  % Refer all data to high side
  Vsc=n*Vsc; Isc=Isc/n;
else; end
if ocside == 'low'
  Voc=n*Voc; Ioc=Ioc/n;
else; end
Req=Psc/Isc^2; R1=Req*R1dc/(R1dc+n^2*R2dc); R2=(Req-R1)/n^2;
Zsc=Vsc/Isc; X1=sqrt(Zsc^2-Req^2)/2; X2=X1/n^2;
Rc=Voc^2/Poc; Xm=Voc/sqrt(Ioc^2-(Voc/Rc)^2);
disp(' '); disp(['   TRANSFORMER EQUIVALENT CIRCUIT PARAMETERS — ', ...
  date]);
disp('              Classical Data Reduction');
disp(' '); disp(' ')
disp([blanks(3) 'R1(ohm)' blanks(8) 'R2(ohm)' blanks(8) 'X1(ohm)' ...
      blanks(8) 'X2(ohm)']);
disp([blanks(3) num2str(R1) blanks(9) num2str(R2) ...
      blanks(8) num2str(X1) blanks(9) num2str(X2)]);
disp(' ');
disp([blanks(3) 'Rc(ohm)' blanks(8) 'Xm']);
disp([blanks(4) num2str(Rc) blanks(8) num2str(Xm)]);
% Refinement of data reduction
thetoc=acos(Poc/Voc/Ioc); Eoc=abs(Voc-(R1+j*X1)*Ioc*exp(-j*thetoc));
Poc=Poc-Ioc^2*R1; Qoc=Voc*Ioc*sin(thetoc)-Ioc^2*X1;
Rc=Eoc^2/Poc; Xm=Eoc^2/Qoc;
thetsc=acos(Psc/Vsc/Isc); Esc=abs(Vsc-(R1+j*X1)*Isc*exp(-j*thetsc));
Psc=Psc-Esc^2/Rc; Qsc=Vsc*Isc*sin(thetsc)-Esc^2/Xm;
Req=Psc/Isc^2; R1=Req*R1dc/(R1dc+n^2*R2dc); R2=(Req-R1)/n^2;
X1=Qsc/Isc^2/2; X2=X1/n^2;
disp(' '); disp(['   TRANSFORMER EQUIVALENT CIRCUIT PARAMETERS — ', ...
  date]);
disp('              Refined Data Reduction');
disp(' '); disp(' ')
disp([blanks(3) 'R1(ohm)' blanks(8) 'R2(ohm)' blanks(8) 'X1(ohm)' ...
      blanks(8) 'X2(ohm)']);
disp([blanks(3) num2str(R1) blanks(9) num2str(R2) ...
      blanks(8) num2str(X1) blanks(9) num2str(X2)]);
disp(' ');
disp([blanks(3) 'Rc(ohm)' blanks(8) 'Xm']);
disp([blanks(4) num2str(Rc) blanks(8) num2str(Xm)]);
```

```
%%%%%%%%%%%%%%%%%%%%%%%%%%%%%%%%%%%%%%%%%%%%%%%%%%%%%%%%%%%%%%%%%%%%%%%%%%%
%
% tranperf1.m - Calculates voltages, currents, regulation and
%               efficiency for a given load with constant load
%               voltage maintained.
%
%%%%%%%%%%%%%%%%%%%%%%%%%%%%%%%%%%%%%%%%%%%%%%%%%%%%%%%%%%%%%%%%%%%%%%%%%%%
clear;
R1=0.06; R2=0.015; X1=0.18; X2=0.045; Xm=400; Rc=1200;
V1=240; V2=120; KVA=5; n=V1/V2; R2p=n^2*R2; X2p=n^2*X2;
R1=0.1067; R2=0.02667; X1=0.3173; X2=0.07933; Xm=386.7; Rc=1694;
V1=480; V2=240; KVA=15; n=V1/V2; R2p=n^2*R2; X2p=n^2*X2;
R1=0.068; R2=0.017; X1=0.14; X2=0.035; Xm=515.4; Rc=2097;
disp('       RATED VOLTAGE LOAD CHARACTERISTIC'); disp(' ');
Load=input('Percent Load( 0->100 ) = ');
PF=input('Load Power Factor( 0->1 ) = ');
if PF ~= 1
   PFS2=input('PF Sense( leading,lagging ) = ','s');
else
   PFS2='unityPF'
end
Pout=1000*KVA*PF*Load/100;
if PFS2 == 'lagging'; thet2=-acos(PF);
elseif PFS2 == 'leading'; thet2=acos(PF);
else thet2=0;
end
I2=KVA*1000/V2*Load/100*exp(j*thet2); E1=(R2p+j*X2p)*I2/n+n*V2;
Im=E1/(j*Xm); Ic=E1/Rc; Io=Ic+Im; I1=I2/n+Io;
V1=(R1+j*X1)*I1+E1;
thet1=angle(V1)-angle(I1); PFin=cos(thet1);
if thet1 > 0; PFS1='lagging';
elseif thet1 < 0; PFS1='leading';
else PFS1='unityPF';end
Pin=real(V1*conj(I1)); eff=Pout/Pin*100; deg=180/pi;
Zm=j*Rc*Xm/(Rc+j*Xm); Znl=R1+j*X1+Zm;
Vnl=abs(V1*Zm/Znl); Reg=(Vnl/n-V2)/V2*100;
deg=180/pi; angV1=angle(V1)*deg; angV2=angle(V2)*deg;
angE1=angle(E1)*deg; angI1=angle(I1)*deg; angI2=angle(I2)*deg;
angIm=angle(Im)*deg; angIc=angle(Ic)*deg;
disp(' '); disp('                 TRANSFORMER PERFORMANCE');
disp(' '); disp(' ')
disp([blanks(3) 'V1 mag' blanks(8) 'V1 ang' blanks(8) 'V2 mag' ...
     blanks(8) 'V2 ang']);
disp([blanks(3) num2str(abs(V1)) blanks(11) num2str(angV1) ...
     blanks(13) num2str(abs(V2)) blanks(9) num2str(angV2)]);
disp(' ');
disp([blanks(3) 'E1 mag' blanks(8) 'E1 ang']);
disp([blanks(4) num2str(abs(E1)) blanks(7) num2str(angE1)]);
disp(' ');
```

```
disp([blanks(3) 'I1 mag' blanks(8) 'I1 ang' blanks(8) 'I2 mag' ...
     blanks(8) 'I2 ang']);
disp([blanks(3) num2str(abs(I1)) blanks(9) num2str(angI1) ...
     blanks(8) num2str(abs(I2)) blanks(9) num2str(angI2)]);
disp(' ');
disp([blanks(3) 'Im mag' blanks(8) 'Im ang' blanks(8) 'Ic mag' ...
     blanks(8) 'Ic ang']);
disp([blanks(3) num2str(abs(Im)) blanks(8) num2str(angIm) ...
     blanks(8) num2str(abs(Ic)) blanks(7) num2str(angIc)]);
disp(' ');
disp([blanks(4) 'PFin' blanks(10) 'PFS1' blanks(10) 'PFout' ...
     blanks(9) 'PFS2']);
disp([blanks(3) num2str(PFin) blanks(8) PFS1 ...
     blanks(9) num2str(PF) blanks(9) PFS2 ]);
disp(' ');
disp([blanks(4) 'Pin' blanks(11) 'Pout' blanks(11) 'eff' ...
     blanks(10) '%Reg']);
disp([blanks(4) num2str(Pin/1000) blanks(10) num2str(Pout/1000) ...
     blanks(10) num2str(eff) blanks(9) num2str(Reg)]);


%%%%%%%%%%%%%%%%%%%%%%%%%%%%%%%%%%%%%%%%%%%%%%%%%%%%%%%%%%%%%%%%%%%%%%%%%%
%
% tranperf2.m - Calculates voltages, currents, regulation and
%               efficiency for a constant input voltage with a
%               specified load impedance.
%
%%%%%%%%%%%%%%%%%%%%%%%%%%%%%%%%%%%%%%%%%%%%%%%%%%%%%%%%%%%%%%%%%%%%%%%%%%
clear;
R1=0.06; R2=0.015; X1=0.18; X2=0.045; Xm=400; Rc=1200;
V1=240; V2=120; KVA=5; n=V1/V2; R2p=n^2*R2; X2p=n^2*X2;
R1=0.1672; R2=0.04179; X1=0.3156; X2=0.0789; Xm=392.4; Rc=1807;
V1=480; V2=240; KVA=15; n=V1/V2; R2p=n^2*R2; X2p=n^2*X2;
% Load characterization at rated load voltage
disp(' DESCRIPTION OF LOAD @ RATED VOLTAGE '); disp(' ');
Load=input(' Percent Load( 0->100 ) = ');
PF=input(' Load Power Factor( 0->1 ) = ');
if PF ~= 1
  PFS2=input(' PF Sense( leading,lagging ) = ','s');
else PFS2='unityPF';
end
V1=input(' Input Voltage( V ) = ');
if PFS2 == 'lagging'; thet2=acos(PF);
elseif PFS2 == 'leading'; thet2=-acos(PF);
elseif PFS2 =='unityPF'; thet2=0;
else; end
ZL=V2^2/KVA/1000/Load*100*exp(j*thet2); Z2=R2p+j*X2p+n^2*ZL;
Zm=j*Rc*Xm/(Rc+j*Xm); Zin=R1+j*X1+Z2*Zm/(Z2+Zm);
thet1=angle(Zin); PFin=cos(thet1);
```

```
if thet1 > 0; PFS1='lagging';
elseif thet1 < 0; PFS1='leading';
else PFS1='unityPF';end
I1=V1/Zin; E1=V1-(R1+j*X1)*I1;
Im=E1/(j*Xm); Ic=E1/Rc;
I2=(I1-Im-Ic)*n; V2=I2*ZL; Pout=real(V2*conj(I2));
Pin=real(V1*conj(I1)); eff=Pout/Pin*100; deg=180/pi;
Znl=R1+j*X1+Zm; Vnl=abs(V1*Zm/Znl);
Reg=(Vnl/n-abs(V2))/abs(V2)*100;
deg=180/pi; angV1=angle(V1)*deg; angV2=angle(V2)*deg;
angE1=angle(E1)*deg; angI1=angle(I1)*deg; angI2=angle(I2)*deg;
angIm=angle(Im)*deg; angIc=angle(Ic)*deg;
disp(' '); disp('              TRANSFORMER PERFORMANCE');
disp(' '); disp(' ')
disp([blanks(3) 'V1 mag' blanks(8) 'V1 ang' blanks(8) 'V2 mag' ...
      blanks(8) 'V2 ang']);
disp([blanks(3) num2str(abs(V1)) blanks(11) num2str(angV1) ...
      blanks(13) num2str(abs(V2)) blanks(9) num2str(angV2)]); disp(' ');
disp([blanks(3) 'E1 mag' blanks(8) 'E1 ang']);
disp([blanks(4) num2str(abs(E1)) blanks(7) num2str(angE1)]); disp(' ');
disp([blanks(3) 'I1 mag' blanks(8) 'I1 ang' blanks(8) 'I2 mag' ...
      blanks(8) 'I2 ang']);
disp([blanks(3) num2str(abs(I1)) blanks(9) num2str(angI1) ...
      blanks(8) num2str(abs(I2)) blanks(9) num2str(angI2)]); disp(' ');
disp([blanks(3) 'Im mag' blanks(8) 'Im ang' blanks(8) 'Ic mag' ...
      blanks(8) 'Ic ang']);
disp([blanks(3) num2str(abs(Im)) blanks(8) num2str(angIm) ...
      blanks(8) num2str(abs(Ic)) blanks(7) num2str(angIc)]); disp(' ');
disp([blanks(4) 'PFin' blanks(10) 'PFS1' blanks(10) 'PFout' ...
      blanks(9) 'PFS2']);
disp([blanks(3) num2str(PFin) blanks(8) PFS1 ...
      blanks(9) num2str(PF) blanks(9) PFS2 ]); disp(' ');
disp([blanks(4) 'Pin' blanks(11) 'Pout' blanks(11) 'eff' ...
 blanks(10) '%Reg']);
disp([blanks(4) num2str(Pin/1000) blanks(10) num2str(Pout/1000) ...
      blanks(10) num2str(eff) blanks(9) num2str(Reg)]);


%%%%%%%%%%%%%%%%%%%%%%%%%%%%%%%%%%%%%%%%%%%%%%%%%%%%%%%%%%%%%%%%%%%%%%%%%%%%
%
% traneff.m - Calculates efficiency vs. load apparent power for
%             constant load voltage & specified range of lagging
%             or leading load power factor.
%
%%%%%%%%%%%%%%%%%%%%%%%%%%%%%%%%%%%%%%%%%%%%%%%%%%%%%%%%%%%%%%%%%%%%%%%%%%%%
clear;
R1=0.09; R2=0.0225; X1=0.18; X2=0.045; Xm=400; Rc=1200;
V1=240; V2=120; KVA=5;
n=V1/V2; R2p=n^2*R2; X2p=n^2*X2;
```

```
nPF=5; PF=linspace(0.2,1,nPF); PFS2='lagging';
if PFS2 == 'lagging'; thet2=-acos(PF);
else; thet2=acos(PF); end
for k=1:nPF
   I2R=KVA*1000/V2*exp(j*thet2(k));       % Rated load current
   nval=150; I2=linspace(0,1.5*I2R,nval); % Specify load values
   for i=1:nval
      E1=(R2p+j*X2p)*I2(i)/n+n*V2;
      Im=E1/(j*Xm); Ic=E1/Rc; Io=Ic+Im; I1=I2(i)/n+Io;
      V1=(R1+j*X1)*I1+E1;
      Pin=real(V1*conj(I1)); Pout=real(V2*conj(I2(i)));
      eff(i)=Pout/Pin*100; maxeff=max(eff);
      load(i)=abs(I2(i))/abs(I2R)*100;
   end
   for i=1:nval
      if eff(i)==maxeff; nmax=i; else; end
   end
   plot(0,0,load,eff,load(nmax),maxeff,'o');
   hold on;
end
title('Efficiency vs. load for constant load voltage');
xlabel('Load apparent power, %'); ylabel('Efficiency, %');
legend('Efficiency', 'Max eff point',4); grid; hold off;
if nPF > 1
text(80,35,['PF range ',num2str(PF(1)),' to ', ...
   num2str(PF(nPF)),', incr = ',num2str(PF(nPF)-PE(nPF-1))]);
text(80,25,['PF sense - ',PFS2]);
%text(80,96.25,['PF range ',num2str(PF(1)),' to ', ...
%    num2str(PF(nPF)),', incr = ',num2str(PF(nPF)-PF(nPF-1))]);
%text(80,25,['PF sense - ',PFS2]);
else; end


%%%%%%%%%%%%%%%%%%%%%%%%%%%%%%%%%%%%%%%%%%%%%%%%%%%%%%%%%%%%%%%%%%%%%%%%%%%%
%
% encost.m - Uses supplied 24-hour load profile to calculate the
%            loss energy cost to operate the load profile at
%            rated output voltage. Also calculates the average
%            or all day efficiency.
%
%%%%%%%%%%%%%%%%%%%%%%%%%%%%%%%%%%%%%%%%%%%%%%%%%%%%%%%%%%%%%%%%%%%%%%%%%%%%
clear;
R1=0.20; R2=0.002; X1=0.60; X2=0.006; Xm=130; Rc=4000;
V1=2400; V2=240; KVA=15; n=V1/V2; R2p=n^2*R2; X2p=n^2*X2;
KWHrcost=0.10;                 % $/kWh cost
LOAD=[55 75 100 90 65 55];     % Percent apparent power load
PF=[0.8 0.85 0.9 0.85 0.77 0.8]; % Power Factor, lagging assumed
HRS=[6 6 4.5 3.5 2 2];         % Hours at load point
n=length(HRS); T(1)=HRS(1); for i=2:n; T(i)=T(i-1)+HRS(i); end
```

```
if T ~= 24; disp('WARNING - 24 hour period not specified'); end
for i=1:n
   I2=KVA*1000/V2*LOAD(i)/100/PF(i)*exp(-j*acos(PF(i)));
   E1=(R2p+j*X2p)*I2/n+n*V2;
   Im=E1/(j*Xm); Ic=E1/Rc; Io=Ic+Im; I1=I2/n+Io;
   V1=(R1+j*X1)*I1+E1;
   thet1=angle(V1)-angle(I1); PFin=cos(thet1);
   Pout=real(V2*conj(I2)); Pin=real(V1*conj(I1));
   eff(i)=Pout/Pin*100; Ploss(i)=Pin-Pout; Po(i)=Pout/1000;
end
cost=0; adeff=0;
for i=1:n
   cost=cost+Ploss(i)/1000*KWHrcost*HRS(i);
   adeff=adeff+eff(i)*HRS(i)/T(n);
end
subplot(2,1,1); stairs(T,Po); grid
title('Power load profile');
ylabel('Load, kW'); xlabel('Time, h');
subplot(2,1,2); stairs(T,eff); grid
title('Efficiency profile');
ylabel('Efficiency, %'); xlabel('Time, h');
disp('  ALL DAY EFFICIENCY STUDY'); disp(' ')
disp([' 24-Hour Energy cost - $', num2str(cost)]); disp(' ');
disp([' All-Day Efficiency(%) - ', num2str(adeff)]);



%%%%%%%%%%%%%%%%%%%%%%%%%%%%%%%%%%%%%%%%%%%%%%%%%%%%%%%%%%%%%%%%%%%%%%%%%
%
% tranreg.m - Calculates the full-load regulation of a transformer
%             as a function of load power factor with constant
%             load voltage applied.
%
%%%%%%%%%%%%%%%%%%%%%%%%%%%%%%%%%%%%%%%%%%%%%%%%%%%%%%%%%%%%%%%%%%%%%%%%%
clear;
R1=0.06; R2=0.015; X1=0.18; X2=0.045; Xm=400; Rc=1200;
V1=240; V2=120; KVA=5;
n=V1/V2; R2p=n^2*R2; X2p=n^2*X2;
npts=100; thet2=linspace(-pi/2,pi/2,npts);
Reg=zeros(1,npts); PF=Reg;
for i=1:npts
   I2=KVA*1000/V2*exp(-j*thet2(i)); E1=(R2p+j*X2p)*I2/n+n*V2;
   Im=E1/(j*Xm); Ic=E1/Rc; Io=Ic+Im; I1=I2/n+Io;
   V1=(R1+j*X1)*I1+E1;
   Zm=j*Rc*Xm/(Rc+j*Xm); Znl=R1+j*X1+Zm;
   Vnl=abs(V1*Zm/Znl);
   Reg(i)=(Vnl/n-V2)/V2*100;
end
thet2=thet2*180/pi;
plot(thet2,Reg); grid;
```

```
title('Regulation vs. power factor');
xlabel('Power factor angle ( lead to lag ), degrees');
ylabel('Regulation, %');


%%%%%%%%%%%%%%%%%%%%%%%%%%%%%%%%%%%%%%%%%%%%%%%%%%%%%%%%%%%%%%%%%%%%%%%%%%%%%
%
% inrush.m - Calculates the inrush current to an unloaded
%            transformer by solution of the nonlinear DEQ.
%
%%%%%%%%%%%%%%%%%%%%%%%%%%%%%%%%%%%%%%%%%%%%%%%%%%%%%%%%%%%%%%%%%%%%%%%%%%%%%
clear;
f=60; R1=0.6; w=2*pi*f; % Frequency, primary resistance
Vm=sqrt(2)*240; zeta=0*pi/180; % Voltage, phase shift
ncyc=12;                % No. source cycles
% No load voltage(RMS) and Magnetizing current(RMS) arrays
Voc=[0 200 240 250 260 270 280 290 300 310 320 480];
Im=[0 0.52 0.65 0.7 0.75 0.85 1 1.5 3 6 12 120]*sqrt(2);
Lam=Voc/4.44/f;         % Flux linkage(max value) array
np=length(Im);
% Vmax=300;             % Activate to check magnetization curve
% for i=1:length(Voc)
%    if Voc(i)>Vmax; nn=i-1; break;
%    else; end
% end
% plot(Im(1:nn),Voc(1:nn)); grid;
% ylabel('Voltage'); xlabel('Magnetizing current'); pause
% plot(Im(1:nn),Lam(1:nn)); grid;
% ylabel('Flux linkage'); xlabel('Magnetizing current'); pause
T=1/f; t0=0; x0=0; x=[x0]; t=[t0];
h=10e-06; incr=T/100; tsav=t0;
for i=1:fix(ncyc*T/h) % Solution for magnetizing current(x)
   if x0 == 0; x0=1e-08; else; end
   if abs(x0) > max(Lam)    % Lam-Im slope
     dLamdi=(Lam(np)-Lam(np-1))/(Im(np)-Im(np-1))/sqrt(2);
   else
     a=interp1(Im,Lam,abs(x0)); b=interp1(Im,Lam,1.01*abs(x0));
     dLamdi=(b-a)/(0.01*abs(x0));
   end
   xx=x0; tt=t0;
   k1=h*(-R1*xx+Vm*sin(w*tt-zeta))/dLamdi;   % Fourth-order
   xx=x0+k1/2;tt=t0+h/2;                      % Runge-Kutta
   k2=h*(-R1*xx+Vm*sin(w*tt-zeta))/dLamdi;   % integration
   xx=x0+k2/2; tt=t0+h/2;                      % with fixed
   k3=h*(-R1*xx+Vm*sin(w*tt-zeta))/dLamdi;   % increment h
   xx=x0+k3; tt=t0+h;
   k4=h*(-R1*xx+Vm*sin(w*tt-zeta))/dLamdi;
   x0=x0+(k1+2*k2+2*k3+k4)/6;
   t0=t0+h;
```

```
   if (t0-tsav) >= incr
      tsav=t0; t=[t t0]; x=[x x0];
   else; end
end
plot(t,x); grid; title('Inrush current');
ylabel('Primary current, A'); xlabel('Time, s');
text(0.7*max(t),0.9*max(x),['Voltage = ',num2str(Vm/sqrt(2))]);
text(0.7*max(t),0.8*max(x),['Phase angle = ',num2str(zeta*180/pi)]);
text(0.7*max(t),0.7*max(x),['No. cycles = ',num2str(ncyc)]);


%%%%%%%%%%%%%%%%%%%%%%%%%%%%%%%%%%%%%%%%%%%%%%%%%%%%%%%%%%%%%%%%%%%%%%%%%
%
% spcore.m - determines d,h & l for core transformer given N
%            and h/w ratio. See data1cor.m for req'd inputs.
%            Also, calculates approx magnetizing current and
%            weights of conductor and core material.
%
%%%%%%%%%%%%%%%%%%%%%%%%%%%%%%%%%%%%%%%%%%%%%%%%%%%%%%%%%%%%%%%%%%%%%%%%%
clear;
global Ggammam GSF Gkl GV1 GI1 Gf Gdelc Grw Grhoc
global GBm GHm Gpm GPm Gkc GN
data1cor,
x0=[GV1*GI1/1000 5 2]'; % Solution is x0 dependent.
delx=zeros(3,1); f=delx; x=x0;
% Newton-Raphson solution, x1=d x2=w x3=l
for i=1:100
   [f,J]=core(x0); delx=-inv(J)*f; x=x0+delx;
      if all(abs(delx)>1e-12);break; else; end
      x0=x;
end
l=x(3); d=x(1); w=x(2); h=Grw*x(2); N=GN;
disp(' '); disp(' ');
disp('    CORE SIZING for CORE-TYPE TRANSFORMER');
if any(x<0); disp('EXTRANEOUS SOLUTION'); pause; else; end
format compact
disp(' '); disp(' ')
disp(['    l (in)' '     d (in)' '     w (in)  ' ...
   '  h (in) ' '  N (turns)  ']);
disp([ l d w h N ]);
% Magnetizing current — A & pu ( Joint/Cross grain neglected )
Im=2*GHm*(h+w+l)/N/sqrt(2); puIm=Im/GI1;
disp(' '); disp(' ')
disp(['    Im (A)     ' '  puIm  ']);
disp([ Im puIm ]);
% Conductor & core weight
Wfe=2*Ggammam*(2*l+(Grw+1)*w)*l*d*GSF;
Wc=2*Gkc*0.322*h*w*(l+d+2*w);
disp(' '); disp(' ')
```

```
disp([' Wfe (lb) ' ' Wc (lb)      ']);
disp([ Wfe Wc ]);


%%%%%%%%%%%%%%%%%%%%%%%%%%%%%%%%%%%%%%%%%%%%%%%%%%%%%%%%%%%%%%%%%%%%%%%%
%
% data1cor.m - input data for spcore.m
%
%%%%%%%%%%%%%%%%%%%%%%%%%%%%%%%%%%%%%%%%%%%%%%%%%%%%%%%%%%%%%%%%%%%%%%%%
% Leading G denotes global variable
GV1=480; KVA=15; PF=0.8; eff=0.97; Gf=60;
Gdelc=1500;          % FL current density - Apsi
Grhoc=1.02e-06;      % Conductor resistivity - Ohm_in
Gkl=1.42;            % Ratio conductor/core losses @ FL
GBm=1.50;            % Max core flux density - T
GHm=0.92*2.021;      % Max core mag fld intensity - A-t/in
Gpm=2.5*0.7;         % Core loss power density - W/lb
Grw=2.0;             % Ratio of window height/width
Ggammam=0.283;       % Core material density - lbs/cu in
GSF=0.95;            % Lamination stacking factor
GN=200;              % Primary coil turns
GI1=1000*KVA/GV1;    % Primary rated current - A
GPm=(1-eff)*1000*KVA*PF/(1+Gkl); % Core losses - W
Gkc=0.35-0.005*sqrt(GV1)+0.0035*(GV1*GI1)^0.25;  % Window fill factor


%%%%%%%%%%%%%%%%%%%%%%%%%%%%%%%%%%%%%%%%%%%%%%%%%%%%%%%%%%%%%%%%%%%%%%%%
%
% core.m - evaluates nonlinear sizing equations and the
%          associated Jacobian for spcore.m
%
%%%%%%%%%%%%%%%%%%%%%%%%%%%%%%%%%%%%%%%%%%%%%%%%%%%%%%%%%%%%%%%%%%%%%%%%
function [f,J] = core(x);
data1cor,
f=zeros(3,1); J=zeros(3,3);
f(1)=2*Ggammam*Gpm*x(1)*x(3)*GSF*(2*x(3)+(Grw+1)*x(2))-GPm;
f(2)=2*Grhoc*Gdelc^2*Gkc*Grw*x(2)^2*(x(3)+x(1)+x(2))-Gkl*GPm;
f(3)=4.44*GN*Gf*GSF*GBm*x(1)*x(3)/(39.37)^2-2*GV1;
J(1,1)=2*Ggammam*Gpm*GSF*(2*x(3)^2+(Grw+1)*x(2)*x(3));
J(1,2)=2*Ggammam*Gpm*GSF*((Grw+1)*x(1)*x(3));
J(1,3)=2*Ggammam*Gpm*GSF*(4*x(1)*x(3)+(Grw+1)*x(1)*x(2));
J(2,1)=2*Grhoc*Gdelc^2*Gkc*Grw*x(2)^2;
J(2,2)=2*Grhoc*Gdelc^2*Gkc*Grw*(2*x(2)*x(3)+2*x(1)*x(2)+3*x(2)^2);
J(2,3)=2*Grhoc*Gdelc^2*Gkc*Grw*x(2)^2;
J(3,1)=4.44*GN*Gf*GSF*GBm/(39.37)^2*x(3);
J(3,2)=0;
J(3,3)=4.44*GN*Gf*GSF*GBm/(39.37)^2*x(1);
```

```
%%%%%%%%%%%%%%%%%%%%%%%%%%%%%%%%%%%%%%%%%%%%%%%%%%%%%%%%%%%%%%%%%%%%%%%%%%
%
% spshell.m - determines d,h & l for shell transformer given N
%             and h/w ratio. See data1sh.m for req'd inputs.
%             Also, calculates approx magnetizing current and
%             weights of conductor and core material.
%
%%%%%%%%%%%%%%%%%%%%%%%%%%%%%%%%%%%%%%%%%%%%%%%%%%%%%%%%%%%%%%%%%%%%%%%%%%
clear;
global Ggammam GSF Gkl GV1 GI1 Gf Gdelc Grw Grhoc
global GBm GHm Gpm GPm Gkc GN
data1sh,
x0=[GV1*GI1/1000 2 2]';    % Solution is x0 dependent.
delx=zeros(3,1); f=delx; x=x0;
% Newton-Raphson solution, x1=d x2=w x3=l
for i=1:100
   [f,J]=shell(x0); delx=-inv(J)*f; x=x0+delx;
     if all(abs(delx)<1e-12);break; else; end
   x0=x;
end
l=x(3); d=x(1); w=x(2); h=Grw*x(2); N=GN;
disp(' '); disp(' ');
disp('    CORE SIZING for SHELL-TYPE TRANSFORMER');
if any(x<0); disp('EXTRANEOUS SOLUTION'); pause; else; end
format compact
disp(' '); disp(' ')
disp(['    l (in)' '    d (in)' '    w (in)  ' ...
   '  h (in) ' ' N (turns) ']);
disp([ l d w h N ]);
% Magnetizing current - A & pu ( Joint/Cross grain neglected )
Im=GHm*(2*h+2*w+2.5*l)/N/sqrt(2); puIm=Im/GI1;
disp(' '); disp(' ')
disp(['    Im (A)   ' ' puIm ']);
disp([ Im puIm ]);
% Conductor & core weight
Wfe=2*Ggammam*(1+w+h)*l*d*GSF;
Wc=2*Gkc*0.322*h*w*(1+d+w);
disp(' '); disp(' ')
disp(['   Wfe (lb) ' ' Wc (lb)     ']);
disp([ Wfe Wc ]);


%%%%%%%%%%%%%%%%%%%%%%%%%%%%%%%%%%%%%%%%%%%%%%%%%%%%%%%%%%%%%%%%%%%%%%%%%%
%
% data1sh.m - input data for spshell.m
%
%%%%%%%%%%%%%%%%%%%%%%%%%%%%%%%%%%%%%%%%%%%%%%%%%%%%%%%%%%%%%%%%%%%%%%%%%%
% Leading G denotes a global variable
GV1=480; KVA=15; PF=0.80; eff=0.97; Gf=60;
```

```
Gdelc=1250;          % FL current density - Apsi
Grhoc=1.02e-06;      % Conductor resistivity - Ohm_in
Gkl=1.42;            % Ratio conductor/core losses @ FL
GBm=1.50;            % Max core flux density - T
GHm=0.92*2.021;      % Max core mag fld intensity - A-t/in
Gpm=2.5*0.7;         % Core loss power density - W/lb
Grw=2.5;             % Ratio of window height/width
Ggammam=0.283;       % Core material density - lbs/cu in
GSF=0.95;            % Lamination stacking factor
GN=160;              % Primary coil turns
GI1=1000*KVA/GV1;    % Primary rated current - A
GPm=(1-eff)*PF*1000*KVA/(1+Gkl); % Core losses - W
Gkc=0.5-0.002*sqrt(GV1)+0.0035*(GV1*GI1)^0.25; % Window fill factor


%%%%%%%%%%%%%%%%%%%%%%%%%%%%%%%%%%%%%%%%%%%%%%%%%%%%%%%%%%%%%%%%%%%%%%%%%
%
% shell.m - evaluates nonlinear sizing equations and the
%           associated Jacobian for spshell.m
%
%%%%%%%%%%%%%%%%%%%%%%%%%%%%%%%%%%%%%%%%%%%%%%%%%%%%%%%%%%%%%%%%%%%%%%%%%
function [f,J] = shell(x);
data1sh,
f=zeros(3,1); J=zeros(3,3);
f(1)=2*Ggammam*Gpm*x(1)*x(3)*GSF*(x(3)+(Grw+1)*x(2))-GPm;
f(2)=2*Grhoc*Gdelc^2*Gkc*Grw*x(2)^2*(x(1)+x(2)+x(3))-Gkl*GPm;
f(3)=4.44*GN*Gf*GSF*GBm*x(1)*x(3)/(39.37)^2-GV1;
J(1,1)=2*Ggammam*Gpm*GSF*(x(3)^2+(Grw+1)*x(2)*x(3));
J(1,2)=2*Ggammam*Gpm*GSF*((Grw+1)*x(1)*x(3));
J(1,3)=2*Ggammam*Gpm*GSF*(2*x(1)*x(3)+(Grw+1)*x(1)*x(2));
J(2,1)=2*Grhoc*Gdelc^2*Gkc*Grw*x(2)^2;
J(2,2)=2*Grhoc*Gdelc^2*Gkc*Grw*(2*x(2)*x(3)+2*x(1)*x(2)+3*x(2)^2);
J(2,3)=2*Grhoc*Gdelc^2*Gkc*Grw*x(2)^2;
J(3,1)=4.44*GN*Gf*GSF*GBm/(39.37)^2*x(3);
J(3,2)=0;
J(3,3)=4.44*GN*Gf*GSF*GBm/(39.37)^2*x(1);


%%%%%%%%%%%%%%%%%%%%%%%%%%%%%%%%%%%%%%%%%%%%%%%%%%%%%%%%%%%%%%%%%%%%%%%%%
%
% joint.m - determines the mmf per joint for a transformer
%           core. Calls Hm7.m where B-H curve is in Lines/
%           sq.in. & Ampturns/in. Material is M-7 grain
%           oriented ESS.
%
%%%%%%%%%%%%%%%%%%%%%%%%%%%%%%%%%%%%%%%%%%%%%%%%%%%%%%%%%%%%%%%%%%%%%%%%%
%clear; clf; % Disable for use with tranmag.m
BFe=linspace(0,200000, 200); n=length(BFe);
del=0.010; % Avg joint length (in.)
for i=1:n
```

```
HFe=Hm7(BFe(i));
FT(i)=HFe*del;
BT(i)=0.5*(BFe(i)/64.52e03+pi*4e-07*HFe*39.37)*64.52;
end
for i=1:n; if BT(i)>=50; m=i; break; end; end
% Suppress plot for use with tranmag.m
%plot(0,0,FT(m:n),BT(m:n)); grid
%title('Joint mmf drop');
%xlabel('Amp-turns per joint');
%ylabel('Flux density, kilolines/sq.in.');
BT=1000*BT; % BT calculated in kilolines/in^2


%%%%%%%%%%%%%%%%%%%%%%%%%%%%%%%%%%%%%%%%%%%%%%%%%%%%%%%%%%%%%%%%%%%%%%%%%%
%
% Hm7.m - B-H interpolation routine
%         Used by joint.m
%
%%%%%%%%%%%%%%%%%%%%%%%%%%%%%%%%%%%%%%%%%%%%%%%%%%%%%%%%%%%%%%%%%%%%%%%%%%
function y = Hm7(Btx)
% B-H values that follow are valid for M-7, 0.014 in. ESS
B=[0 20 30 40 50 60 65 70 80 85 90 95 100 105 110 115 ...
   117 120 125 130 140 210 ]*1000;
H=[ 0 0.4 0.51 0.61 0.71 0.80 0.85 0.90 1.04 1.18 1.35 1.7 2.4 ...
   4.4 9.0 18 26 100 400 1000 10000 100000 ];
% Activate to plot B-H curve
% m=120; plot(H(1:m),B(1:m)); grid; pause;     % Linear plot
% m=21; semilogx(H(2:m),B(2:m));grid; pause;   % Semilog plot
n=length(B); k=0;
if Btx==0; k=-1; y=0; end
if Btx<0; k=-1; y=0; disp('WARNING — Btx < 0, Htx = 0 returned'); end
if Btx>B(n); y=H(n); k=-1; disp('CAUTION — Beyond B-H curve'); end
for i=1:n
if k==0 & (Btx-B(i))<=0; k=i; break; end
end
if k>0;
y=H(k-1)+(Btx-B(k-1))/(B(k)-B(k-1))*(H(k)-H(k-1));
else;
end


%%%%%%%%%%%%%%%%%%%%%%%%%%%%%%%%%%%%%%%%%%%%%%%%%%%%%%%%%%%%%%%%%%%%%%%%%%
%
% tranmag.m - Calculates magnetization curve for core or
%             shell single-phase transformer.
%             Calls sptdata.m type, winding, turns and
%             dimension data.
%             Joint mmf drop calculated by joint.m.
%
%%%%%%%%%%%%%%%%%%%%%%%%%%%%%%%%%%%%%%%%%%%%%%%%%%%%%%%%%%%%%%%%%%%%%%%%%%
```

```
clear;
sptdata;
joint;
if style == 'shell_design';
  A=d*l/2*SF; kphi=2;
  MLP1=2*h+w+3*l/2; MLP2=w+l/2;
elseif style == 'core__design';
  A=d*l*SF; kphi=1;
  MLP1=2*h+3*l+w; MLP2=w+l;
else; end
npts=200; Im=zeros(1,npts+1); V=Im; B=linspace(0,Bmax,npts);
BR=VR/(4.44e-08*N*A*fR*kphi); B=[B';BR];
for i=1:npts+1
  FJ=interp1(BT, FT, B(i)); % Look up joint mmf
  Im(i)=(Hm7(B(i))*MLP1+Hm7cg(B(i))*MLP2+2*FJ)/N;
  V(i)=4.44e-08*N*fR*B(i)*A*kphi;
end
Xm=0.99*VR/Im(npts+1)*sqrt(2);
if style=='shell_design'
  Pcore=2*Pcm7(BR)*2*0.283*(l+h+w)*l*d*SF; % 2 x Epstein losses
elseif style=='core__design'
  Pcore=2.5*Pcm7(BR)*2*0.283*(2*l+w+h)*l*d*SF;
else; end
Rc=(0.99*VR)^2/Pcore;
disp(' '); disp('   MAGNETIZING REACTANCE & CORE LOSS RESISTANCE');
disp(' '); disp([blanks(7) 'Xm (ohms)' blanks(10) 'Rc (ohms)']);
disp([blanks(6) num2str(Xm) blanks(9) num2str(Rc)]);
pause; clf; figure(1);
plot(Im(1,1:npts),V(1,1:npts),Im(1,npts+1),V(1,npts+1),'o');
grid; title('Transformer magnetization curve');
xlabel('Peak value magnetizing current( Im ), A');
ylabel('Primary voltage( V1 ), V');
text(0.50*max(Im), 0.75*max(V),design);
legend('Saturation curve','Rated voltage',4);


%%%%%%%%%%%%%%%%%%%%%%%%%%%%%%%%%%%%%%%%%%%%%%%%%%%%%%%%%%%%%%%%%%%%%%%%%%%%%%
%
% sptdata.m - provides input data for tranmag.m to use in
%             calculation & plot of magnetization curve.
%
%%%%%%%%%%%%%%%%%%%%%%%%%%%%%%%%%%%%%%%%%%%%%%%%%%%%%%%%%%%%%%%%%%%%%%%%%%%%%%
style='shell_design'; % Specify style: shell or core
l=4.75; d=3.0; w=3.5; h=6.0; % Fe core dimensions
Bmax=110000; SF=0.95; % Max flux density desired, stacking factor
VR=480; fR=60; N=160; % Rated voltage & freq, Primary turns
if style=='shell_design'
  design='Shell transformer design';    % Plot label
elseif style=='core__design'
```

```
    design='Core transformer design';
else; end


%%%%%%%%%%%%%%%%%%%%%%%%%%%%%%%%%%%%%%%%%%%%%%%%%%%%%%%%%%%%%%%%%%%%%%%%%%%%
%
% Hm7cg.m - B-H interpolation routine
%           Used by joint.m
%
%%%%%%%%%%%%%%%%%%%%%%%%%%%%%%%%%%%%%%%%%%%%%%%%%%%%%%%%%%%%%%%%%%%%%%%%%%%%
function y = Hm7cg(Btx)
% B-H values that follow are valid for M-7, 0.014 in. ESS
% in the cross-grain region. 90% B value with grain direction.
B=[0 20 30 40 50 60 65 70 80 85 90 95 100 105 110 115 ...
   117 120 125 130 140 210 ]*1000*0.90;
H=[ 0 0.4 0.51 0.61 0.71 0.80 0.85 0.90 1.04 1.18 1.35 1.7 2.4 ...
   4.4 9.0 18 26 100 400 1000 10000 100000 ];
% Activate to plot B-H curve
% m=16; plot(H(1:m),B(1:m)); grid; pause; % Linear plot
% m=21; semilogx(H(2:m),B(2:m));grid; pause; % Semilog plot
n=length(B); k=0;
if Btx==0; k=-1; y=0; end
if Btx<0; k=-1; y=0; disp('WARNING - Btx < 0, Htx = 0 returned'); end
if Btx>B(n); y=H(n); k=-1; disp('CAUTION - Beyond B-H curve'); end
for i=1:n
if k==0 & (Btx-B(i))<=0; k=i; break; end
end
if k>0;
y=H(k-1)+(Btx-B(k-1))/(B(k)-B(k-1))*(H(k)-H(k-1));
else;
end


%%%%%%%%%%%%%%%%%%%%%%%%%%%%%%%%%%%%%%%%%%%%%%%%%%%%%%%%%%%%%%%%%%%%%%%%%%%%
%
% Pcm7.m - Determines Epstein loss given value of B
%           Used by tranmag.m
%
%%%%%%%%%%%%%%%%%%%%%%%%%%%%%%%%%%%%%%%%%%%%%%%%%%%%%%%%%%%%%%%%%%%%%%%%%%%%
function y = Pcm7(Bx)
% B-Wlb values that follow are valid for M-7,0.014",60 Hz
b=[0 10 20 30 40 60 80 100 120 140 180 210]*1000;
wlb=[0 0.0082 0.033 0.073 0.125 0.26 0.455 0.78 1.35 ...
   1.84 3.04 4.13];
B=[0:210]*1000; Wlb=spline(b,wlb,B);
% Activate to plot B-Wlb curve
% m=101; plot(Wlb(1:m),B(1:m)); grid; pause; % Linear plot
% m=101; semilogx(Wlb(2:m),B(2:m)); grid; pause; % Semilog plot
n=length(B); k=0;
```

```
if Bx==0; k=-1; y=0; end
if Bx<0; k=-1; y=0; disp('WARNING - Bx < 0, Wlb = 0 returned'); end
if Bx>B(n); y=Wlb(n); k=-1; disp('CAUTION - Beyond B-Wlb curve');end
for i=1:n
if k==0 & (Bx-B(i))<=0; k=i; break; end
end
if k>0;
y=Wlb(k-1)+(Bx-B(k-1))/(B(k)-B(k-1))*(Wlb(k)-Wlb(k-1));
else;
end


%%%%%%%%%%%%%%%%%%%%%%%%%%%%%%%%%%%%%%%%%%%%%%%%%%%%%%%%%%%%%%%%%%%%%%%%%%%%
%
% leakind.m - Calculates primary & secondary leakage inductances
%             for single-phase shell or core transformer with
%             either vertical or concentric layered windings.
%
%%%%%%%%%%%%%%%%%%%%%%%%%%%%%%%%%%%%%%%%%%%%%%%%%%%%%%%%%%%%%%%%%%%%%%%%%%%%
clear;
n=6;                        % Total no. winding sections
                            % Odd integer >= 3 for shell
                            % 2(k+2) where k odd integer for core
g=0.062;                    % Spacing between wdg layers
N1=160; N2=80;              % Primary & secondary turns
KVA=15; V1=240; f=60;       % Ratings
I1=KVA*1000/V1;
style='shell_type';         % shell_type or core__type
wdgtype='conc_layer';       % vert_layer or conc_layer
h=6.0; w=3.5; d=3; l=4.75;  % Core dimensions
rhoc=0.85e-06;              % Conductor resistivity - Ohm_in
puc=0.8;                    % Per unit portion of conductor in layer
if style=='shell_type'
n1=N1/(n-1); n2=N2/(n-1);   % Unit layer turns(n1,2n2,2n1,...2n2,n1)
   if wdgtype=='vert_layer';
      p=(h-(n-1)*g)/2/(n-1);   %Unit layer thickness
      Ll1=6.384e-08*(n-1)*n1^2*(d+pi*(l+w))/h*(p/3+g/2); % Pri. leakage
      Ll2=6.384e-08*(n-1)*n2^2*(d+pi*(l+w))/h*(p/3+g/2); % Sec. leakage
      Ac=puc*p*(w-g)/n1; MLTc=2*(l+d+w);
   elseif wdgtype=='conc_layer';
      p=(w-(n-1)*g)/2/(n-1);
      Ll1=6.384e-08*(n-1)*n1^2*(d+pi*(l+w))/w*(p/3+g/2);
      Ll2=6.384e-08*(n-1)*n2^2*(d+pi*(l+w))/w*(p/3+g/2);
      Ac=puc*p*(h-g)/n1; MLTc=2*(l+d+w);
   else; end
elseif style=='core__type'
n1=N1/(n-2); n2=N2/(n-2);
   if wdgtype=='vert_layer'
      p=(h-(n/2-1)*g)/(n-2);
```

```
      Ll1=6.384e-08*(n-2)*n1^2*(2*d+pi*(l+w/2))/h*(p/3+g/2);
      Ll2=6.384e-08*(n-2)*n2^2*(2*d+pi*(l+w/2))/h*(p/3+g/2);
      Ac=puc*p*(w-g)/2/n1; MLTc=2*(l+d+w);
    elseif wdgtype=='conc_layer'
      p=(w-(n-1)*g)/2/(n-2);
      Ll1=6.384e-08*(n-2)*n1^2*(2*d+pi*(l+w/2))/w*(p/3+g/2);
      Ll2=6.384e-08*(n-2)*n2^2*(2*d+pi*(l+w/2))/w*(p/3+g/2);
      Ac=puc*p*(h-g)/n1; MLTc=2*(l+d+w);
    else; end
else; end
X1=2*pi*f*Ll1; X2=2*pi*f*Ll2;
R1=1.05*rhoc*N1*MLTc/Ac; R2=(N2/N1)^2*R1; Jc=I1/Ac;
disp(' '); disp('  RESISTANCE & LEAKAGE REACTANCE CALCULATIONS');
disp(' '); disp(' ')
disp(['   Style    '  '   Wndg sects   ' '   Wndg type   ']);
disp(['  ' style '        '   num2str(n) '           ' wdgtype]);
disp(' '); disp(' ')
disp([blanks(3) 'R1 (ohms)' blanks(6) 'R2 (ohms)' blanks(6) ...
   'Jc (apsi)']);
disp([blanks(4) num2str(R1) blanks(9) num2str(R2) blanks(8) ...
   num2str(Jc)]);
disp(' '); disp(' ');
disp([blanks(3) 'Ac1 (sq in)' blanks(4) 'Ac2 (sq in)']);
disp([blanks(4) num2str(Ac) blanks(7) num2str(N1/N2*Ac)]);
disp(' '); disp(' ');
disp([blanks(3) 'L1 (H)' blanks(8) 'L2 (H)' blanks(5) 'X1 (ohms)' ...
   blanks(5) 'X2 (ohms)']);
disp([blanks(2) num2str(Ll1) blanks(6) num2str(Ll2) blanks(5) ...
   num2str(X1) blanks(9) num2str(X2)]);
```

SUMMARY

- Transformers are high-efficiency devices for processing power with an inverse voltage-current relationship ($V_1/I_2 = V_2/I_1$) with frequency unaltered.

- Transformers take advantage of the high magnetic energy density of ferromagnetic material to allow economical device design. However, the nonlinear nature of ferromagnetic material leads to introduction of harmonics in current and/or voltage. The relative magnitude of these harmonics with proper device design is typically small.

- The ferromagnetic core material of the transformer has two distinct loss mechanisms—hysteresis loss and eddy current loss. The first produces losses that vary directly with frequency while the second has losses dependent on frequency squared. Both loss types are almost directly proportional to flux density squared.

- The most common transformer equivalent circuit is the Epstein or tee equivalent circuit. Goodness of performance criteria demand that $(R_1, R_2, X_1, X_2) \ll (R_c, X_m)$. An approximate equivalent circuit with the excitation branch moved to the input terminals can be used for computational advantage with little loss in accuracy of results for many analysis situations.

- The values of the common transformer equivalent circuit can be determined from simple short-circuit and open-circuit test data. R_1, R_2, X_1, and X_2 result from reduction of the short-circuit test data. R_c and X_m are synthesized from the open-circuit test data.

- The nonlinear nature of the transformer magnetizing inductance can lead to a transient inrush current significantly greater than the rated value of current upon energizing a transformer.

- Autotransformers have a significant power advantage over the two-winding transformer. However, conductive isolation is not present and the issue of safety limits their use to particular applications.

- Three-phase transformers function on a per phase basis identically to single-phase transformers.

- Three-phase transformers are commonly connected in one of four arrangements: Y-Y, Δ-Δ, Δ-Y, and Y-Δ. The latter two arrangements result in a phase shift of voltage and current as well as a magnitude adjustment.

- Harmonics that result from magnetic nonlinearity can be managed by connection in three-phase transformers to eliminate their appearance in the connecting lines—an advantage not available with single-phase transformers.

- MATLAB programs are developed to ease the tedious sinusoidal steady-state analysis necessary to assess transformer performance.

- Core and winding loss limitations and induced voltage values place a coupled set of relationships on ferromagnetic core dimensions and winding turn choice. Using these constraints along with practical limits on magnetic flux density and winding current density, procedures are developed for transformer design.

PROBLEMS

4.1   An ideal transformer has an input voltage of 480 V. The output current and voltage are 10 A and 120 V. Determine the value of input current.

4.2   An ideal transformer with a 4:1 turns ratio has a 75-turn secondary coil. If the input voltage is 240 V (rms), 60 Hz, find the peak value of flux through the primary coil.

4.3 A 15-kVA, 2400:240-V, 60-Hz transformer has the following equivalent circuit parameters:

$$R_1 = 2.5\,\Omega \qquad R_2 = 0.025\,\Omega \qquad X_1 = 7.0\,\Omega \qquad X_2 = 0.070\,\Omega$$

$$R_c = 32\,\text{k}\Omega \qquad X_m = 11.5\,\text{k}\Omega$$

If the transformer is supplying a 10-kW, 0.8 PF lagging load at rated voltage, calculate (a) input voltage, (b) input current, and (c) input power factor.

4.4 Determine the efficiency of the transformer of Prob. 4.3 for the given load.

4.5 Let the transformer described in Prob. 4.3 supply rated kVA to a unity PF load at rated voltage. Find (a) the efficiency and (b) regulation.

4.6 The transformer of Prob. 4.3 is connected to a load that would draw rated current at 0.8 PF lagging if the output voltage were rated value. However, the input voltage is rated value at the point of operation. Determine the values of input and output current.

4.7 If short-circuit and open-circuit tests were performed on the transformer of Prob. 4.3 with rated current and rated voltage, respectively, predict the values of P_{SC} and P_{OC} to be recorded.

4.8 A 480:240-V transformer can be modeled by the equivalent circuit of Fig. 4.19. The following data were recorded with the secondary winding open circuit.

$$\overline{V}_1 = 480\angle 0°\,\text{V} \qquad \overline{V}_2 = 239.83\angle 0.01°\,\text{V} \qquad \overline{I}_1 = 1.43\angle -73.08°\,\text{A}$$

Determine R_1 and X_1 and predict the value of R_2 and X_2. Do you recommend this test for determination of R_1 and X_1? Give the reason for your answer.

4.9 Show that although a transformer with a high-voltage primary and a low-voltage secondary may require more ground insulation for the primary, the voltage stress between turns within the secondary winding is comparable with that of the primary.

4.10 Draw the sinusoidal steady-state equivalent circuit of a transformer if the primary and secondary coils are perfectly coupled. Assume that the transformer still has winding and core losses.

4.11 Draw the sinusoidal steady-state equivalent circuit of a transformer if the arbitrary constant of [4.18] to [4.21] is selected as $a = kN_1/N_2$, where k is the coefficient of coupling.

4.12 Use the MATLAB program ⟨imthd.m⟩ to predict the increase in third-harmonic input current for 10 percent load if the load power factor changes from 0.8 lagging to 0.8 leading. Run ⟨hyst.m⟩ for 80 percent maximum flux prior to executing ⟨imthd.m⟩.

4.13 It is discovered that the open-circuit and short-circuit test data for a transformer were both performed on a 90 percent tap of the primary winding. The equivalent circuit parameters have been calculated based on that data.

(*a*) Describe how you could make adjustment on the equivalent circuit parameters to give the values for the 100 percent or full primary winding circuit.

(*b*) What parameters may have error after your adjustment procedure? Justify your answer.

4.14 Execute ⟨hyst.m⟩ for the case of 70 percent maximum flux excitation and determine (*a*) the percentage third-harmonic voltage for the case of fundamental current excitation and (*b*) the percentage third-harmonic current for the case of fundamental voltage excitation.

4.15 Using the approximate equivalent circuit of Fig. 4.25 with V_1 constant in value, show that the maximum efficiency occurs for the point of operation where $V_1^2/R_c = (I_2')^2 R_{eq}$. This can be described as the point where the fixed losses are equal to the variable losses.

4.16 When making efficiency calculations using [4.48], the secondary winding losses are represented as $(I_2')^2 R_2'$. Show that $(I_2')^2 R_2' = I_2^2 R_2$; thus, the representation is correct.

4.17 Show that the rated current short-circuit test losses and the rated voltage open-circuit test losses each have identical value regardless of the side of the transformer selected for test.

4.18 Use ⟨traneff.m⟩ to predict the load point of maximum efficiency for the transformer of Example 4.7 when supplying a 0.8 PF lagging load at rated voltage.

4.19 Use ⟨traneff.m⟩ with the transformer described in the code listing to determine the efficiency of the transformer at 50 percent load with 0.2 PF leading at rated voltage. Compare the result with the 0.2 PF lagging case of Fig. 4.26 and offer an explanation for any difference in efficiency between the two cases.

4.20 Calculate currents $\bar{I}_{21}$ and $\bar{I}_{11}$ of Fig. 4.30 for the residential distribution transformer of Example 4.14.

4.21 The residential distribution transformer of Fig. 4.30 has a voltage rating of 2400:240/120 V. The three loads are pure resistive where L_1, L_2, and L_3 are modeled, respectively, by $R_1 = 10\ \Omega$, $R_2 = 12\ \Omega$, and $R_3 = 20\ \Omega$. Assume an ideal transformer with input voltage $V_{11} = 2400\,\text{V}$. (*a*) Calculate the values of input power P_{11} and input current I_{11}. (*b*) Determine the minimum kVA rating of the transformer for this described application.

4.22 The residential distribution transformer of Fig. 4.30 is rated at 20 kVA, 2400:240/120 V, 60 Hz. Treat the transformer as ideal with rated input voltage. Loads L_1 and L_2 are identical. All three loads have a 0.9 PF lagging. The total transformer load is 15 kW. The 240-V load L_3 forms half the total load. Determine the values of currents I_{21} and I_{11}.

4.23 Determine the value of all-day efficiency (η_{day}) for the two transformers of Example 4.12.

4.24 For the autotransformer of Fig. 4.31*b*, derive an equation for the impedance reflection property. Specifically, consider an impedance Z_H to be connected across the output terminals and determine an expression that describes that impedance in terms of N_1 and N_2 as seen looking into the input terminals.

4.25 Reconnect the autotransformer of Fig. 4.31*b* as a buck autotransformer by placing the dotted terminal of coil N_2 adjacent to the dotted terminal of coil N_1. Derive relationships for $\overline{V}_H/\overline{V}_X$ and $\overline{I}_H/\overline{I}_X$.

4.26 For the step-up autotransformer of Fig. 4.31, add the coil resistance and leakage reactance in series with the "ideal" coils N_1 and N_2. Find a two-winding equivalent circuit that models this nonideal autotransformer.

4.27 Develop the transformation ratios $\overline{V}_H/\overline{V}_X$ and $\overline{I}_H/\overline{I}_X$ for the step-down autotransformer of Fig. 4.32, thus justifying the statement immediately prior to Example 4.15.

4.28 The step-down autotransformer of Fig. 4.32 can be realized as an adjustable voltage source if the coils N_1 and N_2 are one continuous coil and the connection point for V_X is a sliding contact. Assume that such is the case and that $V_H = 125$ V while $V_X = 12.5$ V.

(*a*) If $I_X = 15$ A, determine the currents in each section of the winding (I_1 and I_2).

(*b*) What is the power advantage of this adjustable autotransformer at the adjusted point for V_X when compared to a reconnected two-winding transformer with rating of 125:12.5 V, 200 VA.

4.29 For a three-phase transformer bank connected Δ-Y, the turns ratio is such that the secondary line voltage is larger than the primary line voltage ($V_{A'B'} > V_{AB}$). Show a schematic similar to that of Fig. 4.33*d* where $\overline{V}_{AB}$ properly lags $\overline{V}_{A'B'}$.

4.30 Rework Example 4.17 with the transformer designed as (*a*) Δ-Y and (*b*) Y-Y.

4.31 Work Example 4.18 if the phase sequence is *a·c·b* and all else is unchanged.

4.32 Work Example 4.18 if the transformer is Y-Y and all else is unchanged.

4.33 The Y-Y transformer bank of Fig. 4.42 is supplied from a 7200-V, three-phase distribution system. The secondary supplies a balanced, 208-V, 45-kVA, 0.866 PF lagging load. Assume that $\overline{I}_{A'}$ is on the reference. Determine the values of $\overline{I}_A$, $\overline{I}_{A'}$, $\overline{V}_{AB}$, and $\overline{V}_{A'B'}$.

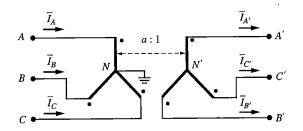

Figure 4.42

4.34 For the Δ-Δ transformer bank of Fig. 4.43, $\bar{I}_A = 10\angle 0°\,$A. Find the values of currents $\bar{I}_{AB}$, $\bar{I}_{A'B'}$, and $\bar{I}_{A'}$ as labeled on the schematic.

4.35 The step-up, Y-Δ transformer bank of Fig. 4.44 is rated for 75 kVA, 208:480 V. It is supplying rated apparent power to a balanced, unity PF load. (*a*) Determine the value of the phase turns ratio *a*. (*b*) If $\bar{V}_{A'B'} = 480\angle 0°\,$V, find the values of currents $\bar{I}_{A'}$ and $\bar{I}_A$.

4.36 A Δ-Δ, three-phase transformer bank is made up of three identical single-phase transformers. However, one of the transformers is presently removed for repair, leaving the so-called open-delta transformer bank of Fig. 4.45. If $\bar{I}_A = 10\angle 0°\,$A and $\bar{V}_{AB} = 240\angle 30°\,$V, determine (*a*) current $\bar{I}_{B'}$ and (*b*) voltage $\bar{V}_{B'C'}$.

4.37 Determine the rating of the open-delta transformer bank of Fig. 4.45 as a percentage of the rating that the transformer bank would have if the third leg were restored.

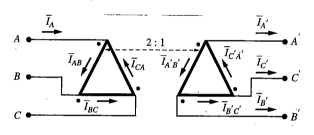

Figure 4.43

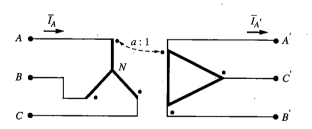

Figure 4.44

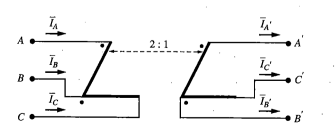

Figure 4.45

4.38 Execute ⟨leakind.m⟩ with the transformer values given in the listed code for the cases of $n = 6$ and $n = 10$. Draw a conclusion about the impact on leakage reactance values due to increasing the number of interleaved primary-secondary winding sections.

4.39 Design a two-winding single-phase transformer that matches the following specifications:

480:240 V   15 kVA   60 Hz

Shell-type   Concentric-layered winding

97% full-load efficiency at 0.8 PF lagging

Air-cooled   Class H insulation

REFERENCES

1. G. R. Slemon, *Magnetoelectric Devices: Transducers, Transformers, and Machines,* John Wiley & Sons, New York, 1966.
2. S. A. Stigant and A. C. Franklin, *The J&P Transformer Book,* 10th ed., Halsted Press, John Wiley & Sons, New York, 1973.
3. W. M. Flanagan, *Handbook of Transformer Design and Applications,* 2d ed., McGraw-Hill, New York, 1993.
4. R. Feinberg (ed.), *Modern Power Transformer Practice,* Halsted Press, John Wiley & Sons, New York, 1980.
5. A. Still and C. S. Siskind, *Elements of Electrical Machine Design,* 3d ed., McGraw-Hill, New York, 1954.
6. J. H. H. Kuhlmann, *Design of Electrical Apparatus,* 3d ed., John Wiley & Sons, New York, 1950.

5

DC MACHINES

5.1 INTRODUCTION

DC machines can be thought of as a dying breed, but death will come slowly. Prior to the development of reliable, high-power solid-state switching devices, the dc motor was the dominant electric machine for all variable-speed motor drive applications. However, the "power electronic revolution" has led to a significant shift from dc motor drives to adjustable-speed induction motor drives in the low integral to midrange horsepower variable-speed applications for new products. The dc motor retains niches on either end of the power spectrum where it is still a machine of choice. For example, the dc motor turns out to be the most economical choice in the automotive industry for cranking motors, windshield wiper motors, blower motors, and power window motors. Certain cost-effective servopositioning systems utilize dc motors such as shown by Fig. 5.1. In high power level applications such as propulsion drives (Fig. 5.2) and steel roll mill drives (Fig. 5.3), the torque density $(N \cdot m/m^3)$ of the dc machine can make it a choice over competing variable-frequency ac motor drives. The dc motor continues to have areas of application choice for new products. DC machines already are installed in areas where they are not presently the choice for new installations, yet still have many years of service life remaining. Thus, study of the dc machine principles is in order for training in the technical field.

5.2 PHYSICAL CONSTRUCTION

Figure 5.4 shows the frame assembly and armature of a six-pole dc motor. A cross-section drawing of a four-pole dc machine is shown by Fig. 5.5 on page 232 to further clarify component configurations. The armature core is an axial stack of circular laminations punched from electric sheet steel that typically range in thickness from 0.35 to 0.6 mm. The field pole and interpole core material may be solid ferromagnetic material since the principal flux of these structures is constant in value and thus does not result in hysteresis and eddy current losses. Fabrication from stacks of laminated pieces is common practice for three reasons. First, milling of the pole structures, especially the field pole core, is more expensive

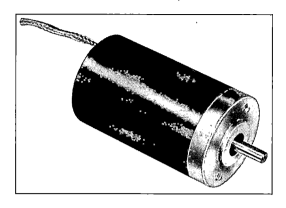

Figure 5.1
DC servomotor with PM field excitation. (*Courtesy of Globe Motors.*™)

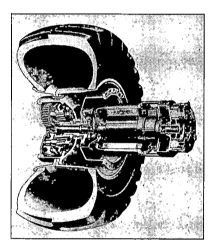

Figure 5.2
DC traction motor for off-highway vehicle. (*Courtesy of LeTourneau, Inc.*)

than punching. Second, there are small tooth ripple magnetic losses in the solid core that can be reduced. Third, during transient conditions, the solid core structure has induced eddy currents that can act to slow the change in flux resulting from a change in coil current.

Typically, a four-pole count is the smallest number for dc machines. A two-pole dc machine results in relatively long magnetic field paths around the stator yoke. In order to avoid excessive field excitation requirements, the stator yoke thickness of a two-pole machine must necessarily have sufficient cross-sectional area to result in low flux density. As a consequence, the power density of a two-pole machine suffers. However, for purpose of introductory study, the practical problem will be ignored and a two-pole machine will be examined for the sake of simplicity.

Figure 5.3
Roll mill dc motor. (*Courtesy of National Electric Coil, Columbus, Ohio. Photo by Chuck Drake.*)

(*a*) (*b*)

Figure 5.4
Six-pole dc motor. (*Courtesy of General Electric Canada Inc.*) (a) Motor frame assembly.
(b) Motor armature.

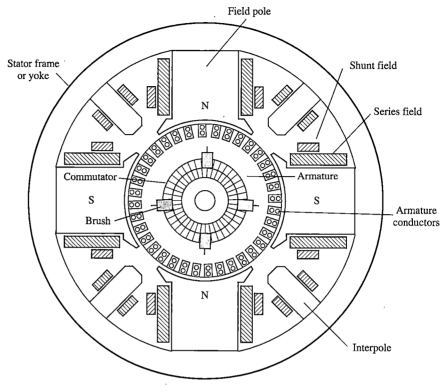

Figure 5.5
Cross-section drawing of four-pole dc machine

5.3 VOLTAGE AND TORQUE PRINCIPLES

Any detrimental effect of the magnetic field produced by the armature conductors when carrying current will be ignored at this point of development of dc machine characteristics. When the armature field impact is established later, it will be seen to simply reduce the flux magnitude along the principal magnetic path and thus does not negate any relationships to be developed in the interim.

5.3.1 NO-LOAD MAGNETIC FIELD

A simple two-pole, radial air gap dc machine is shown in cross section by Fig. 5.6. A single armature coil is present at this point; additional coils will be added later. Treating the air gap as uniform under the field pole arc and the ferromagnetic material as highly permeable, the air gap flux density $B_g(\theta)$, measured in a clockwise sense from the center of the north pole, is displayed by Fig. 5.7. The flat layout is viewed in an axial direction. Except for the polarity transition of the interpolar region, the flux density of the air gap is constant in magnitude.

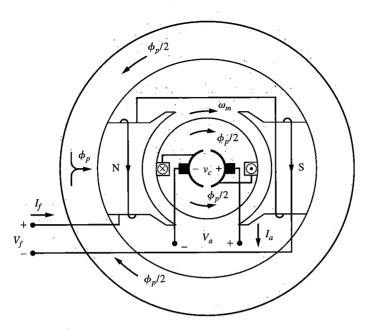

Figure 5.6
Simple two-pole dc generator

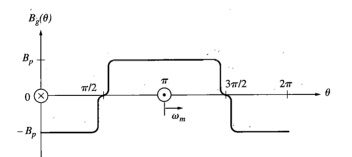

Figure 5.7
Air gap flux density

If the diameter of the armature is large compared to the armature slot depth, and if the slot width is considered negligibly small, then the flux density seen by the armature coil is not significantly different from the air gap flux density. The flat layout, viewed from the surface of the armature toward the shaft, shown by Fig. 5.8 establishes the areas of constant flux density existing in the armature coil environment.

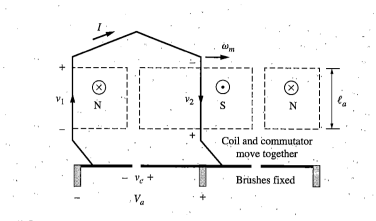

Figure 5.8
Areas of magnetic field influence

5.3.2   ARMATURE COIL INDUCED VOLTAGE

The voltages (v_1, v_2) induced over the armature core length ℓ_a of each side of the coil moving through the time-stationary B-field of Fig. 5.8 are given by the $B\ell v$ rule as

$$v_1 = v_2 = B\ell v = B_p\ell_a r\omega_m \qquad \textbf{[5.1]}$$

where r is the radius measured from the armature shaft centerline to the mean diameter of the armature coil. Polarities of each of the voltages can be verified by Lenz's law. Based on [5.1], KVL gives the coil terminal voltage to be

$$v_c = v_1 + v_2 = 2B_p\ell_a r\omega_m \qquad \textbf{[5.2]}$$

It is apparent from [5.2] that the coil terminal voltage v_c is given by a magnitude scaling of $B_g(\theta)$ where $\theta = \omega_m t$. This voltage is sketched in Fig. 5.9a. From examination of either Fig. 5.6 or Fig. 5.8 while looking through the brush terminals, it can be concluded that as the armature rotates,

$$V_a = v_c \qquad 0 < \omega_m t < \pi/2$$

$$V_a = -v_c \qquad \pi/2 < \omega_m t < 3\pi/2$$

$$V_a = v_c \qquad 3\pi/2 < \omega_m t < 2\pi$$

The resulting brush terminal voltage, commonly called the *armature voltage* V_a, is plotted in Fig. 5.9b. Although the coil voltage is an ac waveform, the voltage seen through the brushes is a unidirectional or dc waveform. Hence, the brush-commutator arrangement acts as a rectifier of the coil voltage, but with polarity discernment. If either the polarity of the B-field or the direction of ω_m are reversed, both v_1 and v_2 change polarities and the resulting plot of V_a is inverted in polarity. If both the B-field and ω_m are reversed, v_1, v_2, and V_a are unchanged from Fig. 5.9.

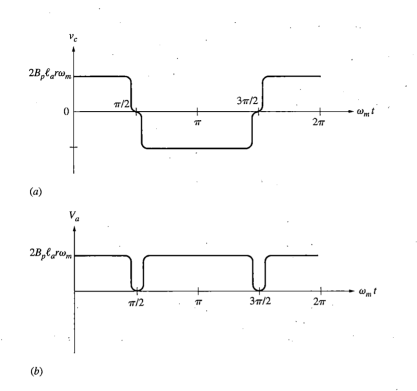

(a)

(b)

Figure 5.9
Simple dc machine voltages. (a) Coil terminal voltage. (b) Voltage view through brush terminals.

One additional observation is that when the brushes span the gaps between commutator segments $(\theta = \pi/2, 3\pi/2)$, the coil terminals are short circuit; however, at this same time, $v = 0$. Consequently, there is no coil voltage available to force current through the temporarily created short circuit.

The armature voltage V_a of Fig. 5.9b is a somewhat crude pulsating dc voltage—a better-quality voltage is desired. To that end, consider the dc machine of Fig. 5.10a which has four coils arranged in a *double-layer winding* that is the typical coil placement scheme for dc machines. The number of commutator segments has been increased to four to accommodate the coil connections. Since the field structure has not been changed from that of Fig. 5.6, the flux density viewed around the air gap looking in the axial direction is unchanged. Except for the shift due to choice of a different reference point from which to measure θ, Figs. 5.7 and 5.10b display the identical flux-density distribution; however, the latter figure serves to show the coil-slot placement. Figure 5.10c presents a flat layout of the coil connections when viewed from the armature surface looking toward the shaft center. The areas of influence for the magnitude B-field, or field pole footprints, are enclosed by the dashed-line rectangles. A general observation concerning dc

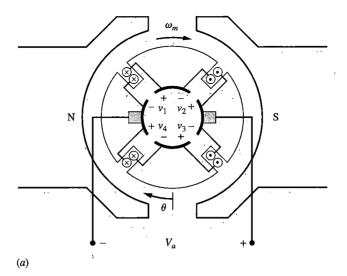

(a)

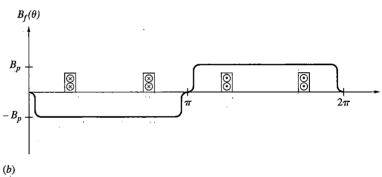

(b)

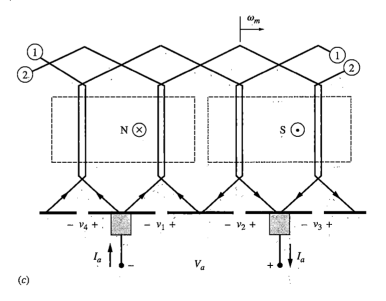

(c)

Figure 5.10
Two-layer winding dc machine with four coils. (a) Cross-section view. (b) Air gap flux density.
(c) Flat layout of coil placement and connection.

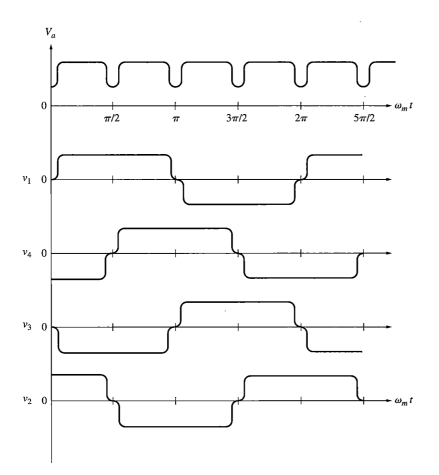

Figure 5.11
Coil and armature voltage for dc machine of Fig. 5.10

machines can be made from this figure: *Current flowing into brushes of one polarity and out of brushes of the opposite polarity divides into paths as it flows through the coils such that current directions in the conductors under a north pole are in one direction while conductor currents under a south pole flow in the opposite direction.*

Since the B-field has not changed, each of the four coil voltages (v_1, v_2, v_3, v_4) of Fig. 5.10 has the identical shape of the coil voltage of Fig. 5.9a; however, each voltage must be properly time-phase shifted. Figure 5.11 displays the coil voltages on a timing diagram. As the armature rotates in the indicated clockwise direction, the armature voltage is given by

$$V_a = v_1 + v_2 \qquad 0 < \omega_m t < \pi/2$$

$$V_a = v_4 + v_1 \qquad \pi/2 < \omega_m t < \pi$$

$$V_a = v_3 + v_4 \qquad \pi < \omega_m t < 3\pi/2$$

$$V_a = v_2 + v_3 \qquad 3\pi/2 < \omega_m t < 2\pi$$

where rotation was assumed to begin with the armature initially 45° counterclockwise from the position shown by Fig. 5.10a. The resulting sketch of V_a is shown at the top of Fig. 5.11, where it is apparent that the quality of the armature voltage has improved over the single-coil case. As the number of coils per pole is increased from this value of 2 to a typical number of 12 to 15, the armature voltage becomes near constant in value, having only a small high-frequency ripple.

The voltage induced over a single-coil side or conductor is given by [5.1]. Define:

Z = number of *conductors* (coil sides that individually conduct coil current) in the armature winding

a = number of *parallel paths* in the armature winding as seen between brush pairs of opposite polarity

If a coil has only one turn, then the coil has two conductors. A two-turn coil would have four conductors. For the winding of Fig. 5.10c, $a = 2$ since there are two parallel paths between the brushes. The conductor count is $Z = 8$. It can be concluded that the number of conductors along any path between the brushes with additive voltages given by [5.1] is Z/a. Thus, the armature voltage for an open-circuit dc machine is

$$E \triangleq V_a|_{I_a=0} = \frac{Z}{a} B_p \ell_a r \omega_m \qquad \textbf{[5.3]}$$

where the no-load voltage E is known as the *counter emf* (cemf). Let Φ_p be the total magnetic flux flowing from a field pole to the armature core; then,

$$B_p = \frac{\Phi_p}{(2\pi r/p)\ell_a} \qquad \textbf{[5.4]}$$

Substitute [5.4] into [5.3] to find

$$E = \frac{pZ}{2\pi a} \Phi_p \omega_m = K\Phi_p \omega_m \qquad \textbf{[5.5]}$$

where

$$K = \frac{pZ}{2\pi a} \qquad \textbf{[5.6]}$$

is known as the *emf constant* or the *torque constant,* the value of which depends only on the machine design. The armature winding of Fig. 5.10c is known as a *lap winding* for which $a = p$ always. Other winding configurations are possible,[1] one of which is known as the *wave winding* with $a = 2$ always.

Example 5.1

A four-pole dc machine armature has 54 slots. It is lap-wound with single-turn coils. (*a*) How many armature coils are required? (*b*) Determine the value of the emf constant.

(*a*) Since there are two coil sides per slot, the number of coils is identical to the number of slots, or 54.

(*b*) $a = p = 4$. $Z = (2 \text{ conductors/slot})(54 \text{ slots}) = 108 \text{ conductors}$. By [5.6],

$$K = \frac{pZ}{2\pi a} = \frac{(4)(108)}{2\pi(4)} = 17.189 \frac{\text{V} \cdot \text{s}}{\text{Wb} \cdot \text{rad}}$$

As the armature rotates, different coils occupy different positions in the conducting paths of the armature; yet there always remains the same number of conductors in series between the brushes. Thus, the resistance of the conducting path between brushes is a constant value. An isolated coil has a small inductance associated with it; however, during the interval of constant coil voltage, a constant value of current flows through the coil so there is no voltage drop associated with this coil inductance. In practical dc machines, the number of armature slots is greater than the presented illustrations. Further, each slot usually holds two to four coil sides. Consequently, during the current transition (*commutation interval*) for a coil, it is mutually coupled with another coil, experiencing a nearly equal and opposite current transition such that $Ldi/dt - Mdi/dt$ is small. Regardless, the coil terminals are shorted by a brush over the commutation interval. Consequently, coil inductance does not produce a voltage seen from a point of view through the brush terminals. Thus, the armature can be modeled by a series connection of a controlled voltage source E given by [5.5] and resistor R_a that includes the resistance of the conducting path through the armature windings between brushes, the carbon brush resistance, and the contact resistance between the brushes and the commutator surface. Figure 5.12 diplays the armature equivalent circuit where the sign convention adopted always results in a positive value for armature current I_a. Power is supplied by the controlled source to the electrical circuit for generator action. Conversely, power is absorbed by the controlled source for motor action.

5.3.3 ELECTROMAGNETIC DEVELOPED TORQUE

The controlled source E of Fig. 5.12 serves as the ideal energy conversion area of the dc machine. For the dc motor armature circuit of Fig. 5.12*b*, the power flowing into controlled source E is identical to the *developed power* P_d associated with the *developed torque* T_d acting to rotate the armature structure in the direction of shaft speed ω_m. Thus, using [5.5],

$$P_d = EI_a = K\Phi_p\omega_m I_a$$

The electromagnetic developed torque follows as

$$T_d = \frac{P_d}{\omega_m} = K\Phi_p I_a \qquad \textbf{[5.7]}$$

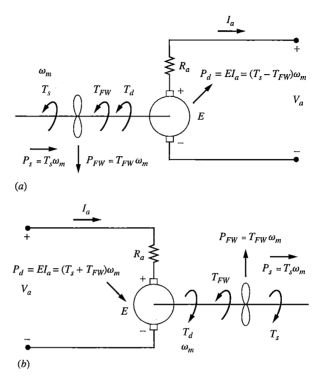

Figure 5.12
DC machine armature equivalent circuit. (a) Generator. (b) Motor.

where T_d acts on the armature in the direction of ω_m in Fig. 5.12b. If this were a generator case, developed torque T_d of Fig. 5.12a taken in the direction of ω_m would be the torque exerted by the mechanical prime mover, less rotational loss torque, required to sustain rotation at the level of electromechanical energy conversion. Figure 5.12 clarifies the relationship between developed torque (T_d), friction and windage torque (T_{FW}), and shaft torque (T_s). For the generator case, T_s is the torque produced by the mechanical prime mover acting to rotate the generator shaft. For the motor case, torque T_s is the torque produced by the motor acting on the coupled mechanical load.

Example 5.2 The dc machine of Example 5.1 runs as a motor with a flux per pole of 0.1 Wb, R_a = 0.14Ω, and an armature terminal voltage of 200 V. While operating at 1100 rpm, determine the value of electromagnetic developed torque.

By [5.5],

$$E = K\Phi_p\omega_m = (17.189)(0.1)(1100\pi/30) = 198.003 \text{ V}$$

From Fig. 5.12b,

$$I_a = \frac{V_a - E}{R_a} = \frac{200 - 198.003}{0.14} = 14.264 \text{ A}$$

Then, using [5.7],

$$T_d = K\Phi_p I_a = (17.189)(0.1)(14.264) = 24.519 \text{ N} \cdot \text{m}$$

5.4 CLASSIFICATION BY FIELD WINDING

All dc machines have the armature equivalent circuit of Fig. 5.12. It is the field winding that establishes the flux per pole Φ_p and offers versatility to the dc machine by producing a variety of voltage-current characteristics for the dc generator and torque-speed characteristics for the dc motor.

5.4.1 BASIC FIELD WINDINGS

The two types of field windings that can be used in design of a dc machine are illustrated for the two-pole machine by Fig. 5.13. A particular machine may have either or both field winding types. The general characteristics of the winding types are as follows:

1. *Shunt Field*—The winding consists of a large number of turns (N_f) of small wire. It is designed to operate in shunt or parallel with the armature circuit, but it may also be connected to an independent source (separately excited).

2. *Series Field*—The winding consists of a small number of turns (N_s) of large wire. It is designed to operate in series with the armature circuit while presenting a small voltage drop.

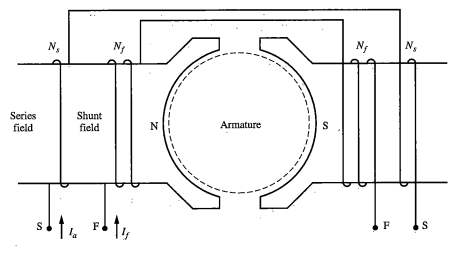

Figure 5.13
Basic field windings

5.4.2 FIELD CONNECTION ARRANGEMENTS

The various dc machine equivalent circuits resulting from use of the two basic field types are shown by Fig. 5.14, where the appropriate armature current (I_a) and line current (I_L) arrow is retained depending on whether the machine acts as a generator or as a motor.

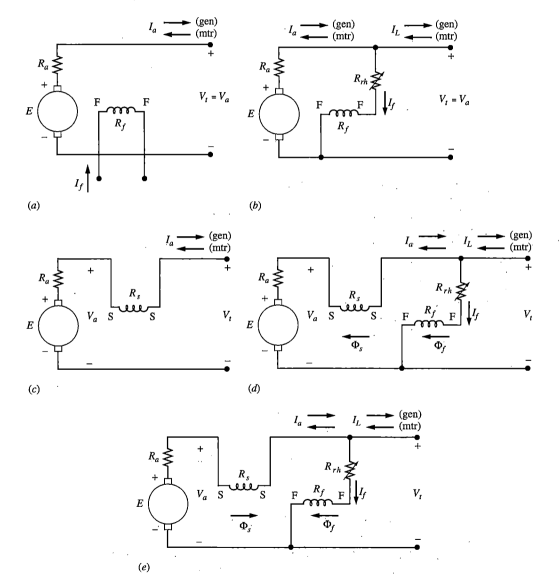

Figure 5.14
DC machine equivalent circuits. (a) Separately excited. (b) Shunt. (c) Series. (d) Cumulative compound. (e) Differential compound.

5.5 Nature and Interaction of Magnetic Fields

The general nature of the magnetic field flux density in the air gap of a dc machine with no current present in the armature coils has been discussed in Sec. 5.3.1; however, flux densities within the ferromagnetic structure have not been addressed. Further, when armature coil current with its associated mmf is introduced, air gap flux density is significantly altered from the $I_a = 0$ case. The impact on performance must necessarily be assessed.

5.5.1 Principal Magnetic Circuit

The analog magnetic circuit for the two-pole dc machine of Fig. 5.6 with $I_a = 0$ is shown by Fig. 5.15, where the shunt field coil forms the mmf source $\mathcal{F}_p = N_f I_f$. If the series field were added, then $\mathcal{F}_p$ would be appropriately altered to include the series field mmf contribution as

$$\mathcal{F}_p = N_f I_f \pm N_s I_a \qquad \textbf{[5.8]}$$

The positive sign applies to a cumulative compound connection and the negative sign to a differential compound connection. For a series dc motor, set $I_f = 0$ and use the positive sign.

The vertical line in the center of the magnetic circuit diagram connects points of equal magnetic potential. Owing to the symmetry of the magnetic circuit, any analysis need only consider the magnetic circuit of one pole that spans the space from the centers of the interpolar regions on either side of the pole. It is with understanding of this symmetry that values of mmf and flux are presented on a per pole basis.

With the exception of $\mathcal{R}_g$ in Fig. 5.15, all reluctances are associated with ferromagnetic material. For small values of Φ_p, the ferromagnetic material will exhibit a high permeability so that $\mathcal{R}_g$ is the dominant value yielding a near linear $\Phi_p - \mathcal{F}_p$ characteristic; however, as Φ_p is further increased by larger values of $\mathcal{F}_p$, the $\Phi_p - \mathcal{F}_p$ characteristic will display magnetic saturation. Since $\mathcal{F}_p$ depends on

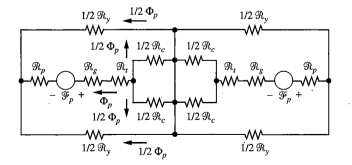

Figure 5.15
DC machine magnetic circuit

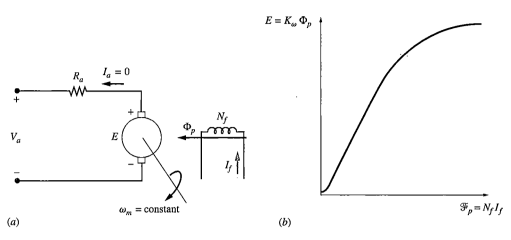

(a) (b)

Figure 5.16
Test determination of OCC. (a) Test setup. (b) Resulting OCC plot.

the field current(s), and since [5.5] shows that E depends directly on Φ_p, any accurate analysis of dc machine performance must consider the nonlinear $\Phi_p - \mathscr{F}_p$ characteristic. Common practice is to handle this consideration through use of the *open-circuit characteristic* (OCC)—a plot of no-load terminal voltage E vs. field current where all data is recorded for a constant value of speed.

Figure 5.16 shows the test setup and the typical shape of an OCC for a dc machine. A small value of E may be recorded for $I_f = 0$ owing to a slight residual magnetism present in the ferromagnetic structure. If the number of turns per pole (N_f) for the field is not known, the horizontal axis is simply plotted as field current (I_f) knowing that the result is only a scaling factor different from $\mathscr{F}_p$. Since speed is held to a constant value for all data points, the vertical axis is only a scaling factor $(K_\omega = K\omega_m)$ different from Φ_p.

Example 5.3   The two plots of Fig. 5.17 show the OCC for a 600-V, 1200 rpm dc machine where the data were recorded at 1200 and 800 rpm. Explain how if only one of the curves were available, the second could be generated.

The ratio of the cemf for two different speeds ω_{m1} and ω_{m2} with values of constant Φ_p gives

$$\frac{E_2}{E_1} = \frac{K\Phi_p\omega_{m2}}{K\Phi_p\omega_{m1}} = \frac{\omega_{m2}}{\omega_{m1}} \qquad \textbf{[1]}$$

The condition of constant Φ_p is met if I_f is unchanged. Thus, at any particular value of field current,

$$E_2 = \frac{\omega_{m2}}{\omega_{m1}}E_1$$

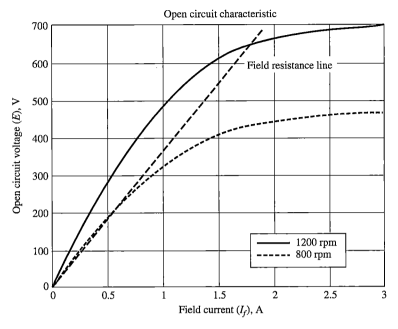

Figure 5.17
Speed adjustment of OCC

If the OCC for 1200 rpm were available, then at each selected value of field current, the value of E for 800 rpm is

$$E_2 = \frac{800}{1200}E_1 = \frac{2}{3}E_1$$

5.5.2   MAGNETIC FIELD INTERACTION

Discussion of magnetic fields of the dc machine to this point has not considered the presence of armature conductor current. The air gap flux density distribution of Fig. 5.10b for the two-pole dc machine of Fig. 5.10a resulted with $I_a = 0$. In order to obtain results for a more practical dc machine, assume that the number of slots is increased to a total of 10 as shown in the flat layout of Fig. 5.18. Ampere's circuital law, given by [3.11], can be applied around closed paths symmetrical about the center of one of the poles where one such path is indicated by dashed lines in Fig. 5.18a. Treating the ferromagnetic material as infinitely permeable and realizing that the air gap path lengths are identical, half of the mmf obtained must equally appear across each air gap. The resulting air gap mmf $\mathscr{F}_A$ due to armature current is shown by Fig. 5.18b. This is a ramped stair-step waveform where the ramps are based on the assumption that the armature conductor current is uniformly distributed over the slot width. As the number of armature slots is

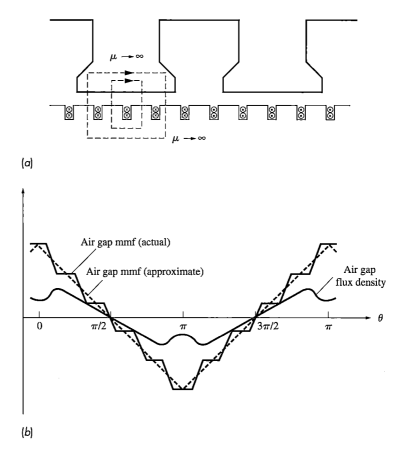

Figure 5.18
Armature only excited. (a) Conductor layout. (b) Air gap fields.

increased, the air gap mmf will approach the triangle waveform indicated in Fig. 5.18b by dashed lines.

Since practical dc machines have mmf waveforms that approach the triangular wave of Fig. 5.18b, it will be used for discussion of air gap flux density B_A due to armature current. Under the field pole face where the reluctance is uniform, the B_A wave is directly proportional to $\mathscr{F}_A$. However, in the interpolar region, the reluctance increases significantly due to the principal flux path lengths now being from the armature surface to the yoke. Although the mmf has a peak value in this region, the reluctance increase is typically more than an order of magnitude. Consequently, the magnitude of $B_A(\theta)$ actually decreases from the value at the pole face tip as indicated by the qualitative sketch of B_A in Fig. 5.18b.

The analysis to this point has assumed that the ferromagnetic material is infinitely permeable. Thus, magnetic linearity exists, and superposition addition of $B_f(\theta)$ from Fig. 5.10b and $B_A(\theta)$ of Fig. 5.18b is justified. Figure 5.19 shows the resulting addition to form the air gap flux density $B_g(\theta)$. It is seen that the flux

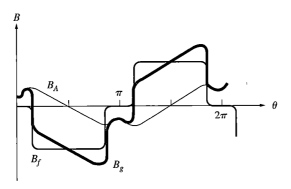

Figure 5.19
Resultant air gap flux density

density at π and 2π is no longer zero with armature current present. If the armature winding of Fig. 5.10c is advanced until one of the coils is shorted by a brush spanning two commutator segments, the sides of the shorted coil are located at π and 2π. Since a nonzero value of voltage is induced behind the low impedance coil, high currents flow across the brush face, leading to destructive heating. Further, the coil current at the negative brush must reverse directions from $-\frac{1}{2}I_a$ to $\frac{1}{2}I_a$ over the interval during which the brush spans the two commutator segments while the coil current at the positive brush reverses from $\frac{1}{2}I_a$ to $-\frac{1}{2}I_a$—a process called *commutation*. If this component of short-circuit current results in the coil current not having a magnitude of $\frac{1}{2}I_a$ at the point that the commutator segment leaves the brush contact, the current must change immediately to the value of $\frac{1}{2}I_a$. This sudden current change through the coil, which has a small but nonzero inductance, results in an induced voltage sufficiently large to arc across the air path between the separating brush and commutator segment.

The ionized air (ozone) resulting from commutator to bar sparking is a good conductor. If a sufficient ozone cloud forms around the commutator periphery so that points of high potential difference are effectively connected together, a line-to-line short circuit is formed—a massive destructive condition known as *commutator flashover*. Even for low-voltage dc machines where commutator flashover is less likely to occur, the additional heating and transfer of material from the commutator bars to brush surface by the arc discharge can damage both the brushes and the commutator. Reliable dc machine operation demands that corrective action be taken.

Interpoles The interpolar region distortion of B_g that precipitates the above discussed commutation problem is a direct result of armature current. In addition, the degree of distortion depends on the magnitude of armature current. The corrective action is to install poles surrounded by windings conducting armature current between the main field poles as illustrated by Fig. 5.20a. These new poles, known as *interpoles* or *commutation poles*, must have the current flow directed so as to nearly zero the interpolar region flux as illustrated by Fig. 5.20b, thereby removing the induced voltage behind the shorted coil undergoing commutation. Figure 5.21c

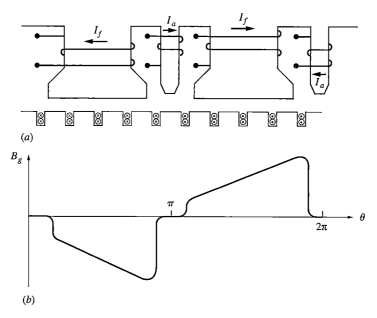

Figure 5.20
Interpoles. (a) Construction. (b) Corrected interpolar flux density.

is an mmf vector plot clarifying the directions and relative strengths of the field mmf ($\mathcal{F}_f$), armature mmf ($\mathcal{F}_A$), and interpole mmf ($\mathcal{F}_I$). The resistance of the interpole winding is lumped into the value of R_a for analysis purposes.

Armature Reaction   The interpoles only affected the air gap flux density in the interpolar region. The B_g waveform under the main field poles is still distorted from the case of $I_a = 0$. The armature mmf $\mathcal{F}_A$ under the pole face region acts to increase the flux density over half of the field pole area and to decrease the flux density over the other half. As long as magnetic linearity exists, the areas A_1 and A_2 are equal as illustrated by Fig. 5.21a. The total flux per pole being the $\int B \, dA$, is unaltered from the case of $I_a = 0$ to $I_a \neq 0$. However, magnetic linearity does not exist for normal conditions where the dc machine is designed to operate near the knee of its $\Phi_p - \mathcal{F}_p$ characteristic. Consequently, $\mathcal{F}_A$ will reduce B_g over half of the pole by a larger amount than the resulting increase over the other pole half as illustrated by Fig. 5.21b where $A_1 > A_2$. Hence, Φ_p is reduced in value—a phenomenon known as *armature reaction*. Typical values for reduction in Φ_p due to armature reaction range from 5 to 10 percent for machines operating near the knee of the $\Phi_p - \mathcal{F}_p$ characteristic. Armature reaction impact decreases as the operating point moves in either direction from the knee of the $\Phi_p - \mathcal{F}_p$ curve. Although the demagnetization of armature reaction may be neglected in approximate dc machine analysis, accurate performance prediction must include an accounting of armature reaction.

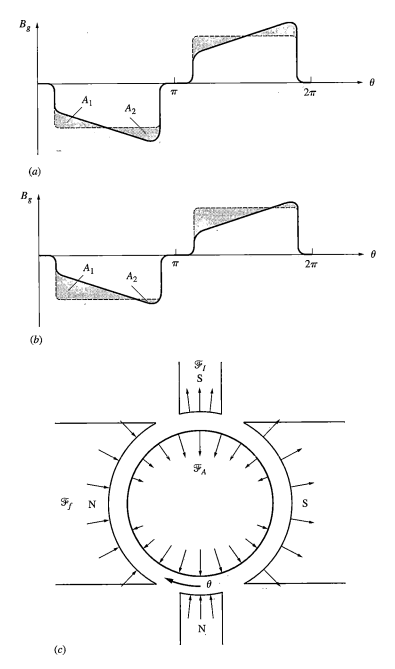

Figure 5.21
Polar region magnetic field. (a) Air gap flux density—magnetically linear. (b) Air gap flux density—magnetically nonlinear. (c) Vector model of acting coil mmf's.

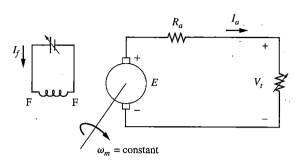

Figure 5.22
Experimental setup for armature reaction determination

Consideration of armature reaction is an analytical challenge owing to the non-linearities involved. Without knowledge of armature winding data, experimental work can be conducted to determine the effects of armature reaction on the cemf E at several load current conditions. Plots can be made to show the effective E vs. I_f with armature reaction present. An experimental setup for armature reaction evaluation is shown in Fig. 5.22. Since armature reaction is negligible for a magnetically linear dc machine, data only need be recorded for I_f greater than about 50 percent of the rated value. A family of curves is to be plotted for selected values of I_a ranging from 0 to rated value. A value of I_f is selected and R_L is adjusted until one of the selected values of I_a is obtained. The actual value of E with armature reaction is determined by

$$E = V_t + I_a R_a \qquad \textbf{[5.9]}$$

Figure 5.23 displays a typical plot of E vs. I_f with I_a as a parameter. Once this graph is completed, the analysis for the dc machine is made by normal procedures except that the value of E is determined from the cemf curve corresponding to the armature current present at the point of study. Reasonable judgment allows linear interpolation between curves.

Compensating Windings The interpoles corrected distortion of the air gap flux density waveform only in the interpolar region. For applications that encounter abrupt changes in load, such as dc roll mill motors, high momentary currents can occur. The resulting extreme distortion of the air gap flux can lead to individual coil voltages that may be sufficiently large to create arc discharge between adjacent commutator segments. Just as in the case of an arcing brush to commutator segment for a commutating coil, the resulting ozone from the arc can lead to commutator flashover. For machines in such applications, windings that conduct armature current are placed in slots distributed over the pole face as illustrated by Fig. 5.24. These windings are known as *compensating windings*. When properly designed, the distortion of the air gap flux density under the pole spans is nearly eliminated; thus, armature reaction is nullified.

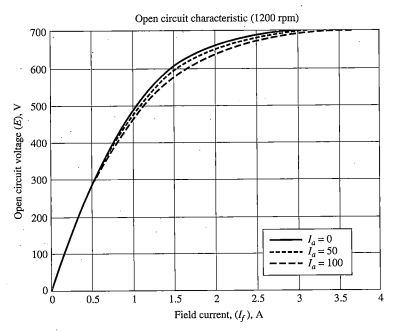

Figure 5.23
OCC showing armature reaction

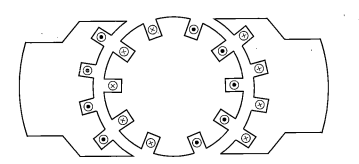

Figure 5.24
Compensating winding

5.6 GENERATOR PERFORMANCE

With the advancement of solid-state rectifying devices, the alternator with either diode bridge or phase-controlled converter rectification is the choice over dc generators as the prime source of dc power in new installations. However, many dc motor drives are controlled for bilateral power flow to allow dynamic braking. In so doing, the dc motor acts as a dc generator; thus, study of the characteristics of a dc generator serves a justifiable purpose.

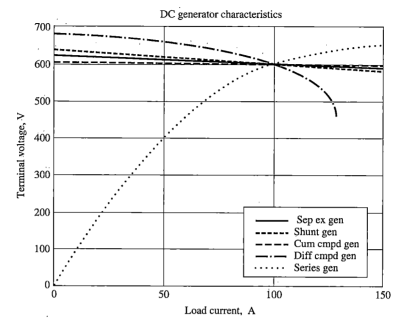

Figure 5.25
DC generator terminal characteristics

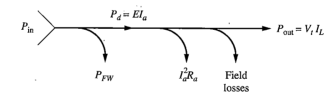

Figure 5.26
Power flow diagram for dc generator

Any one of the five field connection arrangements of Fig. 5.14 can be implemented to form a dc generator, but the resulting terminal voltage-current characteristics differ significantly among the arrangements. Under the assumption that a dc machine intended to operate primarily as a generator will be driven by a speed-regulated mechanical source, the voltage-current characteristics of a dc generator are determined for constant-speed operation. Figure 5.25 shows the terminal volt-ampere characteristics for the various dc generators where all curves pass through the rated voltage-rated load point.

The general power flow diagram for a dc generator is shown by Fig. 5.26. If the generator is self-excited, $P_{in} = P_s$. For the case of a separately excited generator, $P_{in} = P_s + V_f I_f$. Any core losses are lumped with the mechanical rotational losses P_{FW}.

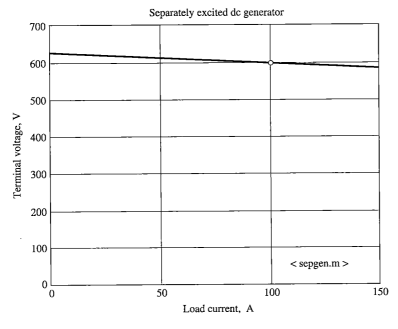

Figure 5.27
Separately excited dc generator

5.6.1 SEPARATELY EXCITED DC GENERATOR

The separately excited dc generator of Fig. 5.14*a* offers close to constant terminal voltage when no control over field current is utilized; however, an independent dc source for field excitation is required. Figure 5.27 shows the terminal characteristics of a 60-kW machine as calculated by the MATLAB program ⟨sepgen.m⟩. The program loads the OCC stored in the file ⟨eif⟩, a two-column ASCII array that contains the E-I_f data of Fig. 5.23. A different file can be constructed with a text editor for loading, or an array could be constructed within ⟨sepgen.m⟩ and the 'load eif' statement removed for analysis of other machines. The effects of armature reaction have been neglected in the formulation of ⟨sepgen.m⟩. If armature reaction were considered, the curve of Fig. 5.27 would be unchanged for zero load current and drop below the present curve slightly as load current increases in value.

Example 5.4

The OCC characteristic of Fig. 5.23 applies to the separately excited dc generator of Fig. 5.27. For the rated condition point of operation (600 V, 100 A), determine the value of field current with and without armature reaction being considered. Let $R_a = 0.25\,\Omega$.

For either case,

$$E = V_t + I_a R_a = 600 + (100)(0.25) = 625\,\text{V}$$

Enter Fig. 5.23 with $E = 625$ to find $I_f \cong 1.6\,\text{A}$ from the $I_a = 0$ OCC curve that neglects armature reaction. With armature reaction considered for $E = 625$, $I_f \cong 1.75\,\text{A}$. The field current requirement is increased 9.4 percent due to armature reaction consideration, which is within the typical range of 5 to 10 percent.

Example 5.5

The separately excited dc generator of Fig. 5.14a has $R_a = 0.4\,\Omega$ and $R_f = 100\,\Omega$. It is known that the rotational losses are 600 W for a speed of 1200 rpm. The OCC of Fig. 5.17 applies to the generator. (a) If the generator is operating at 1500 rpm with no load, determine the terminal voltage. (b) If $I_f = 1.5\,\text{A}$, $n_m = 1200\,\text{rpm}$, and a 15-Ω load is connected across the armature terminals, find the values of armature current (I_a), shaft input power (P_s), efficiency (η), and developed torque (T_d).

(a) Enter the OCC of Fig. 5.17 using the 1200 rpm curve to find $E_{1200} \cong 485\,\text{V}$ for $I_f = 1.0\,\text{A}$. If the generator field current is unchanged as the speed is increased to 1500 rpm, this condition of constant flux per pole allows direct ratio of the generated voltages.

$$V_t = E_{1500} = \frac{1500}{1200}(485) = 606.3\,\text{V}$$

(b) Enter the 1200 rpm OCC of Fig. 5.17 with $I_f = 1.5\,\text{A}$ to find $E \cong 608.5\,\text{V}$. By Ohm's law,

$$I_a = \frac{E}{R_a + R_L} = \frac{608.5}{0.4 + 15} = 39.51\,\text{A}$$

The shaft input power must be the sum of rotational losses and power converted from mechanical to electrical form, or

$$P_s = P_{FW} + EI_a = 600 + (608.5)(39.51) = 24{,}642\,\text{W}$$

This separately excited generator has input power to the field as well as the shaft.

$$P_{in} = P_s + I_f^2 R_f = 24{,}642 + (1.5)^2(100) = 24{,}867\,\text{W}$$

$$P_{out} = I_a^2 R_L = (39.51)^2(15) = 23{,}416\,\text{W}$$

$$\eta = \frac{P_{out}}{P_{in}}(100\%) = \frac{23{,}416}{24{,}867}(100\%) = 94.16\%$$

The developed torque can be determined from the converted power by

$$T_d = \frac{P_d}{\omega_m} = \frac{EI_a}{\omega_m} = \frac{(608.5)(39.51)}{1200(\pi/30)} = 191.32\,\text{N·m}$$

5.6.2   Shunt DC Generator

The shunt dc generator of Fig. 5.14b has the advantage over the separately excited generator in that an independent dc voltage source for field excitation is not required. However, if a controller or voltage regulator is not present to adjust field current, as the terminal voltage decreases with increased $I_a R_a$ voltage drop due to

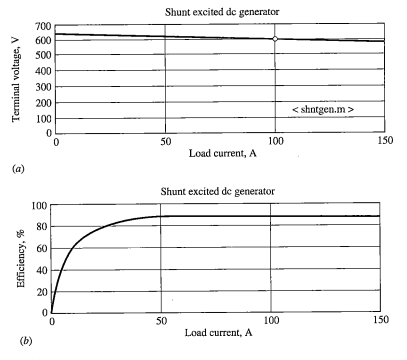

(a)

(b)

Figure 5.28
Shunt excited dc generator. (a) Terminal characteristic. (b) Efficiency.

load current increase, the field current decreases directly with V_t. The decrease in I_f further decreases E; thus, V_t decreases. From an analysis point of view, determination of a value of I_f when only load current is known can be a tedious task; however, the MATLAB program ⟨shntgen.m⟩ can easily pursue this iterative process to determine the resulting field current value. Figure 5.28a shows the terminal characteristic for a 600-V, 60-kW shunt dc generator produced by the MATLAB program. A qualitative assessment of the additional terminal voltage droop when compared with the separately excited dc generator is better made by inspection of Fig. 5.25 where the volt-ampere curves are superimposed.

The shunt dc generator of Fig. 5.14b is operating with no load at 1200 rpm and is characterized by the OCC of Fig. 5.17. $R_a = 0.5\,\Omega$, $R_f = 200\,\Omega$, and the rheostat is set so that $R_{rh} = 160\,\Omega$. (a) Determine the value of terminal voltage. (b) Discuss the nature of terminal voltage if the value of R_{rh} is increased.

Example 5.6

(a) Since the generator has no load $(I_L = 0)$, $I_a = I_f$ and the small voltage drop $I_f R_a$ can be neglected in analysis. Hence, there are two simultaneous conditions that must be satisfied:

1. $V_t = I_f(R_f + R_{rh})$
2. $V_t \cong E = f(I_f)$

The second condition is simply the OCC. If a plot of the first condition (known as the *field resistance line*) is superimposed on the 1200 rpm OCC, the intersection of these two curves is the point of operation. The field resistance line shown in Fig. 5.17 intersects the 1200 rpm OCC at $E \cong V_t = 648$ V. In order for this generator to self-excite and reach this operating point, it was necessary that a small residual voltage be present to initially establish a small $I_f > 0$ so that the generated voltage and field current can bootstrap to the operating point.

(*b*) If R_{rh} is increased, the field resistance line of Fig. 5.17 has a greater slope and, thus, its intersection with the OCC occurs at a smaller value of *E*. If R_{rh} is increased to a large enough value, the generated voltage will decrease to approximately the value of residual voltage.

The MATLAB program ⟨shntgen.m⟩ also determines the efficiency of the shunt generator across the load current range from 0 to 150 percent of rated value. Figure 5.28*b* displays the resulting curve for the example 60-kW dc generator. The typical efficiency-load curve will always have the characteristic shape of Fig. 5.28*b*, wherein the efficiency is zero at no load and rises to a respectable value by approximately 25 percent load. After reaching a maximum value in the 50 to 100 percent load range, the efficiency decreases in the above-rated load region. Efficiency calculations could be added to the other dc generator performance MATLAB code of this text by similar methods.

5.6.3   SERIES EXCITED DC GENERATOR

The series excited dc generator of Fig. 5.14*c* has the typical terminal characteristic of Fig. 5.29. This curve was produced by the MATLAB program ⟨sergen.m⟩ for a 60-kW machine with a 10-turn series field. Owing to the large change in terminal voltage with change in load current, the uncontrolled series dc generator has an unacceptable voltage regulation to be considered for a dc power supply application. However, in certain vehicle propulsion applications, the series dc motor is a configuration of choice. In the dynamic braking mode for such systems, the dc motor acts as a dc series generator. By consecutively adding parallel-connected load resistors across the terminals of the machine as speed decreases, a highly effective vehicle-retarding torque can be obtained at low speeds.

5.6.4   CUMULATIVE COMPOUND DC GENERATOR

The shunt dc generator displays a terminal voltage droop with increase in load current while the series dc generator exhibits a terminal voltage rise for an increase in load current. When properly designed, the cumulative compound dc generator of Fig. 5.14*d* blends the detrimental features of the shunt and series generator to produce a near-constant terminal voltage regardless of load current. Figure 5.30 is the terminal characteristic of a 60-kW cumulative compound dc generator produced by the MATLAB program ⟨cumgen.m⟩ using 750 shunt field turns and 1 series field

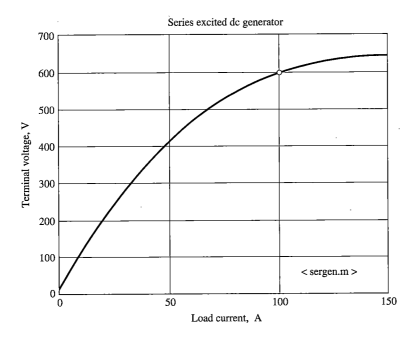

Figure 5.29
Series excited dc generator

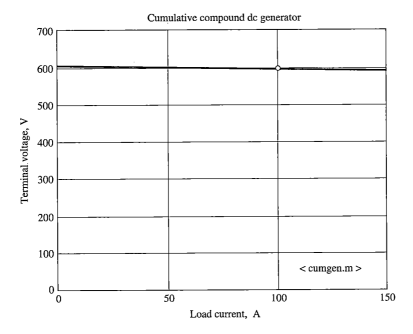

Figure 5.30
Cumulative compound dc generator

turn to produce a near-load independent terminal voltage—a condition known as *flat compounding*.

Example 5.7

The cumulative compound dc generator of Fig. 5.30 has $N_f = 750$ turns, $N_s = 1$ turn, $R_a = 0.25\,\Omega$, and $R_s = 0.03\,\Omega$. Neglect armature reaction and predict the value of field current for $I_L = 100\,A$ if the OCC of Fig. 5.23 is applicable.

From Fig. 5.30, with $I_L = 0$, $E \cong 604\,V$. Entering Fig. 5.23 with $E =. 604\,V$, $I_{fNL} \cong 1.48\,A$. Neglecting the small $I_{fNL}(R_a + R_s)$ voltage drop, the shunt field resistance is

$$R_f = \frac{E}{I_{fNL}} = \frac{604}{1.48} = 408.1\,\Omega$$

The value of field current for rated conditions is then

$$I_f = \frac{V_t}{R_f} = \frac{600}{408.1} \cong 1.47\,A$$

As an alternate approach, the value of the cemf at the load point is

$$E = V_t + I_a(R_a + R_s) = 600 + 100(0.25 + 0.03) = 628\,V$$

Enter Fig. 5.23 at $E = 628\,V$ to see that the equivalent shunt field current is $I_{feq} \cong 1.6\,A$, or the field mmf required is $1.6(750) = 1200$ A-t. Thus,

$$I_f = \frac{1200 - N_s I_a}{N_f} = \frac{1200 - (1)(100)}{750} = 1.47\,A$$

5.6.5 DIFFERENTIAL COMPOUND DC GENERATOR

Based on the observed nature of the cumulative compound dc generator, it would be anticipated that the differential compound dc generator of Fig. 5.14e should display a drooping terminal voltage characteristic as load current increases. Figure 5.31, produced by the MATLAB program ⟨diffgen.m⟩, presents the terminal characteristic for a 60-kW differential compound dc generator with 750 shunt field turns and 3 series field turns. The differential compound dc generator is always set up so that the shunt field is dominant for small values of load current. Although this generator type is not suitable for a constant-voltage dc power supply, it has found extensive application in high-quality arc welding. The machine produces a high voltage when striking an arc and inherently moves to a low-voltage, current-limited performance to sustain a regulated arc.

5.7 MOTOR PERFORMANCE

The five field connection arrangements of Fig. 5.14 are generally reduced to four connections for study when the motor performance is based on a constant value of impressed terminal voltage. In such a case, the field voltage remains constant for

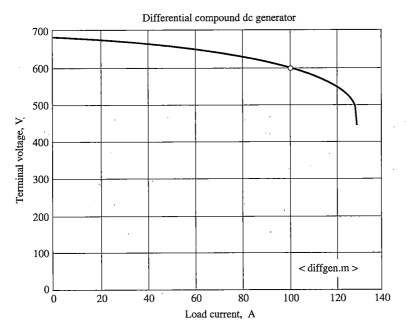

Figure 5.31
Differential compound dc generator

the shunt dc motor so that its field current remains constant, rendering its performance nature identical to the separately excited dc motor.

Applying KVL to the equivalent circuit of Fig. 5.14*b* and solving for speed gives

$$\omega_m = \frac{V_a - I_a R_a}{K\Phi_p} \qquad \textbf{[5.10]}$$

The expression of [5.10] yields the motor shaft speed regardless of the field connection. For the case of a separately excited or a shunt motor, $V_a = V_t$. If a series field is present, the numerator can be replaced by $V_t - I_a(R_a + R_s)$ to express [5.10] in terms of terminal voltage. The motor shaft torque is given by

$$T_s = T_d - T_{FW} = K\Phi_p I_a - P_{FW}/\omega_m \qquad \textbf{[5.11]}$$

Figure 5.32 displays the torque-speed performance characteristics for the various dc motors with constant terminal voltage applied where all configurations produce identical rated shaft torque-speed values. However, in many power conditioned drives, the field winding may be excited from an independent source where the field current is controlled to allow versatility in performance such as shaping torque or power profiles as functions of motor speed.

Figure 5.33 presents the general power flow diagram for a dc motor. If the motor is self-excited, $P_{in} = V_t I_L$. For the case of a separately excited motor, $P_{in} = V_t I_a + V_f I_f$. Any core losses are lumped with the mechanical rotational losses P_{FW}.

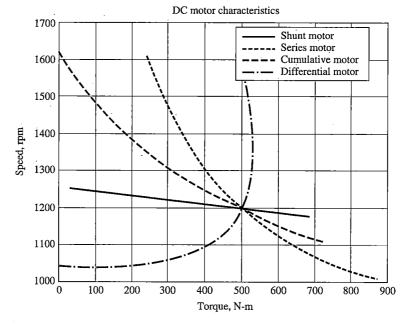

Figure 5.32
DC motor performance characteristics

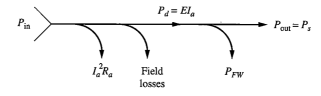

Figure 5.33
Power flow diagram for dc motor

5.7.1   SHUNT EXCITED DC MOTOR

The shunt dc motor of Fig. 5.14*b* exhibits the nearest to a constant speed characteristic of all dc motor configurations when no control of field current is exercised. Figure 5.34*a* presents the torque-speed and load current–speed characteristics for an 80-hp, 600-V shunt dc motor produced by the MATLAB program ⟨shntmtr.m⟩ where it is seen that the speed droop from no load to full load is approximately 50 rpm, giving a speed regulation of

$$SR = \frac{n_{mNL} - n_{mFL}}{n_{mFL}}(100\%) \qquad\qquad \textbf{[5.12]}$$

$$SR = \frac{1255 - 1200}{1200}(100\%) = 4.58\%$$

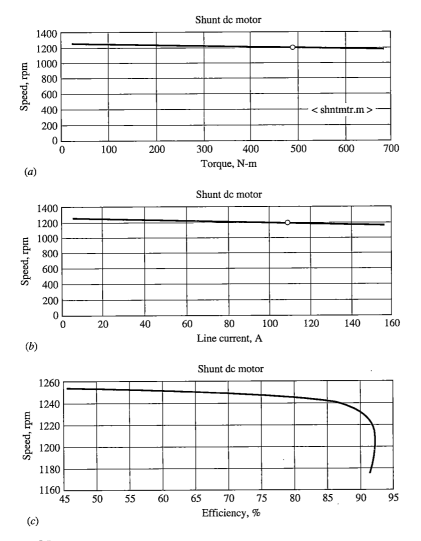

Figure 5.34
Shunt excited dc motor. (a) Torque-speed characteristic. (b) Line current–speed characteristic. (c) Efficiency.

The torque-speed curve of Fig. 5.34b appears to be nearly a straight-line plot. (a) Show that if armature reaction and rotational losses are neglected, the plot is indeed a straight line. (b) Discuss the effect of armature reaction on the torque-speed curve.

    (a) Writing [5.10] in terms of terminal voltage gives

$$\omega_m = \frac{V_t - I_a R_a}{K\Phi_p} \qquad \textbf{[1]}$$

The developed torque expression of [5.7] can be solved for I_a and the result substituted into [1] to yield

Example 5.8

$$\omega_m = \frac{V_t}{K\Phi_p} - \frac{R_a}{(K\Phi_p)^2}T_d \qquad\qquad [2]$$

Since Φ_p is independent of I_a if armature reaction is neglected, equation [2] is a straight line with a vertical axis intercept of $V_t/K\Phi_p$ and a slope of $-R_a/(K\Phi_p)^2$.

(b) Since $I_aR_a \ll V_t$ in [1], it must be concluded that the second term of [2] is an order of magnitude or more smaller in value than the first term. For small values of T_d, I_a is necessarily small. Hence, armature reaction is negligible, and the speed-torque curve of Fig. 5.34a is not impacted. As T_d, thus I_a, increases in value, armature reaction leads to a reduction in Φ_p. Owing to the relative sizes of the two terms of [2], their difference will increase in value, or the motor speed increases for a particular value of T_d. Consequently, the torque-speed curve of Fig. 5.34a would display less speed droop if armature reaction were considered. Stated in terms of speed regulation, armature reaction can reduce the speed regulation of a dc shunt motor. If a shunt motor is operating with a weak shunt field, armature reaction can reduce Φ_p sufficiently so that the loaded speed exceeds the unloaded speed, giving a negative value of speed regulation.

Example 5.9

A shunt dc motor is rated for 230 V, 1350 rpm, 10 hp, $I_L = 37.5$ A, and $I_f = 0.75$ A. It is known that $R_a = 0.35\,\Omega$ and $P_{FW} = 519$ W at rated speed. (a) Determine the rated condition developed torque, cemf, and efficiency. (b) For operation at 230 V with $I_L = 20$ A and a field current $I_f = 0.75$ A, calculate the value of developed torque and speed.

(a)
$$T_d = \frac{P_s + P_{FW}}{\omega_m} = \frac{(10)(746) + 519}{1350(\pi/30)} = 56.44 \text{ N·m}$$

By KCL,

$$I_a = I_L - I_f = 37.5 - 0.75 = 36.75 \text{ A}$$

$$E = V_t - I_aR_a = 230 - (36.75)(0.35) = 217.14 \text{ V}$$

The efficiency is

$$\eta = \frac{P_{out}}{P_{in}}(100\%) = \frac{P_s}{V_tI_L}(100\%) = \frac{(10)(746)}{(230)(37.5)}(100\%) = 86.49\%$$

(b) Let subscript 1 denote rated conditions and subscript 2 denote the 20-A line current condition.

$$I_{a2} = I_{L2} - I_f = 20 - 0.75 = 19.25 \text{ A}$$

By ratio of developed torques,

$$\frac{T_{d2}}{T_{d1}} = \frac{K\Phi_pI_{a2}}{K\Phi_pI_{a1}}$$

or

$$T_{d2} = \frac{I_{a2}}{I_{a1}}T_{d1} = \frac{19.25}{36.75}(56.44) = 29.56 \text{ N·m}$$

$$E_2 = V_t - I_{a2}R_a = 230 - (19.25)(0.35) = 223.26 \text{ V}$$

By ratio of cemfs,

$$n_{m2} = \frac{E_2}{E_1} n_{m1} = \frac{223.26}{217.14}(1350) = 1388.1 \text{ rpm}$$

The MATLAB program ⟨shntmtr.m⟩ also produced the efficiency curve of Fig. 5.34c for the example 80-hp dc shunt motor over the load torque range from a small value to approximately 150 percent of rated torque. The efficiency remains above 90 percent for load torque above approximately 50 percent of rated value but drops off characteristically for light loads as the near constant rotational and field winding losses become comparable to the output power in value.

5.7.2 SERIES EXCITED DC MOTOR

The series dc motor of Fig. 5.14c under constant terminal voltage operation has the torque-speed and line current–speed characteristics of Fig. 5.35. These curves were generated by the MATLAB program ⟨sermtr.m⟩ for a 600-V, 1200 rpm, 80-hp series dc motor with eight series field turns (N_s). The torque increase at low speeds that makes this motor configuration a popular choice for traction applications such

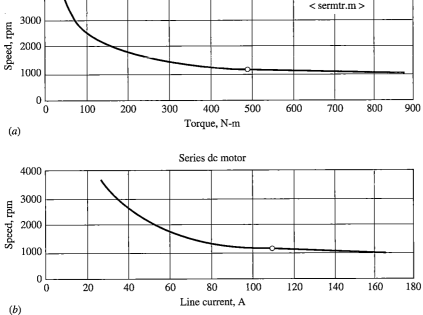

(a)

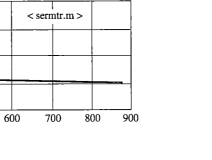

(b)

Figure 5.35
Series excited dc motor. (a) Torque-speed characteristic. (b) Line current–speed characteristic.

as locomotive motors, or for automotive cranking motors, is readily apparent from inspection of the $T_d - n_m$ characteristic.

In traction motor applications, the prime source of power is typically an internal combustion engine that drives a generator to convert energy from mechanical to electrical form. It is desirable to have the traction motors fully utilize the power capacity of the engine over a wide vehicle speed range; thus, the motor output power should be near constant in value over the vehicle speed range until the point is reached at lower speed for which torque limit is necessary to protect drive train components from damage. A constant power condition is described by a hyperbola on a torque-speed plot. Although the torque-speed characteristic of the series motor does appear to asymptotically approach the vertical and horizontal axes, it is not a true hyperbola. In order to approximate the desired constant power performance, *field weakening* control is introduced by discrete step changes in the number of active series field turns as illustrated schematically by Fig. 5.36*a*. The MATLAB program ⟨sermtrfw.m⟩ calculates a family of torque-speed curves according to the values of turns per pole entered in the N_s array. Figure 5.36*b* displays the results for a 600-V, 80-hp dc series motor with $N_s = 4, 5, 7, 10$ turns where the lower curve corresponds to the largest value of N_s. The dashed line superimposed on the family of curves is a torque-speed plot for a constant 80 hp. As the motor increases in speed, the number of active field turns is decreased by switching to give the stepped torque-speed characteristic indicated by the bold trace of Fig. 5.36*b*. Since

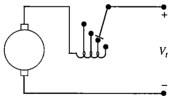

(*a*)

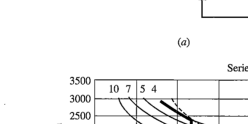

(*b*)

Figure 5.36
Field weakening for dc series motor. (*a*) Schematic. (*b*) $T_d - n_m$ curve ($N_s = 4, 5, 7, 10$ turns).

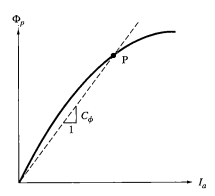

Figure 5.37
Magnetic linearization

practical considerations allow only a small number of steps, an exact tracking of the constant power characteristic is not possible; however, acceptable vehicle performance can usually be obtained.

Although not particularly troublesome to numerical methods, the nonlinear dependence of flux per pole Φ_p on armature current I_a renders closed form analysis of the dc series motor quite cumbersome. With sacrifice in accuracy that must be carefully assessed as the point of operation changes, the dc series motor can be treated as magnetically linear as indicated by the straight dashed-line approximation of Fig. 5.37. Under this approximation, the flux per pole is given by

$$\Phi_p = C_\phi I_a \qquad [5.13]$$

where C_ϕ is the slope of the straight line passing through the origin and the operating point P of the actual $\Phi_p - I_a$ magnetization curve. If P lies along the linear portion of the magnetization curve, [5.13] yields accurate values for the flux per pole. However, if the operating point P is beyond the knee of the magnetization curve, values for Φ_p determined by [5.13] can only be used for small excursions about the operating point to maintain acceptable analytical accuracy.

Using the magnetically linear approximation of [5.13], the respective cemf and developed torque equations of [5.5] and [5.7] become

$$E = K\Phi_p\omega_m = KC_\phi I_a\omega_m = K_\ell I_a\omega_m \qquad [5.14]$$

and

$$T_d = K\Phi_p I_a = KC_\phi I_a^2 = K_\ell I_a^2 \qquad [5.15]$$

where

$$K_\ell = KC_\phi \qquad [5.16]$$

The value of K_ℓ has an obvious dependence on I_a, but it can be treated as a constant as long as [5.13] yields acceptably accurate results. From inspection of [5.14] and [5.15], it is apparent that the magnetic linearization has not changed the dc series motor to a case of algebraic linearity; however, it has resulted in a simple second-degree nonlinearity which is easier to manage analytically than if a higher-degree polynomial in I_a were chosen in lieu of the first-degree curve fit of [5.13].

Example 5.10

Use the magnetically linear approximation of [5.13] to show that the characteristic shape of the speed-torque curve of Fig. 5.35a for the dc series motor is to be expected.

Substitute [5.13] into [2] of Example 5.8 and use [5.15] to find

$$\omega_m = \frac{V_t}{KC_\phi I_a} - \frac{(R_a + R_s)T_d}{(KC_\phi I_a)^2} = \frac{V_t}{K_\ell I_a} - \frac{(R_a + R_s)T_d}{K_\ell^2 I_a^2} \qquad \textbf{[1]}$$

From [5.15],

$$I_a^2 = \frac{T_d}{K_\ell} \qquad \textbf{[2]} \qquad\qquad I_a = \sqrt{\frac{T_d}{K_\ell}} \qquad \textbf{[3]}$$

Use [2] and [3] in [1] and simplify the result to yield

$$\omega_m = \frac{V_t}{\sqrt{K_\ell}\sqrt{T_d}} - \frac{R_a + R_s}{K_\ell} \qquad \textbf{[4]}$$

Equation [4] does have the asymptotic shape of the speed-torque plot of Fig. 5.35a. As $T_d \to 0$, $\omega_m \to \infty$ and as T_d becomes large, ω_m becomes small. The result also shows that $\omega_m \propto 1/\sqrt{T_d}$ rather than $1/T_d$. Thus, the torque-speed curve is not a hyperbola, explaining why field weakening had to be utilized to force the dc series motor to track the constant power hyperbola in Fig. 5.36b.

5.7.3   CUMULATIVE COMPOUND DC MOTOR

The cumulative compound dc motor of Fig. 5.14d under constant terminal voltage operation has the torque-speed and line current–speed characteristic of Fig. 5.38. This set of curves was generated by the MATLAB program ⟨cummtr.m⟩ for a 600-V, 1200 rpm, 80-hp cumulative compound dc motor.

The mmf's produced by the shunt field and the series field of the cumulative compound motor are additive. If the shunt field turns per pole were reduced to zero the machine becomes a series motor. Conversely, if the series field turns per pole were reduced to zero, the motor is a shunt motor. Consequently, the motor torque–speed curve is logically expected to lie somewhere between that of the shunt and that of the series motor. This expectation is validated by inspection of Fig. 5.32, where the torque-speed characteristics of the three motor types are plotted on a common set of axes. The choice of N_s and N_f establishes the relative strengths of shunt field and series field mmf's and thus determines whether the torque-speed characteristic of a particular cumulative compound motor has a behavior closer to that of a series motor or a shunt motor.

The performance attributes of the cumulative compound motor that make it attractive and the dc machine of choice in certain dc drive applications are the favorable features that the hybrid configuration pulls from both the shunt and

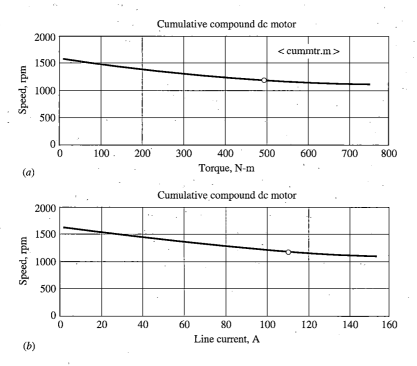

Figure 5.38
Cumulative compound dc motor. (a) Torque-speed characteristic. (b) Line current–speed characteristic.

series motor. Like the series motor, the cumulative compound motor exhibits a high torque at low speed, allowing it to accelerate heavy loads. For an unloaded condition, the shunt field mmf does not vanish; thus, the motor can operate at no load like the shunt motor without dangerous overspeed.

Any attempt to derive an algebraic equation directly showing speed as a function of torque for the cumulative compound motor quickly becomes cumbersome and results in a complex expression from which it is difficult to make a qualitative assessment. It is more convenient to validate the nature of the torque-speed curve of Fig. 5.38a using armature current as an intermediary. Since the mmf's of the shunt and series fields are additive, the flux per pole for the cumulative compound motor is given by

$$\Phi_p = \frac{N_f I_f + N_s I_a}{\mathfrak{R}} \qquad \text{[5.17]}$$

where $\mathfrak{R}$ is the equivalent reluctance per pole of the machine magnetic circuit. Substituting [5.17] into [2] of Example 5.8 and including the series field winding resistance, the speed is given by

$$\omega_m = \frac{V_t - I_a(R_a + R_s)}{K(N_f I_f + N_s I_a)/\mathfrak{R}} \cong \frac{\mathfrak{R} V_t}{K(N_f I_f + N_s I_a)} \qquad \text{[5.18]}$$

where the $I_a(R_a + R_s)$ voltage drop can usually be neglected compared to V_t. Use of [5.17] in [5.7] results in the following expression for developed torque:

$$T_d = \frac{K(N_f I_f + N_s I_a) I_a}{\mathfrak{R}}$$ **[5.19]**

From [5.18] and [5.19], it can be concluded that an increase in armature current I_a results in a decrease in speed ω_m while the torque increases in value. Conversely, a decrease in I_a results in a speed increase and an associated torque decrease. Further, as $I_a \rightarrow 0$, $T_d \rightarrow 0$ and $\omega_m \rightarrow \mathfrak{R} V_t / K N_f I_f$ (a bounded constant value). With these observations, the general nature of the cumulative compound motor torque-speed curve of Fig. 5.38a is seen to be plausible.

Example 5.11 | A cumulative compound motor with $N_f = 750$ turns, $N_s = 3$ turns, $R_a = 0.25\ \Omega$, and $R_s = 0.3\ \Omega$ has the OCC of Fig. 5.23. Neglect armature reaction. If $I_a = 80$ A, $I_f = 1$ A, and $V_t = 600$ V, determine the values of speed and developed torque at the point of operation.

The total mmf per pole is

$$\mathscr{F}_p = N_f I_f + N_s I_a = 750(1) + 3(80) = 990\,\text{A-t}$$

A value of shunt field current exists such that for $N_s = 0$, the identical value of $\mathscr{F}_p$ would be produced. Enter Fig. 5.23 with the equivalent value of shunt field current $I_{feq} = \mathscr{F}_p / N_f = 990/750 = 1.32$ A to determine the cemf as $E_1 \cong 570$ V if the speed were 1200 rpm. Since the OCC of Fig. 5.23 is stored in the array $\langle$eif$\rangle$, a more accurate value can be determined by the MATLAB session below:

```
>> load eif
>> m=length(eif);
>> E=eif(1:m,1);  If=eif(1:m,2);
>> interp1(If, E, 1.32)
ans =
    571.3075
```

For the actual point of operation,

$$E_2 = V_t - I_a(R_a + R_s) = 600 - 80(0.25 + 0.3) = 556.0\ \text{V}$$

Since E_1 for a speed of 1200 rpm was determined at the same value of Φ_p as E_2,

$$n_m = \frac{E_2}{E_1}(1200) = \frac{556.0}{571.3}(1200) = 1167.9\,\text{rpm}$$

The developed torque follows as

$$T_d = \frac{P_d}{\omega_m} = \frac{E I_a}{\omega_m} = \frac{556.0(80)}{1167.9(\pi/30)} = 363.7\,\text{N·m}$$

5.7.4 DIFFERENTIAL COMPOUND DC MOTOR

The differential compound dc motor of Fig. 5.14*e* with a constant terminal voltage impressed has the torque-speed and line current–speed characteristic of Fig. 5.39 as determined by execution of ⟨diffmtr.m⟩. Although the differential compound dc motor can easily be connected by accident if the series coil terminals are reversed when attempting to configure a cumulative compound, its use should be avoided. If the differential compound motor is placed in operation and loaded until the subtractive series field mmf cancels the shunt field mmf, an unstable increase in speed occurs as indicated by Fig. 5.39*a*. The speed expression of [5.18] is applicable if the polarity of the series field mmf is reversed, yielding

$$\omega_m \cong \frac{\mathfrak{R} V_t}{K(N_f I_f - N_s I_a)} \qquad\qquad \textbf{[5.20]}$$

from which it is seen that the speed does increase significantly as $(N_f I_f - N_s I_a) \to 0$. Since the cemf also reduces significantly as mmf cancellation is approached, the armature current increases rapidly. If the protective means such as fuses or overcurrent breakers do not open the motor circuit, destructive speeds can quickly be reached.

Differential compound dc motor

(a)

Differential compound dc motor

(b)

Figure 5.39
Differential compound dc motor. (*a*) Torque-speed characteristic. (*b*) Line current–speed characteristic.

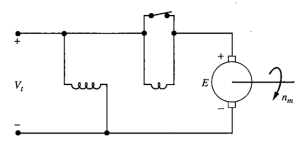

Figure 5.40
Compound-wound dc motor test

Example 5.12 | The unloaded compound-wound dc motor of Fig. 5.40 is energized by a reduced terminal voltage V_t so that it operates at a low speed. The switch that has been shorting the series field is opened, and it is observed that the motor speed increases. Is the motor connected differential or cumulative compound?

Prior to opening the switch, the series field component was zero. After opening the switch, the series field mmf has the value $N_s I_a$. From [5.20], it is seen that the speed would increase if the motor is differential compound. If the motor were connected cumulative compound, [5.18] indicates that the motor speed would decrease. Consequently, it is concluded that the motor is connected differential compound.

5.8 MOTOR CONTROL

Two aspects of dc motor operation and control need to be specifically addressed to complete the introductory study—starting control and speed control. Each of these ideas requires additional hardware to implement.

5.8.1 STARTING CONTROL

If a dc motor is to be connected to a constant voltage supply, external provision must be made to limit the armature current to a safe level until the motor has accelerated to a speed where the cemf has sufficient value to inherently limit armature current. The armature current is given by

$$I_a = \frac{V_a - K\Phi_p\omega_m}{R_a}$$

[5.21]

Upon initial start-up when the speed is zero, the armature current is limited only by the combined brush and winding resistance (R_a); thus, the current can reach damaging levels. The practical solution is to temporarily insert a resistance in series with the armature circuit as illustrated by Fig. 5.41 for the case of a shunt motor. The denominator of [5.21] now becomes $R_a + R_{st}$. The full value of R_{st} is selected to limit armature to an acceptable value when $\omega_m = 0$. R_{st} is reduced to zero as the value of speed, hence E, increases.

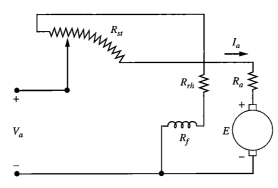

Figure 5.41
Starting resistance for dc motor

Numerous schemes exist for implementation of this starting resistance that vary from a manually operated variable resistor to automated relays that reduce R_{st} based on time, current, or speed. However, the fundamental concept is common to all cases. In addition, a dc motor starter has some control provision to disconnect the armature circuit from the supply mains in the event of a supply voltage interruption. In such case, the start-up sequence must be reinitiated after a power failure. Otherwise, the motor could slow down or stop during a supply interruption and the high current threat returns upon restoration of the supply voltage.

The shunt dc motor of Example 5.9 is to be equipped with the starter of Fig. 5.41 for operation on a 230-V dc bus. Determine the value of R_{st} to assure that the initial armature current does not exceed 150 percent of rated value. | **Example 5.13**

The maximum allowable armature current is

$$I_{a\max} = 1.5(I_{aR} - I_{fR}) = 1.5(37.5 - 0.75) = 55.125 \text{ A}$$

By use of [5.21] with $\omega_m = 0$,

$$R_{st} = \frac{V_t}{I_{a\max}} - R_a = \frac{230}{55.125} - 0.35 = 3.82 \ \Omega$$

5.8.2 SPEED CONTROL

The speed control schemes for dc motors can be placed in three categories:

1. Field control
2. Armature resistance control
3. Armature voltage control.

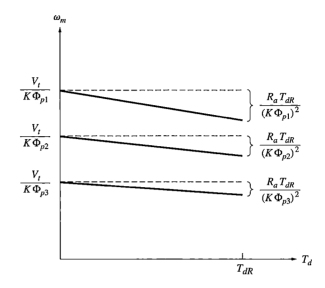

Figure 5.42
Field control of speed

Field Control   For the case of the shunt dc motor, a field rheostat as illustrated by Fig. 5.14b is placed in series with the shunt field so that field current, thus Φ_p, can be varied. The motor speed for the ith setting of field current (I_{fi}) is then given by [2] of Example 5.8 as

$$\omega_m = \frac{V_t}{K\Phi_{pi}} - \frac{R_a}{(K\Phi_{pi})^2}T_d \qquad \textbf{[5.22]}$$

Based on [5.22], the nature of the speed-torque characteristic is shown by Fig. 5.42 for discrete settings of field current.

The field rheostat method of speed control for shunt motors offers the advantages of comparatively low equipment cost and relatively small increase in losses. Its disadvantage lies in the fact that the useful range of speed control is limited. Obviously, speed cannot be controlled to small values since Φ_p cannot be increased without bound.

On the opposite end of the control range Φ_p must be reduced to attain higher speeds. Initially, the decrease in Φ_p leads to increased speed regulation as indicated by the increased negative slope of the speed-torque curves of Fig. 5.42. If Φ_p is decreased sufficiently, the point is reached where the armature reaction due to any load increase can significantly reduce the net flux per pole, resulting in an unstable speed increase with load increase like the differential compound motor. This latter problem can be overcome by addition of a weak series field that is cumulatively connected. Such a motor is commonly known as a *stabilized shunt motor*.

Field control of speed can also be implemented for a series motor by placing a *diverter resistor* in parallel with the series winding. In such an arrangement, cur-

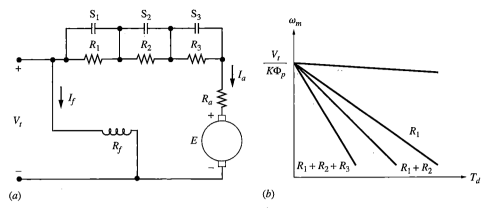

(a) (b)

Figure 5.43
Armature resistance control of speed. (a) Schematic. (b) Speed-torque characteristic.

rent divides between the parallel path through the series field winding and the diverter resistor, resulting in less mmf per ampere of armature current. The performance result is similar to that of field weakening discussed in Sec. 5.7.2, where the number of series field turns was reduced by switch operation. Diverter resistor speed control is sensitive to current division apportionment of current flow through the two parallel paths made up of materials having both different temperature-resistivity characteristics and different temperatures.

Armature Resistance Control If external resistors are added to the armature circuit of a shunt dc motor as illustrated by Fig. 5.43a, the speed equation can be expressed as

$$\omega_m = \frac{V_t}{K\Phi_p} - \frac{R_a + \displaystyle\sum_i R_i}{(K\Phi_p)^2} T_d \qquad \text{[5.23]}$$

The state of the contactors (S_1, S_2, S_3) determines the value of external resistance. For a particular value of shunt field current (I_f), the speed-torque characteristic is made up of the family of curves displayed by Fig. 5.43b. Unlike the field control method, the speed can now be controlled for small values, even down to near zero speed. The armature resistance control of speed is equally applicable to series motors. A disadvantage in any case of application is the high ohmic losses associated with the external resistors that conduct armature current.

The shunt dc motor of Example 5.9 has field current adjusted for rated conditions. If the armature resistance control of Fig. 5.43a is implemented and rated terminal voltage is impressed with all contactors open, determine the value of $R_1 + R_2 + R_3$ to control the motor speed to 100 rpm if the coupled mechanical load requires a torque of 30 N·m.   **Example 5.14**

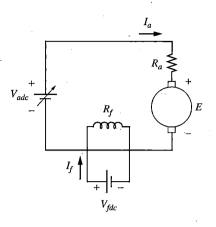

Figure 5.44
Armature voltage control of speed

From Example 5.9, rated field current results in $E = 217.14$ V at 1350 rpm. Hence,

$$K\Phi_p = \frac{E}{\omega_m} = \frac{217.14}{1350(\pi/30)} = 1.536\frac{V \cdot s}{rad}$$

Rearranging [5.23],

$$\sum R_i = R_1 + R_2 + R_3 = \frac{K\Phi_p}{T_d}(V_t - K\Phi_p\omega_m) - R_a$$

$$R_1 + R_2 + R_3 = \frac{1.536}{30}[230 - (1.536)(100)(\pi/30)] - 0.35 = 10.60\ \Omega$$

Armature Voltage Control   Armature voltage control is similar in end result to armature resistance control. The large ohmic losses associated with the no longer present external resistors vanish. However, a high power level dc voltage source to feed the armature circuit is required and a separate dc source for field excitation must be added as indicated by Fig. 5.44. Source V_{adc} may be a dc generator. In such case, the complete arrangement is known as a *Ward-Leonard* system. Or V_{adc} may be a power electronic converter (known as a *phase-controlled converter*) capable of producing a variable-magnitude dc voltage by rectification of ac input voltages. The speed is described by

$$\omega_m = \frac{V_{adc}}{K\Phi_p} - \frac{R_a}{(K\Phi_p)^2}T_d \qquad\qquad \textbf{[5.24]}$$

Example 5.15 | The shunt dc motor of Example 5.9 is operated with the armature voltage control of Fig. 5.44. The source V_{fdc} maintains field current at the rated condition value. Determine the voltage V_{adc} that must be applied if the motor speed is controlled to 100 rpm with a coupled mechanical load that requires 30 N·m.

The value of $K\Phi_p$ from Example 5.14 is valid. Rearranging [5.24],

$$V_{adc} = K\Phi_p\omega_m + \frac{R_aT_d}{K\Phi_p}$$

$$V_{adc} = (1.536)(100)\pi/30 + \frac{0.35(30)}{1.536} = 167.7 \text{ V}$$

5.9 DC MOTOR DESIGN

As the case in other design sections of this book, the scope is limited to allow concise presentation of basic processes and procedures underlying dc motor design. For example, only simplex lap armature windings are considered. Designs of interpole windings and compensating windings are not addressed. References are cited for the reader who wants further details of derivations and exposure to a broader range of design possibilities.

5.9.1 CLASSIFICATIONS AND STANDARDIZATIONS

The National Electrical Manufacturers Association in *NEMA Standard MG-1* presents guidelines for labeling, classifying, rating, and packaging dc motors. These guidelines can be organized into nameplate, environmental, and mechanical subdivisions.

Nameplate The minimum nameplate data for a dc motor includes the specific items:

1. Frame designation (see Table 5.1)
2. Rated output power (hp)
3. Speed at rated load (rpm)
4. Rated-load current (A)
5. Rated voltage (V)
6. Field winding type (shunt, series, . . .)
7. Maximum ambient temperature for rated load
8. Insulation class (see *IEEE Standards 117* and *275*)

Environmental Maximum ambient temperature and altitude derating are addressed. The NEMA styles are discussed—forced-air cooled, self-ventilated, and totally-enclosed. The style has impact on the environment within which the motor can operate and on the power density of the motor.

Mechanical NEMA frame designations standardize envelopes and mounting dimensions, assuring interchangeability of motors. A complete list of frame designations and dimensions is available in *NEMA Standard MG-1*. Figure 5.45 shows

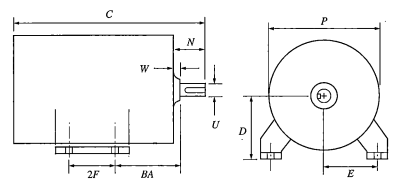

Figure 5.45
Basic NEMA frame dimensions

Table 5.1
NEMA frame dimensions (in)

Frame Designation	C	N-W	U	2F	BA	P	D	E
168AT	15.88	1.75	0.875	8.00	3.25	7.88	4.00	3.25
188AT	19.25	2.25	1.125	9.00	2.75	9.50	4.50	3.75
219AT	22.75	2.75	1.375	11.00	3.50	11.00	5.25	4.25
258AT	26.62	3.25	1.625	12.50	4.25	13.00	6.25	5.00
288AT	30.75	3.75	1.875	14.00	4.75	14.00	7.00	5.50
323AT	27.62	4.25	2.125	9.00	5.25	16.00	8.00	6.25
365AT	33.12	4.75	2.375	12.25	5.88	18.00	9.00	7.00
405AT	36.00	5.25	2.625	13.75	6.62	20.00	10.00	8.00
505AT	50.12	6.50	3.250	18.00	8.50	25.00	12.50	10.00
583AT	53.62	7.50	3.750	16.00	10.00	29.00	14.50	11.50
687AT	68.38	9.50	4.500	32.00	10.00	29.50	14.50	11.50
688AT	72.88	9.50	4.500	36.00	11.50	35.00	17.00	13.50

a partial set of the NEMA frame dimensions. Specific dimension values for a selected group of frames that find use in dc motors over the range from 5 to 800-hp dc motors are found in Table 5.1.

5.9.2   VOLUME AND BORE SIZING

The developed torque produced by a dc machine is given by substitution of [5.6] into [5.7] as

$$T_d = K\Phi_p I_a = \frac{pZ}{2\pi a}\Phi_p I_a \qquad \textbf{[5.25]}$$

If the flux density under a field pole is assumed to have a uniform value B_p, the flux per pole can be written as

$$\Phi_p = B_p \frac{\psi \pi d \ell_a}{p} \qquad \textbf{[5.26]}$$

where ψ is the ratio of pole arc to pole pitch, d is the armature diameter, and ℓ_a is the armature stack length. Substitute [5.26] into [5.25], multiply by $\pi d/\pi d$, and rearrange to yield

$$T_d = \frac{pZ}{2\pi a} B_p \frac{\psi \pi d \ell_a}{p} I_a \frac{\pi d}{\pi d} = \frac{\pi}{2} \psi B_p \left(\frac{ZI_a}{a\pi d} \right) d^2 \ell_a \qquad \textbf{[5.27]}$$

If the slot-embedded armature coils are assumed to lie at the surface of the armature diameter, then the term $ZI_a/a\pi d$ is a surface current density (J_s). Equation [5.27] can then be written as

$$T_d = 2\psi B_p J_s \frac{\pi}{4} d^2 \ell_a \qquad \textbf{[5.28]}$$

It is apparent from [5.28] that the torque developed by a dc machine depends directly on the air gap volume $(\pi/4)d^2\ell_a$, the armature winding current density, and the air gap flux density under the field poles.

In order to minimize size, the values of current density and flux density are maintained at the limits allowed by ability to cool the armature conductors and by saturation limits imposed by the ferromagnetic material. Thus, for a class of dc machines, it is concluded from [5.28] that

$$\frac{d^2\ell_a}{T_d} = \text{constant} = v_T \qquad \textbf{[5.29]}$$

where v_T is a normalized sizing value. By nature, the armature teeth of a dc machine with parallel-sided slots taper to a narrower width at the tooth root than at the tooth tip. As machines become smaller in size, the tooth root flux density has a significant impact on the magnetic circuit design. Consequently, the sizing value v_T must be increased as machine size decreases to allow for acceptable tooth root flux density. Figure 5.46 presents values of v_T that can be used for initial sizing of dc motors depending upon the method of cooling. The graphed values of v_T are valid for motors with rated speed of approximately 1800 rpm. For other speeds, the value of v_T varies inversely proportional to speed. Design engineers usually develop values of v_T based upon successful design history within their organization.

The next step in the initial sizing process of dc machine design is to set the outside frame diameter (D_f). In the case of a NEMA frame designation, $D_f = P$ from Table 5.1. Usually, adequate frame thickness (t_f) and radial depth to accommodate field pole design results if the armature diameter is taken to be

$$0.55D_f \le d \le 0.65D_f \qquad \textbf{[5.30]}$$

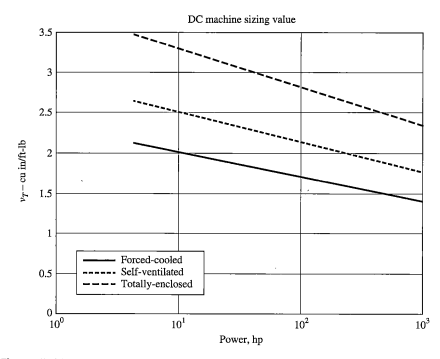

Figure 5.46
DC machine sizing value

If good slot wedging techniques and end-turn banding are used to adequately retain the armature coils against centrifugal force, the armature can operate with surface speed up to 12,000 ft/min. Thus, it should be checked to assure that

$$d \leq \frac{12}{\pi} \frac{12,000}{n_m}$$

[5.31]

With a tentative value for armature diameter (d) selected, the armature stack length follows from [5.29].

$$\ell_a = \frac{v_T T_d}{d^2}$$

[5.32]

Although specific values for d and ℓ_a have been established, some adjustment of the dimension ℓ_a may be necessary as the design is refined. An alternate approach to the stator volume sizing can be found in Ref. 2, pp. 14–19.

5.9.3 Armature Design

The logic flowchart of Fig. 5.47 sets up the iterative procedure to be followed for the armature design process.

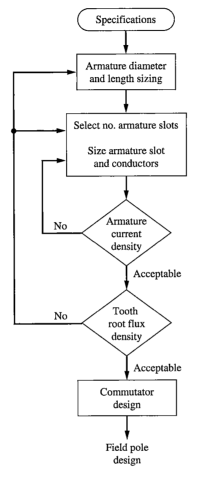

Figure 5.47
Logic flowchart for armature design

Number of Armature Slots In order to minimize flux pulsation along the pole face and produce a slot width that allows for good slot wedge integrity, the armature slot pitch (λ) should typically be in the range

$$1 \leq \lambda = \frac{\pi d}{N} \leq 1.5 \text{ in} \qquad \text{[5.33]}$$

For $d < 20$ in, the lower end of the range usually results in better design.

Since the armature lamination stack experiences a reversing flux, the number of field poles should be selected to give a cyclic frequency in the range of 45 to 70 Hz for operation at rated speed.

$$45 \leq \frac{p n_{mR}}{120} \leq 70 \text{ Hz} \qquad \text{[5.34]}$$

where p must be an even integer. In addition, good design practice is to select the number of armature slots (N) as an integer that falls in the range

$$11.5p \leq N \leq 15.5p \qquad \textbf{[5.35]}$$

Voltage and Torque Constants   This presentation will consider only the case of a simplex lap winding ($a = p$) as illustrated by Fig. 5.10. Integral-horsepower dc machines commonly have a single-turn armature coil with multiple coils per slot (n_c) with two or three coils per slot being typical values. Having selected N and p to satisfy [5.33] to [5.35] and a value for n_c, the total conductors for the armature winding is given by

$$Z = 2n_c N \qquad \textbf{[5.36]}$$

The voltage and torque constants for use with British units (flux in lines, speed in rpm, and torque in ft·lbs) are given by, respectively,

$$k_E = \frac{pZ}{a60 \times 10^8} \qquad \textbf{[5.37]}$$

$$k_T = \frac{pZ}{a8.525 \times 10^8} \qquad \textbf{[5.38]}$$

Rated Current and Flux Per Pole   With the rated power and speed point known, the rated condition shaft torque for a dc motor is

$$T_{sR} = \frac{5250hp_R}{n_{mR}} \text{ ft·lbs} \qquad \textbf{[5.39]}$$

Using the design goal efficiency (η_R) for rated conditions, the rated condition current is

$$I_{aR} = \frac{746hp_R}{\eta_R V_t} \qquad \textbf{[5.40]}$$

If reasonable value rotational losses are 3 percent of the machine power rating, the rated condition flux per pole is

$$\Phi_{pR} = \frac{T_{sR}/0.97}{k_T I_{aR}} \text{ lines} \qquad \textbf{[5.41]}$$

Slot Design   Now that rated armature current and rated flux per pole are established, the armature slot, tooth, and coil can be sized. For good design practice, the stator slot width (b_s) of Fig. 5.48 should be selected such that

$$0.4\lambda \leq b_s \leq 0.5\lambda \qquad \textbf{[5.42]}$$

The slot depth (d_s) usually falls in the range

$$2b_s \leq d_s \leq 4b_s \qquad \textbf{[5.43]}$$

In addition to containment of the two coil sides, the slot must assure adequate insulation between coil sides (coil separator) and accept a means of retaining the

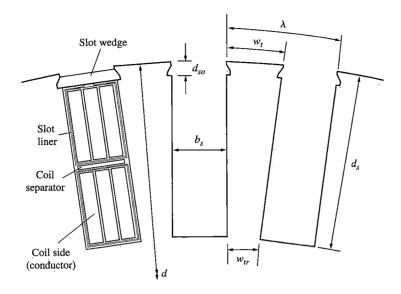

Figure 5.48
Armature slot section view

coil sides (slot wedge) as illustrated by Fig. 5.48. The following material thicknesses are suggested:

1. 250-V motors: Use 0.015-in coil ground insulation and coil separator.
2. 500- to 600-V motors: Use 0.030-in coil ground insulation and coil separator.
3. NEMA 288AT or smaller frames: Use 0.060-in slot wedge.
4. Larger than NEMA 288AT frames: Use 0.125-in slot wedge.

Guided both by the above insulation dimension suggestions and by the tentative values of slot width and slot depth determined by [5.42] and [5.43], the conductor size specification can begin. Allowable conductor current density for successful cooling depends on the flow of air around the coils and the thermal impedance from the point of heat generation within the conductor to the cooling air. Shorter lamination stacks are more easily cooled than longer lamination stacks. For air-cooled machines, typical values of allowable stator current density (Δ_a) are in the range

$$500 \le \Delta_a \le 800 \ \text{A/cm}^2 \quad \text{or} \quad 3200 \le \Delta_a \le 5200 \ \text{A/in}^2 \quad \textbf{[5.44]}$$

The smaller values of Δ_a are typical of totally enclosed machines, whereas the larger values pertain to forced-air cooled machines. Self-ventilated machines would fall in the mid-range.

It is rare in dc machine design to find a standard square wire size compatible with selected slot dimensions; however, magnet wire manufacturers readily supply rectangular conductors drawn to desired dimension specifications and served with film insulation. The required value of stator conductor cross-section area follows as

$$s_a = \frac{I_{aR}}{a\Delta_a} \qquad \text{[5.45]}$$

If a conductor of appropriate cross-section area does not fit the slot well (too large or too small), the iterative process indicated by Fig. 5.47 should be carried out until a satisfactory slot design is attained.

Coil Characterization   The end-turn layout of an armature coil is shown by Fig. 5.49. Since the end-turn projection of the armature coil directly impacts the axial length of the armature assembly, it is desirable to hold the projection to as small a value as practically possible. The design objective is usually accomplished if the following dimensions of Fig. 5.49 are maintained:

$$
\begin{aligned}
s_e &= 0.125 \text{ to } 0.250 \text{ in} \\
b_e &= 0.50 \text{ to } 1.00 \text{ in} \\
g_e &= d_s \\
d_e &= s_e + b_s \\
\lambda_c &= \pi(d - d_s)/N \\
\tau_c &= \text{integer } (N/p)\,\lambda_c
\end{aligned}
\qquad \text{[5.46]}
$$

The smaller values of s_e and b_e apply to armatures of under approximately 24 in in diameter while the larger values pertain to larger-diameter armatures. With the above values selected, the complete end-turn overhang (*OH*) for each end turn is given by

$$\alpha = \sin^{-1}(d_e/\lambda_c) \qquad \text{[5.47]}$$

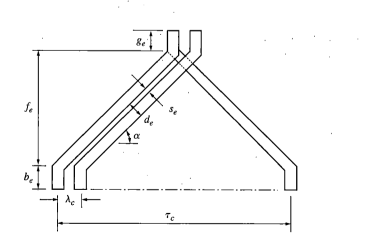

Figure 5.49
Armature coil end turn

$$OH = b_e + f_e + g_e = b_e + \frac{\tau_c}{2}\tan\alpha + g_e \qquad \textbf{[5.48]}$$

The mean length turn (*MLT*) of an armature coil can be determined as

$$MLT_a = 2(\tau_c/\cos\alpha + 2b_e + 2g_e + \ell_a)$$

The value of resistance for the armature circuit is

$$R_a = \rho\frac{MLT_aZ/2}{a^2s_a} \qquad \textbf{[5.49]}$$

where copper resistivity $\rho = 0.69 \times 10^{-6}$ Ω·in at 20°C. The value must be adjusted to the anticipated operating temperature (T) of the armature winding by

$$\rho = \frac{234.5 + T}{254.5}(0.69 \times 10^{-6}) \qquad \textbf{[5.50]}$$

The bulk material of the carbon brushes and their contact interface with the com-mutator typically result in a 2- to 4-V drop. If this brush drop is to be included in the analysis, then the value of R_a should be increased to give proper accounting.

Flux Density Check   Prior to beginning the commutator design, check of the critical flux density for the armature tooth root should be made. The apparent flux density at the tooth root of Fig. 5.48 is given by

$$B_{tra} = \frac{\Phi_{pR}p}{\psi Nw_{tr}\ell_a SF} \qquad \textbf{[5.51]}$$

where

$$w_{tr} = \frac{\pi(d - 2d_s)}{N} - b_s \qquad \textbf{[5.52]}$$

The stacking factor has values $0.94 \leq SF \leq 0.97$ depending on the lamination thickness and the axial assembly pressure of the armature lamination stack. For good design, $B_{tra} < 170$ kilolines/in^2 (2.6 T). If B_{tra} is unacceptably large, action should be taken to decrease the slot width or change the number of armature slots. If an acceptable B_{tra} cannot be so obtained, then ℓ_a must be increased as indicated by the logic flowchart of Fig. 5.47.

Commutator Design   The number of commutator bars (K_c) for a two-layer winding is equal to the number of armature coils. Thus, for n_c coils per slot,

$$K_c = n_cN \qquad \textbf{[5.53]}$$

Methods for mechanical retention of the tapered commutator bars of Fig. 5.50 are beyond the scope of this presentation that addresses only electrical and magnetic design issues. The silver-bearing copper commutator bars are separated by electri-cal insulation sheets with thickness of 0.030 to 0.040 in. Since the full terminal

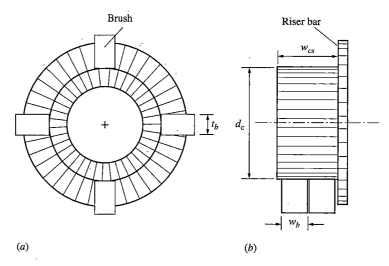

(a)   (b)

Figure 5.50
Commutator. (a) End view. (b) Side view.

voltage of the dc machine appears between adjacent brush sets, it is imperative that sufficient collective interbar insulation exist to assure that bar-to-bar arcing does not occur. The usual assessment is a measure of goodness known as *average volts per bar,* which should generally not exceed 20 V, or

$$e_{cav} = \frac{V_t P}{K_c} \leq 20 \text{ V} \qquad \textbf{[5.54]}$$

If the criterion of [5.54] is not satisfied, an increase in the number of armature slots should be considered.

     The carbon brushes typically offer acceptable service life if the commutator surface speed does not exceed 9000 ft/min. Hence, the commutator brush surface diameter should satisfy

$$d_c \leq \frac{9000(12)}{\pi n_m} \cong \frac{34{,}000}{n_m} \qquad \textbf{[5.55]}$$

The carbon brushes should have a thickness chosen so that the brush spans $n_c +$ 1/2 commutator bars insofar as standard 1/8-in increments in brush dimensions allow. Thus,

$$t_b \cong \frac{\pi d_c}{K_c} \left(n_c + \frac{1}{2} \right) \qquad \textbf{[5.56]}$$

Modern electrographitic brushes exhibit acceptable life expectancy with a current density of $\Delta_b = 80$ A/in². Thus, the total width of the n_b brushes per set is found as

$$n_b w_b = \frac{2 I_{aR}}{p \Delta_b t_b} \qquad \textbf{[5.57]}$$

where a brush holder per pole is assumed. The width of a single brush should lie in the range $1 \leq w_b \leq 2$ in. Once [5.57] is evaluated, then the number of brushes per set (n_b) can be readily determined.

5.9.4 FIELD POLE DESIGN

Prior to design of the actual field pole, two items must be addressed—air gap length and frame thickness. The air gap length (δ) increases with armature diameter and can be decided by

$$\delta \cong 0.0335 \sqrt{d} \qquad \textbf{[5.58]}$$

Due to the reluctance variation of the armature teeth, the effective length (δ_e) of the air gap used in magnetic circuit calculations is greater than the tooth-to-pole face determined by [5.58].

$$\delta_e = \frac{\lambda(5\delta + b_s)}{\lambda(5\delta + b_s) - b_s^2}\delta \qquad \textbf{[5.59]}$$

The frame of a dc machine must carry $1/2\ \Phi_p$ plus any leakage flux that passes between adjacent field pole structures. Reference 3, pp. 338–347, derives expressions for the pole-to-pole permeance allowing calculation of the leakage flux. However, typical values for field pole leakage flux are in the range of 10 to 20 percent of the flux per pole and can be adequately accounted for by use of a leakage factor (LF) with values $1.1 \leq LF \leq 1.2$. In order to avoid an excessive mmf requirement for the frame flux path, the frame flux density (B_f) should not exceed 100 kilolines/in². Consequently, the cross-sectional area of the frame perpendicular to flux flow as illustrated by Fig. 5.51 must be such that

$$A_f = \frac{\frac{1}{2}LF\Phi_p}{B_f} = \frac{LF\Phi_{pR}}{200{,}000} \qquad \textbf{[5.60]}$$

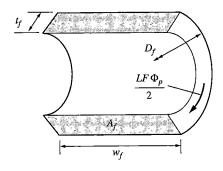

Figure 5.51
Frame section

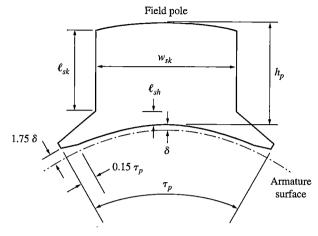

Figure 5.52
Field pole shape

For most dc machines, the frame can maintain a uniform thickness that extends to half of the armature coil overhang. Thus,

$$w_f = \ell_a + OH \qquad \qquad \textbf{[5.61]}$$

The necessary frame thickness is determined as

$$t_f = \frac{A_f}{w_f} = \frac{LF\Phi_{pR}}{200{,}000(\ell_a + OH)} \qquad \qquad \textbf{[5.62]}$$

With δ and t_f now determined, the height of the field pole can be calculated.

$$h_p = \frac{1}{2}(D_f - 2\delta - 2t_f - d) \qquad \qquad \textbf{[5.63]}$$

The pole shank and pole shoe lengths of Fig. 5.52 have the reasonable values of

$$\ell_{sk} \cong 0.9\,h_p \qquad \ell_{sh} \cong 0.1\,h_p \qquad \qquad \textbf{[5.64]}$$

Acceptable magnetic circuit performance and adequate space for the field winding usually results if the pole shank width is sized so that the flux density is 110 kilo-lines/in². Thus,

$$w_{sk} = \frac{LF\Phi_{pR}}{110{,}000\,\ell_a SF} \qquad \qquad \textbf{[5.65]}$$

The per unit field pole arc–to–pole pitch ratio (ψ) should be in the range of 0.65 to 0.70.

$$\psi \cong \frac{\tau_p}{\tau} = \frac{\tau_p}{\pi d/p} \qquad \qquad \textbf{[5.66]}$$

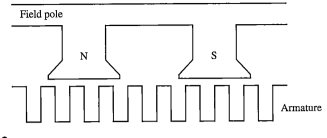

Figure 5.53
Field pole span of teeth

This ratio is stated as approximate in that attention should also be given to the number of armature slots spanned by the pole arc (τ_p). The common design practice is to select a pole arc such that

$$\frac{\tau_p}{\lambda} = m + \frac{1}{2} \qquad \text{[5.67]}$$

where m is an integer. When [5.67] is satisfied, the situation per pair of poles is illustrated by Fig. 5.53, where the reluctance of the magnetic circuit around a complete flux path remains nearly constant regardless of the armature position. Such a case reduces the flux pulsation that would otherwise result, leading to magnetic heating losses in the pole tips. Even with the satisfaction of [5.67], there can be some reluctance variation as the leading edge of a pole tip approaches a tooth while the trailing pole tip exits a tooth, since the slot width and tooth width are not typically equal dimensions. A chamfer of the pole tips as illustrated by Fig. 5.52 is usually successful in significantly diminishing this tooth-transition based flux ripple.

The field pole design per se is generally not an iterative process. The logic flowchart of Fig. 5.54 shows the simple single-pass approach.

5.9.5 MAGNETIC CIRCUIT ANALYSIS

Since the value of Φ_p may well be adjusted in change of operating points for a dc machine, the open-circuit saturation curve (Φ_p vs. mmf_p) must be determined prior to analysis at other than the rated point. Further, the field winding cannot be designed for the rated point of operation until the mmf requirement for that point is known.

Apparent Tooth Density As stated in earlier discussion, the armature teeth are tapered by nature so that the flux density at the tooth root (width w_{tr} of Fig. 5.48) may experience significant saturation. When the tooth root area reaches saturation, flux tends to also travel along a path radially outward and parallel to the tooth sides. Reference 3, p. 69, presents a method to computationally handle the analysis when tooth root saturation occurs. A B_a-H curve for use in the tooth area analysis is constructed that accounts for the parallel permeance path through the

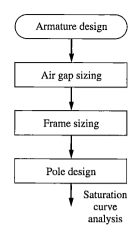

Figure 5.54
Logic flowchart for field pole design

slot area where apparent tooth flux density (B_a) at each point is calculated as though the flux were confined only to the tooth ferromagnetic material by

$$B_a = B + k_t H \qquad\qquad \textbf{[5.68]}$$

The constant k_t is the permeability of the parallel air path given by

$$k_t = 3.2(SF\lambda_3/w_{t3} - 1) \qquad\qquad \textbf{[5.69]}$$

where λ_3 and w_{t3} are, respectively, the slot pitch and tooth width calculated at one-third the tooth depth from the bottom.

The MATLAB program $\langle$dcOCsat.m$\rangle$ has been formulated to calculate the necessary values and plot the magnetization curve (Φ_p vs. mmf_p) for a dc machine with winding and dimensional data read from $\langle$dcdata.m$\rangle$. The program assumes that the field pole and armature ferromagnetic material is M-22, 26-gage ESS and that the frame is AISI 1010 steel by appropriately calling $\langle$Hm22.m$\rangle$ and $\langle$H1010.m$\rangle$ that contain the *B-H* curves of these two materials.

If the resulting magnetization curve displays saturation problems, then the offending portion of the magnetic structure must be identified and corrective action taken as indicated by Fig. 5.55. The final action taken by $\langle$dcOCsat.m$\rangle$ is to form the OCC curve for the dc machine and save the file for later use.

5.9.6   Field Winding Design

Once the magnetization curve has been calculated, the value of mmf_p to produce Φ_{pR} can be determined. At least 1.05 mmf_p should be used as mmf_{pR} in field winding design to allow for armature reaction. Any field winding arrangement desired

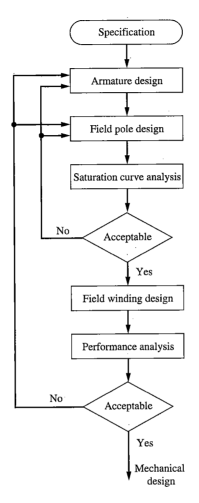

Figure 5.55
Logic flowchart for dc motor design

can be designed to produce mmf_{pR} as long as the winding will physically fit into the available space surrounding the field pole. A layout drawing of the field pole and frame is usually necessary to determine space availability. The space available is further diminished with the addition of the interpoles and their associated windings.

5.9.7 DESIGN REFINEMENT

The MATLAB program ⟨shmtrdes.m⟩ is formulated by modification of ⟨shntmtr.m⟩ to read the OCC that has been saved by the last run of ⟨dcOCsat.m⟩. The program then plots the speed-torque, speed-current, and speed-efficiency curves for the motor design to allow assessment with regard to performance specifications. Should the motor not meet the desired performance, then the design must be iterated as indicated by Fig. 5.55.

If the design were other than a shunt motor, then the appropriate motor performance program introduced earlier in the chapter can be modified using ⟨shmtrdes.m⟩ as a guideline to formulate a performance prediction program for the motor design. Or, if so desired, ⟨dcOCsat.m⟩ could be modified to save the OCC in ⟨eif⟩ overwriting the existing file. Then the performance programs could be used without modification except for editing the data.

5.9.8   SAMPLE DESIGN

The shunt dc motor to be designed is specified as follows:

500 V   400 hp   1750 rpm   583AT frame
Full-load   Efficiency—94%   Self-ventilated
Maximum speed—2500 rpm

From Table 5.1, $P = D_f = 29.00$ in. Using the mean value of [5.30],

$$d = 0.6D_f = 0.6(29.00) = 17.40 \text{ in}$$

Rated developed torque for the motor under the reasonable assumption of 3 percent rotational losses is

$$T_{dR} = \frac{5250P_{sR}}{0.97n_{mR}} = \frac{5250(400)}{0.97(1750)} = 1237 \text{ ft·lb}$$

From Fig. 5.46, $v_T = 1.91$ in³/ft·lb. Since the rated speed is near 1800 rpm, adjustment of v_T is not necessary. Based on [5.29],

$$d^2\ell_a = v_T T_d = (1.91)(1237) \cong 2363 \text{ in}^3$$

Thus,

$$\ell_a = \frac{d^2\ell_a}{d^2} = \frac{2363}{(17.40)^2} = 7.80 \text{ in}$$

If slot pitch $\lambda = 1$, [5.33] gives

$$N = \frac{\pi d}{\lambda} = \frac{\pi(17.40)}{1} = 54.66 \text{ slots}$$

Use $N = 54$ slots. With N chosen,

$$\lambda = \frac{\pi d}{N} = \frac{\pi(17.40)}{54} = 1.012 \text{ in}$$

The choice to design a four-pole machine is made based on [5.34] where

$$\frac{pn_m}{120} = \frac{4(1750)}{120} = 58.33 \text{ Hz}$$

yields a cyclic frequency of the armature flux within the range for acceptable core losses. Also, by [5.35]

$$\frac{N}{p} = \frac{54}{4} = 13.5$$

which lies within the acceptable range for slots per pole.

The common value of $n_c = 3$ coils per slot is chosen for the design trial. Thus, from [5.36], the total armature conductors is

$$Z = 2n_c N = 2(3)(54) = 324 \text{ conductors}$$

The voltage and torque constants follow from [5.37] and [5.38].

$$k_E = \frac{pZ}{a60 \times 10^8} = \frac{4(324)}{4(60 \times 10^8)} = 5.4 \times 10^{-8} \frac{\text{V}}{\text{line·rpm}}$$

$$k_T = \frac{pZ}{a8.525 \times 10^8} = \frac{4(324)}{4(8.525 \times 10^8)} \doteq 3.8 \times 10^{-7} \frac{\text{ft·lb}}{\text{line·A}}$$

The values of rated point current and flux are found from [5.40] and [5.41].

$$I_{aR} = \frac{746P_{sR}}{\eta_R V_t} = \frac{746(400)}{0.94(500)} \cong 635 \text{ A}$$

$$\Phi_{pR} = \frac{T_{dR}}{k_T I_{aR}} = \frac{1237}{(3.8 \times 10^{-7})(635)} = 5.126 \text{ Megalines}$$

Since no radial cooling ducts are planned for the armature stack, the current density will be selected on the low side for self-ventilation. Let $\Delta_a = 4000$ A/in²; then the armature conductor area is

$$s_a = \frac{I_{aR}}{a\Delta_a} = \frac{635}{(4)4000} = 0.0397 \text{ in}^2$$

In an attempt to have an armature slot width in the midrange of [5.42], select an armature conductor with bare copper width $w_c = 0.110$ in. The armature conductor will be insulated with a high-temperature film wrap to a thickness of 0.003 in. Then, the conductor height is

$$d_c = \frac{s_a}{w_c} = \frac{0.0397}{0.110} = 0.361 \text{ in}$$

Under the previously suggested guidelines for ground insulation and slot wedge thickness, the following work determines the slot width and depth:

Slot Width

3(0.110)	conductor
6(0.003)	conductor film
2(0.007)	glass tape
2(0.030)	slot liner
0.422 in	

Allowing for irregularity in materials, use $b_s = 0.450$ in.

Slot Depth

2(0.361)	conductor
6(0.003)	conductor film
4(0.007)	glass tape
2(0.030)	slot liner
0.030	coil separator
0.125	slot wedge
0.062	wedge inset
1.045 in	

Allowing for irregularity in materials, use $d_s = 1.150$ in. The slot depth–to–width ratio is

$$\frac{d_s}{b_s} = \frac{1.150}{0.450} = 2.55$$

which falls within the range suggested by [5.43]. A layout of the slot detail is shown by Fig. 5.56.

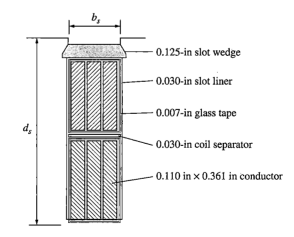

Figure 5.56
Armature slot detail

For the coil end-turn overhang calculations of [5.46] to [5.48], let

$$s_e = 0.125 \text{ in}$$

$$b_e = 0.50 \text{ in}$$

$$g_e = d_s = 1.15 \text{ in}$$

$$d_e = s_e + b_s = 0.125 + 0.45 = 0.575 \text{ in}$$

$$\lambda_c = \frac{\pi(d - d_s)}{N} = \frac{\pi(17.40 - 1.15)}{54} = 0.945 \text{ in}$$

$$\tau_c = \text{integer}\left(\frac{N}{p}\right) \times \lambda_c = \text{integer}\left(\frac{54}{4}\right) \times 0.945 = 12.285 \text{ in}$$

$$\alpha = \sin^{-1}\left(\frac{d_e}{\lambda_c}\right) = \sin^{-1}\left(\frac{0.575}{0.945}\right) = 37.5°$$

$$OH = b_e + \frac{\tau_c}{2}\tan\alpha + g_e = 0.50 + \frac{12.285}{2}\tan(37.5°) + 1.15 = 6.36 \text{ in}$$

The armature coil resistance is found from [5.49] for an estimated average conductor temperature of 150°C.

$$\rho = \frac{234.5 + 150}{254.5}(0.69 \times 10^{-6}) = 1.04 \times 10^{-6} \text{ }\Omega\text{·in}$$

$$MLT_a = 2\left(\frac{\tau_c}{\cos\alpha} + 2b_e + 2g_e + \ell_a\right)$$

$$MLT_a = 2\left[\frac{12.285}{\cos(37.5°)} + 2(0.50) + 2(1.15) + 7.80\right] = 53.17 \text{ in}$$

$$R_a = \frac{\rho MLT_a Z}{2a^2 s_a} = \frac{1.04 \times 10^{-6}(53.17)(324)}{2(4)^2(0.0397)} = 0.0141 \text{ }\Omega$$

The apparent tooth root flux density is checked by use of [5.51] to [5.52] and [5.65] to [5.66].

$$w_{tr} = \frac{\pi(d - 2d_s)}{N} - b_s = \frac{\pi(17.40 - 2 \times 1.150)}{54} - 0.45 = 0.428 \text{ in}$$

Select $\tau_p/\lambda = 9.5$, then

$$\tau_p = 9.5\lambda = 9.5(1.012) = 9.614 \text{ in}$$

$$\psi = \frac{p\tau_p}{\pi d} = \frac{4(9.614)}{\pi(17.40)} = 0.703$$

$$B_{tra} = \frac{p\Phi_{pR}}{\psi N w_{tr}\ell_a SF} = \frac{4(5.126 \times 10^6)}{0.703(54)(0.428)(7.80)(0.96)} = 168.5 \text{ kilolines/in}^2$$

Since B_{tra} is within acceptable limits, no refinement of dimension is required.
The number of commutator bars is

$$K_c = n_c N = 3(54) = 162 \text{ bars}$$

The average volts per commutator bar is determined.

$$e_{cav} = \frac{V_t p}{K_c} = \frac{500(4)}{162} = 12.34 \le 20 \text{ V}$$

Since e_{cav} is well within the acceptable range, no adjustment of armature slot number or coils per slot is necessary.
The commutator and brush dimensions are found by use of [5.55] to [5.57].

$$d_c = \frac{34,000}{n_m} = \frac{34,000}{2500} = 13.60 \text{ in}$$

$$t_b = \frac{\pi d_c}{K_c}\left(n_c + \frac{1}{2}\right) = \frac{\pi(13.60)}{162}\left(3 + \frac{1}{2}\right) = 0.923 \text{ in}$$

As the commutator reaches the wear limit, the brush span of bars increases; thus, the 1/8-in increment below 0.923 in should be selected, or use $t_b = 0.875$ in.

$$n_b w_b = \frac{2I_{aR}}{p\,\Delta_b t_b} = \frac{2(635)}{4(80)(0.875)} = 4.54 \text{ in}$$

Use three brushes per set with $w_b = 1.50$ in. Then, the brush current density for rated load is

$$\Delta_b = \frac{2I_{aR}}{pn_b w_b t_b} = \frac{2(635)}{4(3)(1.50)(0.875)} = 80.6 \text{ A/in}^2$$

Allowing 1/2-in spaces between the riser side and the outside and 1/4 in between brushes, the commutator surface dimension is

$$w_{cs} = 2(0.5) + 2(0.25) + n_b w_b = 2(0.5) + 2(0.25) + 3(1.50) = 6.00 \text{ in}$$

The air gap is sized by [5.58].

$$\delta = 0.0335\sqrt{d} = 0.0335\sqrt{17.40} = 0.1397 \cong 0.140 \text{ in}$$

Since the value of $\psi = 0.703$, the field pole tips are reasonably close so that field pole leakage flux will be on the upper end of typical. Thus, use $LF = 1.2$. By [5.59] to [5.64],

$$A_f = \frac{LF\Phi_{pR}}{200,000} = \frac{1.2(5.126 \times 10^6)}{200,000} \cong 30.75 \text{ in}^2$$

$$w_f = \ell_a + OH = 7.80 + 6.36 = 14.16 \text{ in}$$

Use $w_f = 14.50$ in; then

$$t_f = \frac{A_f}{w_f} = \frac{30.75}{14.50} = 2.12 \text{ in}$$

$$h_p = \frac{1}{2}(D_f - 2\delta - 2t_f - d)$$

$$h_p = \frac{1}{2}(29 - 2 \times 0.140 - 2 \times 2.12 - 17.40) = 3.54 \text{ in}$$

$$\ell_{sk} = 0.9h_p = 0.9(3.54) = 3.19 \text{ in}$$

$$\ell_{sh} = h_p - \ell_{sk} = 3.54 - 3.19 = 0.35 \text{ in}$$

$$w_{sk} = \frac{LF\Phi_{pR}}{110,000\,\ell_a SF} = \frac{1.2(5.126 \times 10^6)}{110,000(7.80)(0.96)} = 7.46 \cong 7.50 \text{ in}$$

After execution of ⟨dcOCsat.m⟩ with the values for the design entered in ⟨dcdata.m⟩, Fig. 5.57 results. It is found that $mmf_p = 9600$ A-t to produce rated

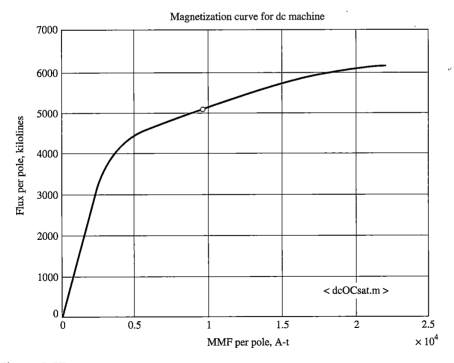

Figure 5.57
Magnetization curve for design example motor

flux per pole. Assuming that armature reaction increases the field mmf requirement by 6 percent, the shunt field should be designed to produce $mmf_{pR} =$ (1.06)9600 = 10,176 A-t. Without the benefit of a layout, it is assumed that a field winding of average width of 2.50 in and a height of $0.85\ell_{sk} \cong 2.71$ in can be fitted into the interpolar space. The area of the field winding cross section is (2.50)(2.71) = 6.77 in². If No. 14 square wire insulated with heavy film over double glass is selected for the field conductor, 1100 turns (N_f) can be fitted into the available area. The mean length turn of the field winding is

$$MLT_f = 2\left[\ell_a + w_{sk} + 2(2.5)\right]$$

$$MLT_f = 2\left[7.80 + 7.50 + 2(2.5)\right] = 40.6 \text{ in}$$

No. 14 square magnet wire has a resistance of 3.25 Ω per 1000 ft for a temperature of 160°C. Thus, the resistance per pole is

$$R_{fp} = \frac{N_f MLT_f}{1000(12)}(3.25) = \frac{1100(40.6)}{1000(12)}(3.25) = 12.09 \text{ } \Omega$$

For a series connection of all field poles,

$$R_f = pR_{fp} = 4(12.09) = 48.36 \text{ } \Omega$$

Full voltage across the field winding yields

$$I_f = \frac{V_{tR}}{R_f} = \frac{500}{48.36} = 10.34 \text{ A}$$

$$mmf_p = N_f I_f = 1100(10.34) = 11{,}374 \text{ A-t}$$

It is concluded that this field winding can adequately excite the motor for the rated point of operation.

Current density of the field winding for the rated point of operation is

$$\Delta_f = \frac{I_{fR}}{s_f} = \frac{10{,}176/1100}{0.00389} = 2378 \text{ A/in}^2$$

If adequate air flow is directed over the field winding, it can be cooled at this level of current density.

The MATLAB program ⟨shmtrdes.m⟩ has been run to analyze the performance of this motor design. Figure 5.58 displays the result, where it is seen that the motor clearly produces an output of 400 hp at the rated speed of 1750 rpm while exhibiting an efficiency slightly above the design goal of 94 percent. The asterisk of Fig. 5.58*d* indicates the 94 percent efficiency goal. For this shunt motor case of near-constant speed and voltage operation, any core losses of the armature have been absorbed in the P_{FW} value.

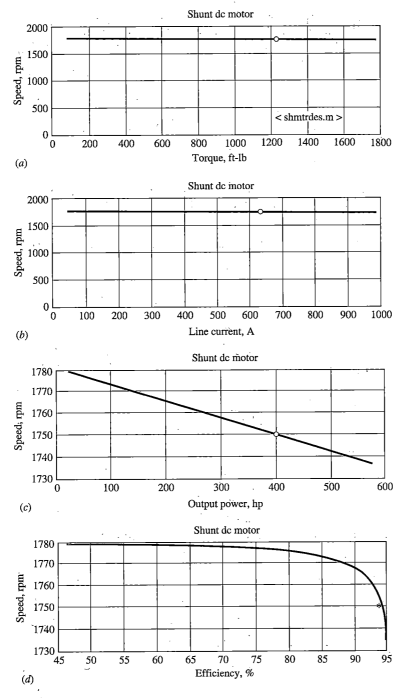

Figure 5.58
Design example 400-hp shunt dc motor. (a) Speed-torque. (b) Speed-current. (c) Speed-output power. (d) Speed-efficiency.

5.10   COMPUTER ANALYSIS CODE

```matlab
%%%%%%%%%%%%%%%%%%%%%%%%%%%%%%%%%%%%%%%%%%%%%%%%%%%%%%%%%%%%%%%%%%%%%%%%%%%
%
% sepgen.m - plots terminal characteristic for separately
%            excited dc generator. Armature reaction neglected.
%
%%%%%%%%%%%%%%%%%%%%%%%%%%%%%%%%%%%%%%%%%%%%%%%%%%%%%%%%%%%%%%%%%%%%%%%%%%%
clear; clf;
VtR=600;       % Rated terminal voltage
PR=60e03;      % Rated output power
Ra=0.25;       % Armature resistance
IaR=PR/VtR;    % Rated output current
load eif       % Load stored OCC
m=length(eif); npts=200;
E=eif(1:m,1); If=eif(1:m,2);
% Determine E & If at rated condition
IfR=interp1( E, If, VtR+IaR*Ra); ER=interp1(If, E, IfR);
Ia=linspace(0,1.5*IaR,npts);
Vt=ER-Ia(1:npts)*Ra;
plot(0,0,Ia,Vt,IaR,VtR,'o'); grid;
title('Separately excited dc generator');
xlabel('Load current, A'); ylabel('Terminal voltage, V');


%%%%%%%%%%%%%%%%%%%%%%%%%%%%%%%%%%%%%%%%%%%%%%%%%%%%%%%%%%%%%%%%%%%%%%%%%%%
%
% shntgen.m - plots terminal characteristic for shunt excited
%            dc generator. Armature reaction neglected.
%
%%%%%%%%%%%%%%%%%%%%%%%%%%%%%%%%%%%%%%%%%%%%%%%%%%%%%%%%%%%%%%%%%%%%%%%%%%%
clear; clf;
VtR=600;       % Rated terminal voltage
PR=60e03;      % Rated output power
Ra=0.25;       % Armature resistance
Pfw=3060;      % F&W losses at speed of analysis (W)
ILR=PR/VtR;    % Rated output current
load eif       % Load stored OCC
m=length(eif); npts=200;
E=eif(1:m,1); If=eif(1:m,2);

% Iterative determination of rated If
IfR=0;
for i=1:50; IfR=interp1( E, If, VtR+(ILR+IfR)*Ra); end

Rfeq=VtR/IfR;     % Total shunt field circuit resistance
IL=linspace(0,1.5*ILR,npts);
```

```
% Iterative determination of Vt-IL
for i=1:npts
   If1=IfR;
   for j=1:100
      Vt1=interp1(If, E, If1)-(IL(i)+If1)*Ra;
      If2=Vt1/Rfeq;
      if abs(If2-If1)<= 0.001; Vt(i)=Vt1; break; end;
      If1=If2;
   end
      eff(i)=Vt(i)*IL(i)/(Vt(i)*IL(i)+(IL(i)-If2)^2*Ra+ ...
         Vt(i)*If2+Pfw)*100;
end
subplot(2,1,1); plot(0,0,IL,Vt,ILR,VtR,'o'); grid;
title('Shunt excited dc generator');
xlabel('Load current, A'); ylabel('Terminal voltage, V');
subplot(2,1,2); plot(IL, eff); grid;
xlabel('Load current, A'); ylabel('Efficiency, %');
title('Shunt excited dc generator');


%%%%%%%%%%%%%%%%%%%%%%%%%%%%%%%%%%%%%%%%%%%%%%%%%%%%%%%%%%%%%%%%%%%%%%%%
%
% sergen.m - plots terminal characteristic for series excited
%            dc generator. Armature reaction neglected.
%
%%%%%%%%%%%%%%%%%%%%%%%%%%%%%%%%%%%%%%%%%%%%%%%%%%%%%%%%%%%%%%%%%%%%%%%%
clear; clf;
VtR=600;     % Rated terminal voltage
PR=60e03;    % Rated output power
nmR=1200;    % Rated speed
Ra=0.25;     % Armature resistance
Rs=0.03;     % Series field resistance
Ns=10;       % Series field turns
IaR=PR/VtR;  % Rated output current

load eif % Load stored OCC(in shunt field amps) for speed nmR
m=length(eif); npts=200;
E=eif(1:m,1); If=eif(1:m,2);
% Create Kphip & mmfs arrays
Kphip=eif(1:m,1)/(nmR*pi/30);
mmfs=eif(1:m,2)*Ns*IaR/(interp1(E,If,VtR+IaR*(Ra+Rs)));
npts=200; Ia=linspace(0,1.5*IaR,npts);
for i=1:npts        % Determine Vt-IL values
   Vt(i)=interp1(mmfs, Kphip, Ns*Ia(i))*nmR*pi/30-Ia(i)*(Ra+Rs);
end

plot(Ia,Vt,IaR,VtR,'o'); grid;
```

```
title('Series excited dc generator');
xlabel('Load current, A'); ylabel('Terminal voltage, V');



%%%%%%%%%%%%%%%%%%%%%%%%%%%%%%%%%%%%%%%%%%%%%%%%%%%%%%%%%%%%%%%%%%%%%%%%%%%%%
%
% cumgen.m - plots terminal characteristic for cumulative
%            compound dc generator. Armature reaction neglected.
%
%%%%%%%%%%%%%%%%%%%%%%%%%%%%%%%%%%%%%%%%%%%%%%%%%%%%%%%%%%%%%%%%%%%%%%%%%%%%%
clear; clf;
VtR=600;        % Rated terminal voltage
PR=60e03;       % Rated output power
Ra=0.25;        % Armature resistance
Rs=0.03;        % Series field resistance
Nf=650;         % Shunt field turns/pole
Ns=1;           % Series field turns/pole
ILR=PR/VtR;     % Rated output current
load eif        % Load stored OCC
m=length(eif); npts=200;
E=eif(1:m,1); If=eif(1:m,2);

% Iterative determination of rated If
IfR=0;
for i=1:50;
IfR=interp1( E, If, VtR+(ILR+IfR)*(Ra+Rs))-Ns/Nf*(ILR-IfR);
end

Rfeq=VtR/IfR;    % Total shunt field circuit resistance
IL=linspace(0,1.5*ILR,npts);

% Iterative determination of Vt-IL
for i=1:npts
  If1=IfR;
  for j=1:100
    Vt1=interp1(If, E, If1+Ns/Nf*(IL(i)-If1))- ...
      (IL(i)+If1)*(Ra+Rs);
    If2=Vt1/Rfeq;
    if abs(If2-If1)<= 0.001; Vt(i)=Vt1; break; end;
    If1=If2;
  end
end
plot(0,0,IL,Vt,ILR,VtR,'o'); grid;
title('Cumulative compound dc generator');
xlabel('Load current, A'); ylabel('Terminal voltage, V');
```

```
%%%%%%%%%%%%%%%%%%%%%%%%%%%%%%%%%%%%%%%%%%%%%%%%%%%%%%%%%%%%%%%%%%%%%
%
% diffgen.m - plots terminal characteristic for differential
%             compound dc generator. Armature reaction neglected.
%
%%%%%%%%%%%%%%%%%%%%%%%%%%%%%%%%%%%%%%%%%%%%%%%%%%%%%%%%%%%%%%%%%%%%%
clear; clf;
VtR=600;       % Rated terminal voltage
PR=60e03;      % Rated output power
Ra=0.25;       % Armature resistance
Rs=0.03;       % Series field resistance
Nf=750;        % Shunt field turns/pole
Ns=3;          % Series field turns/pole
ILR=PR/VtR;    % Rated output current
load eif       % Load stored OCC
m=length(eif); npts=200;
E=eif(1:m,1); If=eif(1:m,2);

% Iterative determination of rated If
IfR=0;
for i=1:50;
IfR=interp1( E, If, VtR+(ILR+IfR)*(Ra+Rs))+Ns/Nf*(ILR-IfR);
end

Rfeq=VtR/IfR;     % Total shunt field circuit resistance
IL=linspace(0,1.5*ILR,npts);

% Iterative determination of Vt-IL
for i=1:npts
  If1=0.75*IfR;
  for j=1:100
    x=If1-Ns/Nf*(IL(i)-If1);
    if x<0; break; end
    Vt1=interp1(If, E, x)- (IL(i)+If1)*(Ra+Rs);
    If2=Vt1/Rfeq;
    if abs(If2-If1)<= 0.001; Vt(i)=Vt1; break; end;
    If1=If2;
  end
end
m=length(Vt);
plot(0,0,IL(1:m),Vt,ILR,VtR,'o'); grid;
title('Differential compound dc generator');
xlabel('Load current, A'); ylabel('Terminal voltage, V');
```

```
%%%%%%%%%%%%%%%%%%%%%%%%%%%%%%%%%%%%%%%%%%%%%%%%%%%%%%%%%%%%%%%%%%%%%%%%%%%%%%%
%
% shntmtr.m - plots developed torque-speed curve for shunt
%             excited dc motor with rated voltage applied.
%             Armature reaction neglected.
%
%%%%%%%%%%%%%%%%%%%%%%%%%%%%%%%%%%%%%%%%%%%%%%%%%%%%%%%%%%%%%%%%%%%%%%%%%%%%%%%
clear; clf;
VtR=600;                        % Rated terminal voltage
PR=80;                          % Rated output horsepower
Ra=0.25;                        % Armature resistance
a=0.2; b=0.6e-5;                % F&W loss equation coefficients
nmR=1200;                       % Rated speed(rpm)
% Rated developed torque & armature current
TdR=PR*746/(nmR*pi/30)+(a+b*nmR^1.7)*30/pi;
IaR=(VtR-sqrt(VtR^2-4*Ra*TdR*nmR*pi/30))/2/Ra;

load eif                        % Load stored OCC of Fig. 5.23
m=length(eif); npts=200;
Kphip=eif(1:m,1)/(nmR*pi/30); If=eif(1:m,2);
% Rated If
IfR=interp1( Kphip, If, (VtR-IaR*Ra)/(nmR*pi/30));
Rfeq=VtR/IfR;    % Total shunt field circuit resistance
npts=25; Ia=linspace(1.5*PR*746/VtR,0,npts);
% Determination of Td-nm
for i=1:npts
  Kphi=interp1(If, Kphip, IfR);
  Td(i)=Kphi*Ia(i);
  wm(i)=VtR/Kphi-Td(i)*Ra/Kphi^2; nm(i)=wm(i)*30/pi;
  Pfw=a*nm(i)+b*nm(i)^2.7;
  eff(i)=(1-(Pfw+Ia(i)^2*Ra+VtR*IfR)/(IfR+Ia(i))/VtR)*100;
  if eff(i)<0; m=i-1; break; end %  F&W over driving
end
subplot(2,1,1); plot(0,0,Td(1:m),nm(1:m),TdR,nmR,'o'); grid;
title('Shunt dc motor');
ylabel('Speed, rpm'); xlabel('Torque, N-m');
subplot(2,1,2); plot(0,0,Ia(1:m)+IfR,nm(1:m),IaR+IfR,nmR,'o'); grid;
title('Shunt dc motor');
ylabel('Speed, rpm'); xlabel('Line current, A');
figure(2)
subplot(2,1,1); plot(eff(1:m),nm(1:m)); grid;
title('Shunt dc motor');
ylabel('Speed, rpm'); xlabel('Efficiency, %');
```

```
%%%%%%%%%%%%%%%%%%%%%%%%%%%%%%%%%%%%%%%%%%%%%%%%%%%%%%%%%%%%%%%%%%%%%%%%%%%
%
% sermtr.m - plots developed torque-speed curve for series
%           excited dc motor with rated voltage applied.
%           Armature reaction neglected.
%
%%%%%%%%%%%%%%%%%%%%%%%%%%%%%%%%%%%%%%%%%%%%%%%%%%%%%%%%%%%%%%%%%%%%%%%%%%%
clear; clf;
VtR=600;                        % Rated terminal voltage
PR=80;                          % Rated output horsepower
Ra=0.25;                        % Armature resistance
Rs=0.03;                        % Series field resistance
Ns=10;                          % Series field turns
nmR=1200;                       % Rated speed(rpm)
% Rated developed torque & armature current assuming 4% F&W losses
TdR=PR*746/(nmR*pi/30)/0.96;
IaR=(VtR-sqrt(VtR^2-4*(Ra+Rs)*TdR*nmR*pi/30))/2/(Ra+Rs);

load eif   % Load stored OCC(in shunt field amps) for speed of nmR
m=length(eif); npts=200;
E=eif(1:m,1); If=eif(1:m,2);
% Create Kphip & mmfs arrays
Kphip=eif(1:m,1)/(nmR*pi/30);
mmfs=eif(1:m,2)*Ns*IaR/(interp1(E,If,VtR-IaR*(Ra+Rs)));
npts=200; Ia=linspace(0.25*IaR,1.5*IaR,npts);

% Determination of Td-nm
for i=1:npts
  Kphi=interp1(mmfs, Kphip, Ns*Ia(i));
  Td(i)=Kphi*Ia(i);
  wm(i)=VtR/Kphi-Td(i)*Ra/Kphi^2;
end
nm=wm*30/pi;
subplot(2,1,1); plot(0,0,Td,nm,TdR,nmR,'o'); grid;
title('Series dc motor');
ylabel('Speed, rpm'); xlabel('Torque, N-m');
subplot(2,1,2); plot(0,0,Ia,nm,IaR,nmR,'o'); grid;
title('Series dc motor');
ylabel('Speed, rpm'); xlabel('Line current, A');


%%%%%%%%%%%%%%%%%%%%%%%%%%%%%%%%%%%%%%%%%%%%%%%%%%%%%%%%%%%%%%%%%%%%%%%%%%%
%
% sermtrfw.m — plots developed torque-speed curve for series
%             excited dc motor with rated voltage applied
%             and series field weakening.
%             Armature reaction neglected.
%
%%%%%%%%%%%%%%%%%%%%%%%%%%%%%%%%%%%%%%%%%%%%%%%%%%%%%%%%%%%%%%%%%%%%%%%%%%%
```

```
clear; clf;
VtR=600;                        % Rated terminal voltage
PR=80;                          % Rated output horsepower
Ra=0.25;                        % Armature resistance
Rs=0.03;                        % Series field resistance
NsT=10;                         % Total series field turns
nmR=1200;                       % Rated speed(rpm)
% Rated developed torque & armature current assuming 4% F&W losses
TdR=PR*746/(nmR*pi/30)/0.96;
IaR=(VtR-sqrt(VtR^2-4*(Ra+Rs)*TdR*nmR*pi/30))/2/(Ra+Rs);

load eif   % Load stored OCC(in shunt field amps) for speed of nmR
m=length(eif); npts=200;
E=eif(1:m,1); If=eif(1:m,2);
% Create Kphip & mmfs arrays
Kphip=eif(1:m,1)/(nmR*pi/30);
mmfs=eif(1:m,2)*NsT*IaR/(interp1(E,If,VtR-IaR*(Ra+Rs)));
npts=200; Ia=linspace(1.5*IaR,0.25*IaR,npts);

% Determination of Td-nm
subplot(2,1,1)
Ns=[4 5 7 10];                  % Series field turns
for n=1:length(Ns)
for i=1:npts
  Kphi=interp1(mmfs, Kphip, Ns(n)*Ia(i));
  Td(i)=Kphi*Ia(i);
  wm(i)=VtR/Kphi-Td(i)*Ra/Kphi^2;
  if wm(i)>3000*pi/30; break; end
end
nm=wm*30/pi;
plot(0,0,Td(1:i),nm(1:i),TdR,nmR,'o');
hold on
title('Series dc motor');
ylabel('Speed, rpm'); xlabel('Torque, N-m');
end
for n=1:i; T(n)=PR*746/0.96/wm(n); end
plot(T,nm,'--'); grid


%%%%%%%%%%%%%%%%%%%%%%%%%%%%%%%%%%%%%%%%%%%%%%%%%%%%%%%%%%%%%%%%%%%%%%%%%%%%%%%%
%
% cummtr.m - plots developed torque-speed curve for cumulative
%            compound excited dc motor with rated voltage applied.
%            Armature reaction neglected.
%
%%%%%%%%%%%%%%%%%%%%%%%%%%%%%%%%%%%%%%%%%%%%%%%%%%%%%%%%%%%%%%%%%%%%%%%%%%%%%%%%
clear; clf;
VtR=600;                        % Rated terminal voltage
PR=80;                          % Rated output horsepower
```

```
Ra=0.25;                          % Armature resistance
Rs=0.03;                          % Series field resistance
Nf=750;                           % Shunt field turns
Ns=3;                             % Series field turns
nmR=1200;                         % Rated speed(rpm)
% Rated developed torque & armature current assuming 4% F&W losses
TdR=PR*746/(nmR*pi/30)/0.96;
IaR=(VtR-sqrt(VtR^2-4*(Ra+Rs)*TdR*nmR*pi/30))/2/(Ra+Rs);

load eif                          % Load stored OCC for speed of nmR
m=length(eif); npts=200;
Kphip=eif(1:m,1)/(nmR*pi/30); If=eif(1:m,2);
% Rated If
IfR=interp1( Kphip, If, (VtR-IaR*(Ra+Rs))/(nmR*pi/30))-Ns/Nf*IaR;
Rfeq=VtR/IfR;     % Total shunt field circuit resistance
npts=200; Ia=linspace(0,1.5*PR*746/VtR,npts);
% Determination of Td-nm
for i=1:npts
   Kphi=interp1(If, Kphip, IfR+Ns/Nf*Ia(i));
   Td(i)=Kphi*Ia(i);
   wm(i)=VtR/Kphi-Td(i)*(Ra+Rs)/Kphi^2;
end
nm=wm*30/pi;
subplot(2,1,1); plot(0,0,Td,nm,TdR,nmR,'o'); grid;
title('Cumulative compound dc motor');
ylabel('Speed, rpm'); xlabel('Torque, N-m');
subplot(2,1,2); plot(0,0,Ia+IfR,nm,IaR+IfR,nmR,'o'); grid;
title('Cumulative compound dc motor');
ylabel('Speed, rpm'); xlabel('Line current, A');


%%%%%%%%%%%%%%%%%%%%%%%%%%%%%%%%%%%%%%%%%%%%%%%%%%%%%%%%%%%%%%%%%%%%%%%%%%
%
% diffmtr.m - plots developed torque-speed curve for differential
%           compound excited dc motor with rated voltage applied.
%           Armature reaction neglected.
%
%%%%%%%%%%%%%%%%%%%%%%%%%%%%%%%%%%%%%%%%%%%%%%%%%%%%%%%%%%%%%%%%%%%%%%%%%%%
clear; clf;
VtR=600;                          % Rated terminal voltage
PR=80;                            % Rated output horsepower
Ra=0.25;                          % Armature resistance
Rs=0.03;                          % Series field resistance
Nf=250;                           % Shunt field turns
Ns=3;                             % Series field turns
nmR=1200;                         % Rated speed(rpm)
% Rated developed torque & armature current assuming 4% F&W losses
TdR=PR*746/(nmR*pi/30)/0.96;
IaR=(VtR-sqrt(VtR^2-4*(Ra+Rs)*TdR*nmR*pi/30))/2/(Ra+Rs);
```

```
load eif                         % Load stored OCC for speed of nmR
m=length(eif); npts=200;
Kphip=eif(1:m,1)/(nmR*pi/30); If=eif(1:m,2);
% Rated If
IfR=interp1( Kphip, If, (VtR-IaR*(Ra+Rs))/(nmR*pi/30))+Ns/Nf*IaR;
Rfeq=VtR/IfR;    % Total shunt field circuit resistance
npts=200; Ia=linspace(0,1.5*PR*746/VtR,npts);
% Determination of Td-nm
for i=1:npts
  Kphi=interp1( If, Kphip, IfR-Ns/Nf*Ia(i));
  Td(i)=Kphi*Ia(i);
  wm(i)=VtR/Kphi-Td(i)*(Ra+Rs)/Kphi^2;
end
nm=wm*30/pi;
subplot(2,1,1); plot(0,0,Td,nm,TdR,nmR,'o'); grid;
title('Differential compound dc motor');
ylabel('Speed, rpm'); xlabel('Torque, N-m');
subplot(2,1,2); plot(0,0,Ia+IfR,nm,IaR+IfR,nmR,'o'); grid;
title('Differential compound dc motor');
ylabel('Speed, rpm'); xlabel('Line current, A');


%%%%%%%%%%%%%%%%%%%%%%%%%%%%%%%%%%%%%%%%%%%%%%%%%%%%%%%%%%%%%%%%%%%%%%%%%%%
%
% dcOCsat.m - plots the magnetization curve for a dc
%             machine. Calls dcdata.m for input data.
%
%%%%%%%%%%%%%%%%%%%%%%%%%%%%%%%%%%%%%%%%%%%%%%%%%%%%%%%%%%%%%%%%%%%%%%%%%%%
clear;
dcdata,
kE=1.666667e-10*p*N^2*nc/a; kT=1.173e-9*p*N^2*nc/a;
phipR=TdR/kT/IaR;
% Build apparent tooth flux density array
Bt=linspace(0,180000,100);
for i=1:length(Bt); Ht(i)=hm22(Bt(i)); end
lam3=pi*(d-4/3*ds)/N;
kt=3.2*(SF*lam3/(lam3-bs)-1);
for i=1:length(Ht)
  Ba(i)=Bt(i)+kt*interp1(Bt,Ht,Bt(i));
end
taut=fix(psi*N/p)+0.5;    % Teeth per pole span
At3=SF*la*(lam3-bs)*taut;    % Total tooth area @ 1/3 depth
rc=(d-2*ds-dshft)/2;
Ac=2*SF*la*rc;              % Rotor core area
Af=wf*tf;                   % Frame area
taup=taut*p/N*pi/4*(d+2*del);% Pole arc
Ask=SF*(la+0.125)*wsk;      % Pole shank area
Ash3=SF*(la+0.125)*(2*wsk+taup)/3;% Shoe area @ 1/3 depth
lam=pi*d/N;                 % Slot pitch
```

```
qty=lam*(5*del+bs); ks=qty/(qty-bs^2);  % Carter coefficient
phip=linspace(0,1.2*phipR,200);
phip=[phip phipR]; m=length(phip);
for i=1:m
  ATt=interp1(Ba,Ht,phip(i)/At3)*ds;
  ATc=hm22(phip(i)/Ac)*pi/2/p*(d-ds-rc/2);
  ATf=h1010(LF*phip(i)/Af/2)*pi/2/4*(ODf-tf);
  ATsh=hm22(LF*phip(i)/Ash3)*lsh;
  ATsk=hm22(LF*phip(i)/Ask)*(ODf-d-2*del-2*lsh-2*tf)/2;
  ATg=p*ks*del*phip(i)/pi/0.665/d/la/3.2;
  ATp(i)=ATg+ATsk+ATsh+ATf+ATc+ATt;
end
plot(ATp(1:m-1),phip(1:m-1)/1000,ATp(m),phip(m)/1000,'o');
title('Magnetization curve for dc machine'); grid
xlabel('MMF per pole, A-t');
ylabel('Flux per pole, Kilolines');
% Generate & save OCC for use in performance study
EIfdes=[kE*nmR*phip(1:m-1)' ATp(1:m-1)'/Nf];
save EIfdes EIfdes -ascii -double


%%%%%%%%%%%%%%%%%%%%%%%%%%%%%%%%%%%%%%%%%%%%%%%%%%%%%%%%%%%%%%%%%%%%%%%%%%%%%%%
%
% dcdata. m - provides input data for dcOCsat.m . All
%             dimensions in inches & areas in sq. in.
%
%%%%%%%%%%%%%%%%%%%%%%%%%%%%%%%%%%%%%%%%%%%%%%%%%%%%%%%%%%%%%%%%%%%%%%%%%%%%%%%
% Rated conditions
PsR=400; nmR=1750;
etaR=0.94; VtR=500;
TdR=5250*PsR/0.97/nmR;
IaR=746*PsR/etaR/VtR;

% Winding data
p=4; N=54; nc=3; a=4;
Nf=1100; Ra=0.0141; Rf=48.36;

% Slot dimensions
ds=1.15; bs=0.450;

% Armature dimensions
la=7.80; d=17.40; SF=0.96;
del=0.140; dshft=4.0;

% Field pole dimensions
wsk=7.50; lsh=0.35;
LF=1.20; psi=0.703;

% Frame dimensions
ODf=29.00; tf=2.12; wf=14.50;
```

```
%%%%%%%%%%%%%%%%%%%%%%%%%%%%%%%%%%%%%%%%%%%%%%%%%%%%%%%%%%%%%%%%%%%%%%
%
% Hm22.m - B-H interpolation routine
%
%%%%%%%%%%%%%%%%%%%%%%%%%%%%%%%%%%%%%%%%%%%%%%%%%%%%%%%%%%%%%%%%%%%%%%
function y = Hx(Bx)
% B-H values that follow are valid for M-22, 26 ga. ESS
B=[0 0.4 0.8 2 8 9.2 11 12.5 13.8 15.2 16.5 18 19 19.6 ...
     19.8 20 20.4 20.6 21.7 28]*6.45e3;        % Lines/sq.in.
H=[0 0.18 0.26 0.38 1.4 1.8 3 5 9.5 28 70 160 260 370 ...
     420 520 825 1000 2000 8000]*2.021;  % A-t/in.
% Activate to plot B-H curve
% m=20; plot(H(1:m),B(1:m)); grid; pause; % Linear plot
% m=20; semilogx(H(2:m),B(2:m));grid; pause; % Semilog plot
n=length(B); k=0;
if Bx==0; k=-1; y=0; end
if Bx<0; k=-1; y=0; disp('WARNING — Bx < 0, Hx = 0 returned'); end
if Bx>B(n); y=H(n); k=-1; disp('CAUTION — Beyond B-H curve'); end
for i=1:n
if k==0 & (Bx-B(i))<=0; k=i; break; end
end
if k>0;
y=H(k-1)+(Bx-B(k-1))/(B(k)-B(k-1))*(H(k)-H(k-1));
else;
end



%%%%%%%%%%%%%%%%%%%%%%%%%%%%%%%%%%%%%%%%%%%%%%%%%%%%%%%%%%%%%%%%%%%%%%
%
% H1010.m - B-H interpolation routine
%
%%%%%%%%%%%%%%%%%%%%%%%%%%%%%%%%%%%%%%%%%%%%%%%%%%%%%%%%%%%%%%%%%%%%%%
function y = H1010(Bx)
% B-H values that follow are valid for 1010 steel plate
B=[0 2.6 5.2 7.7 10.3 12.9 18.1 24.5 31 38.7 51.6 64.5 ...
     71 77 83.8 90.3 97 103 110 116 122 129 135 142 ...
     150 260]*1000;      % Kilolines/sq_in
H=[0 0.6 1.3 2 2.5 2.8 3.4 4 4.7 5.5 6.9 8.5 9.9 11.9 ...
     14.3 19.2 28.3 46.5 86.9 155.6 242.5 444.6 647 ...
     950 2021 4e4];    % A_t/in
% Activate to plot B-H curve
% m=26; plot(H(1:m),B(1:m)); grid; % Linear plot
% m=26; semilogx(H(2:m),B(2:m));grid; % Semilog plot
n=length(B); k=0;
if Bx==0; k=-1; y=0; end
if Bx<0; k=-1; y=0; disp('WARNING — Bx < 0, Hx = 0 returned'); end
if Bx>B(n); y=H(n); k=-1; disp('CAUTION — Beyond B-H curve'); end
for i=1:n
if k==0 & (Bx-B(i))<=0; k=i; break; end
```

```
end
if k>0;
y=H(k-1)+(Bx-B(k-1))/(B(k)-B(k-1))*(H(k)-H(k-1));
else;
end


%%%%%%%%%%%%%%%%%%%%%%%%%%%%%%%%%%%%%%%%%%%%%%%%%%%%%%%%%%%%%%%%%%%%%%%%%%%%%%
%
% shmtrdes.m - plots developed torque-speed curve for shunt
%              excited dc motor with rated voltage applied
%              for the example design motor of Chapter 5.
%              Uses OCC calculated & saved by dcOCsat.m.
%              Armature reaction neglected.
%
%%%%%%%%%%%%%%%%%%%%%%%%%%%%%%%%%%%%%%%%%%%%%%%%%%%%%%%%%%%%%%%%%%%%%%%%%%%%%%
clear; clf;
VtR=500;                        % Rated terminal voltage
PsR=400;                        % Rated output horsepower
Ra=0.0141;                      % Armature resistance
a=0.4; b=1.2e-5;                % F&W loss equation coefficients
nmR=1750;                       % Rated speed(rpm)
% Rated developed torque & armature current
TdR=PsR*746/(nmR*pi/30)+(a+b*nmR^1.7)*30/pi;
IaR=(VtR-sqrt(VtR^2-4*Ra*TdR*nmR*pi/30))/2/Ra;

load eifdes; eif=eifdes;    % Load stored OCC from dcOCsat.m run
m=length(eif); npts=200;
Kphip=eif(1:m,1)/(nmR*pi/30); If=eif(1:m,2);
% Rated If
IfR=interp1(Kphip, If, (VtR-IaR*Ra)/(nmR*pi/30));
Rfeq=VtR/IfR;    % Total shunt field circuit resistance
npts=25; Ia=linspace(1.5*PsR*746/VtR,0,npts);
% Determination of Td-nm
for i=1:npts
  Kphi=interp1(If, Kphip, IfR);
  Td(i)=Kphi*Ia(i);
  wm(i)=VtR/Kphi-Td(i)*Ra/Kphi^2; nm(i)=wm(i)*30/pi;
  Pfw=a*nm(i)+b*nm(i)^2.7;
  eff(i)=(1-(Pfw+Ia(i)^2*Ra+VtR*IfR)/(IfR+Ia(i))/VtR)*100;
  Ps(i)=(Td(i)*wm(i)-Pfw)/746;
  if eff(i)<0; m=i-1; break; end  % F&W over driving
end
subplot(2,1,1); plot(0,0,0.7376*Td(1:m),nm(1:m), ...
   0.7376*TdR,nmR,'o'); grid; title('Shunt dc motor');
ylabel('Speed, rpm'); xlabel('Torque, ft-lb');
subplot(2,1,2); plot(0,0,Ia(1:m)+IfR,nm(1:m),IaR+IfR,nmR,'o'); grid;
title('Shunt dc motor');
ylabel('Speed, rpm'); xlabel('Line current, A');
```

```
figure(2)
subplot(2,1,1); plot(Ps(1:m),nm(1:m),PsR,nmR,'o'); grid;
title('Shunt dc motor');
ylabel('Speed, rpm'); xlabel('Output power, hp');
subplot(2,1,2); plot(eff(1:m),nm(1:m),94,nmR,'*'); grid;
title('Shunt dc motor');
ylabel('Speed, rpm'); xlabel('Efficiency, %');
```

SUMMARY

- DC machines range in size from small subfractional-hp servomotors in office products and automobiles to large multi-thousand-hp roll mill motors.

- DC machines actually have ac currents flowing in the armature coils, but these currents are rectified by the commutator structure so that external currents flowing to or from the armature brush terminals are dc currents.

- The induced voltage or cemf appearing external at the armature brush terminals is a dc voltage given by $E = K\Phi_p\omega_m$.

- The developed torque acting to rotate a dc motor is described by the simple relationship $T_d = K\Phi_p I_a$. .

- The magnetic field produced by the armature windings of a dc machine is ideally oriented 90° electrical from the magnetic field established by the field pole windings—a condition that produces a maximum torque for particular values of field strengths.

- The armature winding mmf acts to distort the air gap flux produced by the field windings so that the voltage behind the terminals of armature coils that are shorted by brushes during the commutation process is nonzero. Consequently, interpoles conducting armature current must be installed between the main field poles to cancel the distorted interpolar region flux and reduce the induced voltage behind coils undergoing commutation to near zero to prevent destructive brush sparking.

- The armature winding mmf is by nature a cross-magnetizing mmf that acts to reduce the flux under the main poles except at conditions of magnetic linearity— a phenomenon known as *armature reaction.*

- DC machine field windings are of two types—series windings that conduct armature current and shunt windings that conduct a current independent of armature current. A particular machine may have either or both of these two winding types.

- DC machines are classified or identified according to the field winding connection.

- Significantly different voltage-current characteristics are produced by dc generators depending on the field winding connection configuration implemented.

- Significantly different torque-speed characteristics are exhibited by dc motors depending on the field winding connection configuration chosen.

- Torque produced by a dc machine depends on the air gap volume. Hence, air gap volume sizing is the starting point of dc machine design.

PROBLEMS

5.1 If the air gap magnetic field of Fig. 5.8 were time-varying, that is described by $B_g(\theta, t)$, extend the $B\ell v$ rule to give the coil voltage analogous to [5.2].

5.2 Determine the value of the emf constant K for a 4-pole, 45-slot, wave-wound dc machine with two conductors per slot.

5.3 If a dc machine is to be analyzed in British units (flux in lines, speed in rpm, torque in ft·lb), determine (a) the emf constant k_E such that $E = k_E \Phi_p n_m$ and (b) the torque constant k_T such that $T_d = k_T \Phi_p I_a$.

5.4 The table below gives data for the OCC of a dc machine measured at 1500 rpm.

E	5.00	62.50	375.00	500.00	625.00	687.50	700.00
I_f	0.00	0.10	0.70	1.05	1.60	2.50	3.00

On a common graph, plot the OCC of this machine for speeds of 1500 and 1000 rpm.

5.5 A shunt dc generator has the OCC described by Prob. 5.4. The generator is operating open-circuit with a total shunt circuit resistance (R_f + series connected rheostat) of 400 Ω. The small $I_f R_a$ voltage drop can be considered negligible. Determine the terminal voltage of the dc generator (a) for a speed of 1500 rpm and (b) for a speed of 1250 rpm.

5.6 For the shunt dc generator of Prob. 5.5, the field resistance is $R_f = 375\Omega$. Determine the value to which the rheostat connected in series with the shunt field must be adjusted if the generator is to produce an open-circuit voltage of 500 V at a speed of 1500 rpm.

5.7 A shunt dc generator has the following nameplate data:

      230 V   12 kW   1200 rpm   1.5 A field current

From test measurement, it is known that $R_f = 100\Omega$ and $R_a = 0.296\Omega$. The generator is operating at nameplate rated conditions. Armature reaction can be considered negligible. Determine the values of (a) armature current and (b) cemf. (c) If the rotational losses are known to be 231 W at 1200 rpm, determine the full-load efficiency. (d) Find the resistance value of the field rheostat that must have been present.

5.8 The shunt dc generator of Prob. 5.7 is operated with the load resistance unchanged. However, the field is now separately excited with $I_f = 1.5$ A and the speed $n_m = 600$ rpm. (a) Determine the value of current $I_L = I_a$ supplied to the load. (b) If the rotational losses are given by $P_{FW} = 0.0833 n_m + C_m n_m^{2.8}$ (W), calculate the efficiency of the generator at this point of operation.

5.9 A cumulative compound dc generator is operating at 1200 rpm and described by the OCC of Fig. 5.59. For the point of operation, $I_L = 280$ A and $I_f = 1.4$ A. It is known that $N_f = 1000$ turns per pole and $N_s = 3$ turns per pole. (a) Neglecting armature reaction, find the cemf E. (b) If armature reaction is known to be 100 A-t at 100 A and can be considered to vary directly with I_a over the region of operation interest, find the value of E.

5.10 Repeat Prob. 5.9 if the generator connection is changed to differential compound and armature reaction at the point of operation is 50 A-t. All else is unchanged.

5.11 A separately excited dc generator with $R_a = 0.25\Omega$ is operating at 1200 rpm and has the OCC of Fig. 5.23. At the point of operation, $I_a = 100$ A and $I_f = 1.5$ A. Determine the terminal voltage (a) with and (b) without armature reaction considered.

5.12 Modify the MATLAB program ⟨cumgen.m⟩ to calculate and plot efficiency across the load current range using a technique similar to that of ⟨shntgen.m⟩.

5.13 Given a shunt dc motor for which $V_t = 550$ V, $I_a = 35$ A, $R_a = 0.7\Omega$, and $R_f = 400\Omega$. The machine does not have a shunt field rheostat installed. Determine the value of (a) cemf E and (b) line current I_L.

5.14 Given a shunt dc motor for which $R_a = 0.08\Omega$. For a light load operating point, it is known that $V_t = 230$ V, $I_a = 5$ A, and $n_m = 1190$ rpm. If the terminal voltage remains constant but the coupled mechanical load is increased until $I_a = 50$ A, find the values of (a) speed, and (b) developed torque. (c) Determine the percentage speed regulation if the 5-A point is no-load and the 50-A point is full-load condition.

5.15 For the shunt dc motor of Prob. 5.14, the rotational losses are known to be 500 W at 1200 rpm and can be considered to vary as speed squared over the region of interest. Also, it is known that $R_f + R_{rh} = 115\ \Omega$. Calculate the value of efficiency for the motor at the $I_a = 50$ A point of operation.

5.16 A 20-hp, 230-V, 1150 rpm shunt dc motor has four poles, four parallel paths in the armature winding, and 882 armature conductors. $R_a = 0.188\Omega$. At rated conditions, $I_a = 73$ A and $I_f = 1.6$ A. Determine for rated conditions the values of (a) developed torque, (b) flux per pole, (c) rotational losses, and (d) efficiency.

5.17 A shunt dc motor with negligible rotational losses is coupled to a mechanical load characterized by a constant torque regardless of speed. $R_a = 0.2\Omega$. At a particular point of operation, it is known that $V_t = 230$ V, $I_a = 70$ A, and $n_m = 1150$ rpm. If a rheostat in series with the field winding is adjusted so

Open-circuit characteristic

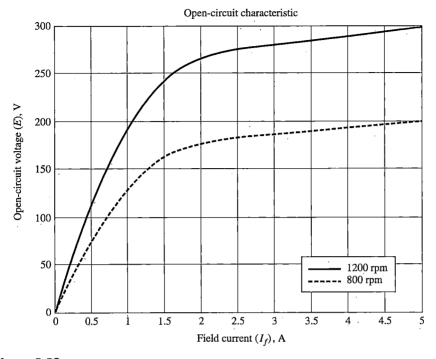

Figure 5.59
OCC for dc machine

that the flux per pole is reduced to 75 percent of the value at the known data point and V_t is unchanged, determine the values of (a) current and (b) speed that now exist.

5.18 A series dc motor runs at 300 rpm and draws a current of 75 A from a 500-V dc source. $R_s + R_a = 0.4\,\Omega$. The motor is coupled to a mechanical load that requires a constant torque regardless of speed. Rotational losses and armature reaction can be neglected. Magnetic linearity is assumed to exist. If the terminal voltage is increased to 600 V, determine the new steady-state values of (a) armature current and (b) speed.

5.19 Rework Prob. 5.18 except assume that the coupled mechanical load has a square-law torque ($T_s = k_T \omega_m^2$).

5.20 A cumulative compound dc motor is operated from a 230-V dc bus. The OCC is given by Fig. 5.59. At the point of operation, $I_a = 50$ A and $I_f = 1.5$ A. It is known that $N_f = 200$ turns per pole, $N_s = 2$ turns per pole, $R_f = 100\,\Omega$, $R_s = 0.05\,\Omega$, and $R_a = 0.3\,\Omega$. Determine (a) speed and (b) developed torque. Assume that armature reaction is negligible.

5.21 Rework Prob. 5.20 except the motor is connected differential compound and all else is unchanged.

5.22 Show that the line current–speed characteristic of a shunt dc motor is a straight line that can be written in the form $n_m = I_{a0} + mI_a$ if constant terminal voltage is applied and armature reaction is negligible.

5.23 Determine the value of $K\Phi_p$ for the shunt dc motor of Fig. 5.32 if the constant value of terminal voltage is $V_t = 600$ V and rotational losses are considered negligible.

5.24 Equation [2] of Example 5.8 shows that the developed torque-speed characteristic for a shunt dc motor is a straight line. Present a convincing argument that the load torque-speed characteristic is also nearly a straight line if armature reaction is neglected.

5.25 Given a 600-V, 80-hp, 1200 rpm shunt dc motor with $R_a = 0.25\Omega$. Rotational losses at rated speed are known to be 2300 W. The OCC of Fig. 5.23 is applicable. Determine the field current required to produce rated torque for $I_a = 100$ A with and without armature reaction considered.

5.26 The dc motor described by the data of ⟨shntmtr.m⟩ is placed in a control system that has an independent power supply feeding the field winding so that the machine is now a separately excited dc motor. The controller maintains $I_f = 65I_a$; thus, the motor now has a characteristic similar to a series dc motor. Modify ⟨shntmtr.m⟩ to analyze the machine and produce the resulting torque-speed curve.

5.27 Use a magnetically linear assumption and an approach similar to Example 5.10 to show that the line current-speed curve of Fig. 5.35b is actually the sum of a hyperbola and a constant value. Specifically, show that $\omega_m = c_1/I_a - c_2$ where c_1 and c_2 are positive valued constants.

5.28 A 250-V shunt dc motor with $R_a = 0.13\Omega$ has a no-load speed of 1200 rpm, a no-load line current of 2.5 A, and a no-load field current of 1 A. When operated at full load with terminal voltage and field current unchanged, $I_L = 100$ A. (*a*) Determine the speed regulation if armature reaction is neglected. (*b*) If it is known that armature reaction reduces Φ_p by 5 percent at full load, assume that the developed torque required is unchanged and find the value of speed regulation.

5.29 A sinusoidal voltage source is on the terminals of a dc motor. (*a*) If the motor is a series dc motor, will the motor run? (*b*) If the motor is a shunt dc motor, will the motor run?

5.30 Two identical dc machines are connected as shown in Fig. 5.60. At all times, the left machine operates as a generator while the right machine operates as a motor. Rotational losses are negligible. The mechanical load coupled to the motor is described by $T_L = k_T \omega_{mM}$. (*a*) For any point of operation with $\omega_{mG} > 0$, $\omega_{mM} > 0$, and $I_f \neq 0$, is $\omega_{mG} > \omega_{mM}$, $\omega_{mG} < \omega_{mM}$, or $\omega_{mG} = \omega_{mM}$? (*b*) If V_B is increased in value and ω_{mG} is unchanged, does the value of ω_{mM} increase, decrease, or remain unchanged? (*c*) If ω_{mG} is increased, but ω_{mM} is unchanged,

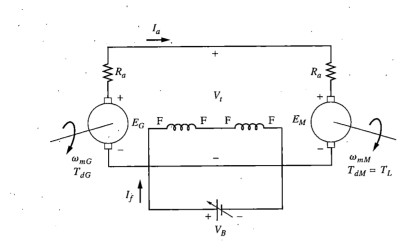

Figure 5.60

was the value of V_B increased or decreased? (*d*) If V_B is unchanged while ω_{mG} is increased, will ω_{mM} increase, decrease, or remain unchanged?

5.31 For the two identical dc machines of Fig. 5.60, the left machine is a generator operating with $\omega_{mG} = 200$ rad/s. The right machine is a motor that drives a coupled mechanical load characterized by constant torque. Rotational losses are negligible. (*a*) If V_B is decreased in value, does ω_{mM} increase, decrease, or remain constant? (*b*) If V_B is increased, does I_a increase, decrease, or remain constant?

5.32 In Example 5.13, the full value of starting resistance of Fig. 5.41 was sized to limit the initial current for the shunt motor of Example 5.9 to 150 percent of rated value. Assume that the starter is designed for four discrete steps over the range of $R_{st} = 3.82\Omega$. The first step is full value of starter resistance ($R_{st1} = 3.82\Omega$) for zero speed. The final step is complete removal of the starter resistance ($R_{st4} = 0$) when speed reaches 1200 rpm. The two intermediate steps (R_{st2}, R_{st3}) are activated at 400 and 800 rpm, respectively, with values of resistance sized so that the armature current never exceeds 150 percent of rated value. Determine the values for R_{st2} and R_{st3}. It is known that $R_f + R_{rh} \gg R_{st1}$.

5.33 Size resistors R_1 and R_2 for the armature resistance control of Example 5.14 so that the motor can operate at the two points ($T_d = 38$ N·m, $n_m = 500$ rpm) and ($T_d = 45$ N·m, $n_m = 750$ rpm).

5.34 Design a dc shunt motor that satisfies the following specifications:

> 250 V   50 hp   1150 rpm   365AT frame
>
> Full-load efficiency—91%   Self-ventilated
>
> Maximum speed 2000 rpm

REFERENCES

1. C. S. Siskind, *Direct-Current Machinery,* McGraw-Hill, New York, 1952.
2. J. H. H. Kuhlmann, *Design of Electrical Apparatus,* 3d ed., John Wiley & Sons, New York, 1950.
3. A. F. Puchstein, *The Design of Small Direct-Current Motors,* John Wiley & Sons, New York, 1961.

chapter

6

INDUCTION MOTORS

6.1 INTRODUCTION

Robust construction, relatively low manufacturing cost, and ease of control have resulted in the induction motor being the most populous of all electric machines. Whereas the dc motors of Chap. 5 and the synchronous motors of Chap. 7 require two excitation connections (*doubly excited*), the induction motor has only one excitation connection (*singly excited*). Currents that flow in the second winding of the induction motor are established by the process of magnetic *induction* through coupling with the singly excited winding, from whence the name *induction motor* is derived.

In applications where power requirement is small and suited to single-phase distribution, the induction motor is available in single-phase versions. Many domestic appliances such as washers, dryers, fans, and air conditioning units use single-phase induction motors. However, common industrial applications use the three-phase induction motor in integral-horsepower ratings with typical voltage ratings in the United States ranging from 230 to 4160 V. Variations in torque-speed characteristics are available with classification set by the National Electrical Manufacturers Association (NEMA), which will be addressed in Sec. 6.7.4. Among other things, NEMA also maintains a list of standard frame sizes voluntarily adhered to by most manufacturers, thereby assuring physical interchangeability between motors of competing manufacturers.

The three-phase induction motor theory and performance characteristics will be developed first in this chapter. The single-phase induction motor will then be discussed through constraint and extension of the three-phase induction motor theory.

6.2 CLASSIFICATION AND PHYSICAL CONSTRUCTION

The *stator* or *primary* (stationary portion) of an induction motor consists of a frame that houses a magnetically active, annular cylindrical structure (*stator lamination stack*) punched from electrical steel sheet with a three-phase winding set embedded in evenly spaced, internal slots. The individual coils of this electrical

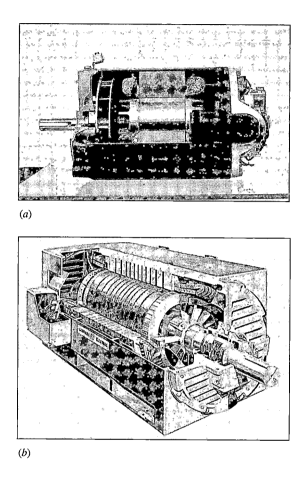

(a)

(b)

Figure 6.1
Squirrel-cage induction motors. (*Courtesy of General Electric Company.*) (a) Random-wound
stator coils. (b) Form-wound stator coils.

winding are *random-wound* for smaller motors and *form-wound* for larger motors
as illustrated by Fig. 6.1.

The *rotor* or *secondary* of an induction motor is made up of a shaft-mounted,
magnetically active, cylindrical structure (*rotor lamination stack*) also constructed
from electrical steel sheet punchings with evenly spaced slots located around the
outer periphery to accept the conductors of the rotor winding. The rotor winding
may be either of two types: *squirrel-cage* or *wound-rotor*.

A squirrel-cage winding can be formed by either die casting or fabrication
process. It is the more robust of the two windings, consisting of aluminum (or copper) bars embedded in the rotor slots and shorted at both ends by aluminum (or
copper) end rings. The rotors of both induction motors of Fig. 6.1 are squirrel-cage
types. For clarity, a view of a squirrel-cage rotor is shown by Fig. 6.2. Variation of
the rotor bar design is a primary method used to alter torque-speed characteristics.

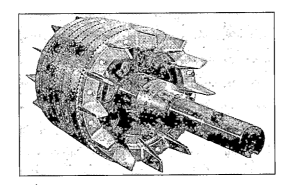

Figure 6.2
Squirrel-cage rotor. (*Courtesy of General Electric Company.*)

The wound-rotor winding for an induction motor is of form similar to the stator winding. Typically, the winding is wye-connected with the three-phase line leads connected to rotor-mounted slip rings. External connection to a three-phase resistor bank is made through stationary brushes that maintain sliding contact with the slip rings. Section 6.7.3 discusses modifying the torque-speed characteristic by varying the rotor circuit resistance.

6.3 STATOR WINDING AND MMF

6.3.1 STATOR WINDINGS

A form-wound stator coil for insertion into the slots of the stator structure is shown in Fig. 6.3a. A random-wound coil can be of this same basic design, or it may be wound by automation in situ rather than being formed externally for insertion in the stator slots as an individual unit. The form-wound coil is made up of rectangular conductors, yielding a large percentage fill of conductor in the slot as illustrated by Fig. 6.3b, while the random-wound coil is suited for use in partially closed slots and results in a smaller percentage of conductor fill, as is apparent from the voids between conductors as seen in Fig. 6.3c.

Typically, the stator winding is a *double-layer winding* (two coil sides per slot) as illustrated by the simple two-slot lamination of Fig. 6.4 showing both the end view of the actual construction and a flat layout of the panoramic view as seen from the inner diameter of the stator lamination stack. The complete interconnection scheme of the two coils is readily apparent from the flat layout.

6.3.2 WINDING MMF

Ampere's circuital law, given by equation [3.11], can be applied around closed paths, such as A or B indicated in Fig. 6.4a, wherein the total enclosed current of the two N-turn coil sides having i A per conductor is $2Ni$. Under the practical

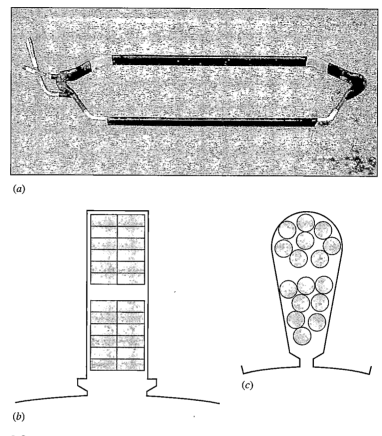

(a)

(b)

(c)

Figure 6.3
Stator coil. (a) Complete coil. (*Courtesy of National Electric Coil, Columbus, Ohio. Photo by Jane Hutt.*) (b) Slot section of form-wound coil. (c) Slot section of random-wound coil.

assumption that the permeability of the stator and rotor ferromagnetic material is high, the mmf drop in the magnetic core material can be neglected with conclusion that the mmf drop across each uniform air gap is of value Ni with variation in space as shown by Fig. 6.5a. The associated Fourier spectrum of this square wave, determined by the MATLAB program ⟨sqmmf.m⟩, is depicted by Fig. 6.5b. The spectrum plot has been normalized by the magnitude of the fundamental mmf component. From Fig. 6.5b, it is apparent that the resulting square-wave mmf contains significant odd harmonic content. The flux produced in a linear magnetic core would have identical harmonic content. As a consequence, increased core losses, detrimental harmonic torques, and excess noise would result. These prob-

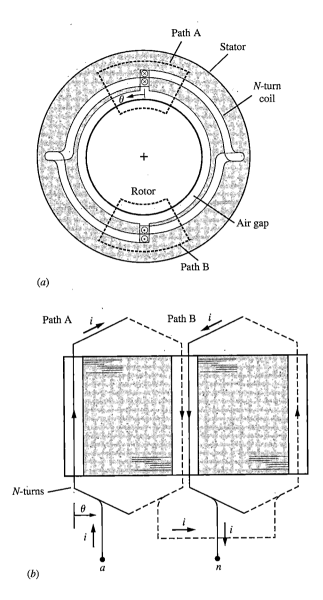

(a)

(b)

Figure 6.4
Two-slot, double-layer stator winding. (a) Actual construction. (b) Flat layout.

lems are diminished significantly when the mmf wave contains only the fundamental frequency of the applied voltage source.

During the design stage, three mechanisms are introduced to reduce the objectionable harmonic content of the mmf waveform:

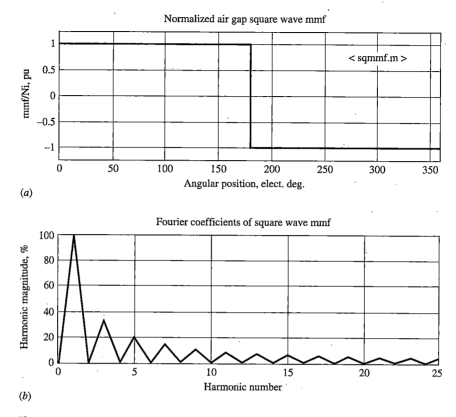

Figure 6.5
Air gap mmf wave for winding of Fig. 6.4. (a) MMF waveform. (b) Fourier spectrum.

1. *Coil Pitch.* Opposite sides of each individual coil are placed to span less than a full magnetic *pole pitch* $\tau = \pi D/p$, where D is the stator bore diameter and p is the number of magnetic poles.

2. *Winding Distribution.* Rather than concentrate the windings in a single slot, the coil is subdivided into coils of lesser turns and the resulting coils are placed in adjacent slots.

3. *Slot Skew.* Slots of either the stator or rotor lamination stack are designed to traverse the axial length of the magnetic core at an angle from the true axial direction. Hence, the average value of the mmf wave over the axial length of the magnetic core is described by a ramped transition, rather than the step transition of Fig. 6.5a, as Ampere's circuital law is applied around closed paths over the circumferential portion of air gap spanning a slot transition.

Figure 6.6 presents the flat layout of a practical winding incorporating pitch, distribution, and skew. The winding is that of one phase for a two-pole, three-phase machine using laminations with $S_1 = 18$ stator slots. The 180° pole pitch

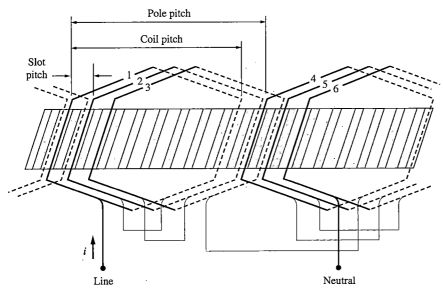

Figure 6.6
Flat layout of winding for one phase (18 slots, 2-pole, three-phase machine).

spans 9 slots for this two-pole machine. Notice that coil 1 spans only 8 slots, or the *angular coil pitch* $\rho = (8/9)(180°) = 160°$. The *angular slot pitch* $\gamma = p180°/S_1 = 20°$ per slot. The skew angle is such that the axial coil-side transition spans one angular slot pitch.

Coils 1 and 4 of Fig. 6.6 form a coil pair that produces a trapezoidal mmf. The MATLAB program ⟨phasemmf.m⟩ generates the component mmf for each coil of the pair and the resultant or sum. Figure 6.7 shows the plots along with the associated Fourier spectrum of the resultant mmf. Comparing the Fourier spectrum of Figs. 6.7 and 6.8, it is seen that skew and pitch have reduced the harmonic content of the coil pair mmf, with the greatest reduction observed for the case of harmonics of order five and above. The program ⟨phasemmf.m⟩ also forms the mmf of the other two coil pairs (2 and 5, 3 and 6) of Fig. 6.6 to yield the air gap mmf for the complete phase winding. Coil pair mmf's, phase winding mmf, and the associated Fourier spectrum of the phase winding mmf are displayed in Fig. 6.8. It is apparent that the harmonic components of order five and above have been practically eliminated. Although the third harmonic has been reduced over the case of a square-wave mmf, it has not been eliminated. However, its existence is not a concern as long as three-wire, wye-connected windings are used. The induced third-harmonic voltages that may be present in a coupled winding can produce no third-harmonic currents since the third-harmonic currents of all phase windings are necessarily in phase and constrained to sum to zero at the neutral point.

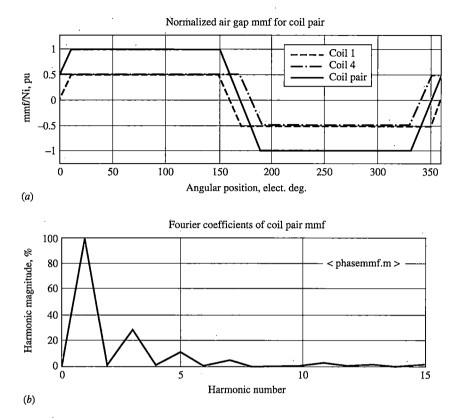

(a)

(b)

Figure 6.7
Coil pair mmf for winding of Fig. 6.6. (a) mmf waveform. (b) Fourier spectrum.

Example 6.1

Run the MATLAB program (phasemmf.m) for the winding of Fig. 6.6 except with a coil pitch of 6 slots to determine the percentage of third, fifth, seventh, and ninth harmonics.

The interactive input values are 1-slot skew, 6-slot pitch, 2 poles, and 18 slots. By either reading from the plot or polling the y vector, the results are as follows:

Third harmonic   —   0 percent

Fifth harmonic   —   4 percent

Seventh harmonic   —   2 percent

Ninth harmonic   —   0 percent

This 2/3-coil pitch has eliminated all harmonics that are multiples of 3 (called *triplen harmonics*). It is a favored coil pitch for delta-connected windings where a path for triplen harmonics exists. However, the fundamental component has also been significantly reduced over the full-pitch coil case.

Although the foregoing work has used the two-pole winding for a specific example, it is equally applicable for a *p*-pole machine. As a direct consequence of the work, use of the fundamental component of the air gap mmf is justified and is

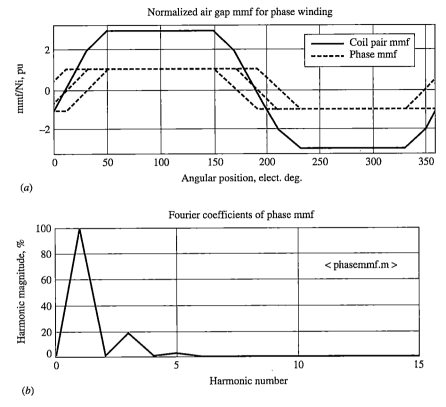

(a)

(b)

Figure 6.8
Phase winding mmf. (a) MMF waveform. (b) Fourier spectrum.

adopted from here onward. Having established that only the fundamental component of the air gap mmf is of concern, formulas are derived in Appendix A for the *pitch factor* (k_p) and the *distribution factor* (k_d) such that the phase winding per pole pair can be considered a single-layer winding (one coil side per slot) concentrated in a pair of slots located a full pole pitch apart with an effective number of turns calculated by

$$N_{1eff} = 2k_p k_d N_{S1} N \qquad [6.1]$$

where $N_{S1} = S_1/pm$ is the slots per pole per phase and N is the turns per actual two-layer coil. N_{S1} has been used rather than N_S, since the secondary will also be seen to have a three-phase winding and the extra subscript is necessary to avoid any ambiguity.

For the winding of Fig. 6.6 let the turns per coil be 10 and determine the effective number of turns for a concentrated, full-pitch coil that establishes an identical component mmf as the actual phase winding for this 2-pole or 1-pole-pair machine.

Example 6.2

Using [A.11] and [A.14] of Appendix A,

$$N_{S1} = \frac{S_1}{pm} = \frac{18}{(2)(3)} = 3 \text{ slots per pole per phase}$$

$$k_d = \frac{\sin(N_{S1}\gamma/2)}{N_{S1}\sin(\gamma/2)} = \frac{\sin(3 \times 20/2)}{3\sin(20/2)} = 0.9598$$

$$k_p = \sin(p/2) = \sin(160/2) = 0.9848$$

Applying [6.1],

$$N_{1eff} = 2k_p k_d N_{S1} N = (2)(0.9848)(0.9598)(3)(10)$$

$$= 56.42 \text{ turns}$$

Thus, the 6 coil sides each having 10 conductors in this two-layer phase winding arrangement with pitch and distribution can be replaced in concept by a single side per slot coil of full pitch with 56.42 turns.

6.3.3 STATOR AIR GAP TRAVELING WAVE

From knowledge of Fourier series, the square wave, air gap mmf of a N_{1eff}-turn, full-pitch, concentrated coil representing the phase winding of phase a can be written as

$$\mathscr{F}_{ga} = \frac{4}{\pi}\frac{N_{1eff}}{2}i_a \sum_{n=1}^{\infty} \frac{1}{n}\sin\left(n\frac{p}{2}\theta\right) \qquad n = 1,3,5,\ldots \qquad \textbf{[6.2]}$$

where p = number of poles and θ is the angular position in mechanical degrees with reference shown in Fig. 6.4. The fundamental component of $\mathscr{F}_{ga}$ follows from evaluation of [6.2] for $n = 1$.

$$\mathscr{F}_{ga1} = \frac{2}{\pi}N_{1eff}i_a\sin\left(\frac{p}{2}\theta\right) \qquad \textbf{[6.3]}$$

This is a standing wave with amplitude variation at any point in space determined by the nature of current i_a.

Since the windings of phases b and c are respectively displaced in space by $-/+$ 120 electrical degrees from phase a, the fundamental component air gap mmf waves can be expressed as

$$\mathscr{F}_{gb1} = \frac{2}{\pi}N_{1eff}i_b\sin\left(\frac{p}{2}\theta - 120°\right) \qquad \textbf{[6.4]}$$

$$\mathscr{F}_{gc1} = \frac{2}{\pi}N_{1eff}i_c\sin\left(\frac{p}{2}\theta + 120°\right) \qquad \textbf{[6.5]}$$

The individual phase currents for a balanced case with $a \cdot b \cdot c$ phase sequence are

$$i_a = I_m \sin(\omega t) \tag{6.6}$$

$$i_b = I_m \sin(\omega t - 120°) \tag{6.7}$$

$$i_c = I_m \sin(\omega t + 120°) \tag{6.8}$$

Substitution of these currents into [6.3] to [6.5] and adding gives the total air gap mmf wave by superposition.

$$\mathcal{F}_g = \mathcal{F}_{ga1} + \mathcal{F}_{gb1} + \mathcal{F}_{gc1}$$

$$= \frac{2}{\pi} N_{1eff} I_m \left[\sin(\omega t) \sin\left(\frac{p}{2}\theta\right) + \sin(\omega t - 120°) \sin\left(\frac{p}{2}\theta - 120°\right) \right.$$

$$\left. + \sin(\omega t + 120°) \sin\left(\frac{p}{2}\theta + 120°\right) \right] \tag{6.9}$$

The trigonometric identity $\sin A \sin B = \frac{1}{2}\cos(A - B) - \frac{1}{2}\cos(A + B)$ can be applied to [6.9] and the result simplified to yield

$$\mathcal{F}_g = \frac{3}{\pi} N_{1eff} I_m \cos\left(\omega t - \frac{p}{2}\theta\right) \tag{6.10}$$

Equation [6.10] is the end result of the preceding exercise in study of the winding arrangement. It completely characterizes the total stator mmf produced by the set of three-phase windings that acts to drive flux around the magnetic circuit of the induction motor.

6.3.4 SYNCHRONOUS SPEED

Equation [6.10] describes a constant-amplitude traveling wave. Its peak amplitude is 3/2 the peak amplitude of the individual phase mmf standing waves. The angular mechanical speed of this traveling wave follows from time differentiation of the argument.

$$\frac{d}{dt}\left(\omega t - \frac{p}{2}\theta\right) = 0$$

or

$$\omega_s = \frac{d\theta}{dt} = \frac{2}{p}\omega \text{ rad/s} \tag{6.11}$$

The angular mechanical speed given by [6.11] is determined precisely by the angular frequency (ω) of the electrical source currents; thus, it is called *synchronous speed*. Synchronous speed can be expressed in British units by conversion.

$$n_s = \frac{60}{2\pi}\omega_s = \frac{60}{2\pi}\left(\frac{2}{p}\right)2\pi f = \frac{120f}{p} \tag{6.12}$$

where f is the cyclic frequency in Hz of the electrical source currents. *As the study of the induction motor proceeds, it will be seen that this concept of synchronous speed plays a central role in all analysis work.*

Example 6.3 | Calculate the synchronous speeds for machines having 2, 4, 6, or 8 poles if operated from a 60-Hz source.

By [6.12] with $f = 60$ Hz,

$$n_s = \frac{120(60)}{p} = \frac{7200}{p}$$

Substitution of the various values of p yields the results of Table 6.1.

Table 6.1
Synchronous speeds (60 Hz)

Poles—p	Speed—n_s (rpm)
2	3600
4	1800
6	1200
8	900

The MATLAB program ⟨fwave.m⟩ visualizes the air gap traveling mmf wave for any even-integer number of poles of choice. The results are displayed in a strobe manner for one magnetic pole pair of movement. Figure 6.9 displays the final frame of the animated simulation for a 12-pole machine.

6.4 ROTOR ACTION AND SLIP

6.4.1 ROTOR COIL INDUCED VOLTAGES

The wound rotor has a set of three-phase windings embedded in slots that are similar to the stator windings. Distribution and pitch are implemented to diminish harmonics just as in the case of the stator windings. Figure 6.10 shows the concentrated coil representation of the wound rotor. Typically, the phase windings are wye-connected with the three line connections available for external connection through a set of slip rings as illustrated by Fig. 6.10*b*.

The squirrel-cage rotor is shown in flat layout with an end ring removed by Fig. 6.11*a*. The embedded individual rotor bars form the sides of single-turn coils where the end ring forms a common end-turn path for all rotor bar currents. From this point of view, the squirrel-cage rotor also forms a set of three-phase windings as shown schematically in Fig. 6.11*b*.

It is concluded that for both the wound and squirrel-cage rotors, the rotor windings form a three-phase set of phase windings. At this stage of development, these windings are treated as open-circuit. Further, the stator windings, when supplied by

Air gap mmf traveling wave

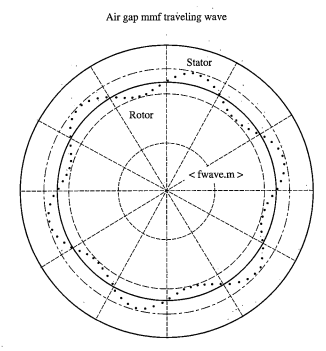

Figure 6.9
Air gap traveling mmf wave simulation

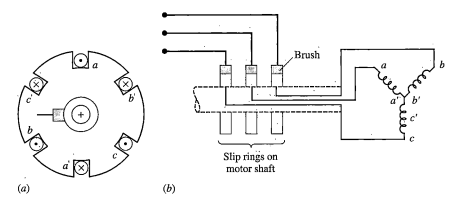

Figure 6.10
Wound-rotor windings. (a) Cross section of rotor. (b) Coil connections to slip rings.

a set of balanced three-phase currents, were shown in the previous section to produce a constant-amplitude traveling mmf wave. As a consequence, a constant-amplitude traveling flux wave is present in the air gap. The stator could be conceptually replaced by a permanent-magnet structure revolving at synchronous speed and surrounding the rotor, as illustrated by Fig. 6.12. If the rotor revolves at the mechanical speed ω_m (or n_m), a normalized speed difference, called *slip* (*s*) can be defined as

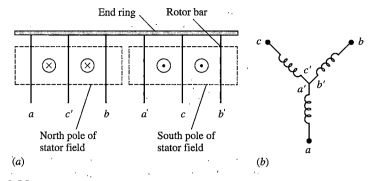

Figure 6.11
Squirrel-cage rotor windings. (a) Flat layout with end ring removed. (b) Schematic representation.

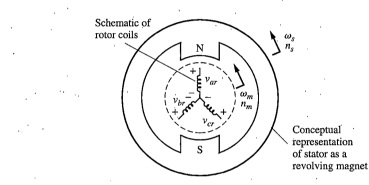

Figure 6.12
Rotor windings in presence of stator field

$$s = \frac{\omega_s - \omega_m}{\omega_s} = \frac{n_s - n_m}{n_s} \qquad \textbf{[6.13]}$$

where ω_s and ω_m are in units of rad/s while n_s and n_m are in units of rpm.

From a reference point of one of the rotor coils, it follows from [6.13] that the sinusoidally distributed stator traveling flux wave appears to have a speed of $s\omega_s = \omega_s - \omega_m$. By Faraday's law, voltages varying sinusoidally with time are induced behind the terminals of each rotor coil described by

$$v_{ar} = V_{rm}\cos(\omega_r t) \qquad \textbf{[6.14]}$$

$$v_{br} = V_{rm}\cos(\omega_r t - 120°) \qquad \textbf{[6.15]}$$

$$v_{cr} = V_{rm}\cos(\omega_r t + 120°) \qquad \textbf{[6.16]}$$

For a p-pole machine, the radian frequency of the rotor coil voltages depends on the mechanical speed at which the stator flux wave passes a rotor coil. Its value is given by

$$\omega_r = \frac{p}{2}(s\omega_s) = s\left(\frac{p}{2}\omega_s\right) = s\omega \qquad \textbf{[6.17]}$$

The cyclic frequency (f_r) of rotor voltages is related to the cyclic frequency (f) of the stator coil currents by

$$f_r = \frac{\omega_r}{2\pi} = \frac{s\omega}{2\pi} = sf \qquad \text{[6.18]}$$

6.4.2 ROTOR AIR GAP TRAVELING WAVE

If the balanced set of rotor windings with a balanced set of induced voltages are shorted as indicated in Fig. 6.13, a balanced set of three-phase currents flow that lag behind the voltages by a phase shift angle ϕ.

$$i_{ar} = I_{rm}\cos(s\omega t - \phi) \qquad \text{[6.19]}$$

$$i_{br} = I_{rm}\cos(s\omega t - \phi - 120°) \qquad \text{[6.20]}$$

$$i_{cr} = I_{rm}\cos(s\omega t - \phi + 120°) \qquad \text{[6.21]}$$

Like the stator coils, the rotor coils being displaced by 120° in space and conducting a balanced set of three-phase currents will establish a rotor mmf wave in the air gap that is constant in amplitude and traveling with respect to the rotor at a mechanical speed of

$$\omega_{rr} = \frac{2}{p}\omega_r = \frac{2}{p}(s\omega) = s\left(\frac{2}{p}\omega\right) = s\omega_s \qquad \text{[6.22]}$$

However, since the rotor is mechanically revolving at speed ω_m, the mechanical speed of the rotor revolving field as viewed from the stator reference is

$$\omega_{rs} = s\omega_s + \omega_m = \omega_s - \omega_m + \omega_m = \omega_s \qquad \text{[6.23]}$$

Specifically, the rotor field as seen from the stator travels at the same speed as the stator field. Although not necessarily in phase, the air gap of the induction motor has two constant-amplitude traveling mmf waves. Both mmf waves revolve at synchronous speed ω_s when viewed from the fixed stator reference. The action of these two fields attempting to align produces a torque on the rotor structure known as the *electromagnetic torque* or *developed torque* (T_d).

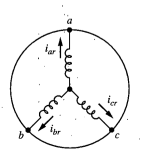

Figure 6.13
Shorted rotor coil set

The MATLAB program ⟨rotmmf.m⟩ simulates in a strobe manner the rotor air gap mmf wave as seen from the stator reference for each 10° (electrical) of movement over the span of one magnetic pole pair. For the specified value of slip, the rotor position can be observed by tracking the rotor bars indicated by small circles. Figure 6.14 displays the final frame of the animated simulation for a 12-pole machine with slip $s = 0.25$.

Example 6.4

A 4-pole, 60-Hz, three-phase induction motor is operating at a shaft speed of 1750 rpm. Determine: (a) speed of the stator air gap mmf wave; (b) frequency of the rotor-induced currents; and (c) the speed of the rotor mmf wave with respect to the rotor.

(a) The air gap mmf wave travels at synchronous speed as given by [6.11] or [6.12].

$$\omega_s = \frac{2}{p}\omega = \frac{2}{p}(2\pi f) = \frac{2}{4}(2\pi \times 60) = 188.49 \text{ rad/s}$$

$$n_s = \frac{120f}{p} = \frac{120(60)}{4} = 1800 \text{ rpm}$$

(b) Frequency of the rotor currents is found by [6.18]. By [6.13],

Rotor produced mmf wave and rotor position

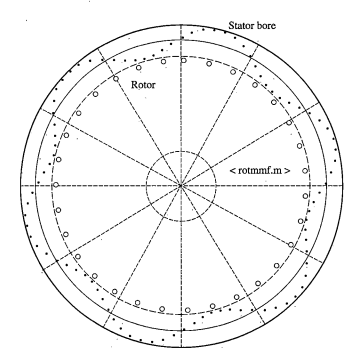

Figure 6.14
Rotor traveling wave simulation

$$s = \frac{n_s - n_m}{n_s} = \frac{1800 - 1750}{1800} = 0.02778$$

Thus,

$$f_r = sf = (0.02778)(60) = 1.6667 \text{ Hz}$$

(c) By use of [6.22],

$$\omega_{rr} = s\omega_s = (0.02778)(188.49) = 5.236 \text{ rad/s}$$

Expressed in rpm,

$$n_{rr} = sn_s = (0.02667)(1800) = 50 \text{ rpm}$$

Note that $n_m + n_{rr} = 1750 + 50 = 1800 \text{ rpm} = n_s$, verifying the conclusion that the rotor field as seen from the stator reference travels at the same speed as the stator field.

6.5 EQUIVALENT CIRCUIT

Similarities between the induction motor and the transformer of Chap. 4 are readily apparent. Power from a sinusoidal source is supplied to a stator winding or primary. This winding establishes a flux that mutually couples a rotor winding or secondary. The cyclic mutual flux traverses a ferromagnetic material that gives rise to eddy current and hysteresis losses. Not all of the flux established by the primary winding necessarily links the secondary winding, suggesting that leakage reactance exists. Any current that flows in the secondary winding acts to oppose changes in the primary generated mutual flux, requiring the existence of an mmf balance. As a result of these similarities, the practical transformer equivalent circuit seems like a suitable candidate for modeling the induction motor. However, there are three notable dissimilarities between the induction motor and the transformer that would be expected to impact direct application of the practical transformer model. The mutual flux path must cross the high-reluctance air gap, resulting in reduced magnetizing reactance and increased leakage reactance for the induction motor as compared to a transformer. Also, there is relative motion between the induction motor primary and secondary coils. Additionally, the secondary winding has its terminals shorted.

Figure 6.15 depicts an adaptation of the three-phase transformer to primary and secondary winding of an induction motor. Focusing on the arrangement of phase a, the rms value of the secondary induced voltage of frequency $f_r = sf$ follows from [4.7] as

$$E_{2s} = 4.44 f_r N_{2eff} \Phi_{ma} = 4.44 sf N_{2eff} \Phi_{ma} \qquad \textbf{[6.24]}$$

Letting $L_{\ell 2}$ represent the secondary leakage inductance, the secondary leakage reactance is given by

$$X_{2s} = \omega_r L_{\ell 2} = 2\pi f_r L_{\ell 2} = 2\pi sf L_{\ell 2} \qquad \textbf{[6.25]}$$

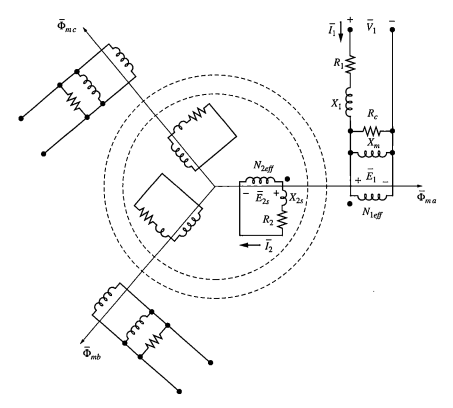

Figure 6.15
Induction motor modeled as a transformer

Applying KVL around the secondary loop,

$$\bar{E}_{2s} = 4.44 s f N_{2eff} \bar{\Phi}_{ma} = R_2 \bar{I}_2 + j2\pi s f L_{\ell 2} \bar{I}_2 \qquad \text{[6.26]}$$

Rearrange [6.26] to yield

$$s\bar{E}_2 = R_2 \bar{I}_2 + jsX_2 \bar{I}_2 \qquad \text{[6.27]}$$

where $E_2 = 4.44 f N_{2eff} \Phi_{ma}$ and $X_2 = 2\pi f L_{\ell 2}$; the values of secondary induced voltage and leakage reactance are measured at blocked rotor ($\omega_m = 0$, $s = 1$) conditions where $f_r = f$.

Dividing both sides of [6.27] by s gives

$$\bar{E}_2 = \frac{R_2}{s}\bar{I}_2 + jX_2\bar{I}_2 \qquad \text{[6.28]}$$

Equation [6.28] suggests that if the secondary resistance were replaced by a slip-dependent value, then the model could be treated as though there were no relative motion between the primary and secondary coils. Figure 6.16 presents this model

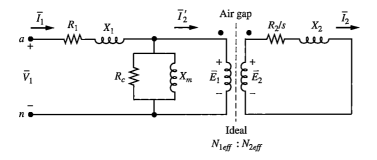

Figure 6.16
Exact per phase equivalent circuit for three-phase induction motor

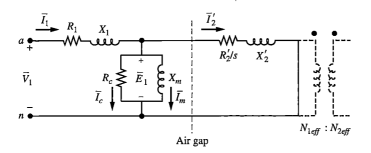

Figure 6.17
Exact per phase equivalent circuit for three-phase induction motor (stator reference)

for phase a. Since the actual rotor variables are of little concern in performance analysis, the ideal transformer can be reflected to the extreme right in the circuit of Fig. 6.16 to give the equivalent circuit for phase a of the three-phase induction motor displayed by Fig. 6.17. Using $a = N_{1eff}/N_{2eff}$, impedance reflections require that

$$\frac{R_2'}{s} = a^2\frac{R_2}{s}, \quad X_2' = a^2X_2 \quad \text{and} \quad \bar{I}_2' = \frac{\bar{I}_2}{a}$$

The location of the air gap, immediately to the right of the excitation branch ($R_c \| jX_m$), has been retained for future reference.

6.6 TEST DETERMINATION OF PARAMETERS

Prior to application of the per phase equivalent circuit in any performance calculations, establishment of test procedures to yield values for the elements of the equivalent circuit is in order. Although it was remarked in the previous section that magnetizing inductance (X_m) is reduced and that leakage reactances (X_1, X'_2) are increased when compared to a transformer on a per unit basis, this change is not sufficient to negate the relative comparison of R_c and X_m with R_1, X_1, R'_2, and X'_2 for a transformer made in Sec. 4.4.3. Consequently, goodness of performance criteria still require that $\{R_c, X_m\} \gg \{R_1, X_1, R'_2, X'_2\}$.

6.6.1 BLOCKED ROTOR TEST

For the condition of blocked rotor $(\omega_m = 0, s = 1)$, the rotor is mechanically constrained from rotation while excited by a balanced three-phase source. The reflected secondary resistance of Fig. 6.17 becomes $R'_2/s = R'_2$. Since $|R'_2 + jX'_2| \ll |R_c\|jX_m|$ the excitation branch can be neglected. Hence, the per phase input impedance for blocked rotor operation is

$$Z_{br} \cong R_1 + R'_2 + j(X_1 + X'_2) = R_{eq} + jX_{eq} \qquad \textbf{[6.29]}$$

This condition is directly analogous to the short-circuit test of the transformer discussed in Sec. 4.5.2.

    The motor is instrumented to record input current $(I_1 = I_{br})$, input voltage $(V_1 = V_{br} = V_L/\sqrt{3})$, and per phase input power $(P_{br} = P_T/3)$. The test should be performed at near rated current. Owing to unequal current distribution in rotor bars due to skin effect, the secondary resistance R'_2 is frequency-dependent. It is common practice to perform the blocked rotor test with reduced excitation frequency (f_{br}) in order to obtain a value of R'_2 that is typical of rotor frequency for rated slip operation. From this data, the values of R_{eq} and X_{br} are straightway synthesized.

$$R_{eq} = \frac{P_{br}}{I_{br}^2} \qquad \textbf{[6.30]}$$

$$Z_{br} = \frac{V_{br}}{I_{br}} \qquad \textbf{[6.31]}$$

$$X_{br} = (Z_{br}^2 - R_{eq}^2)^{1/2} \qquad \textbf{[6.32]}$$

    Simple halving of R_{eq} and X_{br} to give R_1, R'_2 and X_1, X'_2, as in the case of the transformer, is not justified. An additional test is performed by placing a controllable dc voltage source (V_{dc}) across two input lines and measuring the resulting dc current (I_{dc}). The resistance seen by this test is simply $2R_1$. The secondary current is zero, since there is no time-varying coupling flux and reactances X_1 and X_m act as short circuits in the dc steady state. Thus,

$$R_1 = \frac{1}{2}\frac{V_{dc}}{I_{dc}} \qquad \textbf{[6.33]}$$

The value of R_1 as found by [6.33] is typically increased by 3 to 5 percent for use in the equivalent circuit to account for any increased conductor resistance due to

Table 6.2
Empirical distribution of X_{eq}

NEMA Design	X_1	X_2'
A	$0.5X_{eq}$	$0.5X_{eq}$
B	$0.4X_{eq}$	$0.6X_{eq}$
C	$0.3X_{eq}$	$0.7X_{eq}$
D	$0.5X_{eq}$	$0.5X_{eq}$
Wound rotor	$0.5X_{eq}$	$0.5X_{eq}$

unequal current distribution attributable to small skin effect and cross-slot magnetic fields during ac operation. With R_1 determined,

$$R_2' = R_{eq} - R_1 \qquad \textbf{[6.34]}$$

Stator and rotor slots differ in number and shape. Further, the end-turn areas are not mirror images. Consequently, the coefficient of coupling for the stator winding is not necessarily equal to the coefficient of coupling of the rotor winding. Thus, the justification of $X_1 = X_2'$ as presented for the transformer in Sec. 4.4.3 is not valid. The apportionment of X_{br} to give X_1 and X_2' in the absence of design calculations is commonly carried out according to empirical methods that are based on vast experience and documented in *IEEE Standard 112*. The relationships presented in Table 6.2 depend upon the NEMA design (listed on the motor nameplate) to be discussed in Sec. 6.7. Before entering Table 6.2, the leakage reactance must be adjusted to rated frequency (f) by

$$X_{eq} = \frac{f}{f_{br}}X_{br} \qquad \textbf{[6.35]}$$

6.6.2 No-Load Test

The no-load test is performed with the motor shaft free of any coupled mechanical load ($\omega_m \cong \omega_s, s \cong 0$) with rated and balanced three-phase excitation applied to the motor input terminals. Since s is small, the value of R_2'/s in Fig. 6.17 is large. Consequently, I_2' is small, but not zero, as power supplied to the secondary must account for rotor ohmic losses and friction and windage losses. The motor is instrumented to record input current ($I_1 = I_{n\ell}$), input voltage ($V_1 = V_{n\ell} = V_L/\sqrt{3}$), per phase input power ($P_{n\ell} = P_T/3$), and motor shaft speed ($n_{n\ell}, \omega_{n\ell}$). Rated frequency excitation is used to correctly yield the frequency-sensitive value of R_c.

The phase angle by which $\bar{I}_{n\ell}$ lags $\bar{V}_{n\ell}$ follows as

$$\theta_{n\ell} = \cos^{-1}\left[\frac{P_{n\ell}}{V_{n\ell}I_{n\ell}}\right] \qquad \textbf{[6.36]}$$

KVL applied to the equivalent circuit of Fig. 6.17 gives

$$\bar{E}_1 = V_1 \angle 0° - I_{n\ell} \angle -\theta_{n\ell}(R_1 + jX_1) \qquad \text{[6.37]}$$

With $\bar{E}_1$ determined by [6.37], the reflected secondary current is found.

$$\bar{I}_2 = \frac{\bar{E}_1}{\dfrac{R'_2}{s_{n\ell}} + jX'_2} \qquad \text{[6.38]}$$

where

$$s_{n\ell} = \frac{\omega_s - \omega_{n\ell}}{\omega_s} = \frac{n_s - n_{n\ell}}{n_s} \qquad \text{[6.39]}$$

Power dissipated by R_c is calculated by

$$P_c = P_{n\ell} - I_1^2 R_1 - (I'_2)^2 \frac{R'_2}{s} \qquad \text{[6.40]}$$

where the last term of [6.40] accounts for both the secondary ohmic losses and the friction and windage losses on a per phase basis. The value of R_c can be synthesized directly.

$$R_c = \frac{E_1^2}{P_c} \qquad \text{[6.41]}$$

The reactive power (Q_m) flowing to X_m is determined and used to evaluate X_m.

$$Q_m = V_{n\ell} I_{n\ell} \sin(\theta_{n\ell}) - I_{n\ell}^2 X_1 - (I'_2)^2 X'_2 \qquad \text{[6.42]}$$

$$X_m = \frac{E_1^2}{Q_m} \qquad \text{[6.43]}$$

Example 6.5 | A three-phase induction motor nameplate has the following information:

10 hp   60 Hz   230 V   27 A   1755 rpm

215T frame   Design B   Code F   SF 1.15

Recorded test data are known where voltages are measured line-to-line and power values are total three-phase:

DC Test	Blocked Rotor Test	No-Load Test
10.8 V	22.26 V	230 V
25 A	27 A	12.9 A
	770 W	791 W
	15 Hz	1798 rpm

Determine the values of all parameters for the equivalent circuit of Fig. 6.17.
    Using equations [6.30] to [6.43],

$$R_{eq} = \frac{P_{br}}{I_{br}^2} = \frac{770/3}{(27)^2} = 0.3521 \ \Omega$$

$$Z_{br} = \frac{V_{br}}{I_{br}} = \frac{22.26/\sqrt{3}}{27} = 0.4760 \ \Omega$$

$$X_{br} = (Z_{br}^2 - R_{eq}^2)^{1/2} = (0.4760^2 - 0.3521^2)^{1/2} = 0.3203 \ \Omega$$

Using 5 percent increase to account for ac conditions,

$$R_1 = (1.05)\frac{1}{2}\frac{V_{dc}}{I_{dc}} = (1.05)\frac{1}{2}\frac{10.8}{25} = 0.2268 \ \Omega$$

$$R_2' = R_{eq} - R_1 = 0.3521 - 0.2268 = 0.1253 \ \Omega$$

$$X_{eq} = \frac{f}{f_{br}}X_{br} = \frac{60}{15}(0.3203) = 1.2812 \ \Omega$$

Based on Table 6.2,

$$X_1 = 0.4X_{eq} = 0.4(1.2812) = 0.5125 \ \Omega$$

$$X_2' = 0.6X_{eq} = 0.6(1.2812) = 0.7687 \ \Omega$$

$$\theta_{n\ell} = \cos^{-1}\left[\frac{P_{n\ell}}{V_{n\ell}I_{n\ell}}\right] = \cos^{-1}\left[\frac{791/3}{(230/\sqrt{3})(12.9)}\right] = 81.14°$$

$$\overline{E}_1 = \frac{230}{\sqrt{3}} \angle 0° - 12.9\angle -81.14°(0.2268 + j0.5125) = 125.82\angle 0.85°$$

$$s_{n\ell} = \frac{n_s - n_{n\ell}}{n_s} = \frac{1800 - 1798}{1800} = 0.001111$$

$$\overline{I}_2' = \frac{\overline{E}_1}{\dfrac{R_2'}{s} + jX_2'} = \frac{125.82\angle 0.85°}{\dfrac{0.1253}{0.001111} + j0.7687} = 1.1156\angle 0.46°$$

$$P_c = P_{n\ell} - I_1^2 R_1 - (I_2')^2 \frac{R_2'}{s}$$

$$= \frac{791}{3} - (12.9)^2(0.2268) - (1.1156)^2\left(\frac{0.1253}{0.001111}\right) = 85.56 \ \text{W}$$

$$R_c = \frac{E_1^2}{P_c} = \frac{(125.82)^2}{85.56} = 185.02 \ \Omega$$

$$Q_m = V_{n\ell}I_{n\ell}\sin(\theta_{n\ell}) - I_{n\ell}^2 X_1 - (I_2')^2 X_2'$$

$$= \left(\frac{230}{\sqrt{3}}\right)(12.9)\sin(81.14°) - (12.9)^2(0.5125) - (1.1156)^2(0.7687) = 1606.37 \ \text{VARs}$$

$$X_m = \frac{E_1^2}{Q_m} = \frac{(125.82)^2}{1606.37} = 9.855 \ \Omega$$

Although not difficult in nature, the above example shows that determination of the induction motor equivalent circuit parameters is a somewhat tedious task. Round-off error can accumulate if precision of the answers is not guarded. The MATLAB program ⟨imeqckt.m⟩ can quickly calculate the equivalent circuit parameters from test data while maintaining double-precision accuracy. The code is edited to reflect the necessary 13 data entries (p, f, design, V_{dc}, I_{dc}, V_{br}, I_{br}, P_{br}, f_{br}, $V_{n\ell}$, $I_{n\ell}$, $P_{n\ell}$, and $n_{n\ell}$). The data presently in the code is that of Example 6.5, where the results displayed to the screen are as follows:

```
Parameters  =
   R1         R2pr        X1          X2pr        Rc          Xm
ans  =
2.2680e-001  1.2528e-001  5.1252e-001  7.6878e-001  1.8510e+002  9.8554e+000
```

6.7 PERFORMANCE CALCULATIONS AND NATURE

6.7.1 INDUCTION MOTOR POWER FLOW

Since there are no electrical sources connected directly to the rotor of an induction motor, the power that crosses the air gap in magnetic form before being converted to electrical form is the only source of power supplied to the rotor. Specifically, the average power that crosses the air gap must be equal to the sum of the rotor coil ohmic losses dissipated as heat and the power converted to mechanical form. The per phase average power that crosses the air gap for the per phase equivalent circuit of Fig. 6.17 is given by

$$P_g = (I_2')^2 \frac{R_2'}{s} = (I_2')^2 R_2' + (I_2')^2 \frac{(1-s)}{s} R_2' \qquad \textbf{[6.44]}$$

The first term of the right-hand expression of [6.44] is obviously the per phase ohmic losses of the rotor coils. Consequently, the unfamiliar second term must necessarily be the power converted to mechanical form, or the *developed power* P_d. From [6.44], it can be seen that the per phase rotor ohmic losses P_{rcu} and the per phase developed power P_d can be respectively expressed as

$$P_{rcu} = (I_2')^2 R_2' = sP_g \qquad \textbf{[6.45]}$$

$$P_d = (I_2')^2 \frac{(1-s)}{s} R_2' = (1-s)P_g \qquad \textbf{[6.46]}$$

With the mechanical power conversion term identified in [6.46], the per phase equivalent circuit of Fig. 6.17 can be redrawn to delineate the electromagnetic power conversion. Figure 6.18 results, where the *per phase friction and windage losses* ($P_{FW}/3$) along with the *per phase output shaft power* ($P_s/3$) and the *per phase output shaft torque* ($T_s/3$) have been added. Shaft power (P_s), shaft torque

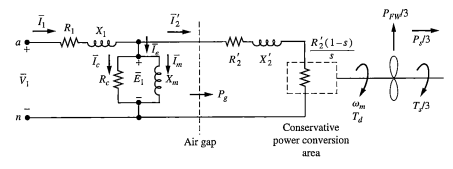

Figure 6.18
Per phase equivalent circuit illustrating electromechanical power conversion

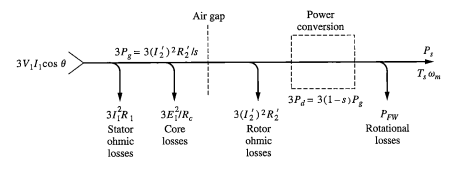

Figure 6.19
Total average power flow diagram for a three-phase induction motor

(T_s), and friction and windage losses (P_{FW}) are inherently total entities for the motor; however, the quantities can analytically be divided by 3 for association with the per phase equivalent circuit power flow calculations. An average power flow diagram for the total three-phase motor is presented in Fig. 6.19 to clarify the complete average power flow process.

6.7.2 DEVELOPED TORQUE DETERMINATION

Certainly one can impress phase voltage V_1 on the equivalent circuit for a particular value of speed (hence slip) and determine the value of I_2' by circuit analysis methods. Air gap power P_g and developed power P_d can follow by application of [6.44] and [6.45], whence the developed torque can be evaluated as $T_d = P_d/\omega_m$. However, it is convenient to develop a formula to yield the developed torque in a single-step calculation.

The developed torque can be determined by dividing [6.46] by shaft speed ω_m.

$$T_d = \frac{P_d}{\omega_m} = \frac{(1-s)}{\omega_m}(I_2')^2 \frac{R_2'}{s} \qquad \textbf{[6.47]}$$

Equation [6.13] can be rearranged to give

$$\frac{1}{\omega_s} = \frac{1-s}{\omega_m}$$

which when substituted into [6.47] yields

$$T_d = \frac{(I_2')^2 \dfrac{R_2'}{s}}{\omega_s} \qquad \textbf{[6.48]}$$

In order to attain the desired torque formula, an expression for $(I_2')^2$ is required. Referring to Fig. 6.17, a Thevenin's equivalent circuit looking to the left from the air gap line can be determined.

By voltage division,

$$\overline{V}_{Th} = \frac{R_c \| jX_m}{R_1 + jX_1 + R_c \| jX_m}\overline{V}_1 = \frac{jR_cX_m}{(R_1R_c - X_1X_m) + j(R_1X_m + R_cX_1 + R_cX_m)}\overline{V}_1$$

$$\textbf{[6.49]}$$

The Thevenin's impedance is

$$\begin{aligned}
\mathbf{Z}_{Th} &= R_{Th} + jX_{Th} = R_c \| jX_m \| (R_1 + jX_1) \\[2mm]
&= \frac{R_cR_1X_m(R_cX_m + R_1X_m + R_cX_1) - R_cX_mX_1(R_cR_1 - X_mX_1)}{(R_cR_1 - X_mX_1)^2 + (R_cX_m + R_1X_m + R_cX_1)^2} \\[2mm]
&\quad + j\frac{R_cX_mX_1(R_cX_m + R_1X_m + R_cX_1) + R_cR_1X_m(R_cR_1 - X_mX_1)}{(R_cR_1 - X_mX_1)^2 + (R_cX_m + R_1X_m + R_cX_1)^2} \quad \textbf{[6.50]}
\end{aligned}$$

Although the expressions for $\overline{V}_{Th}$ and $\mathbf{Z}_{Th}$ are cumbersome, they are not speed (or slip) dependent. Thus, the expressions need only be evaluated once for a particular motor. The per phase equivalent circuit of Fig. 6.17 can now be redrawn as shown by Fig. 6.20. An expression for $(I_2')^2$ follows directly from this circuit.

$$(I_2')^2 = \frac{V_{Th}^2}{\left(R_{Th} + \dfrac{R_2'}{s}\right)^2 + (X_{Th} + X_2')^2} \qquad \textbf{[6.51]}$$

Substitution of [6.51] into [6.48] leads to the desired formula for per phase developed torque. After multiplication by 3 to yield the total developed torque, the result is

$$3T_d = \frac{3V_{Th}^2 \dfrac{R_2'}{s}}{\omega_s\left[\left(R_{Th} + \dfrac{R_2'}{s}\right)^2 + (X_{Th} + X_2')^2\right]} \qquad \textbf{[6.52]}$$

The total developed torque given by [6.52] is obviously a function of speed through the slip variable s; however, the nature of the resulting speed-torque curve is not readily apparent by inspection. The MATLAB program ⟨Te_wm.m⟩ has the capability to generate an array of torque values given by [6.52] for a specified range of slip (s) and then plot the results against the values of speed corresponding to the slip values. The resulting torque-speed curve for the motor of Example 6.5 is shown in Fig. 6.21. The program also computes the value of the Thevenin voltage $\overline{V}_{Th} = V_{Th}\angle\theta$ and the Thevenin impedance $\mathbf{Z}_{Th} = R_{Th} + jX_{Th}$. The Thevenin

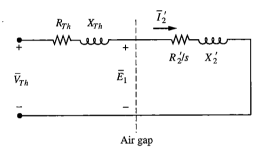

Figure 6.20
Per phase Thevenin equivalent circuit for three-phase induction motor

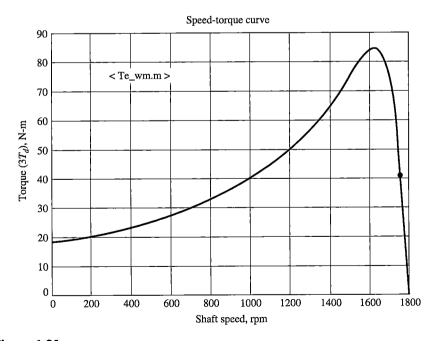

Figure 6.21
Total electromagnetic torque for motor of Example 6.5

equivalent values for the motor of Example 6.5 as displayed to the screen are located below:

```
Thevenin_Values:
     VTh           Theta          Rth           XTh
ans  =
1.2606e+002   1.1011e+000   2.0591e-001   4.9058e-001
```

These values are convenient to use in any longhand calculations relating to the equivalent circuit of Fig. 6.20 or equation [6.52], thus circumventing the cumbersome evaluations of [6.49] and [6.50].

Example 6.6

The speed-torque curve of Fig. 6.21 applies to the motor mode of operation when both torque and speed act in the same angular direction. Two other modes of operation are possible. If the speed were greater than synchronous speed, the torque acts opposite to speed direction, giving generator mode of operation. If speed is opposite to the direction of air gap field rotation, the torque acts to reduce the speed of the motor, giving braking or plugging mode of operation. Use ⟨Te_wm.m⟩ with the s vector edited to give the range $-1 \leq s \leq 2$ to generate a torque-speed curve that displays the three modes of induction machine operation using the machine of Example 6.5.

The plot of Fig. 6.22 displays the resulting plot where the three modes have been indicated.

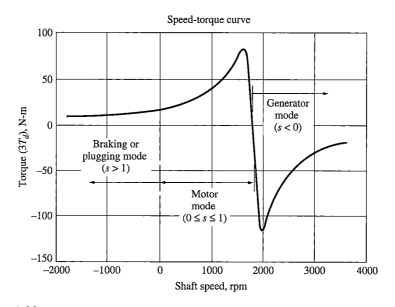

Figure 6.22
Modes of induction machine operation

A three-phase, 230-V, 60-Hz, 10-hp, 1138 rpm induction motor has the following equivalent circuit values:

Example 6.7

$$R_1 = 0.15\,\Omega \qquad R_2' = 0.3\,\Omega \qquad R_c = 225\,\Omega$$

$$X_1 = 0.5\,\Omega \qquad X_2' = 0.5\,\Omega \qquad X_m = 25\,\Omega$$

If $\bar{V}_1 = 230/\sqrt{3}\angle 0°$, [6.49] and [6.50] yield $\bar{V}_{Th} = 130.1\angle 0.21°$, $R_{Th} = 0.145\,\Omega$, and $X_{Th} = 0.49\,\Omega$. For rated load conditions, determine (a) total developed torque $(3T_d)$, (b) friction and windage losses (P_{FW}), (c) reflected secondary current (I_2'), (d) input current (I_1), (e) input power factor (PF), and (f) efficiency (η).

(a) The nameplate rated speed must occur for a small value of slip. Since 1138 rpm is a small percentage less than $n_s = 1200\,\text{rpm}$, the motor is a six-pole machine. Hence, synchronous speed in mks units is

$$\omega_s = \frac{2}{p}\,(\omega) = \frac{2}{6}\,(120\pi) = 40\pi \text{ rad/s}$$

$$s = \frac{n_s - n_m}{n_s} = \frac{1200 - 1138}{1200} = 0.052$$

By [6.52],

$$3T_d = \frac{3V_{Th}^2\,\dfrac{R_2'}{s}}{\omega_s\left[\left(R_{Th} + \dfrac{R_2'}{s}\right)^2 + (X_{Th} + X_2')^2\right]}$$

$$3T_d = \frac{3(130.1)^2\,\dfrac{0.3}{0.052}}{40\pi\left[\left(0.145 + \dfrac{0.3}{0.052}\right)^2 + (0.49 + 0.5)^2\right]} = 64.83 \text{ N·m}$$

(b) Recalling that the motor is operating at rated load,

$$P_{FW} = 3T_d\omega_m - P_s = 64.83(1138\pi/30) - 10(746) = 265.86 \text{ W}$$

(c) From Fig. 6.20,

$$\bar{I}_2' = \frac{\bar{V}_{Th}}{\left(R_{Th} + \dfrac{R_2'}{s}\right) + j(X_{Th} + X_2')}$$

$$= \frac{130.1\angle 0.21°}{\left(0.145 + \dfrac{0.3}{0.052}\right) + j(0.49 + 0.5)} = 21.7\angle -9.29° \text{ A}$$

or

$$I_2' = 21.7 \text{ A}$$

(d) From Fig. 6.20,

$$\overline{E}_1 = \overline{I}_2'\left(\frac{R_2'}{s} + jX_2'\right) = (21.7\angle-9.29°)\left(\frac{0.3}{0.052} + j0.5\right) = 125.64\angle-4.34° \text{ V}$$

Using Fig. 6.18,

$$\overline{I}_1 = \overline{I}_2' + \frac{\overline{E}_1}{R_c} + \frac{\overline{E}_1}{jX_m}$$

$$\overline{I}_1 = 21.7\angle-9.29° + \frac{125.64\angle-4.34°}{225} + \frac{125.64\angle-4.34°}{25\angle90°} = 23.22\angle-21.6° \text{ A}$$

or

$$I_1 = 23.22 \text{ A}$$

(e) The input power factor is

$$PF = \cos(\angle\overline{V}_1 - \angle\overline{I}_1) = \cos(0° + 21.6°) = 0.93 \text{ lagging}$$

(f) The efficiency is found as the percentage ratio of output and input power.

$$P_T = 3V_1 I_1 PF = 3(230/\sqrt{3})(23.22)(0.93) = 8602.7 \text{ W}$$

$$\eta = \frac{P_s}{P_T}(100\%) = \frac{10(746)}{8602.7}(100\%) = 86.7\%$$

Example 6.8 | For the motor of Example 6.7, determine the rated load efficiency using $\eta = (1 - \text{losses}/P_T)\cdot$ 100%.

The value of $P_{FW} = 265.86$ W is known from Example 6.7. The balance of the losses are found as

$$3P_c = \frac{3E_1^2}{R_c} = \frac{3(125.64)^2}{225} = 210.5 \text{ W}$$

$$3I_1^2 R_1 + 3(I_2')^2 R_2' = 3(23.22)^2(0.15) + 3(21.7)^2(0.3) = 666.4 \text{ W}$$

Thus,

$$\text{Losses} = P_{FW} + 3P_c + 3I_1^2 R_1 + 3(I_2')^2 R_2' = 265.86 + 210.5 + 666.4 = 1142.76 \text{ W}$$

and

$$\eta = \left(1 - \frac{\text{losses}}{P_T}\right)100\% = \left(1 - \frac{1142.76}{8602.7}\right)100\% = 86.7\%$$

Example 6.9 | Determine the value of (a) starting (stall) current and (b) starting PF for the motor of Example 6.7.

(a) From Fig. 6.20 with $s = 1$,

$$\bar{I}_2' = \frac{\bar{V}_{Th}}{(R_{Th} + R_2') + j(X_{Th} + X_2')}$$

$$= \frac{130.1 \angle 0.21°}{(0.145 + 0.3) + j(0.49 + 0.5)} = 119.9 \angle -65.59° \text{ A}$$

$$\bar{E}_1 = \bar{I}_2'(R_2' + jX_2') = (119.9 \angle -65.59°)(0.3 + j0.5) = 69.9 \angle -6.55° \text{ V}$$

Based on Fig. 6.18,

$$\bar{I}_1 = \bar{I}_2' + \frac{\bar{E}_1}{R_c} + \frac{\bar{E}_1}{jX_m} = 119.9 \angle -65.59° + \frac{69.9 \angle -6.55°}{225} + \frac{69.9 \angle -6.55°}{25 \angle 90°}$$

$$\bar{I}_1 = 122.47 \angle -66.1°$$

or

$$I_1 = 122.47 \text{ A}$$

(b) $$PF = \cos(\angle \bar{V}_1 - \angle \bar{I}_1) = \cos(0 + 66.1°) = 0.4 \text{ lagging}$$

Example 6.10

The motor of Example 6.7 is operating in the braking mode with $n_m = -100$ rpm . Determine the value of total developed torque acting to brake the motor.

The slip for this braking condition is

$$s = \frac{n_s - n_m}{n_s} = \frac{1200 - (-100)}{1200} = 1.083$$

By [6.52],

$$3T_d = \frac{3V_{Th}^2 \dfrac{R_2'}{s}}{\omega_s \left[\left(R_{Th} + \dfrac{R_2'}{s} \right)^2 + (X_{Th} + X_2')^2 \right]}$$

$$= \frac{3(130.1)^2 \dfrac{0.3}{1.083}}{40\pi \left[\left(0.145 + \dfrac{0.3}{1.083} \right)^2 + (0.49 + 0.5)^2 \right]}$$

$$3T_d = 96.64 \text{ N·m}$$

Example 6.11

The induction machine of Example 6.7 is connected to an infinite bus with $\bar{V}_1 = 230/\sqrt{3} \angle 0°$ with a frequency of 60 Hz. It is driven by a mechanical source at a speed of 1224 rpm; thus, it is operating as an induction generator. Find the value of power being converted from mechanical to electrical form.

The slip is

$$s = \frac{n_s - n_m}{n_s} = \frac{1200 - 1224}{1200} = -0.02$$

From Fig. 6.20,

$$\bar{I}'_2 = \frac{\bar{V}_{Th}}{\left(R_{Th} + \dfrac{R'_2}{s}\right) + j(X_{Th} + X'_2)}$$

$$= \frac{130.1 \angle 0.21°}{\left(0.145 + \dfrac{0.3}{-0.02}\right) + j(0.49 + 0.5)} = 8.74 \angle -175.97° \text{ A}$$

Based on [6.46],

$$3P_d = 3(I'_2)^2 R'_2 \frac{(1-s)}{s} = 3(8.74)^2(0.3)\frac{(1+0.02)}{-0.02} = -3506.2 \text{ W}$$

The negative sign arises since the energy conversion resistor of Fig. 6.18 is supplying power to the secondary circuit rather than absorbing power.

6.7.3 DEVELOPED TORQUE NATURE

Now being aware of the characteristic shape of the speed-torque curve over a wide speed range, the following specific items can be examined:

1. Effect of voltage variation
2. Condition for zero torque
3. Small slip approximation
4. Critical points of torque
5. Rotor coil resistance impact

Voltage Variation Substitution of [6.11] into [6.13] gives

$$s = \frac{\omega_s - \omega_m}{\omega_s} = \frac{\dfrac{2}{p}\omega - \omega_m}{\dfrac{2}{p}\omega} = \frac{\dfrac{2}{p}2\pi f - \omega_m}{\dfrac{2}{p}2\pi f} \qquad \textbf{[6.53]}$$

For operation at constant stator frequency (f), it is apparent from [6.53] that if speed (ω_m) were held constant, then slip (s) is a constant. It can then be concluded from [6.52] that for a particular value of speed, the total developed torque varies as V_{Th}^2. The torque produced by an induction motor is quite sensitive to changes in terminal voltage.

Condition for Zero Torque Multiply both the numerator and denominator of [6.52] by s^2 to yield

$$3T_d = \frac{3sV_{Th}^2 R'_2}{\omega_s\left[(sR_{Th} + R'_2)^2 + s^2(X_{Th} + X'_2)^2\right]} \qquad \textbf{[6.54]}$$

Setting $3T_d = 0$ in [6.54] leads to

$$0 = 3sV_{Th}^2 R_2'$$

Considering $R_2' = 0$ unrealistic and $V_{Th} = 0$ of no concern, it is apparent that $3T_d = 0$ for $s = 0$ ($\omega_m = \omega_s$). The fact that torque is zero at synchronous speed is corroborated by inspection of Fig. 6.21.

Small Slip Approximation Consider s small, but not zero, in [6.54]. Then,

$$3T_d \cong \frac{3sV_{Th}^2}{\omega_s R_2'} \qquad \qquad \textbf{[6.55]}$$

The speed-torque curve is nearly a straight-line segment for small values of slip, that is, near synchronous speed. Practically, the error in torque determined by [6.55] is less than 2 percent for $-0.02 < s < 0.02$.

Extend [6.55] to estimate the speed at which the induction motor of Example 6.7 produces an output power of 5 hp. Neglect P_{FW} in the prediction determination. | **Example 6.12**

Since $\omega_m = (1 - s)\omega_s$, [6.55] can be written as

$$3T_d \omega_m \cong \frac{3sV_{Th}^2}{\omega_s R_2'}(1 - s)\omega_s \cong \frac{3sV_{Th}^2(1 - s)}{R_2'}$$

The value of slip is small enough that $(1 - s) \cong 1$, or

$$P_s \cong 3T_d \omega_m \cong \frac{3sV_{Th}^2}{R_2'}$$

If the power is to be halved from 10 to 5 hp, then the slip must be reduced to half of the rated load value.

$$n_m \cong (1 - s)n_s = \left(1 - \frac{0.052}{2}\right)1200 = 1169 \text{ rpm}$$

Critical Points of Torque The maximum and minimum points for the speed-torque curve follow from differentiating [6.54] with respect to s and equating the result to zero.

$$\frac{\partial(3T_d)}{\partial s} = 0 = \frac{\partial}{\partial s}\left[\frac{3sV_{Th}^2 R_2'}{\omega_s[(sR_{Th} + R_2')^2 + s^2(X_{Th} + X_2')^2]}\right]$$

$$= \frac{\partial}{\partial s}\left[\frac{s}{(sR_{Th} + R_2')^2 + s^2(X_{Th} + X_2')^2}\right]$$

Carrying out the indicated differentiation and multiplying through by the resulting denominator yields

$$0 = (sR_{Th} + R_2')^2 + s^2(X_{Th} + X_2')^2 - s[2R_{Th}(sR_{Th} + R_2') + 2s(X_{Th} + X_2')^2]$$

After simplification of the above quadratic equation in s,

$$0 = s^2[R_{Th}^2 + (X_{Th} + X_2')^2] - (R_2')^2$$

or

$$s_{max} = s = \pm \frac{R_2'}{[R_{Th}^2 + (X_{Th} + X_2')^2]^{1/2}} \qquad \textbf{[6.56]}$$

Obviously, the positive value of s_{max} (s_{max+}) corresponds to the speed-torque curve in the motor mode of Fig. 6.22 while the negative value of s_{max} (s_{max-}) applies to the generator mode.

Equation [6.54] can be rearranged to yield

$$3T_d = \frac{3sV_{Th}^2 R_2'/\omega_s}{s^2[R_{Th}^2 + (X_{Th} + X_2')^2] + 2sR_{Th}R_2' + (R_2')^2} \qquad \textbf{[6.57]}$$

Substituting s_{max+} and s_{max-} as given by [6.56] into [6.57], respectively, and simplifying the two results give the motor mode and generator mode critical torque points as

$$3T_{dmax+} = \frac{3V_{Th}^2}{2\omega_s[R_{Th} + [R_{Th}^2 + (X_{Th} + X_2')^2]^{1/2}]} \qquad \textbf{[6.58]}$$

$$3T_{dmax-} = \frac{3V_{Th}^2}{2\omega_s[R_{Th} - [R_{Th}^2 + (X_{Th} + X_2')^2]^{1/2}]} \qquad \textbf{[6.59]}$$

The denominator of [6.59] has a negative value resulting in $3T_{dmax-} < 0$ as expected from Fig. 6.22. Further, the magnitude of the denominator of [6.59] is less than the denominator of [6.58]; consequently, $|3T_{dmax-}| > 3T_{dmax+}$, which is consistent with the observed maximum and minimum values of Fig. 6.22.

Example 6.13

For the induction motor of Example 6.7, determine the maximum torque that can be developed in the motor mode and the associated speed.

By [6.58],

$$3T_{dmax+} = \frac{3V_{Th}^2}{2\omega_s[R_{Th+}[R_{Th}^2 + (X_{Th} + X_2')^2]^{1/2}]}$$

$$3T_{dmax+} = \frac{3(130.1)^2}{2(40\pi)[0.145 + [(0.145)^2 + (0.49 + 0.5)^2]^{1/2}]} = 176.4 \text{ N·m}$$

From use of [6.56],

$$s_{max+} = \frac{R_2'}{[R_{Th}^2 + (X_{Th} + X_2')^2]^{1/2}} = \frac{0.3}{[(0.145)^2 + (0.49 + 0.5)^2]^{1/2}} = 0.301$$

$$n_{mmax} = (1 - s_{max+})n_s = (1 - 0.301)(1200) = 838.8 \text{ rpm}$$

Rotor Coil Resistance Impact Inspection of [6.58] shows that the value of maximum torque ($3T_{d\max+}$) is independent of rotor coil resistance R_2'. However, the value of $s_{\max}$ from [6.56] is directly dependent on R_2'. Thus, it is concluded that an increase in R_2' results in a direct increase in the value of slip at which maximum torque occurs, but the value of maximum torque itself is unchanged.

Example 6.14

Using the MATLAB program ⟨Te_wm.m⟩, plot the speed-torque curves for the induction motor of Example 6.5 with values of R_2' being 0.12528 Ω and 1.2528 Ω.

  Several approaches are possible by modification of the code. A reasonably simple approach is chosen and implemented as follows. Edit ⟨Te_wm.m⟩ to remove the statement defining "R2pr = 0.12528" as this value will be defined from a screen session. Comment out (add leading %) to "clear;" and the four lines of code to display Thevenin values to the screen. Then invoke MATLAB and carry out the screen session shown below. The second plot to appear (see Fig. 6.23) will display the two speed-torque curves corresponding to the two values of R_2'.

```
    R2pr=0.12528;
    Te_wm
    hold

  Current plot held

    R2pr=10*R2pr;
    Te_wm
```

  From Fig. 6.23, it is apparent that increasing the value of R_2' by a factor of 10 has increased the value of $s_{\max+}$ by an order of magnitude as clearly predicted by [6.56]. Although Example 6.7 must be considered an impractical exercise in that rotor coil resistance of a squirrel-cage induction motor cannot be operationally altered except for temperature effects, practical realization of effective rotor coil resistance can be implemented for the slip-ring induction motor by adding external resistance as illustrated by Fig. 6.24. For fixed voltage-frequency operation of an induction motor, increased effective rotor coil resistance allows the motor to produce a larger torque at low speeds in applications such as cranes or hoists; as the motor accelerates, the external resistance is reduced, allowing the motor to operate at increased speed while carrying the high torque load. However, present day technology with variable-voltage, variable-frequency inverters can produce high torque at low speed using a squirrel-cage induction motor. Consequently, the market for wound-rotor induction motors is vanishing.

6.7.4   FREQUENCY SENSITIVITY OF ROTOR PARAMETERS

NEMA design classifications are assigned to induction motors according to the minimum allowable ratios of breakdown (maximum) torque to rated torque and

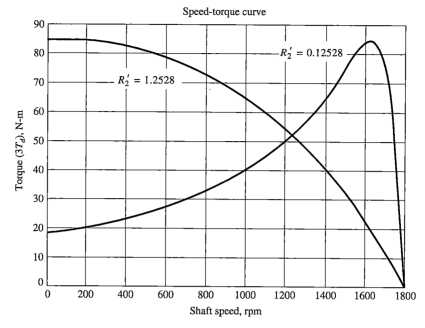

Figure 6.23
Effect of increasing R_2'

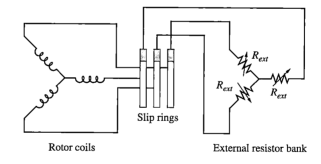

Figure 6.24
External adjustment of effective R_2'

starting torque to rated torque. The ratio values can be found in the *NEMA Standard MG-1* and vary some according to poles and frequency. Typical values are listed in Table 6.3. In addition, NEMA design A, B, and C motors must have a full-load slip of less than 5 percent; full-load slip for NEMA design D must be 5 percent or more.

The NEMA design B motor of Example 6.5 has the speed-torque curve as calculated in Example 6.6. Referring to the associated plot of Fig. 6.21, the 10-hp rated torque of 40.6 N-m has been indicated by a dot. It is seen that the maximum torque satisfies NEMA design B; however, the value of starting torque $(n_m = 0)$ is less than 50 percent of the rated torque value. The disparity stems

Table 6.3 NEMA design torques		
NEMA Design	Max. Rated	Starting Rated
A	225%	150%
B	200%	150%
C	190%	200%
D	225%	225%

from the fact that the analysis of Example 6.6 treated the rotor parameters as constant values when, in reality, R_2' and X_2' are sensitive to changes in rotor frequency. Values determined from the blocked rotor test at a reduced frequency of 15 Hz are not significantly different from dc values; however, as rotor electrical frequency f_r increases due to decreasing rotor mechanical speed, the combination of skin effect and cross-slot flux can significantly alter the portions of R_2' and X_2' associated with the large cross-section rotor bar. This phenomenon will be specifically addressed in Sec. 6.11, but for the present, an empirical adjustment is introduced as follows that is suitable for use with NEMA design B motors:

$$R_2' = \left[0.5 + 0.5\left(\frac{s}{s_{max}}\right)^{1/2}\right]R_{20}' \qquad \textbf{[6.60]}$$

$$X_2' = \left[0.4 + 0.6\left(\frac{s_{max}}{s}\right)^{1/2}\right]X_{20}' \qquad \textbf{[6.61]}$$

where R_{20}' and X_{20}' are the low-frequency values from the blocked rotor test and s_{max} is determined from [6.56] using $X_2' = X_{20}'$ and $R_2' = R_{20}'$.

The MATLAB code of ⟨Te_wm.m⟩ has a section that is commented out to implement the above frequency sensitivity of R_2' and X_2'. Activating this section of code and executing ⟨Te_wm.m⟩ produces the speed-torque curve of Fig. 6.25 for the motor of Example 6.5. Inspection of Fig. 6.25 shows that the resulting increase of R_2' and reduction of X_2' at higher values of rotor electrical frequency f_r has produced the 150 percent value of starting torque necessary to satisfy NEMA design classification.

6.7.5 MACHINE PERFORMANCE EVALUATION

The Thevenin equivalent circuit of Fig. 6.20 was developed in Sec. 6.7.2 to ease the task of repetitive longhand torque calculations. Inherent in the development, the phase input terminals lose their identity. Hence, input current, input power factor, and segregated loss calculations are not possible directly from the Thevenin equivalent circuit. Although cumbersome to execute by longhand procedure, such

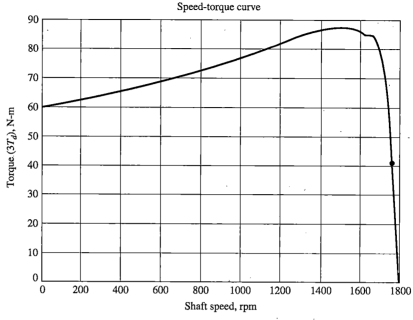

Figure 6.25
Speed-torque curve with frequency sensitivity of R'_2 and X'_2

calculations are best carried out using the equivalent circuit of Fig. 6.17. However, MATLAB based analyses using the circuit of Fig. 6.17 are quite simple. The MAT-LAB program ⟨im_perf.m⟩ is a code to calculate the values of input current, input power factor, total developed torque, output power, and efficiency based on the per phase equivalent circuit of Fig. 6.17 over the speed range from $\omega_m = 0$ to $\omega_m \cong \omega_s$. Empirical adjustment of R'_2 and X'_2 for changes in rotor frequency is included. The particular equation set implemented is as follows.

Input impedance:

$$\mathbf{Z}_{in} = R_1 + jX_1 + R_c \| jX_m \| \left(\frac{R'_2}{s} + jX'_2 \right) \qquad \textbf{[6.62]}$$

Input current:

$$\bar{I}_1 = \frac{\bar{V}_1}{\mathbf{Z}_{in}} \qquad \textbf{[6.63]}$$

Rotor current:

$$\bar{I}'_2 = \frac{R_c \| jX_m}{R_c \| jX_m + \left(\dfrac{R'_2}{s} + jX'_2 \right)} \bar{I}_1 \qquad \textbf{[6.64]}$$

Total input power:

$$P_T = 3V_1 I_1 \cos\left[\angle V_1 - \angle I_1\right] \qquad \text{[6.65]}$$

Total developed torque:

$$3T_d = \frac{3(I_2')^2 \dfrac{R_2'}{s}}{\omega_s} \qquad \text{[6.66]}$$

Shaft speed:

$$n_m = (1 - s)n_s \qquad \text{[6.67]}$$

F and W losses:

$$P_{FW} = P_{FWs}\left(\frac{n_m}{n_s}\right)^n \qquad \text{[6.68]}$$

where P_{FWs} = the friction and windage losses at synchronous speed
n = a suitable exponent (2.5 to 3)

Total output power:

$$P_s = 3T_d \omega_m - P_{FW} \qquad \text{[6.69]}$$

Execution of ⟨im_perf.m⟩ with the equivalent circuit parameter values of Example 6.5 yields the family of plots shown by Fig. 6.26. Rated operation has been indicated on each curve by a dot. At least three typical characteristics of the induction motor can be observed by inspection of these plots:

1. Starting current is commonly 5 to 6 times nameplate rated value.

2. Maximum efficiency occurs at near rated load and reduces significantly for light load.

3. Power factor is near peak value at rated load and reduces significantly for light load.

6.8 REDUCED VOLTAGE STARTING

Input current of five or more times the nameplate rated value is common when an induction motor is started with rated voltage applied. Figure 6.26 corroborates this claim for the example 10-hp motor. The encountered excessive current can lead to damaging heat buildup in the motor when the combination of inertia and coupled load characteristics extends the acceleration interval. However, the more objectionable result is the adverse effect on neighboring equipment due to increased voltage drop in the feeder lines. For a large induction motor that can require several seconds

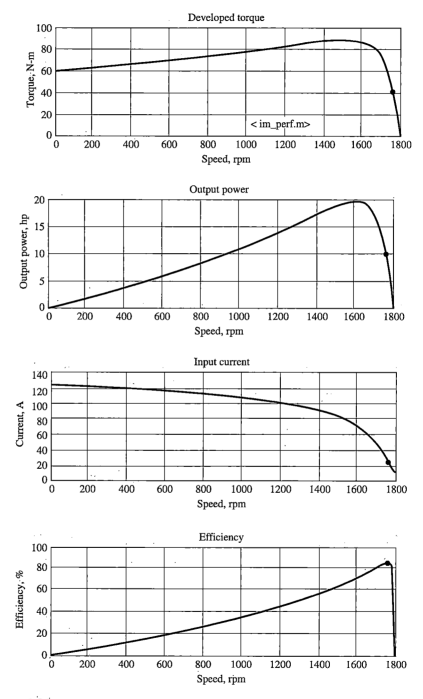

Figure 6.26
Induction motor performance curves

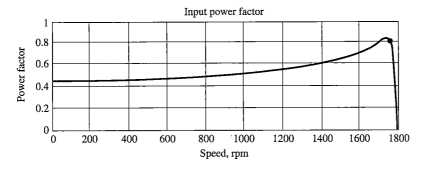

Figure 6.26 (Continued)
Induction motor performance curves

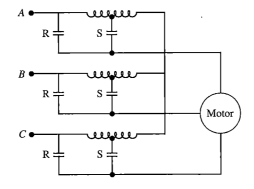

Figure 6.27
Autotransformer starting

to accelerate to operating speed, reduced voltage starting schemes are commonly implemented. Although numerous approaches are found in practice, only the two methods that do not depend on a particular machine winding or access to the individual phase windings will be presented.

6.8.1 AUTOTRANSFORMER STARTING

The autotransformer is suitable for use in the reduced voltage starting of induction motors since it is a high power density device and there is no issue of personal safety in the event of a winding failure for this application. Figure 6.27 schematically shows the connection scheme. Initially, contactors S are closed so that a step-down voltage is applied to the motor. At reduced voltage, the motor torque is reduced over the value available for full voltage and the acceleration time is thus increased. The effective turns ratio of the autotransformer is selected as a compromise that may result in current above the desired level at low speeds and below the desired level at higher speeds. However, the current is reduced to an acceptable level to avoid detrimental voltage dip of the feeder lines. After the motor accelerates to near operating speed, contactors S are opened and contactors R are closed, impressing rated voltage across the motor terminals. The schematic shown leaves the autotransformer energized. If the motor is started infrequently, an extra set of contactors may be installed to de-energize the autotransformer for the purpose of eliminating its core loss energy usage during its period of inactivity.

6.8.2 SOLID-STATE STARTER

With advancement in power semiconductor technology, the solid-state motor starter of Fig. 6.28 has become increasingly popular. The thyristors are unidirectional conduction devices that enter conduction state only when forward biased and a gating signal is applied from control circuitry. By delay of the gating signal, the thyristors conduct only over a portion of the available supply voltage cycle; thus, the magnitude of the voltage impressed on the motor terminals is reduced. With current sensing used to control the degree of delay in the gating signal, the motor current can be controlled to any desired value over the entire acceleration interval of the motor.

6.9 SPEED CONTROL

The developed torque-speed curve of an induction motor fed by constant-frequency voltage is anchored to the zero crossing point at synchronous speed ($\omega_s = (2/p)\omega = 4\pi f/p$). Limited speed changes can be accomplished by altering the slope of the torque-speed curve between synchronous speed and peak torque; however, wide-range speed control can be implemented only by change of the synchronous speed by varying the frequency (f) of the impressed voltage or altering the pole count (p). For simplicity in the discussion that follows, it will be assumed that $3T_d \cong T_s$.

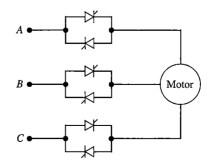

Figure 6.28
Solid-state starter

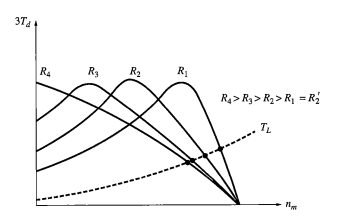

Figure 6.29
Rotor resistance control of speed

6.9.1 ROTOR RESISTANCE CONTROL

An increase in the value of rotor circuit resistance is practical only for a wound-rotor induction motor wherein external resistance can be introduced through the slip-ring connection. This concept was introduced in Example 6.14, but the result focused on increase of low-speed torque. Figure 6.29 presents the torque-speed characteristics of a wound-rotor induction motor determined by increasing the rotor circuit resistance from the value of R_2'. The motor can operate at the speed determined by the intersection of a particular torque-speed characteristic and the superimposed load torque curve labeled T_L.

Although the speed of the wound-rotor induction motor can be controlled by rotor resistance change, it is not without penalty. Equation [6.45] shows that rotor ohmic losses are given by the product of slip and air gap power. Although the increased rotor circuit resistance leads to a decrease in air gap power, the slip (s) increases with increase in rotor circuit resistance. It must be concluded that a larger portion of the air gap power is dissipated as ohmic losses when rotor circuit resistance is increased to control speed and a smaller portion is converted to developed power. Consequently, the efficiency of the motor decreases as the speed is reduced by means of rotor circuit resistance control.

6.9.2 VOLTAGE CONTROL

Small-range speed control of the squirrel-cage induction motor can be implemented by decrease in the magnitude of input voltage. In Sec. 6.7.3, it has been concluded that the induction motor developed torque is proportional to the square of input voltage magnitude at any particular speed. Figure 6.30 is a plot of the family of torque-speed curves that result if the voltage is reduced from rated value. A load torque characteristic T_L has been superimposed on the family of curves

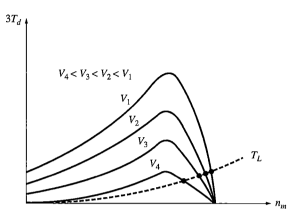

Figure 6.30
Voltage control of speed

where it is clearly seen that limited range speed control results. However, it is noted that the slip increases as speed is decreased just as in the case of rotor resistance control. Hence, efficiency decreases with decrease in speed, and overheating of the motor is a potential problem.

Voltage control of speed can be implemented by supplying the stator winding from the secondary of an adjustable autotransformer. Present-day implementation, however, commonly applies the same circuit technology of the solid-state starter of Fig. 6.28. In this case, the thyristor circuitry is known as a *voltage controller* and the gating signals are controlled to yield a command voltage rather than maintain a current limit.

6.9.3 Pole Changing

Changing the number of poles provides a means to set a different synchronous speed and thus significantly alter the torque-speed characteristic of a squirrel-cage induction motor. The basic concept of the commonly implemented pole-changing scheme is one that simply switches the connection pattern of the phase-groups of coils from alternating magnetic polarity to a pattern that establishes a single magnetic polarity. Since lines of flux must close on themselves, magnetic poles of opposite polarity (called *consequent poles*) appear midway between each coil phase-group as illustrated by Fig. 6.31. The squirrel-cage rotor inherently establishes a number of poles equal to the stator. The consequent pole connection has twice the number of magnetic poles; thus, its synchronous speed is half of that of the normal alternating polarity connection. Obviously, there is close to a factor of 2 in the motor shaft speed between the two connections.

The consequent pole scheme illustrated by Fig. 6.31 has a single path per phase for both connections. The consequent pole case has some increase in leakage reactance over the alternating polarity connection. As a result, the maximum

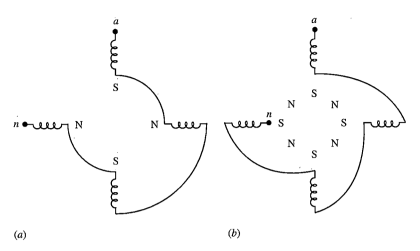

Figure 6.31
Pole changing. (a) Conventional connection. (b) Consequent connection.

torque value will be slightly, but not significantly, less for the consequent pole connection. Since the two maximum torque values are close to the same, the scheme is known as a *constant-torque pole-changing connection.* Two other consequent pole schemes are possible—series to parallel wye connection change (variable torque) and wye to delta connection change (constant horsepower).[3]

6.9.4   FREQUENCY CONTROL

Variable stator frequency control offers the option of changing synchronous speed continuously rather than the discrete single step resulting from pole changing. In addition, frequency control allows low-speed operation. Further, if properly implemented it does not result in the efficiency penalty that arises with both voltage and rotor circuit resistance methods of speed control.

An issue arises in frequency control that has not been a concern in the speed control methods to this point. The potential exists to raise the flux density to high saturation levels, thereby resulting in a large magnetizing current. Due to saturation of the ferromagnetic material, the air gap flux does not increase proportional to the increase in magnetizing current. Consequently, the developed torque does not increase significantly; however, the increased magnetizing current does increase the ohmic losses of the stator windings, usually leading to overheating. Operation of an induction motor with significant saturation is a condition to be avoided.

Insight into the potential saturation problem with frequency control is gained by considering the approximate equivalent circuit of Fig. 6.32. This circuit is obtained from Fig. 6.17 by neglecting the core loss resistor, moving the magnetizing reactance to the input terminals, and replacing any X by ωL. The magnetizing current I_m establishes the mutual flux that links both stator and rotor coils. An

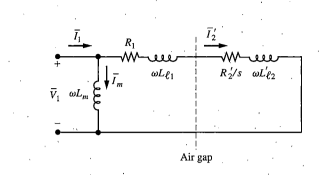

Figure 6.32
Approximate per phase equivalent circuit of induction motor

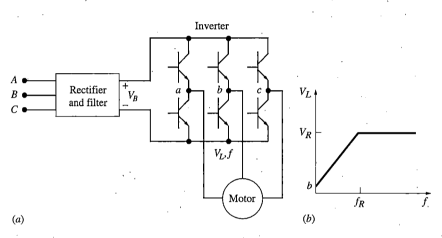

(a)   (b)

Figure 6.33
Variable frequency speed control. (a) Inverter. (b) V_L-f control.

increase in I_m above the value that exists for rated voltage and frequency acts to increase the air gap flux density above the value for which the motor has been designed. The magnitude of the magnetizing current in Fig. 6.32 is

$$I_m = \frac{V_1}{\omega L_m} = \frac{V_L/\sqrt{3}}{2\pi f L_m} = \frac{0.092}{L_m}\left(\frac{V_L}{f}\right)$$ **[6.70]**

whence it is apparent that as frequency is changed, the ratio V_L/f should not exceed the ratio for rated voltage and frequency if magnetic saturation is to be avoided. Further, if V_L/f is maintained at the value for rated conditions, the air gap flux will remain at the rated value, assuring that the motor torque-producing capability is not diminished. Attainment of this goal means that with introduction of frequency control, the ability to control the magnitude of the motor terminal voltage must also be implemented.

The schematic of Fig. 6.33a introduces the circuit topology of a three-phase inverter for variable frequency control. The transistors are operated in the switch

Table 6.4
Six-step voltages

v_{ab}	v_{bc}	v_{ca}
V_B	0	$-V_B$
0	V_B	$-V_B$
$-V_B$	V_B	0
$-V_B$	0	V_B
0	$-V_B$	V_B
V_B	$-V_B$	0

mode. By selection of various conduction states for the six transistors, two popular output voltage waveforms can be generated—six-step and sinusoidal pulse-width modulation (SPWM). In both cases, the rate at which the transistor base drive circuitry passes through a complete cycle of conduction events establishes the frequency of the output voltage waveforms.

The six-step inverter control maintains the conduction states for equal increments per cycle to produce the output line voltages of Table 6.4. Notice that this line voltage set has positive peak values that are offset by $T/6$ or 120°. Figure 6.34 displays a plot of the line voltage v_{ab} from Table 6.4 along with the associated phase voltage and the Fourier spectrum of the phase voltage. It is seen that this phase voltage has fifth and seventh harmonics that may be of concern; however, the coil pitch and distribution factors are commonly chosen to minimize the effect of fifth and seventh harmonics. The rectifier of Fig. 6.33a must be a phase-controlled converter so that V_B can be varied to control the magnitude of the six-step inverter output voltages.

The SPWM inverter control switches the transistors so that the potential of node a of Fig. 6.33a has the bistate values indicated by Fig. 6.35. Nodes b and c are switched to form identical waveforms except for respective shifts of 120° and 240°. The phase voltage associated with this node switching control is shown in Fig. 6.36 along with the resulting Fourier spectrum. The SPWM output voltage has an interesting spectral characteristic. It is essentially void of harmonics of order less than $2N - 1$, where N is the number of pulses per half cycle of the node switching pattern. Although there are significant voltage harmonics in the neighborhood of $2N$, if N is chosen large enough, the motor windings present a sufficiently high impedance to any voltage of this frequency that negligible current results.

The switching points of Fig. 6.35 are determined by crossing points of a triangular waveform with frequency $2Nf$ and a reference sinusoidal waveform with frequency f, where f is the frequency of the output voltages. The crossing points may be determined by analog signals or by an algorithm. In either case, the triangular wave is constant in amplitude A_c. The reference sine wave is variable in amplitude A. The ratio A/A_c is known as the *modulation index* (MI). Change in

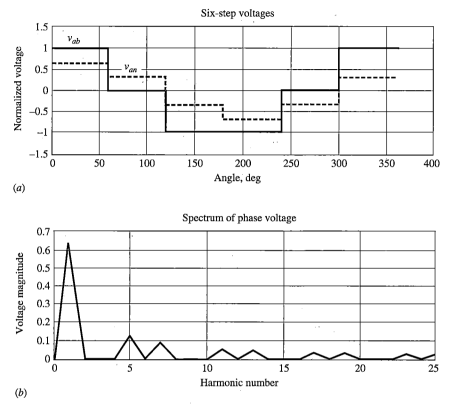

Figure 6.34
Six-step voltages

the modulation index alters the crossing points of the triangular wave and the carrier wave. If MI decreases, the width of the pulses of Fig. 6.35 decreases. Consequently, the width of the pulses of the phase voltage waveform of Fig. 6.36 decreases and the magnitude of the fundamental component of phase voltage decreases. The necessary voltage magnitude control is attained by this means, so that the dc voltage V_B of Fig. 6.33a can be constant in value. The rectifier in this case is simply a diode bridge.

The MATLAB program ⟨Td_nmvf.m⟩ has been formulated to plot the speed-torque curve and rated conditions and then to superimpose plots of the stable portion of the speed-torque curves resulting from variable voltage, variable frequency operation. Control of the ratio V_L/f is provided in the program; however, at low frequency values where R_1 is comparable in value to $\omega L_{\ell 1}$, the value of V_L/f has to be increased above the ratio of rated values or else the developed torque decreases significantly. The practical V_L/f control is that shown by Fig. 6.33b. As a consequence of this control, the air gap flux density at low speeds may exceed design value. Typically, the motor only operates at this con-

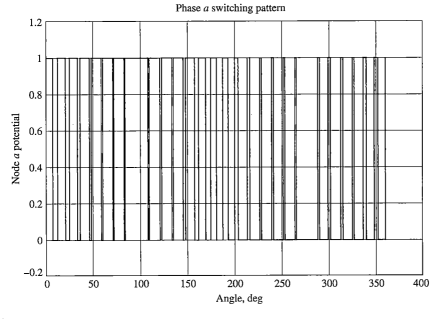

Phase *a* switching pattern

Figure 6.35
Bi-state switching

dition during start-up, so the overheating of stator coils due to the increased magnetizing current is not an issue. It is also noted from Fig. 6.33*b* that the voltage V_L is not allowed to exceed rated value if frequency were raised beyond rated value. This is to avoid dielectric stress of the insulation. The curves of Fig. 6.37 show the resulting speed-torque curve for operation of an induction motor defined by the supplied data while operating over a range of frequency from 5 to 70 Hz.

A load requiring 100 N·m torque at 1500 rpm is to be driven by an induction motor under volts per hertz control with the performance characteristic of Fig. 6.37. Assume that rotational losses are negligible. Determine (*a*) the value of stator frequency and (*b*) the value of input line voltage if the control is described by Fig. 6.33*b*, where *b* is 6 percent of rated line voltage. **| Example 6.15**

(*a*) Locate the load point for $n_m = 1500$ and $3T_d = 100$ on Fig. 6.37. It is judged that this point lies approximately 40 percent of the distance between the 50-Hz and 55-Hz speed-torque curves; thus,

$$f = 50 + 0.4(5) = 52 \text{ Hz}$$

(*b*) From the input data of ⟨Td_nmvf.m⟩, it is found that rated voltage and frequency are given by $V_R = 460$ and $f_R = 60$. The equation taken from the MATLAB code that describes the voltage-frequency control of Fig. 6.33*b* is used to yield

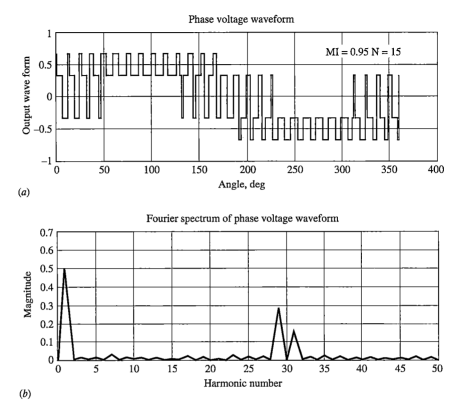

Figure 6.36
Phase voltage for SPWM

$$V_L = \frac{(V_R - b)f}{f_R} + b = \frac{(460 - 0.06 \times 460)(52)}{60} + 0.06(460)$$

$$V_L = 402.35 \text{ V}$$

6.10   SINGLE-PHASE MOTORS

The single-phase, squirrel-cage induction motor with distributed stator windings is structurally similar to the three-phase induction motor. The significant difference is in the stator winding arrangement. The single-phase motor stator windings are effectively distributed, pitched, and skewed to produce a sinusoidal mmf in space. However, a single-layer, concentric coil arrangement is typically used because of its suitability for automated winding. Further, there is only one phase winding per pole; consequently, the air gap mmf is a standing wave as described by [6.3].

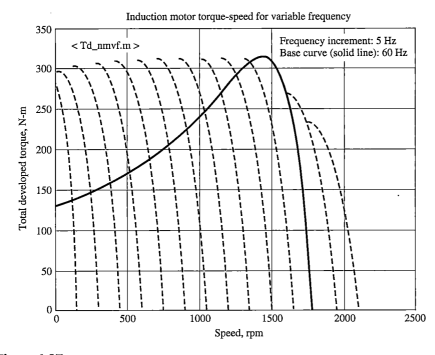

Induction motor torque-speed for variable frequency

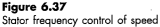

Frequency increment: 5 Hz
Base curve (solid line): 60 Hz

Figure 6.37
Stator frequency control of speed

6.10.1 AIR GAP FIELD

If the current $i_a = I_m \sin(\omega t)$ is substituted into [6.3], the stator-produced (rotor electrically inactive) air gap mmf of the single-phase induction motor becomes

$$\mathcal{F}_1 = \frac{2}{\pi} N_{1eff} I_m \sin(\omega t) \sin\left(\frac{p}{2}\theta\right) = \mathcal{F}_m \sin(\omega t) \sin\left(\frac{p}{2}\theta\right) \quad \textbf{[6.71]}$$

By use of the trigonometric identity $\sin x \sin y = \frac{1}{2}\cos(x - y) - \frac{1}{2}\cos(x + y)$, [6.71] can be written as

$$\mathcal{F}_1 = \mathcal{F}_f + \mathcal{F}_b = \frac{1}{2}\mathcal{F}_m \cos\left(\omega t - \frac{p}{2}\theta\right) - \frac{1}{2}\mathcal{F}_m \cos\left(\omega t + \frac{p}{2}\theta\right) \quad \textbf{[6.72]}$$

Interpreting [6.72] in light of [6.10] and [6.11], it is concluded that the stator-produced standing mmf wave has been decomposed into two traveling mmf waves with amplitudes equal to half the peak value of the standing wave. Further, the traveling waves have angular speeds that are equal in magnitude, but opposite in direction. This decomposition into two oppositely revolving mmf waves is often referred to as *double revolving field theory*. The angular speeds of $\mathcal{F}_f$ and $\mathcal{F}_b$ are, respectively,

$$\omega_{sf} = \frac{2}{p}\omega \qquad \omega_{sb} = -\frac{2}{p}\omega \qquad\qquad \textbf{[6.73]}$$

The MATLAB program ⟨spwave.m⟩ simulates the individual forward and backward air gap traveling waves and sums the two waves to produce the resulting pulsating wave for any even integer number of poles of choice. The results are displayed in a strobe manner for one magnetic pole-pair pitch of movement. Figure 6.38 shows the final frame of the animated simulation for an eight-pole machine.

6.10.2 EQUIVALENT CIRCUIT

There is only a single mmf established by the excited stator coil and, thus, only a single pulsating flux exists. However, one-half of the mmf, hence one-half of the turns, are associated with each of the forward and backward mmf components of [6.72]. Consequently, one-half of the flux linkages of the coil are associated with each $\mathscr{F}_f$ and $\mathscr{F}_b$. If $\mathscr{F}_f$ and $\mathscr{F}_b$ are thought of as mmf's of two separate, series-connected coils, the stator of the single-phase induction motor can be modeled as

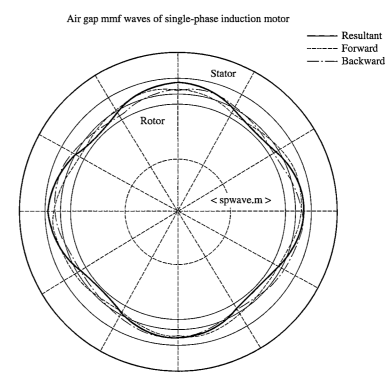

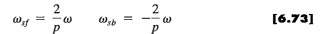

Figure 6.38
Simulation of air gap traveling mmf waves of single-phase induction motor

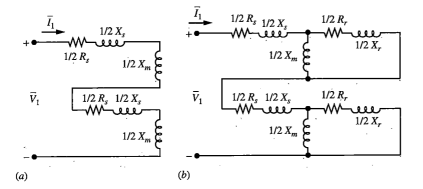

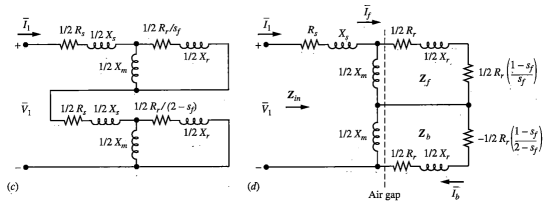

Figure 6.39
Single-phase induction motor equivalent circuits. (a) Rotor electrically inactive. (b) Blocked rotor. (c) Any rotor speed. (d) Rearrangement of (c).

shown in Fig. 6.39a. For the case of half the rotor turns carrying the full rotor flux and acting as shorted transformer secondaries coupled to the two stator coils, the blocked rotor equivalent circuit of Fig. 6.39b results.

A set of slips can be defined for both the forward- and backward-revolving fields as

$$s_f = \frac{\omega_{sf} - \omega_m}{\omega_{sf}} \qquad [6.74]$$

$$s_b = \frac{\omega_{sb} - \omega_m}{\omega_{sb}} = \frac{-\omega_{sf} - \omega_m}{-\omega_{sf}} \qquad [6.75]$$

Equations [6.74] and [6.75] can each be solved for ω_m and the resulting expressions equated to yield

$$s_b = 2 - s_f \qquad [6.76]$$

By analogy with the work of Sec. 6.5, the upper and lower values of $\frac{1}{2}R_r$ in Fig. 6.39b can be replaced, respectively, by $\frac{1}{2}R_r/s_f$ and $\frac{1}{2}R_r/s_b = \frac{1}{2}R/(2 - s_f)$ to yield the equivalent circuit of Fig. 6.39c that is valid for any speed ω_m. Rearrangement of the circuit results in the common equivalent circuit of the single-phase induction motor shown in Fig. 6.39d.

6.10.3 PERFORMANCE NATURE

The developed torque can be calculated directly from the equivalent circuit as the power delivered to the energy conversion resistances divided by mechanical speed, giving

$$T_d = T_{df} + T_{db} = \frac{\frac{1}{2}I_f^2 R_r \left(\dfrac{1 - s_f}{s_f}\right)}{\omega_m} - \frac{\frac{1}{2}I_b^2 R_r \left(\dfrac{1 - s_f}{2 - s_f}\right)}{\omega_m} \qquad \textbf{[6.77]}$$

The first term on the right-hand side of [6.77] is the torque (T_{df}) produced by the forward-revolving field while the second term is the torque (T_{db}) resulting from the backward-revolving field. Or, the developed torque can alternately be found as the sum of the power across the air gap divided by the associated synchronous speed.

$$T_d = T_{df} + T_{db} = \frac{\frac{1}{2}I_f^2 R_r/s_f}{\omega_{sf}} + \frac{\frac{1}{2}I_b^2 R_r/(2 - s_f)}{\omega_{sb}} \qquad \textbf{[6.78]}$$

For both cases, the second term (T_{db}) is a negative quantity reflecting the fact that the backward-revolving field results in a torque that acts against the direction of rotation.

The expressions for developed torque above were derived from average values of power. Hence, the expressions for torque yield average values. Although the details are beyond the scope of this book, it is stated that the single-phase induction motor produces an instantaneous torque that is time-varying.

The MATLAB program ⟨torq1ph.m⟩ calculates the values of T_{df}, T_{db}, and T_d using the equivalent circuit parameters. Figure 6.40 displays the plotted results for a 4-pole, 120-V, 60-Hz single-phase induction motor with the following equivalent circuit values: $R_s = 2.02\,\Omega$, $R_r = 4.12\,\Omega$, $X_s = 2.79\,\Omega$, $X_r = 2.12\,\Omega$, and $X_m = 106.8\,\Omega$. It is readily apparent from the graph that the single-phase induction motor has no starting torque; however, if the motor were to attain a nonzero speed, it does display an average torque. Further, the motor torque-speed curve displays symmetry between the first and third quadrants, leading to the conclusion that operation in either direction of rotation is possible.

Analysis of the single-phase induction motor is slightly more cumbersome than for the three-phase motor case due to a second rotor loop in the equivalent

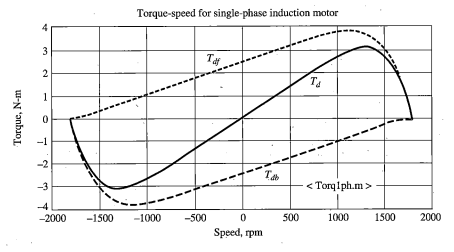

Figure 6.40
Developed torque for a single-phase induction motor

circuit of Fig. 6.39*d*. However, the concept is simple enough as outlined by the following procedure:

$$\mathbf{Z}_f = j\frac{X_m}{2} \,\|\, \left(\frac{1}{2}\frac{R_r}{s_f} + j\frac{1}{2}X_r \right) \qquad \textbf{[6.79]}$$

$$\mathbf{Z}_b = j\frac{X_m}{2} \,\|\, \left(\frac{1}{2}\frac{R_r}{2 - s_f} + j\frac{1}{2}X_r \right) \qquad \textbf{[6.80]}$$

$$\mathbf{Z}_s = R_s + jX_s \qquad \textbf{[6.81]}$$

$$\mathbf{Z}_{in} = \mathbf{Z}_s + \mathbf{Z}_f + \mathbf{Z}_b \qquad \textbf{[6.82]}$$

$$\bar{I}_1 = \frac{\bar{V}_1}{\mathbf{Z}_{in}} \qquad \textbf{[6.83]}$$

By current division,

$$\bar{I}_f = \frac{j\frac{1}{2}X_m}{\mathbf{Z}_f + jX_m}\bar{I}_1 \qquad \textbf{[6.84]}$$

$$\bar{I}_b = \frac{j\frac{1}{2}X_m}{\mathbf{Z}_b + jX_m}\bar{I}_1 \qquad \textbf{[6.85]}$$

With I_f and I_b determined, the developed torque (T_d) can be readily calculated by [6.77] or [6.78]. The MATLAB program ⟨sp_perf.m⟩ carries out the above calculations for an array of speed values and plots the resulting developed electromagnetic

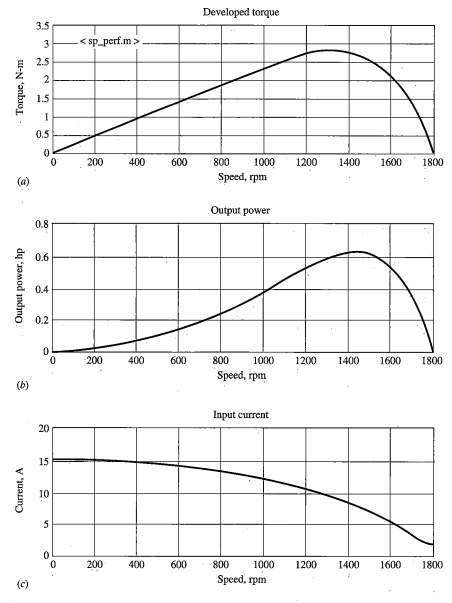

Figure 6.41
Performance of single-phase induction motor

torque, output power, input current, efficiency, and power factor. The resulting performance nature is displayed by Fig. 6.41 for a 4-pole, 120-V, 60-Hz single-phase induction motor characterized by $R_s = 2.02\Omega$, $R_r = 4.12\Omega$, $X_s = 2.79\Omega$, $X_r = 2.12\Omega$, and $X_m = 106.8\Omega$.

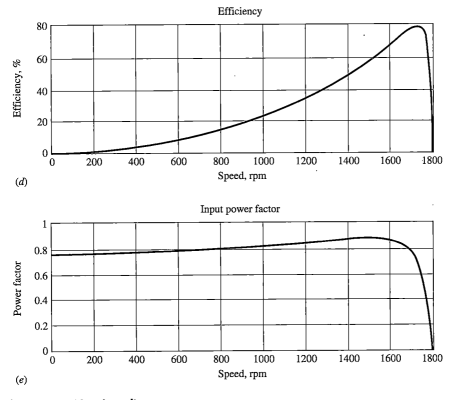

Figure 6.41 (Continued)
Performance of single-phase induction motor

6.10.4 AUXILIARY STARTING WINDING

It was concluded above that the single-phase induction motor has zero starting torque from inspection of Fig. 6.40. The same conclusion is drawn if $s_f = 1$ ($\omega_m = 0$) is substituted into [6.77]. Some method must be devised to accelerate the single-phase motor to a speed where it has sufficient torque to function on its own.

The polyphase induction motor has a nonzero starting torque. It is then logically natural to explore the possibility of using the polyphase principles to initially accelerate the single-phase motor. Economic considerations dictate that the most simple polyphase concept be implemented.

Consider the two-phase wound stator of Fig. 6.42 with N_{eff} turns per coil. The β coil is displaced 90° in space from the α coil. By analogy with the work of Sec. 6.3.3, the fundamental components of air gap mmf for the α and β coils can be written as

$$\mathcal{F}_{g\alpha 1} = \frac{2}{\pi} N_{eff} i_\alpha \cos\left(\frac{p}{2}\,\theta\right)$$

[6.86]

$$\mathcal{F}_{g\beta 1} = \frac{2}{\pi} N_{eff} i_\beta \sin\left(\frac{p}{2}\,\theta\right)$$

[6.87]

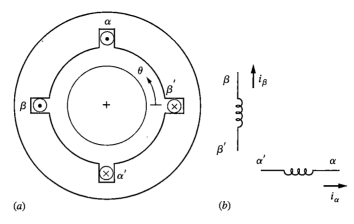

Figure 6.42
Two-phase winding. (a) Physical arrangement. (b) Schematic representation.

If currents of equal amplitude, but with 90° time phase separation, are injected into the coils, the total air gap mmf is given by

$$\mathcal{F}_g = \mathcal{F}_{g\alpha1} + \mathcal{F}_{g\beta1} = \frac{2}{\pi} N_{eff} I_m \left[\cos(\omega t)\cos\left(\frac{p}{2}\theta\right) + \sin(\omega t)\sin\left(\frac{p}{2}\theta\right) \right]$$

[6.88]

After applying the trigonometric identity $\cos(x - y) = \cos x \cos y + \sin x \sin y$ and defining $\mathcal{F}_m = (2/\pi)N_{eff}I_m$, the total air gap mmf can be expressed as

$$\mathcal{F}_g = \mathcal{F}_m \cos\left(\omega t - \frac{p}{2}\theta\right)$$

[6.89]

When [6.89] is compared with [6.10], it is apparent that the two-phase winding with balanced two-phase excitation has produced an air gap mmf that is a constant-amplitude traveling wave. Just as in the case of the three-phase winding, this traveling mmf wave will interact with the rotor coils to produce a nonzero starting torque.

Since the starting winding is needed only during initial acceleration of the single-phase motor, it is designed for intermittent duty and rendered open-circuit by a centrifugal switch that is typically activated in the range of 60 to 70 percent of synchronous speed. The starting winding (intermittent duty) is located with 90° displacement in space from the main winding (continuous duty). A quasi 90° time-phase difference in the currents of the two windings is established by either design with significantly different reactance to resistance ratios, or by insertion of a series-connected capacitor in the starting winding. The former one is known as a *split-phase* motor while the latter is called a *capacitor-start* motor.

6.11   INDUCTION MOTOR DESIGN

Consistent with other design sections of this book, the scope is narrowed to allow a concise presentation of the basic processes and procedures underlying the design of induction motors. As specific examples, only open slots with form-wound coils and integer number of slots per pole per phase are considered. References are cited that give the reader omitted details of derivations and guidance into less constrained design arrangements than allowed by this narrow scope.

6.11.1 CLASSIFICATIONS AND STANDARDIZATIONS

The National Electrical Manufacturers Association in *NEMA Standard MG-1* presents numerous guidelines for labeling, classification, rating, and packaging of induction motors. These guidelines can be loosely organized into nameplate, environmental, electrical, and mechanical categories.

Nameplate The minimum nameplate data given for a squirrel-cage induction motor includes the following specific items:

1. Frame designation (see Table 6.5)
2. Rated output power (hp)
3. Speed at rated load (rpm)
4. Rated-load current (rms, line current)
5. Rated voltage (rms, line voltage)
6. Rated frequency (Hz)
7. Number of phases
8. Locked-rotor kVA code letter (see Table 6.4)
9. Design letter (see Table 6.3)

Table 6.5
Locked-rotor kVA/hp code letters

Code Letter	Locked-rotor kVA/hp
B	3.15–3.55
C	3.55–4.0
D	4.0–4.5
E	4.5–5.0
F	5.0–5.6
G	5.6.–6.3

10. Maximum ambient temperature for rated load (40°C typical)
11. Time rating (continuous, 30 minutes, etc.)
12. Insulation class (see *IEEE Standards 117* and *275*)

Environmental Maximum ambient temperature and altitude ratings are among the items covered. An important item is specification of the NEMA style—open machine and totally-enclosed machine with each style having subcategories that specify the motor's tolerance to a hostile environment. From a design point of view, totally-enclosed machines must have an intermediary heat exchange mechanism rather than directly rejecting losses to the ambient air. As a direct consequence, the totally-enclosed machine must be designed for lower losses per unit of volume than for open machines. Design of only open machines will be considered in this book.

Electrical The electrical quantities listed on the nameplate generally stand as self-explanatory with the exception of locked-rotor kVA code. This specification serves the purpose of indirectly giving a range for the rated-voltage starting current of the induction motor by the ratio (locked-rotor, three-phase apparent power)/(rated output horsepower), whence the locked-rotor current is determined as

$$I_{\text{locked}-\text{rotor}} = \frac{1000(\text{locked} - \text{rotor kVA/hp})(\text{rated hp})}{\sqrt{3}(\text{rated } V_L)} \qquad \textbf{[6.90]}$$

A partial list of the locked-rotor kVA code letters is presented by Table 6.5.

Mechanical NEMA frame designations standardize outside dimensions to assure interchangeability of motors. For a comprehensive list of frame designations and a complete set of dimensions for each, the reader should consult *NEMA Standard MG-1*. In order to clarify the nature of the frame designation, Fig. 6.43 shows a partial set of dimension symbols where the specific values for a selected group of frames are found in Table 6.6.

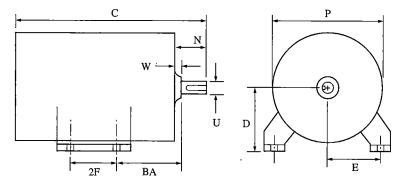

Figure 6.43
Basic NEMA frame dimensions

Table 6.6
NEMA frame dimensions (in)

Frame Designation	C	N-W	U	2F	BA	P	D	E
145T	11.69	2.25	0.875	5.00	2.25	6.88	3.50	2.75
184T	13.56	2.75	1.125	5.50	2.75	8.00	4.50	3.75
215T	17.19	3.38	1.375	7.00	3.50	9.38	5.25	4.25
256T	22.25	4.00	1.625	10.00	4.25	10.78	6.25	5.00
284T	23.38	4.62	1.875	9.50	4.75	14.12	7.00	5.50
326T	27.50	5.25	2.125	10.50	5.25	16.12	8.00	6.25
365T	29.62	5.88	2.375	11.25	5.88	18.62	9.00	7.00
404T	32.38	7.25	2.875	12.25	6.62	21.12	10.00	8.00
445T	39.56	8.50	3.375	16.50	7.50	23.62	11.00	9.00

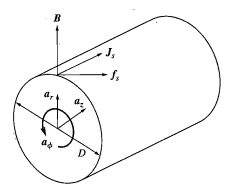

Figure 6.44
Fields and force at stator bore.

6.11.2 VOLUME AND BORE SIZING

If the slot-embedded stator coils are assumed to lie at the stator bore diameter (D) of a p-pole machine, then the three-phase currents of frequency ω flowing in the three-phase winding set can be modeled as a traveling current sheet with density

$$\mathbf{J}_s = J_{sm}\cos\left(\omega t - \frac{p}{2}\theta\right)\mathbf{a}_z \qquad \textbf{[6.91]}$$

The rotor-induced currents produce a traveling flux density wave, not in phase with $\mathbf{J}_s$, that passes through the stator bore diameter surface with description

$$\mathbf{B}_r = B_{rm}\sin\left(\omega t - \frac{p}{2}\theta - \alpha\right)\mathbf{a}_r \qquad \textbf{[6.92]}$$

The stator current density wave and the rotor flux density wave interact, as indicated in Fig. 6.44, to produce a surface force density given by

$$\mathbf{f}_s = \mathbf{J}_s \times \mathbf{B}_r \qquad \textbf{[6.93]}$$

Since J_s and B_r are orthogonal, the magnitude of the resulting force density is

$$f_s = -J_s B_r = -J_{sm} B_{rm} \cos\left(\omega t - \frac{p}{2}\theta\right) \sin\left(\omega t - \frac{p}{2}\theta - \alpha\right) \quad \textbf{[6.94]}$$

where the negative sign indicates that the force density is opposite to the direction of unit vector a_θ.

The force resulting from the field interaction acts at a distance $D/2$ to develop a torque described by

$$T = \frac{D}{2} \int_{\theta=0}^{2\pi} \int_{\ell=0}^{\ell_a} f_s d\left(\frac{D}{2}\theta\right) d\ell = \frac{D^2 \ell_a}{4} \int_0^{2\pi} f_s \, d\theta \quad \textbf{[6.95]}$$

If the trigonometric identity $\cos x \sin y = \frac{1}{2}\sin(x+y) - \frac{1}{2}\sin(x-y)$ is applied to [6.94], the result is substituted into [6.95], and the integration is performed, the torque is determined as

$$T = \frac{\pi}{4} D^2 \ell_a J_{sm} B_{rm} \sin\alpha \quad \textbf{[6.96]}$$

It is apparent from [6.96] that the torque developed by an induction motor depends directly on the air gap volume $(\pi/4)D^2\ell_a$, the stator winding current density, and the air gap flux density.

In order to minimize size, the values of current density and flux density are respectively maintained at the limits allowed by ability to cool and by saturation limits imposed by the ferromagnetic material. The angle α is directly related to the rotor power factor and thus tends to fall into a somewhat typical narrow range. Consequently, for a class of induction motors, it can be concluded from [6.96] that

$$\frac{D^2 \ell_a}{T} \cong \text{constant} = v_T \quad \textbf{[6.97]}$$

For open-style, NEMA design B motors rated for 10 hp or more, a reasonable value for the normalized volume constant of [6.97] is $v_T = 5$ to 6 in³/ft·lb (1.0 to 1.2×10^{-5} m³/N·m). For motors less than 10 hp, the ground and layer insulation and the slot wedge become a larger percent of the slot cross section, forcing increase of the normalized volume constant value to attain acceptable design. As the power rating decreases to 1 hp, the v_T should increase to a value of 9 to 10 in³/ft·lb. Design engineers usually develop values of v_T based on successful design history within their organization.

As the next step in the initial sizing of the ferromagnetic material envelope, the stator lamination stack outside diameter (D_o) should be selected. The target outside diameter of the frame (D_f) is assumed to be known. In the case of a NEMA frame designation, $D_f = P$ from Table 6.6. Selecting a wall thickness (t_f) to give adequate structural integrity for the frame, the stator lamination outside diameter is

$$D_o = D_f - 2t_f \quad \textbf{[6.98]}$$

With the tentative value of D_o decided, the specification of the stator bore diameter must be reconciled with the number of magnetic poles. For machines with smaller pole count, the circumferential flux path along the stator magnetic core between the bottom of the stator slots and the stator lamination outside diameter is longer. In order to avoid an excessive magnetizing mmf requirement, the stator core flux density for machines with a smaller number of poles must be less than for machines with a larger number of poles. An empirical formula from Ref. 1 is helpful in tentative sizing of the stator bore diameter:

$$D = \frac{D_o - 0.647}{1.175 + 1.03/p} \qquad \textbf{[6.99]}$$

The stator lamination stack length then follows from [6.97] as

$$\ell_a = \frac{v_T}{D^2} \qquad \textbf{[6.100]}$$

Although specific values for D and ℓ_a have been established, some adjustment of dimension ℓ_a may be necessary as the design is refined. An alternate approach to the stator volume sizing can be found in Ref. 2, pp. 296–305.

6.11.3 STATOR DESIGN

The above stator volume and bore sizing obviously began the stator design. Insofar as possible, however, the dimensional selections that were made are held outside of the principal iterative stator design process for determination of slot number, slot dimensions, and conductor sizing that is set forth in the logic flowchart of Fig. 6.45.

Number of Stator Slots The industry standard double-layer winding necessarily requires that the number of conductors per slot (C_s) be an even integer. The narrow scope of an integer multiple of 3 slots per pole limits the number of stator slots (S_1) selection to the choices of Table 6.7.

The stator slot pitch is the span of a stator tooth and slot measured at the stator bore diameter D (see Fig. 6.46), or

$$\lambda_1 = \frac{\pi D}{S_1} \qquad \textbf{[6.101]}$$

Table 6.7
Stator slot selection

Poles (p)	No. Stator Slots (S_1)											
2	6	12	18	24	30	36	42	48	54	60	66	72
4		12		24		36		48		60		72
6			18			36			54			72
8				24				48				72

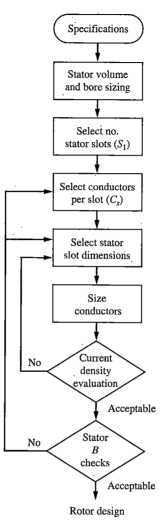

Figure 6.45
Logic flowchart for stator design

Good design practice places the value of stator slot pitch in the range from 0.75 to 1.5 in. Hence, the choice of S_1 from Table 6.7 should be such that

$$0.75 \leq \frac{\pi D}{S_1} \leq 1.5 \qquad \qquad \textbf{[6.102]}$$

Conductors per Slot The air gap traveling mmf wave of [6.10] logically produces an air gap traveling flux wave. Thus, the N_{1eff}-turn, full-pitch, concentrated coil of a pole-pair phase winding sees a flux over its pole pitch described by

$$\phi_p = \Phi_m \sin(\omega t) \qquad \qquad \textbf{[6.103]}$$

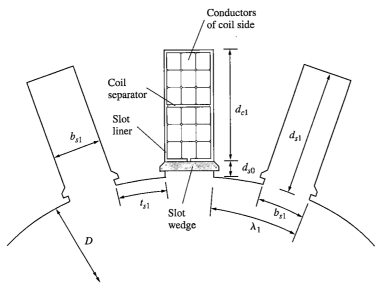

Figure 6.46
Stator slot section view

where Φ_m is the peak value of flux that is established by the stator three-phase coil set and ϕ_p mutually couples the stator and rotor coils. Any stator coil leakage flux is not included in this description.

Using [6.103], the instantaneous value of phasor voltage ($\overline{E}_1$) per phase from Fig. 6.16 is found for a p-pole with a parallel winding path by use of Faraday's law.

$$e_1 = \frac{p}{2} \frac{N_{1eff}}{a} \frac{d\phi_p}{dt} = \omega N_{\phi eff} \Phi_m \cos(\omega t) \qquad \textbf{[6.104]}$$

where the effective series turns per phase have been defined as

$$N_{\phi eff} = \frac{p}{2} \frac{N_{1eff}}{a} \qquad \textbf{[6.105]}$$

The rms phasor magnitude of [6.104] is given by

$$E_1 = \frac{\omega N_{\phi eff} \Phi_m}{\sqrt{2}} = 4.44 N_{\phi eff} f \Phi_m \qquad \textbf{[6.106]}$$

If a typical voltage drop across $R_1 + jX_1$ of Fig. 6.16 is assumed so that $E_1 \cong 0.97V_1$, then the effective series turns per phase follows from [6.106] as

$$N_{\phi eff} \cong \frac{0.97 V_1}{4.44 f \Phi_m} \qquad \textbf{[6.107]}$$

By use of [6.1] and [6.105], the conductors per stator slot can be determined from [6.107] as

$$C_s = 2N \cong \frac{0.97 a V_1}{2.22 k_d k_p p N_{s1} f \Phi_m} \qquad \textbf{[6.108]}$$

In order to complete a conductors-per-slot decision, an estimate of the maximum value of air gap flux must be made. A variation of an empirical formula from p. 306 of Ref. 2 is useful.

$$\Phi_m = \left(0.00145 + \frac{0.00300}{p} \right) \left[\frac{60}{f} P_s \right]^{1/2} \qquad \textbf{[6.109]}$$

P_s is the rated motor output power in horsepower. The value of C_s must be rounded to the nearest even integer.

Slot Design   For good design practice, the stator slot width (b_{s1}) of Fig. 6.46 should be selected such that

$$0.5\lambda_1 \le b_{s1} \le 0.6\lambda_1 \qquad \textbf{[6.110]}$$

The slot depth (d_{s1}) usually falls in the range

$$2b_{s1} \le d_{s1} \le 4b_{s1} \qquad \textbf{[6.111]}$$

In addition to containment of the two coil sides, the slot must provide ground insulation (slot liner), assure adequate insulation between coil sides (coil separator), and accept a means of retaining the coil sides (slot wedge) as illustrated by Fig. 6.46. The following material thicknesses are suggested:

1. 230-V motors: Use 0.015-in slot liner and coil separator.
2. 460-V motors: Use 0.030-in slot liner and coil separator.
3. NEMA 284T or smaller frames: Use 0.060-in slot wedge.
4. Larger than NEMA 284T frames: Use 0.125-in slot wedge.

Also, it is common practice to wrap the coil sides with a layer of 0.005- to 0.007-in thick fiberglass tape to avoid conductor damage during handling and to provide space for varnish impregnating of the finished stator winding.

Guided both by the above insulation dimension suggestions and by the tentative values of slot width and slot depth determined by [6.110] and [6.111], the conductor size specification is begun. Allowable conductor current density for successful cooling depends on the flow of air around the coils and the thermal impedance from the point of heat generation within the conductor to the cooling air. Shorter lamination stacks are more easily cooled than longer lamination stacks. For air-cooled machines, typical values of allowable stator current density (Δ_1) are in the range

$$500 \le \Delta_1 \le 800 \, \text{A/cm}^2 \quad \text{or} \quad 3200 \le \Delta_1 \le \text{A/in}^2 \qquad \textbf{[6.112]}$$

Initially, standard square wire sizes of Table 6.8 should be considered for conductor choice. However, the slot dimensions may not be compatible with these val-

Table 6.8
Film insulated square magnet wire

AWG Size	Bare Width (in)	Overall Width (in)	Area (in²)
0	0.3249	0.3329	0.10420
1	0.2893	0.2972	0.08232
2	0.2576	0.2652	0.06498
3	0.2294	0.2367	0.05125
4	0.2043	0.2113	0.04037
5	0.1819	0.1887	0.03171
6	0.1620	0.1686	0.02536
7	0.1443	0.1507	0.01994
8	0.1285	0.1348	0.01563
9	0.1144	0.1205	0.01251
10	0.1019	0.1079	0.00980
11	0.0907	0.0967	0.00788
12	0.0808	0.0868	0.00619
13	0.0720	0.0780	0.00496
14	0.0641	0.0701	0.00389

Table 6.9
Efficiency

P_s (hp)	η (%)
1–3	75
5–40	85
50–200	90

ues to allow the desired conductor stacking arrangement. In such case, magnet wire manufacturers can supply rectangular conductors drawn to desired dimension specifications.

The value of full-load current is needed to size the conductor cross-section area A_1. Although not known exactly until the design is completed, a slightly conservative value of full-load current can be calculated by assuming a full-load power factor of 0.85 and a full-load efficiency (η) according to Table 6.9. The estimated full-load current is then

$$I_1 = \frac{P_s(746)}{\sqrt{3}\,V_L(0.85)\eta} \qquad \textbf{[6.113]}$$

where P_s is full-load output power in horsepower and V_L is rated line voltage. The required value of the stator conductor cross-section area follows as

$$s_{a1} = \frac{I_1}{a\Delta_1} \qquad \textbf{[6.114]}$$

If a conductor of appropriate cross-section area does not fit the slot well (too large or too small), the iterative process indicated by Fig. 6.45 should be carried out until a satisfactory slot design is attained.

Flux Density Checks Prior to beginning the rotor design, critical flux density checks should be made. The maximum apparent stator tooth flux density is found as

$$B_{stm} = \frac{\dfrac{\pi}{2}\Phi_m}{\dfrac{S_1}{p}t_{s1}\ell_a SF} = \frac{\pi p \Phi_m}{2 S_1 t_{s1} \ell_a SF} \qquad [6.115]$$

where SF is a stacking factor with values $0.94 \leq SF \leq 0.97$ depending on the lamination thickness and axial assembly pressure. For good design, $90 \leq B_{stm} \leq 120$ kilolines/in^2 $(1.4 \leq B_{stm} \leq 1.9$ T).

The maximum value of stator core flux density is given by

$$B_{scm} = \frac{\Phi_m/2}{\left(\dfrac{D_o - D}{2} - d_{s1}\right)\ell_a SF} = \frac{\Phi_m}{(D_o - D - 2d_{s1})\ell_a SF} \qquad [6.116]$$

For small pole-number machines, the stator core path is long. Consequently, B_{scm} must be small enough to avoid an mmf requirement that results in excessive magnetizing current. Good design places $50 \leq B_{scm} \leq 95$ kilolines/in^2 $(0.5 \leq B_{scm} \leq 0.95$ T), where the lower end of the range applies to two-pole machines.

If the flux density checks show that either B_{stm} or B_{scm} is high while the other is low, corrective action may be possible by reapportionment of slot dimensions b_{s1} and d_{s1}. Otherwise, adjustment of the conductors per slot will be necessary as indicated by the flowchart of Fig. 6.45.

6.11.4 ROTOR DESIGN

With a stator design characterized by both acceptable conductor current density and flux density, it would be a rare occurrence to not be able to fit a rotor design with acceptable current density and flux density into the defined cylindrical stator bore envelope. However, no assurance of satisfying critical criteria of the NEMA design letter is guaranteed. Common practice is to design the rotor to fit the stator bore envelope according to the logic process set up by Fig. 6.47, assuring that conductor current density and flux densities are acceptable. After both stator and rotor candidate designs are completed, the equivalent circuit parameters are evaluated, specific performance deficiencies are identified, and iteration of the complete design, if necessary, is made to target the desired performance.

Air Gap Sizing The reluctance associated with the air gap forms a significant portion of the total effective reluctance of the induction motor magnetic circuit. Thus, the air gap length is a dominant factor in determination of the required

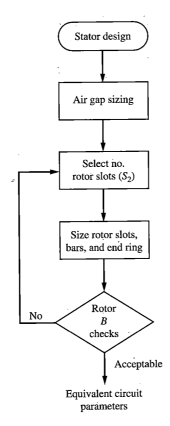

Figure 6.47
Logic flowchart for rotor design

magnetizing current. Consequently, it should be as short as practically possible. An empirical relationship from Ref. 1 is helpful in setting the air gap length (δ).

$$\delta = 0.0016D + 0.001\ell_a + 0.0072 \qquad \textbf{[6.117]}$$

where all dimensions are in inches. Round off to the nearest 0.005 in is good practice. With the air gap length decided, the rotor diameter (D_r) is

$$D_r = D - 2\delta \qquad \textbf{[6.118]}$$

Number of Rotor Bars When selecting the number of rotor bars (S_2), it is pointed out in Ref. 2 (p. 318) that certain combinations can produce mmf wave harmonics that lead to detrimental cusps in the speed-torque curve. To be specifically avoided are

$$S_1 - S_2 = \pm p, \pm 3p, \pm 5p \qquad S_1 - S_2 = \pm 3kp \qquad (k = 1, 2, 3, \dots)$$

$$\textbf{[6.119]}$$

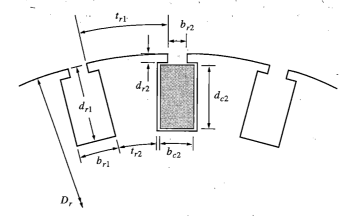

Figure 6.48
Rotor slot section view

Certain other combinations result in motors with increased mechanical noise. Their avoidance depends on the application.

Slot, Bar, and End-Ring Design The narrow presentation will consider only the rectangular rotor bar design of Fig. 6.48. A rotor bar depth-to-width ratio, $(d_{r1} - d_{r2})/b_{r1}$, in the range of 4 to 6 is typically satisfactory for NEMA design B motors. Partially closed rotor slots are necessary to aid in mechanical retention of the bar against centrifugal force. However, the slot opening must be sized to avoid excessive rotor leakage reactance. Empirical relationships from Ref. 1 are suggested for use in sizing the rotor slot lip.

$$b_{r2} \cong 0.01D + 0.045 \qquad \textbf{[6.120]}$$

$$d_{r2} \cong 0.677 b_{r2} \qquad \textbf{[6.121]}$$

All dimensions are in units of inches.

The rotor slot must be sized so that the rotor bar cross-section area satisfies allowable current density. Without ground insulation to thermally degrade, rotor bars can operate at higher temperature than stator conductors. The end rings, having better exposure to cooling air flow, can be operated at even slightly higher temperatures than rotor bars. The following guidelines for rotor bar current density (Δ_b) and end ring current density (Δ_e) are suggested for copper:

$$700 \le \Delta_b \le 775 \text{ A/cm}^2 \quad \text{or} \quad 4500 \le \Delta_b \le 5000 \text{ A/in}^2$$

$$\textbf{[6.122]}$$

$$775 \le \Delta_e \le 930 \text{ A/cm}^2 \quad \text{or} \quad 5000 \le \Delta_e \le 6000 \text{ A/in}^2$$

$$\textbf{[6.123]}$$

If aluminum is used, multiply values by 0.61. These values of Δ_b and Δ_e may be lowered in the final iterative stages of design if a decrease in rotor resistance is indicated to meet performance criteria.

In order to proceed with the rotor bar design process, the rotor bar current and end ring current must be related to the stator coil current. Owing to the reluctance of the air gap, about 10 percent of the full-load stator mmf is required to establish the mutual flux linking both the rotor and the stator coils. The balance of the stator mmf is opposed by the rotor mmf, or

$$0.9 N_{1eff} \frac{I_1}{a} = N_{2eff} I_b \qquad \textbf{[6.124]}$$

where I_b is the rms value of rotor bar current and N_{2eff}, yet to be determined, is the effective number of turns per pole per phase for the rotor.

Define rotor slots per pole per phase (N_{s2}) and rotor slot angle (γ_2) as

$$N_{s2} = \frac{S_2}{mp} \qquad \textbf{[6.125]}$$

$$\gamma_2 = \frac{\pi}{S_2/p} \qquad \textbf{[6.126]}$$

The angular rotor coil pitch is the angular distance between two bars closest to a pole pitch; thus,

$$\rho_r = \frac{fix(S_2/p)}{S_2/p} \pi \qquad \textbf{[6.127]}$$

The rotor distribution factor (k_{dr}) and pitch factor (k_{pr}) follow based on [A.11] and [A.14] of Appendix A.

$$k_{dr} = \frac{\sin(N_{s2}\gamma_2/2)}{N_{s2}\sin(\gamma_2/2)} \qquad \textbf{[6.128]}$$

$$k_{pr} = \sin(\rho_r/2) \qquad \textbf{[6.129]}$$

Realizing that the rotor has only one conductor per slot (C_{s2}), the rotor effective turns per pole per phase can now be written by analogy with [6.1] giving

$$N_{2eff} = k_{pr} k_{dr} N_{s2} C_{s2} = k_{pr} k_{dr} \frac{S_2}{mp} \qquad \textbf{[6.130]}$$

Substituting [6.1] and [6.130] in [6.124] and rearranging determines the rotor bar current in terms of the stator coil current.

$$I_b = \frac{1.8 N k_p k_d S_1}{a k_{pr} k_{dr} S_2} I_1 \qquad \textbf{[6.131]}$$

The current from each rotor bar enters an end ring and divides with half of the current flowing clockwise in the end ring while the other half flows in a counter-clockwise direction. Over a pole pitch, S_2/p rotor bars inject current into an end ring. Each bar current varies sinusoidally in time with a phase shift of γ_2 from the current in an adjacent bar. If the number of rotor bars is large enough that γ_2 is a small angle, then the end ring current (I_e) is simply the half cycle average of a sine wave of amplitude $\frac{1}{2}I_b$, or

$$I_e = \frac{1}{\pi} \int_0^\pi \frac{S_2}{p} \frac{I_b}{2} \sin(\theta) \, d\theta = \frac{S_2}{p\pi} I_b \qquad \textbf{[6.132]}$$

Use of [6.131] and [6.132] along with the current density values of [6.122] and [6.123] allows sizing of the cross-section area of the rotor bar and the end ring, respectively.

$$s_b = \frac{I_b}{\Delta_b} = \frac{1.8 N k_p k_d S_1}{a k_{pr} k_{dr} S_2} \frac{I_1}{\Delta_b} \qquad \textbf{[6.133]}$$

$$s_r = \frac{I_e}{\Delta_e} = \frac{S_2}{p\pi} \frac{I_b}{\Delta_e} = \frac{1.8 N k_p k_d S_1}{p\pi a k_{pr} k_{dr}} \frac{I_1}{\Delta_e} \qquad \textbf{[6.134]}$$

The units of s_b and s_r will be in^2 if Δ_b and Δ_e have units of A/in^2. Choosing the bar depth-to-width ratio in the range of $3 \le rdw \le 6$, the rotor bar dimensions are determined as

$$b_{r1} = \left(\frac{s_b}{rdw}\right)^{1/2} \qquad \textbf{[6.135]}$$

$$d_{r1} - d_{r2} = (rdw s_b)^{1/2} \qquad \textbf{[6.136]}$$

Flux Density Checks   Critical flux checks should be made to assure that acceptable rotor slot proportions have been made. The maximum rotor tooth flux density exists at the root of the rotor tooth with dimension t_{r2} as indicated in Fig. 6.48:

$$B_{rtm} = \frac{\frac{\pi}{2} \Phi_m}{\frac{S_2}{p} t_{r2} \ell_a SF} = \frac{\pi p \Phi_m}{2 S_2 t_{r2} \ell_a SF} \qquad \textbf{[6.137]}$$

As in the case of B_{stm}, good design requires that $90 \le B_{rtm} \le 120$ kilolines/in^2 ($0.9 \le B_{rtm} \le 1.2$ T).

Under the conservative assumption that no flux passes through the shaft of diameter D_{sh}, the maximum value of rotor core flux density is found as

$$B_{rcm} = \frac{\Phi_m/2}{\left(\dfrac{D_r - D_{sh}}{2} - d_{r1}\right) \ell_a SF} = \frac{\Phi_m}{(D_r - D_{sh} - 2d_{r1}) \ell_a SF} \qquad \textbf{[6.138]}$$

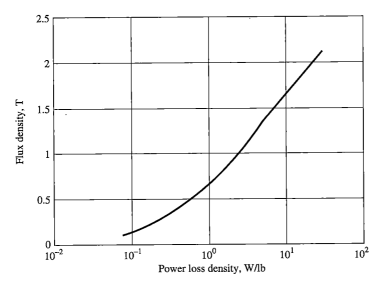

Figure 6.49
Power loss density (60 Hz)

B_{rcm} should not exceed 100 kilolines/in^2 (1 T). If axial ventilation holes pass through the rotor core, the reduction in the area for flux flow should be considered in [6.138]. Should the flux density checks show excessive values for either B_{rtm} or B_{rcm}, adjustment of the number of rotor slots or the rotor slot proportions should be considered as indicated in Fig. 6.47.

6.11.5 EQUIVALENT CIRCUIT PARAMETERS

Inclusion of complete derivation details for the six equivalent circuit parameters of Fig. 6.17 is beyond the scope of this book. The procedure for determination of core loss resistance R_c is covered in detail. An abbreviated derivation for the magnetizing reactance X_m is presented. Formulas for the four remaining parameters are simply given. The reader with an in-depth interest should pursue the cited references.

Core Loss Resistance R_c Determination of R_c requires evaluation of the frequency and flux density sensitive core losses that it models. Core loss graphical data is available from the electric sheet steel manufacturers in power density form (W/lb) as a function of flux density with frequency as a parameter. Figure 6.49 shows a typical core loss power density plot.

The core loss power density is measured by an Epstein test fixture that does not reflect the assembled lamination stack pressure, motor operating temperature, and burrs created by punching dies. Consequently, the loss densities determined by the Epstein test procedure are significantly lower than the values that result in application. Multiplication of values by a factor k_c of 2.5 to 4 is necessary to obtain values typical of actual losses.

When evaluating R_c for steady-state, low-slip operating conditions, any rotor core losses are considered negligible as the frequency of rotor cyclic fluxes is small. For purposes of calculation of stator core losses, the teeth and core losses are handled separately since the flux densities of these two regions may be significantly different. The stator tooth flux density (B_{t13}) at one-third of tooth depth and the stator core flux density (B_{scc}) at midpath of the core are suitable choices for loss calculations.

$$B_{t13} = \frac{\pi p \Phi_m}{2 \left[\pi \left(D + \frac{2}{3} d_{s1} \right) - S_1 b_{s1} \right] \ell_a SF} \qquad \textbf{[6.139]}$$

$$B_{scc} = \frac{0.866 \Phi_m}{(D_o - D - 2d_{s1}) \ell_a SF} \qquad \textbf{[6.140]}$$

Power loss densities (p_{t13}, p_{scc}) are determined from Fig. 6.49 for the two levels of flux density (B_{t13}, B_{scc}). After weights of the teeth and stator core material (W_t, W_c) are calculated, the total stator core losses (P_c) are evaluated as

$$P_c = k_c (W_t p_{t13} + W_c p_{scc}) \qquad \textbf{[6.141]}$$

Using $E_1 \cong 0.97 V_1$ in Fig. 6.17,

$$R_c \cong \frac{(0.97 V_1)^2}{P_c/3} \qquad \textbf{[6.142]}$$

Magnetizing Reactance X_m The mmf drop across the air gap of an induction motor is typically 80 to 85 percent of the total mmf requirement to sustain the mutual flux linking the stator and rotor coils. Hence, an accurate determination of its value is critical in evaluation of the magnetizing inductance X_m.

The larger portion of the flux leaving the rotor surface enters the tips of the stator teeth at the bore diameter D after traversing the air gap width δ. However, a small portion of the flux enters the stator teeth along the sides of the slot. A similar flux pattern exists as the flux leaves the rotor structure. Consequently, the air gap reluctance must reflect an effective air gap length (δ_e) that is slightly greater than the actual air gap length (δ). Factors to account for the increase in effective air gap length due to an open stator slot and a partially closed rotor slot, respectively, are given on p. 328 in Ref. 2.

$$k_s = \frac{\lambda(5\delta + b_{s1})}{\lambda(5\delta + b_{s1}) - b_{s1}^2} \qquad \textbf{[6.143]}$$

$$k_r = \frac{\lambda(4.4\delta + 0.75 b_{r2})}{\lambda(4.4\delta + 0.75 b_{r2}) - b_{r2}^2} \qquad \textbf{[6.144]}$$

The effective air gap is then found as

$$\delta_e = k_s k_r \delta \qquad \textbf{[6.145]}$$

The magnetizing reactance (X_m) is given by

$$X_m = \omega L_m \qquad \text{[6.146]}$$

where the magnetizing inductance is calculated as flux linkages per ampere, or

$$L_m = \frac{N_{\phi eff}\phi_p}{I_m} = \frac{N_{\phi eff}^2 \frac{2}{\pi}\Phi_m}{N_{\phi eff}I_m} = \frac{\frac{2}{\pi}N_{\phi eff}^2}{\mathcal{R}} \qquad \text{[6.147]}$$

$N_{\phi eff}$ is the total effective turns in series per phase, ϕ_p is the total flux per pole, and $\mathcal{R}$ is the reluctance seen by the flux of a pole. If good design practice is maintained, the total reluctance is about 20 percent greater than the air gap reluctance; thus,

$$\mathcal{R} \cong 1.2\mathcal{R}_g = 1.2\frac{1}{\mu_o}\frac{2\delta_e}{\frac{\pi D}{p}\ell_a} = \frac{2.4p\,\delta_e}{\mu_o\pi D\ell_a} \qquad \text{[6.148]}$$

Use of [6.147] and [6.148] in [6.146] leads to the desired expression for magnetizing inductance.

$$X_m = \omega\frac{\frac{2}{\pi}\left(\frac{k_p k_d S_1 N}{am}\right)^2}{\frac{2.4p\delta_e}{\mu_o\pi D\ell_a}} = 1.6713 \times 10^{-7}\left(\frac{fD\ell_a}{p\delta_e}\right)\left(\frac{k_p k_d S_1 N}{am}\right)^2 \qquad \text{[6.149]}$$

The units for D, ℓ_a, and δ_e are in inches.

Stator Coil Resistance R_1 The stator coil resistance per phase for a temperature of 140°C with copper conductor material is given by

$$R_1 = \frac{1.056 \times 10^{-6}C_\phi MLT}{as_{a1}} \qquad \text{[6.150]}$$

where MLT is the mean length turn of a stator coil (in) and s_{a1} is the cross-section area of a coil conductor (in^2). C_ϕ is the total series turns per phase given by

$$C_\phi = \frac{S_1 N}{am} \qquad \text{[6.151]}$$

where N is the actual number of turns per coil. Common practice is to increase the calculated value of R_1 by 3 to 5 percent to account for increase in resistance due to unequal current distribution.

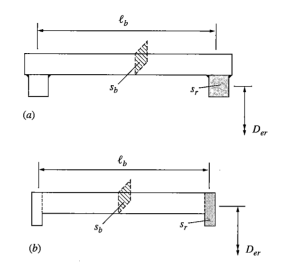

Figure 6.50
Rotor cage dimension symbols. (a) Fabricated cage. (b) Die-cast cage.

Rotor Resistance R_2' Define the following dimension symbols that are illustrated by Fig. 6.50:

s_b = cross-section area of a rotor bar (in^2)

s_r = cross-section area of an end ring (in^2)

ℓ_b = length of rotor bar conducting current (in)

D_{er} = mean diameter of an end ring (in)

The rotor resistance for copper material at 180°C is then given by

$$R_2' = 1.169 \times 10^{-6} m \left(\frac{k_d k_p C_\phi}{k_{dr} k_{pr}}\right)^2 \left[\frac{\ell_b K_1}{S_2 s_b} + \frac{2}{\pi} \frac{D_{er} K_{ring}}{p^2 s_r}\right] \qquad \textbf{[6.152]}$$

If the rotor cage is manufactured from aluminum, R_2' should be multiplied by 1.64 to account for the higher resistivity of aluminum. The factor $K_1 \geq 1$ adjusts the bar portion of R_2' to account for resistance increase due to skin effect, as discussed on pp. 272–277 of Ref. 3.

$$K_1 = \frac{\alpha_d \left[\sinh(2\alpha_d) + \sin(2\alpha_d)\right]}{\cosh(2\alpha_d) - \cos(2\alpha_d)} \qquad \textbf{[6.153]}$$

$$\alpha_d = k_\alpha d_{c2} \left(\frac{sfb_{c2}}{b_{r1}}\right)^{1/2} \qquad \textbf{[6.154]}$$

where d_{c2}, b_{c2}, and b_{r1} are defined in Fig. 6.48 and k_α is 0.3505 for copper bars and 0.2700 for aluminum bars. Since α_d depends on rotor frequency ($f_r = sf$), K_1 must be evaluated each time that motor speed changes.

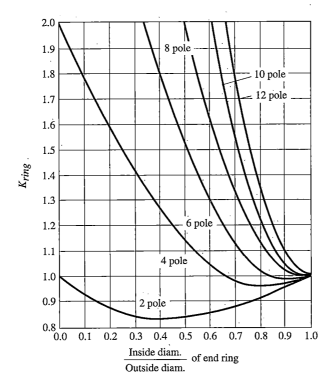

Figure 6.51
End-ring resistance adjustment

The factor K_{ring}, derived in Ref. 4, accounts for the nonuniform distribution of current over the cross section of the end ring. Its value is taken from the graph of Fig. 6.51.

Stator Coil Leakage Reactance X_1 Stator coil leakage reactance consists of four component reactances: skew reactance (X_{sk}); slot reactance (X_{ss}); zigzag reactance (X_{z1}); and end-turn reactance (X_{es}). These component reactances sum to yield the stator coil leakage reactance. All linear dimensions are in units of inches.

$$X_1 = X_{sk} + X_{ss} + X_{z1} + X_{es} \qquad \textbf{[6.155]}$$

$$X_{sk} = \frac{\theta_{sk}^2}{12} X_m \qquad \textbf{[6.156]}$$

where θ_{sk} is the skew angle in radians.

$$X_{ss} = 2.0055 \times 10^{-7} \frac{mfC_\phi^2 \ell_a K_s}{S_1} \left[\frac{d_{c1}}{3b_{s1}} + \frac{d_{s0}}{b_{s1}} \right] \qquad \textbf{[6.157]}$$

where the symbols are clarified in Fig. 6.46. K_s is a factor, introduced in Ref. 5, to account for reduction in leakage reactance due to certain slots having coil sides from different phases if the coil pitch is less than full:

If $0 < \rho \leq \pi/2$, $\qquad\qquad\qquad K_s = \rho/\pi$ [6.158]

If $\pi/2 < \rho \leq 2\pi/3$, $\qquad\qquad K_s = -0.25 + 1.5\rho/\pi$ [6.159]

If $2\pi/3 < \rho \leq \pi$, $\qquad\qquad K_s = 0.25 + 0.75\rho/\pi$ [6.160]

$$X_{z1} = 2.0055 \times 10^{-7} \frac{mfC_\phi^2 \ell_a K_s}{S_1}\left[\frac{a_1}{4\delta} + \frac{d_{r2}(1 - \frac{1}{2}k_{s1})}{b_{r2}}\right]$$ [6.161]

$$a_1 = \frac{1}{2}|\lambda - b_{s1} - b_{r2}|$$ [6.162]

If $S_2 < S_1$, $\qquad\qquad\qquad k_{s1} = 1$ [6.163]

If $S_2 > S_1$, $\qquad\qquad\qquad k_{s1} = \frac{S_1}{S_2}$ [6.164]

$$X_{es} = 0.3192 \times 10^{-7} \frac{fN_\phi^2 \ell_e K_s}{2p}\left[1 + 4\ln\left(\frac{3.94D}{S_1\sqrt{b_{s1}d_{s1}}}\right)\right]$$ [6.165]

where ℓ_e is the length of a coil end turn measured along the coil from lamination stack to lamination stack.

Rotor Leakage Reactance X_2' The rotor coil leakage reactance is subdivided into three component reactances: slot reactance (X_{rs}'); zigzag reactance (X_{z2}'); and end-ring reactance (X_{re}'). The sum of the component reactances forms the rotor leakage reactance. All linear dimensions are in units of inches.

$$X_2' = X_{rs}' + X_{z2}' + X_{re}'$$ [6.166]

$$X_{rs}' = 2.0055 \times 10^{-7} \frac{mfN_\phi^2 \ell_a J_1}{(k_{dr}k_{pr})^2 S_2}\left[\frac{d_{r1} - d_{r2}}{3b_{r1}} + \frac{d_{r2}}{b_{r2}}\right]$$ [6.167]

J_1 is a factor to adjust the bar portion of X_2' to account for inductance decrease due to skin effect. See pp. 272–277 of Ref. 3 for the derivation.

$$J_1 = \frac{3[\sinh(2\alpha_d) - \sin(2\alpha_d)]}{2\alpha_d[\cosh(2\alpha_d) - \cos(2\alpha_d)]}$$ [6.168]

where α_d is determined from [6.154]. J_1 must be re-evaluated if speed changes.

$$X_{z2}' = 2.0055 \times 10^{-7} \frac{mfN_\phi^2 \ell_a}{(k_{rp}k_{rd})^2 S_2}\left[\frac{a_2}{4\delta} + \frac{d_{s0}(1 - \frac{1}{2}k_{r1})}{b_{s1}}\right]$$ [6.169]

$$a_2 = \frac{1}{2} \left| \frac{\pi D_r}{S_2} - b_{s1} - b_{r2} \right| \qquad\qquad [6.170]$$

If $S_1 < S_2$, $$k_{r1} = 1 \qquad\qquad [6.171]$$

If $S_1 > S_2$, $$k_{r1} = \frac{S_2}{S_1} \qquad\qquad [6.172]$$

$$X'_{er} = 7.98 \times 10^{-9} \frac{fN_\phi^2}{p^2} \left\{ \frac{S_2 D_{er}}{mp} \left[1 + 4 \ln\left(\frac{3.94 D_r}{S_2 \sqrt{s_r}} \right) \right] \right.$$

$$\left. + 4 b_r p \left[1 + 4 \ln\left(\frac{3.94 D_r}{S_2 \sqrt{s_b}} \right) \right] \right\} \qquad\qquad [6.173]$$

6.11.6 DESIGN REFINEMENT

With the equivalent circuit parameters known, performance of the induction motor can be predicted across the speed range to compare the designed motor against specification. If desired performance specifications are not met, then design refinement can be implemented as indicated by the logic flowchart of Fig. 6.52. The MATLAB program ⟨imeval.m⟩ can alleviate much of the tedious work of a design refinement. This program reads a data file program called ⟨imdesign.m⟩ that contains general descriptive information on the motor as well as dimensional data. The program calls ⟨pc.m⟩ to determine power loss density values wherein loss data for electric steel sheet is stored.

6.11.7 SAMPLE DESIGN

The induction motor to be designed is specified as follows:

460 V   60 Hz   50 hp   1175 rpm

NEMA design B   365T frame

From Table 6.6, $P = D_f = 18.62$ in. Allowing a frame thickness of 0.50 in, the lamination stack outside diameter is

$$D_o = D_f - 2t_f = 18.62 - 2(0.50) = 17.62 \text{ in}$$

Rated torque for the motor is determined as

$$T_s = \frac{5250(P_s)}{n_m} = \frac{5250(50)}{1175} = 223.4 \text{ ft·lb}$$

Using $v_T = 5$ in³/ft·lb in [6.97],

$$D^2 \ell_a = v_T T_s = (5)(223.4) = 1117 \text{ in}^3$$

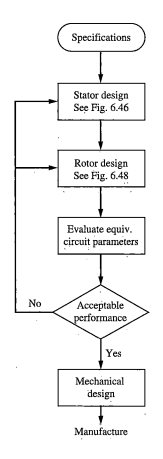

Figure 6.52
Logic flowchart for induction motor design

From [6.99],

$$D = \frac{D_o - 0.647}{1.175 + 1.03/p} = \frac{17.62 - 0.647}{1.175 + 1.03/6} = 12.60 \text{ in}$$

Hence,

$$\ell_a = \frac{D^2 \ell_a}{D^2} = \frac{1117}{(12.60)^2} = 7.03 \cong 7.00 \text{ in}$$

Selecting $S_1 = 54$ from Table 6.7,

$$\lambda_1 = \frac{\pi D}{S_1} = \frac{\pi (12.60)}{54} = 0.733 \text{ in}$$

This value of stator slot pitch is only slightly below the minimum range of 0.75 in and is judged to be acceptable.

The maximum value of flux per pole is estimated from [6.109].

$$\Phi'_m = \left(0.00145 + \frac{0.00300}{p} \right) \left[\frac{60}{f} P_s \right]^{1/2}$$

$$= \left(0.00145 + \frac{0.00300}{6} \right) \left[\frac{60}{60}(50) \right]^{1/2} = 0.0137 \, \text{Wb}$$

The winding factors are determined by formulas from Appendix A.

$$N_{s1} = \frac{S_1}{pm} = \frac{54}{(6)(3)} = 3 \text{ slots per pole per phase}$$

There are $S_1/p = 54/6 = 9$ slots per pole. Choose a pitch of 8 stator slots.

$$\rho = \frac{8}{9}(180°) = 160°$$

$$k_p = \sin(\rho/2) = \sin(160°/2) = 0.9848$$

$$\gamma_1 = \frac{p180°}{S_1} = \frac{6(180°)}{54} = 20° \text{ per slot}$$

$$k_d = \frac{\sin(N_{s1}\gamma_1/2)}{N_{s1}\sin(\gamma_1/2)} = \frac{\sin(3 \times 20°/2)}{3\sin(20°/2)} = 0.9598$$

$$k_{w1} = k_p k_d = (0.9848)(0.9598) = 0.9452$$

Based on [6.108] and choosing a single circuit winding,

$$C_s = \frac{0.97aV_1}{2.22k_{w1}pN_{s1}f\Phi'_m}$$

$$= \frac{0.97(1)(460/\sqrt{3})}{2.22(0.9452)(6)(3)(60)(0.0137)} = 8.29 \text{ conductors per slot}$$

Since the conductors per slot must be an even integer, choose $C_s = 8$. The actual value of maximum flux per pole can now be computed from [6.108].

$$\Phi_m = \frac{0.97aV_1 \times 10^8}{2.22k_{w1}pN_{s1}fC_s} = \frac{0.97(1)(460/\sqrt{3}) \times 10^8}{2.22(0.9452)(6)(3)(60)(8)} = 1.421 \times 10^6 \text{ lines}$$

Full-load current can be estimated by [6.113].

$$I_1 = \frac{P_s(746)}{\sqrt{3}V_L(0.85)\eta} = \frac{(50)(746)}{\sqrt{3}(460)(0.85)(0.90)} = 61.19 \, \text{A}$$

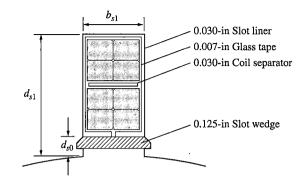

Figure 6.53
Stator slot detail

For a current density of $4000 \ A/in^2$, the stator conductor size is given by [6.114].

$$s_{a1} = \frac{I_1}{a\Delta_1} = \frac{61.19}{(1)(4000)} = 0.0153 \ in^2$$

From Table 6.8, it is seen that No. 8 square magnet wire is an acceptable choice for the stator conductors.

A layout of the chosen slot detail is shown in Fig. 6.53. Allowing a 10 percent variation in conductor dimensions to account for departure from flatness, the slot depth and width are determined.

Slot Depth	(4)(1.1)(0.1348)	conductor
	(4)(0.007)	glass tape
	(2)(0.030)	slot liner
	0.030	coil separator
	0.125	slot wedge
	0.060	wedge inset
	0.8961 in	

Use $d_{s1} = 0.900$ in and $d_{s0} = 0.125 + 0.060 = 0.185$ in.

Slot Width	(2)(1.1)(0.1348)	conductor
	(2)0.007	glass tape
	(2)0.030	slot liner
	0.3706 in	

Use $b_{s1} = 0.375$ in, giving $d_{s1}/b_{s1} = 0.900/0.375 = 2.4$. This shallow-slot design will have a lower leakage reactance than a deep-slot design. Consequently, meeting the low-slip (2.1 percent) requirement will be eased.

Flux density checks are now made using [6.115] and [6.116].

$$t_{s1} = \lambda_1 - b_{s1} = 0.733 - 0.375 = 0.358 \text{ in}$$

$$B_{stm} = \frac{\pi p \Phi_m}{2 S_1 t_{s1} \ell_a SF} = \frac{\pi(6)(1.421 \times 10^6)}{2(54)(0.358)(7)(0.96)} = 103.1 \text{ kilolines/in}^2$$

$$B_{scm} = \frac{\Phi_m}{(D_o - D - 2d_{s1})\ell_a SF}$$

$$= \frac{1.421 \times 10^6}{(17.62 - 12.60 - 2 \times 0.90)(7)(0.96)} = 65.7 \text{ kilolines/in}^2$$

Since both values are within the acceptable range, iterative changes are not necessary. The air gap is sized by [6.117].

$$\delta = 0.0016D + 0.001\ell_a + 0.0072$$

$$= (0.0016)(12.60) + (0.001)(7) + 0.0072 = 0.03436$$

Use $\delta = 0.035$ in.

The number of rotor slots is selected as $S_2 = 51$. Thus, $S_1 - S_2 = 54 - 51 = 3$, which avoids the undesirable combinations specified by [6.119]. The rotor slot lip is sized in accordance with [6.120] and [6.121].

$$b_{r2} = 0.01D + 0.045 = 0.01(12.60) + 0.045 = 0.171 \text{ in}$$

$$d_{r2} = 0.677b_{r2} = 0.677(0.171) = 0.116 \text{ in}$$

The rotor winding factors are needed to calculate bar and end-ring currents.

$$N_{s2} = \frac{S_2}{mp} = \frac{51}{(3)(6)} = 2.833 \text{ slots per pole per phase}$$

$$\gamma_2 = \frac{p 180°}{S_2} = \frac{6(180°)}{51} = 21.18° \text{ per rotor slot}$$

$$\rho_r = \frac{fix(S_2/p)}{S_2/p} 180° = \frac{fix(51/6)}{51/6} 180° = 169.41°$$

$$k_{pr} = \sin(\rho_r/2) = \sin(169.41°/2) = 0.9957$$

$$k_{dr} = \frac{\sin(N_{s2}\gamma_2/2)}{N_{s2}\sin(\gamma_2/2)} = \frac{\sin(2.833 \times 21.18°/2)}{2.833\sin(21.18°/2)} = 0.9604$$

$$k_{w2} = k_{pr}k_{dr} = (0.9957)(0.9604) = 0.9563$$

$$I_b = \frac{1.8Nk_{w1}S_1}{ak_{w2}S_2}I_1 = \frac{1.8(4)(0.9452)(54)}{(1)(0.9563)(51)}61.19 = 461.07 \text{ A}$$

$$I_e = \frac{S_2}{p\pi}I_b = \frac{51}{6\pi}461.07 = 1247.49 \text{ A}$$

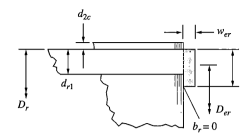

Figure 6.54
Rotor bar end-ring detail

The rotor bars and end rings are to be die-cast aluminum with $\Delta_b = 0.61(5000)$ $= 3050$ A/in^2 and $\Delta_e = 0.61(5500) = 3355$ A/in^2.

$$s_b = \frac{I_b}{\Delta_b} = \frac{461.07}{3050} = 0.1512 \text{ in}^2$$

$$s_r = \frac{I_e}{\Delta_e} = \frac{1247.49}{3355} = 0.3718 \text{ in}^2$$

The bar depth-to-width ratio is chosen as $rdw = 3$. Then,

$$b_{r1} = \left(\frac{s_b}{rdw}\right)^{1/2} = \left(\frac{0.1512}{3}\right)^{1/2} = 0.2245 \text{ in}$$

Use a bar width of $b_{r1} = 0.250$ in.

$$d_{r1} - d_{r2} = (rdws_b)^{1/2} = (3 \times 0.1512)^{1/2} = 0.673 \text{ in}$$

Use a bar depth of $d_{r1} - d_{r2} = 0.700$ in, whence the rotor slot depth is $d_{r1} = 0.700 + d_{r2} = 0.700 + 0.116 = 0.816$ in. The rotor bar–end ring detail chosen is displayed by Fig. 6.54. Since the rotor cage is die-cast, there is no rotor bar over-hang ($b_r = 0$) and the rotor bar completely fills the rotor slot ($d_{2c} = d_{r1} - d_{r2}$, $b_{2c} = b_{r1}$). With the decision to make the end-ring depth the same as the rotor-bar depth, $d_{er} = 0.700$ in. Thus, the end-ring width is

$$w_{er} = \frac{s_r}{d_{er}} = \frac{0.3718}{0.700} = 0.531 \text{ in}$$

The flux density checks can now be made by use of [6.137] and [6.138].

$$D_r = D - 2\delta = 12.60 - 2(0.035) = 12.530 \text{ in}$$

$$t_{r2} = \frac{\pi(D_r - 2d_{r1})}{S_2} - b_{r1} = \frac{\pi(12.530 - 2 \times 0.816)}{51} - 0.250 = 0.421 \text{ in}$$

$$B_{rtm} = \frac{\pi p \Phi_m}{2 S_2 t_{r2} \ell_a SF} = \frac{\pi(6)(1.421 \times 10^6)}{2(51)(0.421)(7)(0.96)} = 92.8 \text{ kilolines/in}^2$$

Choosing a 4-in diameter shaft ($D_{sn} = 4$) and using 1-in diameter axial cooling air holes in the core area ($d_c = 1.00$),

$$B_{rcm} = \frac{\Phi_m}{(D_r - D_{sh} - 2d_c - 2d_{r1})\ell_a SF}$$

$$= \frac{1.421 \times 10^6}{(12.53 - 4.00 - 2 \times 1.00 - 2 \times 0.816)(7)(0.96)} = 43.2 \text{ kilolines/in}^2$$

Apparently the values of rotor tooth and core flux densities are low. Consequently, rotor slot size can be increased if needed in the design refinement stage.

When the values from the above work are entered in the data file ⟨imdesign.m⟩ and ⟨imeval.m⟩ is executed, the rated condition results shown below are displayed to the screen:

```
FLUX  DENSITY  CHECKS  -  kilolines/sq.  in
Bstm  =  103.1
Bscm  =  65.67
Brtm  =  116.9
Brcm  =  47.01


RATED  CONDITIONS
Speed(rpm)   =  1159
Current(A)   =  60.69
Torque(N_m)  =  307.5
Efficiency(%)  =  88.03
Power  factor  =  0.8753
```

Inspection of the results shows that the designed rated speed of 1175 rpm is not met. Hence, design refinement is necessary. Reduced slip operation can be accomplished by reduction of the rotor resistance R_2' with minimal difficulty. After some trial-and-error work, the following rotor dimensions are found to yield the desired full-load speed goal:

$$d_{2c} = 0.900 \text{ in} \qquad b_{2c} = 0.3125 \text{ in}$$

$$d_{er} = 0.900 \text{ in} \qquad w_{er} = 0.750 \text{ in}$$

The resulting screen display from execution of ⟨imeval.m⟩ with an updated ⟨imdesign.m⟩ is as follows:

```
FLUX  DENSITY  CHECKS  -  kilolines/sq.  in
Bstm  =  103.1
Bscm  =  65.67
Brtm  =  116.9
Brcm  =  47.01

RATED  CONDITIONS
Speed(rpm)   =  1176
Current(A)   =  59.87
Torque(N_m)  =  302.9
Efficiency(%) =  89.23
Power  factor  =  0.8738
```

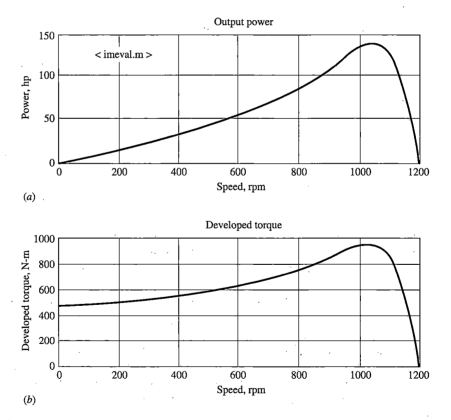

(a)

(b)

Figure 6.55
Performance curves for sample design motor

Final motor performance curves produced by ⟨imeval.m⟩ are shown as Fig. 6.55*a* to *e*.

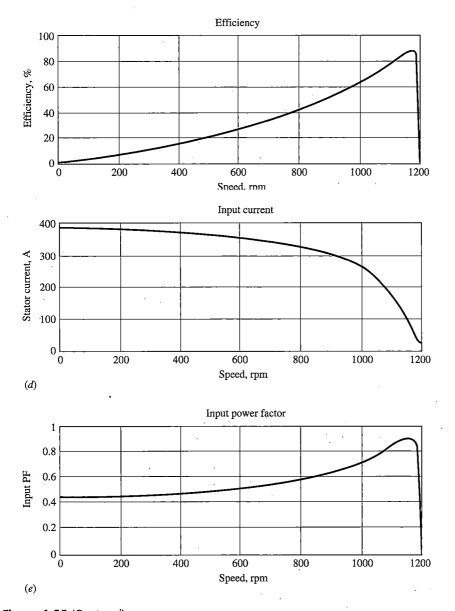

(d)

(e)

Figure 6.55 (Continued)
Performance curves for sample design motor

6.12 Computer Analysis Code

```
%%%%%%%%%%%%%%%%%%%%%%%%%%%%%%%%%%%%%%%%%%%%%%%%%%%%%%%%%%%%%%%%%%%%%%%%%%%%%%%
%
% sqmmf.m - Generates a normalized square wave mmf and its
%           associated Fourier spectrum.
%%%%%%%%%%%%%%%%%%%%%%%%%%%%%%%%%%%%%%%%%%%%%%%%%%%%%%%%%%%%%%%%%%%%%%%%%%%%%%%
clear; clf;
% Construct & plot square wave.
n=1024; x=ones(1,n); x(n/2:n)=-x(n/2:n);ang=linspace(0,360,n);
subplot(2,1,1); axis([0,360,-1.25,1.25]); plot(ang,x); grid;
title('Normalized air gap square wave mmf');
xlabel('Angular position, elect. deg'); ylabel('mmf/Ni, pu');
% Determine Fourier coefficients & plot in percent.
x=abs(fft(x)); x=x/max(x)*100;
subplot(2,1,2); axis; plot([0:25],x(1:26)); grid;
title('Fourier coefficients of square wave mmf');
xlabel('Harmonic number'); ylabel('Harmonic magnitude, %');



%%%%%%%%%%%%%%%%%%%%%%%%%%%%%%%%%%%%%%%%%%%%%%%%%%%%%%%%%%%%%%%%%%%%%%%%%%%%%%%
%
% phasemmf.m - Determines air gap mmfs for a p-pole, 3-phase
%              stator winding with integer slots/pole/phase.
%              Values are normalized by Ni.First set of plots
%              are mmfs for one coil pair. Second set of plots
%              are mmfs for phase winding.
%
%%%%%%%%%%%%%%%%%%%%%%%%%%%%%%%%%%%%%%%%%%%%%%%%%%%%%%%%%%%%%%%%%%%%%%%%%%%%%%%
clear;
skew=input('Skew in no. slots (0-1)   ');
pitch=input('Coil pitch in no. slots  ');
p=input('No. poles   ');
S1=input('No. stator slots  '); clf;
x=ones(1,1024)/2; y=zeros(1,1024); z=y;
% Generate quasi-trapezoidal mmf of single coil.
m=p/2*round(skew/S1/2*1024); n=p/2*round(pitch/S1*1024);
x(n:1024)=-x(n:1024);
for i=1:m; if m==0; break; end
x(i)=(i-1)/m/2; x(n+1-i)=x(i); x(n+i)=-x(i); x(1025-i)=-x(i);
end
y=-shift(x,-0.5); % Shift for mmf of second coil.
% Add mmfs to give coil pair resultant mmf.
ang=linspace(0,360,1024); z=x+y;
% Plot component & resultant mmfs.
subplot(2,1,1); plot(ang,x,'--',ang,y,'-.',ang,z);
axis([0 360 -1.25 1.25]); grid;
title('Normalized air gap mmf for coil pair');
```

```
xlabel('Angular position, elect. deg'); ylabel('mmf/Ni, pu');
legend('Coil 1', 'Coil 4', 'Coil pair',0);
z=abs(fft(z)); z=z/max(z)*100; % Determine Fourier coefficients.
subplot(2,1,2); axis; plot([0:15],z(1:16)); grid
title('Fourier coefficients of coil pair mmf');
xlabel('Harmonic number'); ylabel('Harmonic magnitude, %');
pause; figure;
% Construct mmf resulting from all coil pairs of a phase.
Nspp=fix(S1/2/3); z=x+y; y=z; vertscale=Nspp*max(z)+0.25;
subplot(2,1,1); plot(ang,z,'--');
axis([0,360,-vertscale,vertscale]); grid; hold;
title('Normalized air gap mmf for phase winding');
xlabel('Angular position, elect. deg'); ylabel('mmf/Ni, pu');
for i=2:Nspp; if Nspp==1; break; end
x=shift(z,-1/S1); if Nspp<=4; plot(ang,x,'--'); else; end
y=y+x; z=x; end
plot(ang,y,'b');
legend('Coil pair mmf', 'Phase mmf');
% Determine phase mmf Fourier coefficients.
y=abs(fft(y)); y=y/max(y)*100;
subplot(2,1,2);hold; plot([0:15],y(1:16)); grid;
title('Fourier coefficients of phase mmf');
xlabel('Harmonic number'); ylabel('Harmonic magnitude, %');



%%%%%%%%%%%%%%%%%%%%%%%%%%%%%%%%%%%%%%%%%%%%%%%%%%%%%%%%%%%%%%%%%%%%%%%%%%%%%%%%
%
% fwave.m - simulates air gap traveling mmf wave for one
%           magnetic pole-pair pitch of movement in 10
%           magnetic (electric) degree increments.
%
%%%%%%%%%%%%%%%%%%%%%%%%%%%%%%%%%%%%%%%%%%%%%%%%%%%%%%%%%%%%%%%%%%%%%%%%%%%%%%%%
clear; colordef white;
p=input('How many poles    ');    % No. of poles
theta=linspace(0,2*pi,90); n=length(theta);
ro=6; rs=5; rr=4;          % Outside, bore, & rotor radii
rgr=4.5;                   % Air gap center grid radius
Ro=ro*ones(1,n);Rs=rs*ones(1,n);Rgr=rgr*ones(1,n);Rr=rr*ones(1,n);
F=zeros(1,n); phi=0; cla;
for i=1:37; pause(0.001);
    for j=1:n; F(j)=rgr+0.45*sin(p/2*theta(j)-phi); end
polar(theta,Ro,'-.');hold on; polar(theta,Rs,'-.');
polar(theta,Rgr,'m'); polar(theta,Rr,'g-.');
text(0.9,rs+0.30,'Stator');text(-1.5,rr-0.90,'Rotor');
name=['Air gap mmf traveling wave';
    '                          ';
    '                          ']; title(name);
polar(theta,F,'r.'); hold off; phi=phi+pi/18;
end;
```

```
%%%%%%%%%%%%%%%%%%%%%%%%%%%%%%%%%%%%%%%%%%%%%%%%%%%%%%%%%%%%%%%%%%%%%%%%%%%%%%%
%
% rotmmf.m - simulates rotor-produced air gap traveling mmf
%            wave for one magnetic pole-pair pitch of movement
%            in 10 magnetic (electric) degree increments. Rotor
%            position is indicated by cage.
%
%%%%%%%%%%%%%%%%%%%%%%%%%%%%%%%%%%%%%%%%%%%%%%%%%%%%%%%%%%%%%%%%%%%%%%%%%%%%%%%
clear; colordef white;
p=input('How many poles     ');        % No. of poles
s=input('Rotor slip(per unit)    ');   % Rotor slip
nb=input('How many rotor bars     ');  % No. of rotor bars
theta=linspace(0,2*pi,90); n=length(theta);
rs=5; rr=4;                            % Bore & rotor radii
rgr=4.5;                               % Air gap center grid radius
Rs=rs*ones(1,n);Rgr=rgr*ones(1,n);Rr=rr*ones(1,n); cla;
F=zeros(1,n); phi=0; rb=0.96*rr; rotbar=[]; rotang=[];
for k=1:nb; rotbar=[rotbar;rb]; rotang=[rotang;0]; end
for i=1:37; pause(0.001);
   for j=1:n; F(j)=rgr+0.45*sin(p/2*theta(j)-phi); end
ang=(1-s)*2/p*phi; rotang=[];
   for k=1:nb; rotang=[rotang;ang+(k-1)*2*pi/nb]; end
polar(theta,Rs,'-.'); hold on;
polar(theta,Rgr,'m');polar(theta,Rr,'g-.');
polar(rotang,rotbar,'go'); polar(theta,F,'r.');
text(1.6,rs+0.45,'Stator bore');text(-1.6,rr-0.9,'Rotor');
name=['Rotor produced mmf wave & rotor position';
'                                             ';
'                                           ']; title(name);
hold off; phi=phi+pi/18;
end;


%%%%%%%%%%%%%%%%%%%%%%%%%%%%%%%%%%%%%%%%%%%%%%%%%%%%%%%%%%%%%%%%%%%%%%%%%%%%%%%
%
% imeqckt.m - determines parameters for the equivalent circuit of a
%             three-phase induction motor from test data.
%
%%%%%%%%%%%%%%%%%%%%%%%%%%%%%%%%%%%%%%%%%%%%%%%%%%%%%%%%%%%%%%%%%%%%%%%%%%%%%%%
clear;
p=4; f=60; design='B';         % No. poles, frequency, NEMA design
Vdc=10.8; Idc=25.0;            % dc voltage,current data (terminal-terminal)
% Blocked rotor data (phase voltage and power, line current, br freq.)
Vbr=22.26/sqrt(3); Ibr=27.0; Pbr=770/3; fbr=15;
% No-load data (phase voltage and power, line current, speed-rpm)
Vnl=230/sqrt(3); Inl=12.9; Pnl=791/3; nnl=1798;
Req=Pbr/Ibr^2; Zbr=Vbr/Ibr; Xeq=f/fbr*sqrt(Zbr^2-Req^2);
R1=1.05*Vdc/Idc/2; R2pr=Req-R1;  % 5% adjustment for ac value of R1
if design=='B'; X1=0.4*Xeq; X2pr=0.6*Xeq;
```

```
elseif design=='C'; X1=0.3*Xeq; X2pr=0.7*Xeq;
else; X1=0.5*Xeq; X2pr=0.5*Xeq;
end
thetanl=acos(Pnl/Vnl/Inl); ns=120*60/p; s=(ns-nnl)/ns;
E1=Vnl-Inl*exp(-j*thetanl)*(R1+j*X1); I2pr=E1/(R2pr/s+j*X2pr);
E1=abs(E1); I2pr=abs(I2pr);
Pc=Pnl-Inl^2*R1-I2pr^2*R2pr/s; Rc=E1^2/Pc;
if Pc<0; disp('CHECK FOR DATA ERROR — ESPECIALLY NO-LOAD SPEED'); end
Qm=Vnl*Inl*sin(thetanl)-Inl^2*X1-I2pr^2*X2pr; Xm=E1^2/Qm;
Parameters=[ '          R1             ' 'R2pr          ' 'X1            '...
             'X2pr          ' 'Rc            ' 'Xm           ' ]
format short e
[ R1 R2pr X1 X2pr Rc Xm ]


%%%%%%%%%%%%%%%%%%%%%%%%%%%%%%%%%%%%%%%%%%%%%%%%%%%%%%%%%%%%%%%%%%%%%%%%%%%%%%%
%
% Te_wm.m - calculates total developed torque ( 3Td )
%           using (6.52). Also determines VTh, RTh, & XTh.
%           Assumes V1 on reference. Requires equivalent
%           circuit parameter values.
%
%%%%%%%%%%%%%%%%%%%%%%%%%%%%%%%%%%%%%%%%%%%%%%%%%%%%%%%%%%%%%%%%%%%%%%%%%%%%%%%
clear;
p=4; f=60;        % Poles, frequency
R1=0.2268; R2pr=0.12528; X1=0.51252;          % Equiv. ckt. values
X2pr=0.76878;Rc=185.1; Xm=9.8554;
V1=230/sqrt(3);                               % Phase voltage
RcXm=Rc*j*Xm/(Rc+j*Xm);
VTh=RcXm*V1/(R1+j*X1+RcXm); ang=angle(VTh)*180/pi; VTh=abs(VTh);
ZTh=RcXm*(R1+j*X1)/(RcXm+R1+j*X1); RTh=real(ZTh); XTh=imag(ZTh);
Thevenin_Values=[ '       VTh' '         Theta'...
    '        RTh' '          Xth' ]
format short e
[ VTh ang RTh XTh ]
pause;
npts=99; s=linspace(0.00001,1,npts); s=fliplr(s); ws=2/p*2*pi*f;
R2pr0=R2pr; X2pr0=X2pr; smax=R2pr/sqrt(RTh^2+(XTh+X2pr)^2);
for i=1:npts
% Activate indented code below to frequency-adjust R2pr & X2pr.
%    if i ==1; disp('R2pr & X2pr empirically adjusted for fr'); end
%    if s(i)>=smax
%    R2pr=(0.5+0.5*sqrt(s(i)/smax))*R2pr0;
%    X2pr=(0.4+0.6*sqrt(smax/s(i)))*X2pr0;
%    else; R2pr=R2pr0; X2pr=X2pr0;
%    end
TTd(i)=3*VTh^2*R2pr/s(i)/ws/((RTh+R2pr/s(i))^2+(XTh+X2pr)^2);
nm(i)=(1-s(i))*ws*30/pi;
end
```

```
plot(nm,TTd);grid;
title('Speed-torque curve');
xlabel('Shaft speed, rpm'); ylabel('Torque (3Td), N-m');


%%%%%%%%%%%%%%%%%%%%%%%%%%%%%%%%%%%%%%%%%%%%%%%%%%%%%%%%%%%%%%%%%%%%%%%%%%%
%
% im_perf.m - calculates induction motor performance ( I1, PF,
%             3Td, Ps, efficiency ) based on equivalent circuit
%             of Fig. 6.17. Reads equivalent circuit parameters from im_data.m
%             Assumes F&W losses vary as nth power of speed.
%
%%%%%%%%%%%%%%%%%%%%%%%%%%%%%%%%%%%%%%%%%%%%%%%%%%%%%%%%%%%%%%%%%%%%%%%%%%%%
clear; clf;
im_data,
V1=230/sqrt(3); f=60; p=6; % Phase voltage, frequency, poles
Pfw=224; n=2.8;      % Total F&W losses at syn. speed, speed dependence
npts=200; s=linspace( 0.00001,1,npts); s=fliplr(s);
I1=zeros(1,npts); TTd=I1; PF=I1; Ps=I1; eff=I1; nm=I1;
ws=2/p*2*pi*f; ns=120*f/p;    % Synchronous speed
% Empirical adjustment of R2pr & X2pr for fr variation
R2pr0=R2pr; X2pr0=X2pr; smax=R2pr/sqrt(R1^2+(X1+X2pr)^2);
for i=1:npts
   %if s(i) > smax
   %R2pr=(0.5+0.5*sqrt(s(i)/smax))*R2pr0;
   %X2pr=(0.4+0.6*sqrt(smax/s(i)))*X2pr0;
   %else; R2pr=R2pr0; X2pr=X2pr0; end
Z2=R2pr/s(i)+j*X2pr; Zm=j*Rc*Xm/(Rc+j*Xm);
Zin=R1+j*X1 + Z2*Zm/(Z2+Zm);
I11=V1/Zin; I1(i)=abs(I11); PF(i)=cos(angle(I11));
I2pr=abs(Zm/(R2pr/s(i)+j*X2pr+Zm)*I11); TPin=3*V1*I1(i)*PF(i);
TTd(i)=3*I2pr^2*R2pr/s(i)/ws; nm(i)=(1-s(i))*ns;
PPo=TTd(i)*(1-s(i))*ws-Pfw*(nm(i)/ns)^n;
if PPo<0; break; else;
Ps(i)=PPo; eff(i)=100*Ps(i)/TPin; end
end
subplot(2,1,1), plot(nm,TTd); grid; title('Developed torque');
xlabel('Speed, rpm'); ylabel('Torque, N-m');
subplot(2,1,2), plot(nm,Ps/746); grid; title('Output power');
xlabel('Speed, rpm'); ylabel('Output power, hp');
figure(2); subplot(2,1,1);
plot(nm,I1); grid; title('Input current');
xlabel('Speed, rpm'); ylabel('Current, A');
subplot(2,1,2); plot(nm,eff); grid; title('Efficiency');
xlabel('Speed, rpm'); ylabel('Efficiency, %');
figure(3); subplot(2,1,1);
plot(nm,PF); grid; title('Input power factor');
xlabel('Speed, rpm'); ylabel('Power factor');
```

```
%%%%%%%%%%%%%%%%%%%%%%%%%%%%%%%%%%%%%%%%%%%%%%%%%%%%%%%%%%%%%%%%%%%%%%%%%%%
%
% im_data.m - contains induction motor equivalent circuit
%             parameter values to be read by calling program.
%
%%%%%%%%%%%%%%%%%%%%%%%%%%%%%%%%%%%%%%%%%%%%%%%%%%%%%%%%%%%%%%%%%%%%%%%%%%%%
R1=0.2268;  R2pr=0.12528;
X1=0.51252; X2pr=0.76878;
Rc=185.1; Xm=9.8554;


%%%%%%%%%%%%%%%%%%%%%%%%%%%%%%%%%%%%%%%%%%%%%%%%%%%%%%%%%%%%%%%%%%%%%%%%%%%%
%
% spwave.m - simulates the component forward & backward
%            air gap mmf waves of a single-phase induction
%            motor for one magnetic pole-pair pitch of
%            movement in 10 magnetic (electric) degree
%            increments. Components are then added to give
%            resultant air gap standing wave.
%
%%%%%%%%%%%%%%%%%%%%%%%%%%%%%%%%%%%%%%%%%%%%%%%%%%%%%%%%%%%%%%%%%%%%%%%%%%%%
clear; colordef white;
p=input('How many poles     ');    % No. of poles
theta=linspace(0,2*pi,50); n=length(theta);
ro=6; rs=5; rr=4;               % Outside, bore, & rotor radii
rgr=4.5;                        % Air gap center grid radius
Ro=ro*ones(1,n);Rs=rs*ones(1,n);Rgr=rgr*ones(1,n);Rr=rr*ones(1,n);
F=zeros(1,n); F1=F; F2=F; phi=pi/4;
for i=1:37; pause(0.001);
   polar(theta,Ro,'-.'); hold on; polar(theta,Rs,'-.');
   polar(theta,Rgr,'m'); polar(theta,Rr,'g-.');
for j=1:n;
F1(j)=Rr(j)+0.5+0.225*cos(p/2*theta(j)-phi);
F2(j)=Rr(j)+0.5+0.225*cos(p/2*theta(j)+phi);
F(j)=F1(j)+F2(j)-Rr(j)-0.5; end
polar(theta,F,'r'); polar(theta,F1,'y:'); polar(theta,F2,'c-.');
text(0.9,rs+0.25,'Stator');text(-1.5,rr-0.75,'Rotor');
text(5.5,5,'.. Forward');text(5.5,4,'-. Backward');
text(5.5,6,'- Resultant');
name=['Air gap mmf waves of single-phase induction motor';
   '                                                      ';
   '                                                      ']; title(name);
phi=phi+pi/18; hold off;
end


%%%%%%%%%%%%%%%%%%%%%%%%%%%%%%%%%%%%%%%%%%%%%%%%%%%%%%%%%%%%%%%%%%%%%%%%%%%%
%
% torq1ph.m - calculates torque-speed curve for
%             single-phase induction motor.
%
%%%%%%%%%%%%%%%%%%%%%%%%%%%%%%%%%%%%%%%%%%%%%%%%%%%%%%%%%%%%%%%%%%%%%%%%%%%%
```

```
clear; clf;
p=4; f=60; Vm=120*sqrt(2); % p=poles, f=frequency
Rs=2.02; Xs=2.79; Xm=106.8; Rr=4.12; Xr=2.12;
smax=2; npts=100;    % smax=max slip, npts=points for curve
ws=2/p*2*pi*f;        % Synchronous speed
Zs=Rs+Xs*j; Zm=0+Xm/2*j;
s=zeros(1,npts); dels=smax/npts;
for i=2:npts-1; s(i)=(i-1)*dels; end
s(1)=0.001; s(npts)=smax-0.001;
Td=zeros(1,npts); Tdf=Td; Tdb=Td; wm=Td;
for i=1:npts  % Loop to calculate torque points
ZU=Rr/2/s(i)+Xr/2*j; ZL=Rr/2/(2-s(i))+Xr/2*j;
Zf=Zm*ZU/(Zm+ZU); Zb=Zm*ZL/(Zm+ZL);
Is=Vm/sqrt(2)/(Zs+Zf+Zb);
Irf=abs(Is*Zm/(Zm+ZU)); Irb=abs(Is*Zm/(Zm+ZL));
wm(i)=(1-s(i))*ws;
if wm(i) == 0
Tdf(i)=0.5*Irf^2*Rr/s(i)/ws; Tdb(i)=-0.5*Irb^2*Rr/(2-s(i))/ws;
else;
Tdf(i)=0.5*Irf^2*Rr*(1-s(i))/wm(i)/s(i);
Tdb(i)=0.5*Irb^2*Rr*(s(i)-1)/wm(i)/(2-s(i));
end
Td(i)=Tdf(i)+Tdb(i);
nm=wm*30/pi;
end
plot(nm,Td,nm,Tdf,'--',nm,Tdb,'-.');grid
title('Torque-speed for single-phase induction motor');
xlabel('Speed, rpm'); ylabel('Torque, N-m');


%%%%%%%%%%%%%%%%%%%%%%%%%%%%%%%%%%%%%%%%%%%%%%%%%%%%%%%%%%%%%%%%%%%%%%%%%
%
% sp_perf.m - calculates single-phase induction motor performance
%            I1,PF,Td, Ps, efficiency ) based on equivalent circuit
%            of Fig. 6.39.
%            Assumes F&W losses vary as nth power of speed.
%
%%%%%%%%%%%%%%%%%%%%%%%%%%%%%%%%%%%%%%%%%%%%%%%%%%%%%%%%%%%%%%%%%%%%%%%%%
clear;
V1=120; f=60; p=4;  % Phase voltage, frequency, poles
Rs=2.02; Xs=2.79; Xm=106.8; Rr=4.12; Xr=2.12;
Pfw=10.5; n=2.8;     % Total F&W losses at syn. speed, speed dependence
npts=100; s=linspace( 0.0001,1,npts); s=fliplr(s);
I1=zeros(1,npts); Td=I1; PF=I1; Ps=I1; eff=I1; nm=I1;
ws=2/p*2*pi*f; ns=120*f/p;   % Synchronous speed
Zs=Rs+Xs*j; Zm=0+Xm/2*j;
for i=1:npts  % Loop to calculate torque points
ZU=Rr/2/s(i)+Xr/2*j; ZL=Rr/2/(2-s(i))+Xr/2*j;
Zf=Zm*ZU/(Zm+ZU); Zb=Zm*ZL/(Zm+ZL);
```

```
Is=V1/(Zs+Zf+Zb); I1(i)=abs(Is); nm(i)=(1-s(i))*ns;
Irf=abs(Is*Zm/(Zm+ZU)); Irb=abs(Is*Zm/(Zm+ZL));
PF(i)=cos(angle(Is)); Pin=V1*I1(i)*PF(i);
Td(i)=0.5*Irf^2*Rr/s(i)/ws-0.5*Irb^2*Rr/(2-s(i))/ws;
if Td(i)<0; Td(i)=0; end
TPs=Td(i)*(1-s(i))*ws - Pfw*(nm(i)/ns)^n;
if TPs<0; break; else;
Ps(i)=TPs; eff(i)=100*Ps(i)/Pin; end
end
subplot(2,1,1), plot(nm,Td); grid; title('Developed torque');
xlabel('Speed, rpm'); ylabel('Torque, N-m');
subplot(2,1,2), plot(nm,Ps/746); grid; title('Output power');
xlabel('Speed, rpm'); ylabel('Output power, hp');
figure(2);
subplot(2,1,1); plot(nm,I1); grid; title('Input current');
xlabel('Speed, rpm'); ylabel('Current, A');
subplot(2,1,2); plot(nm,eff); grid; title('Efficiency');
xlabel('Speed, rpm'); ylabel('Efficiency, %');
figure(3);
subplot(2,1,1); plot(nm,PF); grid; title('Input power factor');
xlabel('Speed, rpm'); ylabel('Power factor');


%%%%%%%%%%%%%%%%%%%%%%%%%%%%%%%%%%%%%%%%%%%%%%%%%%%%%%%%%%%%%%%%%%%%%%%%%%
%
% imeval.m - Calculates induction motor performance
%            given winding & dimensional data
%            Calls imdesign.m for input data
%            Calls Pc.m during Rc calculation
%
%%%%%%%%%%%%%%%%%%%%%%%%%%%%%%%%%%%%%%%%%%%%%%%%%%%%%%%%%%%%%%%%%%%%%%%%%%
clear;
imdesign,     % imdesign.m contains dimension, winding, & rating data
V1=VL/sqrt(3)+0*j; ws=2/p*2*pi*f;
Ns1=S1/p/m; Cphi=Ns1*Cs*p/a; del=(D-Dr)/2;
lam1=pi*D/S1; lam2=pi*Dr/S2;
% Fringing coefficients
qty=lam1*(5*del+bs1); ks=qty/(qty-bs1^2);
qty=lam2*(4.4*del+0.75*br2); kr=qty/(qty-br2^2);

% Stator winding factor
gam=p*pi/S1; kd=sin(Ns1*gam/2)/Ns1/sin(gam/2);
rho=pitch*p/S1*pi; kp=cos(pi/2-rho/2); kw1=kd*kp;

% Flux density checks
phim=0.97e08*a*V1/2.22/kd/kp/p/Ns1/f/Cs;
ts1=lam1-bs1; Bstm=pi*p*phim/2/S1/ts1/la/SF;
Bscm=phim/(Do-D-2*ds1)/la/SF;
tr2=pi*(Dr-2*dr1)/S2-br1; Brtm=pi*p*phim/2/S2/tr2/la/SF;
```

```
Brcm=phim/(Dr-Dsh-2*dc-2*dr1)/la/SF;
clc; disp(' '); disp(' ');
disp('FLUX DENSITY CHECKS - kilolines/sq. in'); disp(' ');
disp(['Bstm = ', num2str(Bstm/1000)]);
disp(['Bscm = ', num2str(Bscm/1000)]);
disp(['Brtm = ', num2str(Brtm/1000)]);
disp(['Brcm = ', num2str(Brcm/1000)]);
% R1 calculation
alf=asin((bs1+0.050)/lam1); tau=pi*D/p;
lalf=rho*tau/(2*pi*cos(alf)); lc=la+1+2*(lalf+ds1);
R1=1.056e-06*lc*Cphi/a/sa1; R1=1.05*R1; % ac value

% Xm calculation
dele=ks*kr*del;
Xm=1.6713e-07*f*D*la/p/dele*(kw1*S1*Cs/2/a/m)^2;

% Rc calculation

Bt13=pi*p*phim/2/(pi*(D+2*ds1/3)-S1*bs1)/la/SF;
Bscc=0.866*phim/(Do-D-2*ds1)/la/SF;
Wt=(pi/4*((D+2*ds1)^2-D^2)-S1*bs1*ds1)*la*SF*0.283;
Pt=Wt*Pc(Bt13)*4;
Wy=pi/4*(Do^2-(D+2*ds1)^2)*la*SF*0.283;
Py=Wy*Pc(Bscc)*4;
Rc=3*(0.97*VL/sqrt(3))^2/(Pt+Py);

% X1 components calculation
thetask=p*pi/S1;
Xsk=Xm*thetask^2/12;
pupitch=p*pitch/S1;        % Phase factor
if pupitch <= 0.5; Ks=pupitch;
elseif pupitch <= 2/3; Ks=-0.25+1.5*pupitch;
else; Ks=0.25+0.75*pupitch; end
Fs=(ds1-ds0)/3/bs1+ds0/bs1;
Xss=2.0055e-07*m*f*Cphi^2*la*Ks*Fs/S1;
if S2<S1; ks1=1; else; ks1=S1/S2; end
a1=abs(pi*D/S1-bs1-br2)/2;
Fz=a1/4/del+dr2*(1-ks1/2)/br2;
Xz1=2.0055e-07*m*f*Cphi^2*la*Ks*Fz/S1;
Fes=1+4*log(3.94*D/S1/sqrt(bs1*ds1));
Xes=0.3192e-07*f*(kw1*Cphi)^2*(0.5+lalf)*Ks*Fes/p;

% X2 components calculation
if S1<S2; kr1=1; else; kr1=S2/S1; end
a2=abs(pi*Dr/S2-bs1-br2)/2;
q2=S2/m/p; gamr=p*pi/S2; kdr=sin(q2*gamr/2)/q2/sin(gamr/2);
rhor=fix(S2/p)*p*pi/S2; kpr=cos((pi-rhor)/2); kwr=kdr*kpr;
Fr=(dr1-dr2)/3/br1+dr2/br2;
```

```
Xrs=2.0055e-07*m*f*(kw1*Cphi)^2*la*Fr/kwr^2/S2;
Fz2=a2/4/del+ds0*(1-kr1/2)/bs1;
Xz2=2.0055e-07*m*f*(kw1*Cphi)^2*la*Fz2/kwr^2/S2;
fer1=S2*Der/m/p*(1+4*log(3.94*Dr/S2/sqrt(wer/der)));
fer2=4*br*p*(1+4*log(3.94*Dr/S2/sqrt(br1*dr1)));
Xer=7.98e-09*f*(kw1*Cphi)^2/p^2*(fer1+fer2);

I1m=zeros(1,100); PF=I1m; TTd=I1m; eta=I1m; Ps=I1m;
wm=linspace(0,ws-0.01,100); npts=length(wm);
for i=1:npts;      % Speed loop - performance calculations
s=(ws-wm(i))/ws; fr=s*f;
if krc==1; kalf=0.3505; kbet=1; else; kalf=0.2700; kbet=1.64; end
alfd=kalf*sqrt(fr*b2c/br1)*d2c;
K1=alfd*(sinh(2*alfd)+sin(2*alfd))/(cosh(2*alfd)-cos(2*alfd));
J1=3*(sinh(2*alfd)-sin(2*alfd))/(2*alfd*(cosh(2*alfd)-cos(2*alfd)));
% R2 calculation with skin effect adjustment
R2=kbet*1.169e-06*m*(kw1*Cphi/kwr)^2*(lb*K1/S2/sb+2*Der*Kring/pi/p^2/sr);
X1=Xsk+Xes+Xss+Xz1;
X2=Xer+Xz2+Xrs*J1;       % Adjustment for skin effects
Z1=R1+X1*j; Z2=R2/s+X2*j; ZM=(0+Rc*Xm*j)/(Rc+Xm*j);
Z=Z1*Z2+Z1*ZM+Z2*ZM; I1=V1*(Z2+ZM)/Z; I1m(1,i)=abs(I1);  % Stator current
I2=ZM/(Z2+ZM)*I1; E=I2*Z2;
PF(1,i)=cos(atan2(imag(Z/(Z2+ZM)),real(Z/(Z2+ZM)))); % Power factor
TTd(1,i)=3*abs(I2)^2*R2/s/ws;                % Developed torque
Pfw=0.03*HP*746*(wm(i)/ws)^(2.8);  % F&W losses
Ps(1,i)=TTd(1,i)*wm(i)-Pfw; if Ps(1,i)<0; Ps(1,i)=0; end
Pin=3*real(V1*conj(I1)); eta(1,i)=Ps(1,i)/Pin*100;    % Efficiency
end;
nm=wm*30/pi; Ps=Ps/746;
% Plot performance
figure(1);
subplot(2,1,1), plot(nm,Ps); grid; title('Output power');
xlabel('Speed, rpm'); ylabel('Power, hp');
subplot(2,1,2), plot(nm,TTd);grid; title('Developed torque');
xlabel('Speed, rpm'); ylabel('Developed torque, N-m');
figure(2);
subplot(2,1,1), plot(nm,I1m);grid; title('Input current');
xlabel('Speed, rpm'); ylabel('Stator current, A');
subplot(2,1,2), plot(nm,PF);grid; title('Input power factor');
xlabel('Speed, rpm'); ylabel('Input PF');
figure(3);
subplot(2,1,1), plot(nm,eta);grid; title('Efficiency');
xlabel('Speed, rpm'); ylabel('Efficiency, %');
% Linear interpolation for rated conditions
for i=1:npts-1; if Ps(npts-i) >= HP; k=npts-i; break; end; end;
if i==npts-1; disp('DESIGN CANNOT PRODUCE RATED OUTPUT');
else;
nmR=nm(k+1)-(HP-Ps(k+1))*(nm(k+1)-nm(k))/(Ps(k)-Ps(k+1));
```

```
delnm=nm(k+1)-nm(k); delnm1=nm(k+1)-nmR;
I1R=I1m(k+1)+(I1m(k)-I1m(k+1))*delnm1/delnm;
TTdR=TTd(k+1)+(TTd(k)-TTd(k+1))*delnm1/delnm-0.03*HP*746* ...
(nmR*pi/30)^(1.8)/ws^(2.8);
etaR=eta(k+1)+(eta(k)-eta(k+1))*delnm1/delnm;
PFR=PF(k+1)+(PF(k)-PF(k+1))*delnm1/delnm;
disp(' '); disp(' ');disp('RATED CONDITIONS');disp(' ');
disp(['Speed(rpm) = ',num2str(nmR)]);
disp(['Current(A) = ',num2str(I1R)]);
disp(['Torque(N_m) = ',num2str(TTdR)]);
disp(['Efficiency(%) = ',num2str(etaR)]);
disp(['Power factor = ',num2str(PFR)]);
end



%%%%%%%%%%%%%%%%%%%%%%%%%%%%%%%%%%%%%%%%%%%%%%%%%%%%%%%%%%%%%%%%%%%%%%%%%%%%%%
%
% imdesign.m - Provides input data for imeval.m
%              Limited to parallel-sided, open stator slots
%              and rectangular rotor bars.
%              ( All dimensions in inches & areas in sq. in.)
%
%%%%%%%%%%%%%%%%%%%%%%%%%%%%%%%%%%%%%%%%%%%%%%%%%%%%%%%%%%%%%%%%%%%%%%%%%%%%%%
% Rotor conductor material; Copper assumed for stator conductors.
krc=2; % 1=copper; 2=aluminum

% Winding data
VL=460; f=60; p=6; m=3; HP=50;
S1=54; S2=51; Cs=8; pitch=8; a=1;

% Conductor data
sa1=0.01563; sb=0.250*0.700; sr=0.700*0.531;
d2c=0.673; b2c=0.250;

% General dimensions
D=12.600; Dr=12.530; Do=17.62;
Dsh=4.00; dc=1.00; la=7.00; SF=0.96;

% Stator slot dimensions
bs1=0.375; ds0=0.185; ds1=0.900;

% Rotor slot dimensions
br2=0.171; br1=0.3125;
dr2=0.116; dr1=1.016;

% Squirrel cage data
Der=11.398; der=0.700; wer=0.531;
lb=7.00; br=0.00; Kring=1.00;
```

```
%%%%%%%%%%%%%%%%%%%%%%%%%%%%%%%%%%%%%%%%%%%%%%%%%%%%%%%%%%%%%%%%%%%%%%%%%%%%%%
%
% Pc.m - Determines Epstein loss given value of B
%        Used by imperf.m
%
%%%%%%%%%%%%%%%%%%%%%%%%%%%%%%%%%%%%%%%%%%%%%%%%%%%%%%%%%%%%%%%%%%%%%%%%%%%%%%
function y = Pc(Bx)
% B-Wlb values that follow are valid for M-27,24 ga,60 Hz
B=[0 10 19 24 30 35 40 45 50 55 60 64 70 75 80 85 90 ...
95 100 105 110 115 120 125 130 135 138 165 210 ]*1000;
Wlb=[0 0.08 0.15 0.195 0.27 0.35 0.43 0.52 0.625 0.76 ...
0.89 1.0 1.2 1.35 1.55 1.75 2.0 2.25 2.55 2.8 3.15 3.5 ...
3.85 4.2 4.7 5.2 5.6 11.0 30.0 ];
% Activate to plot B-Wlb curve
% m=29; plot(Wlb(1:m),B(1:m)); grid; pause; % Linear plot
% m=29; semilogx(Wlb(2:m),B(2:m)); grid; pause; % Semilog plot
n=length(B); k=0;
if Bx==0; k=-1; y=0; end
if Bx<0; k=-1; y=0; disp('WARNING - Bx < 0, Wlb = 0 returned'); end
if Bx>B(n); y=Wlb(n); k=-1; disp('CAUTION - Beyond B-Wlb curve');end
for i=1:n
if k==0 & (Bx-B(i))<=0; k=i; break; end
end
if k>0;
y=Wlb(k-1)+(Bx-B(k-1))/(B(k)-B(k-1))*(Wlb(k)-Wlb(k-1));
else;
end


%%%%%%%%%%%%%%%%%%%%%%%%%%%%%%%%%%%%%%%%%%%%%%%%%%%%%%%%%%%%%%%%%%%%%%%%%%%%%%
%
% Td_nmvf.m - Plots total developed torque (TTd) vs speed (nm)
%             for variable-frequency induction motor operation.
%
%%%%%%%%%%%%%%%%%%%%%%%%%%%%%%%%%%%%%%%%%%%%%%%%%%%%%%%%%%%%%%%%%%%%%%%%%%%%%%
clear; clf;
R1=0.2; R2=0.3; X1=0.754; X2=0.754;
Rc=110; Xm=33.9;
p=4; fR=60; VR=460; nmLim=2100;   % nmLim is max allowable speed

% Thevenin resistance & inductance
Zm=j*Xm*Rc/(Rc+j*Xm);
ZTh=Zm*(R1+j*X1)/(R1+j*X1+Zm);
RTh=real(ZTh); LTh=imag(ZTh)/2/pi/fR;
L1=X1/2/pi/fR; L2=X2/2/pi/fR; Lm=Xm/2/pi/fR;

% Frequencies for analysis in addition to rated
nf=14; f=linspace(nmLim/nf,nmLim,nf)*p/120;
```

```
% Rated frequency calculations( Base curve )
npts=50; ws=2/p*2*pi*fR; wm=linspace(0,ws-0.01,npts);
w=2*pi*fR; VTh=abs(VR/sqrt(3)*Zm/(R1+j*X1+Zm));
for i=1:npts
   s=(ws-wm(i))/ws;
   TTd(i)=3*VTh^2*R2/s/ws/( (RTh+R2/s)^2+w^2*(LTh+L2)^2 );
end
plot(wm*30/pi,TTd);grid;
title('Induction motor torque - speed for variable frequency');
xlabel('Speed, rpm'); ylabel('Total developed torque, N-m');
fdev=f(2)-f(1);
text(0.65,0.95,['Frequency increment: ',num2str(fdev),' Hz'],'sc')
text(0.65,0.92,['Base curve(solidline): ',num2str(fR),' Hz'],'sc')
hold on;  % Hold base curve for overplotting

for k=1:nf  % Other than rated frequency calculations
ws=2/p*2*pi*f(k); wm=linspace(0,ws-0.01,npts);
w=2*pi*f(k);

% Set voltage-frequency control
if f(k)/fR <= 1; b=0.06*VR; VL=(VR-b)*f(k)/fR+b; else; VL=VR; end

% Empirical adjustment of core loss resistance
Rcf=0.5*Rc*(VR/VL*f(k)/fR)^2*(fR/f(k)+(fR/f(k))^2);
Zm=j*w*Lm*Rcf/(Rcf+j*w*Lm);
ZTh=Zm*(R1+j*w*L1)/(R1+j*w*L1+Zm);
RTh=real(ZTh); LTh=imag(ZTh)/2/pi/f(k);
VTh=abs(VL/sqrt(3)*Zm/(R1+j*w*L1+Zm));
for i=1:npts
   s=(ws-wm(i))/ws;
   TTd(i)=3*VTh^2*R2/s/ws/( (RTh+R2/s)^2+w^2*(LTh+L2)^2 );
end

% Determination of points above nmmax for plot
smax=R2/sqrt(R1^2+w^2*(LTh+L2)^2); wmmax=(1-smax)*ws;
for m=1:npts; if wm(m)>=wmmax; break; end; end
plot(wm(m:npts)*30/pi, TTd(m:npts),'--');
end
% Activate to superimpose load torque plot
%TL=50+0.004052847*wm .^2;
%plot(wm*30/pi, TL,'-.');
hold off;
```

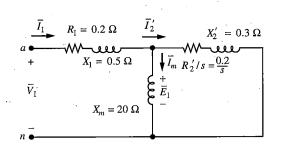

Figure 6.56

$\bar{V}_1 = 220/\sqrt{3}\angle 0°$, it is known that $\bar{V}_{Th} = 123.9\angle 0.56°$ V. Also, $Z_{Th} = 0.19 + j0.49\ \Omega$. The friction and windage losses are given by $P_{FW} = 77.2 \times 10^{-6}n_m^2$ where P_{FW} is in W if n_m is in units of rpm. If the motor is operating at maximum torque condition, determine (a) the total developed torque ($3T_{dmax+}$), (b) the value of input current (I_1), (c) the input power factor (PF), and (d) the value of output power (P_s).

6.13 Rework Prob. 6.12 if the motor is operating at 5 percent slip with all else unchanged.

6.14 The motor of Prob. 6.12 is driven by a mechanical source so that $n_m = n_s$ and all else is unchanged. Determine (a) the value of input current (I_1), (b) the input power factor (PF), and (c) the air gap power ($3P_g$).

6.15 For the induction motor of Example 6.7 operating at rated conditions, calculate (a) the value of total power flowing across the air gap ($3P_g$) and (b) the value of total power converted from electrical to mechanical form ($3P_d$).

6.16 Figure 6.24 shows a wound-rotor motor with a balanced external resistor set connected in the secondary circuit. Discuss the impact on maximum torque ($3T_{dmax+}$) and the speed at which it occurs if R_{ext} were replaced by a L_{ext}.

6.17 Repeat Prob. 6.16 if R_{ext} is replaced by a C_{ext}. Your answer should consider both the case of a small C_{ext} and a large C_{ext}.

6.18 Figure 6.13 shows a shorted rotor coil set with the short connected from a to b to c. Show that for the case of balanced currents, the connection acts no different than if each of the individual phase coil terminals were shorted given $v_a = v_b = v_c = 0$.

6.19 Derive a formula to determine the value of slip that exists for $3P_{dmax}$, the maximum value of power converted from electrical to mechanical form for a three-phase induction motor. (*Hint:* $3P_d = (1 - s)\omega_s 3T_d$)

6.20 A small slip approximation for developed torque is given by [6.55]. Modify ⟨Te_wm.m⟩ to plot the results of [6.55] on the same axes as the exact value of developed torque. You may want to limit the range on slip to $0.1 \le s \le 0$. For the motor of Prob. 6.4, determine the speed range for which [6.55] yields a value of developed torque with no more than 3 percent error.

6.21 The test data that follows were taken on a 4-pole, 3-phase, 230-V, 60-Hz, NEMA design B induction motor:

Blocked rotor: 26.9 V, 68.5 A, 15 Hz, 2570 W

DC stator excitation: 14.5 V, 68.3 A

No-load: 230 V, 60 Hz, 22 A, 1310 W, 1797 rpm

By use of ⟨imeqckt.m⟩, determine the equivalent circuit parameters for this induction motor.

6.22 Use ⟨phasemmf.m⟩ to determine the Fourier spectrum of the phase mmf for the sample design motor of Sec. 6.11.7.

6.23 Calculate the expected full-voltage locked-rotor current for the Code F induction motor of Example 6.5 and compare the result with the calculated value of Fig. 6.26.

6.24 A 120-V, 60-Hz, 4-pole, 1/4-hp single-phase induction motor is described by the equivalent circuit parameters:

$$R_s = 2.5 \ \Omega \qquad R_r = 2.4 \ \Omega$$

$$X_s = 2.8 \ \Omega \qquad X_r = 1.7 \ \Omega$$

$$X_m = 60 \ \Omega$$

The synchronous speed rotational losses are 25 W and vary with speed to the 2.5 power. Use ⟨sp_perf.m⟩ to determine the full-load speed, input current, power factor, and efficiency.

6.25 Design a three-phase induction motor that matches the following specifications:

460 V   60 Hz   15 hp   1170 rpm

NEMA design B   284T frame

REFERENCES

1. T. C. Lloyd and H. B. Stone, "Some Aspects of Polyphase Motor Design," *Trans. AIEE,* vol. 65, December 1946, pp. 812–818.
2. J. H. H. Kuhlmann, *Design of Electrical Apparatus,* 3d ed., John Wiley & Sons, New York, 1950.
3. P. L. Alger, *Induction Machines,* 2d ed., Gordon and Breach, New York, 1970.
4. P. H. Trickey, "Induction Motor Resistance Ring Width," *Electrical Engineering,* February 1936, pp. 144–150.
5. P. L. Alger, "The Calculation of Armature Reactance in Synchronous Machines," *Trans. AIEE,* vol. 47, February 1928, pp. 493–513.

SYNCHRONOUS MACHINES

7.1 INTRODUCTION

The synchronous machine in steady-state operation is characterized by the mechanical speed of the movable member. This speed is identical in value to the average speed of the magnetic field established by the phase winding set that conducts alternating currents. As a direct consequence, the precise mechanical speed of a synchronous machine in steady-state operation is determined by or synchronized with the frequency of the phase winding alternating currents. The collective installed power rating of synchronous machines exceeds that of induction motors, but since a significant percentage of the synchronous machines are large in power rating, the population count is less than for induction motors.

The single-phase synchronous machine is available for low-power, precise-speed applications for operation directly from distribution level voltages. The electric clock is one such application that utilizes hysteresis motors (Fig. 7.1)—a synchronous machine type that is beyond the scope of this text. Another low-power synchronous machine application of great importance is computer disk drives. Although the personal computer is supplied by single-phase power, the disk drive motors are typically brushless dc motors fed from conditioned power rather than directly fed from distribution level voltage. It will be seen in Sec. 7.8 that the brushless dc motor (Fig. 7.2), as well as the switched reluctance motor, is actually a polyphase machine.

Common industrial motor applications use the three-phase synchronous machine (Fig. 7.3) with typical voltage ratings in the United States ranging from 230 to 4160 V. Frequently, the synchronous motor is installed in near constant load applications such as air compressors where the field excitation can be suitably adjusted to allow the motor to operate at a leading power factor, thereby offering improvement in the overall plant power factor.

An all-important usage of the synchronous machine is its application in three-phase ac power generation. The synchronous machine in generator mode operation is commonly called an *alternator*. This population of synchronous machines includes the small automotive alternators (Fig. 7.4), the diesel-electric standby generators (Fig. 7.5), the waterwheel generators (Fig. 7.6), and the turboalternators (Fig. 7.7) of the electric utility power plants.

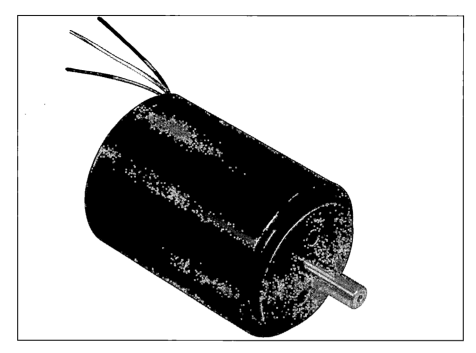

Figure 7.1
Hysteresis motor. (*Courtesy of Globe Motors™.*)

Fractional- and integral-horsepower synchronous motors and standby synchronous generators manufactured in the United States are offered in standard NEMA frame sizes, thereby assuring uniformity in overall dimensions, mounting patterns, and shaft configurations. The large hydraulic-turbine driven, steam-turbine driven, and combustion-turbine driven synchronous generators utilized by the electric utility companies are custom-designed machines that do not adhere to standardized frame sizes; however, certain degrees in commonality of ratings, protective features, and safety requirements for these machines are maintained through guidelines provided by the American National Standards Institute (ANSI) in their ANSI C50 series of publications.

7.2 CLASSIFICATION AND PHYSICAL CONSTRUCTION

The introductory discussion has established by implication the dominance of the three-phase synchronous machine. Consequently, the study will be limited to synchronous machines with a three-phase ac winding set. The time-static magnetic field (*excitation field*) necessary for synchronous machine operation can be established either by coils (*wound field*) conducting dc current or by permanent magnets. By nature, the former arrangement allows adjustment of the excitation field amplitude while the latter provides a fixed excitation field strength.

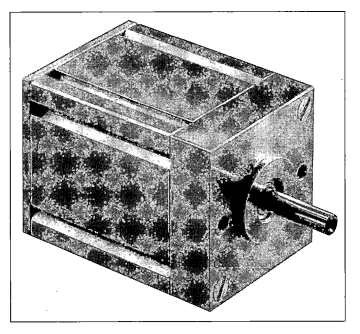

Figure 7.2
Permanent magnet synchronous machine for brushless dc motor. (*Courtesy of Globe Motors™.*)

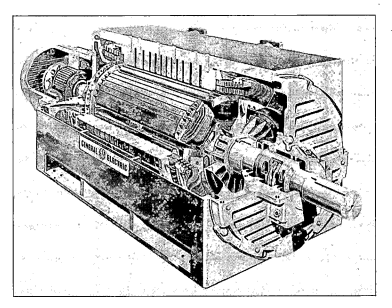

Figure 7.3
Large, three-phase synchronous motor with integral exciter. (*Courtesy of General Electric Company.*)

Figure 7.4
Automotive alternator. (*Courtesy of Prestolite Electric.*)

Figure 7.5
Diesel-electric standby generator set. (*Courtesy of Cummins Engine Company, Inc.*)

7.2.1   STATOR WINDING AND MMF

Without loss of generality, the more common case of the ac winding set being situated on the stator member (often referred to as the *armature*) will be adopted for all discussion. Consequently, the discussions and results of Sec. 6.3 regarding the induction motor windings and subsequently produced mmf wave are equally applicable to

Figure 7.6
Waterwheel or hydraulic turbine driven synchronous generator. (*Courtesy of General Electric Canada Inc.*)

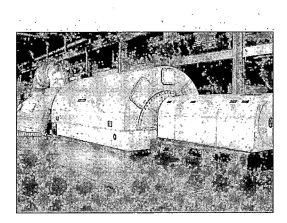

Figure 7.7
Turboalternator or steam turbine driven synchronous generator. (*Courtesy of Kentucky Utilities Company.*)

the synchronous machine. It is pointed out that although a constant amplitude mmf wave that revolves at synchronous speed ($\omega_s = 2\omega/p$) is established by the three-phase winding set, the air gap flux density wave will be of identical form only if the reluctance is uniform around the air gap periphery. The case of nonuniform air gap reluctance and the applicable analysis techniques will be addressed in Sec. 7.7.

7.2.2 ROTOR WINDINGS AND AIR GAP FIELDS

Synchronous generators in electric utility fossil fuel or nuclear power plants are directly driven by steam turbines. The emerging peaking generators installed by the electric utilities and independent power suppliers are driven by combustion turbine engines. In either case, a high-speed synchronous generator is desired for compatibility with the turbine prime mover. The two-pole machine is the more common configuration, although some four-pole machines are utilized. The rotor peripheral velocity of these turboalternators is high. In order to minimize windage losses, the rotors are of the *cylindrical rotor* or *round rotor* (RR) design shown by Fig. 7.8. From inspection of the illustration, it is apparent that if the slots are neglected, the air gap is uniform in width, giving a constant reluctance to flux flowing from the rotor to stator regardless of the point of evaluation. Further, the field windings embedded in the rotor slots are not uniformly spaced over a rotor magnetic pole. The designer chooses a distribution of the rotor windings so that the air gap flux density distribution (B_r) due to only the field winding conducting dc current is approximately

$$B_r = B_{mm}\sin\left(\frac{p}{2}\theta_r\right) \qquad\qquad \textbf{[7.1]}$$

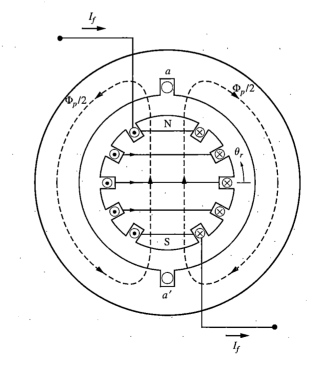

Figure 7.8
Two-pole round rotor synchronous machine

where θ_r is the mechanical angle measured along the rotor periphery with reference from the midpoint between a pair of magnetic poles as indicated in Fig. 7.8 and B_{mrm} is the maximum value of flux density established by rotor excitation that mutually couples with the stator coils.

Example 7.1

Run the MATLAB program ⟨rdrotmmf.m⟩ to determine the harmonic nature of the air gap mmf wave of a round rotor synchronous machine with two poles (p), a typical value of 20 rotor slots (S_2), a 2-in rotor slot width or breadth (b_f), and a 41-in rotor diameter (D_r). Since the air gap is uniform in width, the normalized flux density waveform will be identical; thus, the graph axes are labeled as flux density.

The screen display below shows the total harmonic distortion to be 6.907 percent, indicating a reasonably low harmonic content in the rotor produced mmf wave, although intuition might lead one to be skeptical that such a discretely distributed winding could have resulted in less than significant harmonic content.

```
TOTAL  HARMONIC  DISTORTION(%)  =  6.907
```

Figure 7.9a shows the actual or total air gap mmf wave where the ramp between various discrete levels results from the uniform change in current enclosed over the span of each rotor slot. The fundamental component of the mmf waveform is superimposed to illustrate how closely the actual waveform approaches the desired ideal waveform. Fig. 7.9b displays the Fourier spectrum of the actual mmf waveform.

Synchronous machines in hydraulic turbine or standby generator applications and large synchronous motors typically operate at lower peripheral speeds than turboalternators. Consequently, centrifugal forces and windage losses are not such an important concern, allowing use of the *salient-pole* (SP) rotor design shown by Fig. 7.10. The concentrated field winding arrangement of the salient-pole rotor results in a rotor that is less costly to manufacture than the round-rotor design. The designer uses shaping of the field pole periphery to accomplish a rotor-produced air gap flux density that is approximated by [7.1].

Example 7.2

A common field pole shaping technique for synchronous machines is to select the center of the pole arc (span of the pole head) nonconcentric with the rotor center so that the air gap increases in length toward the pole edges. The MATLAB program ⟨sprBfld.m⟩ forms the normalized permeance ($\mathscr{P} = 1/\mathscr{R}$) variation for a nonconcentric pole arc over the span of two pole pitches. Since the field mmf is constant in value over a pole pitch (complete span of a magnet pole), the flux density varies directly with permeance; thus, the resulting normalized plot can justifiably be labeled flux density. Run ⟨sprBfld.m⟩ for a salient-pole machine with a 15-in diameter rotor (D_r), a 0.125-in pole center air gap length (δ_c), and a 2/3 pole arc–to–pole pitch ratio.

The screen display below shows the total harmonic distortion of the air gap flux density to be relatively low.

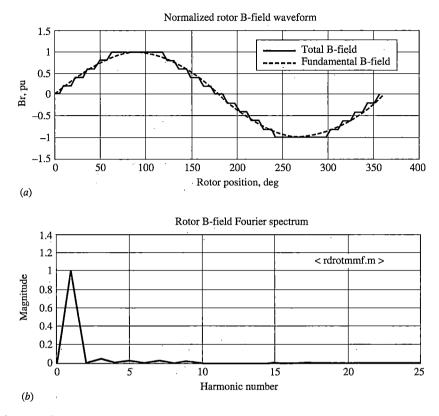

(a)

(b)

Figure 7.9
Round rotor–produced field mmf. (*a*) Actual mmf wave with fundamental component superimposed. (*b*) Fourier spectrum of mmf wave.

TOTAL HARMONIC DISTORTION(%) = 8.299

Fig. 7.11*a* shows the total air gap flux density wave where the reasonable assumption of a parabolic decrease in the flux density from the pole edge to the center point between poles has been implemented in ⟨sprBfld.m⟩ The fundamental component of the air gap *B*-field is superimposed on the plot to allow visual assessment of the closeness of the actual waveform to the desired ideal case. Fig. 7.11*b* displays the Fourier spectrum of the total air gap flux density waveform.

7.3 GENERATED VOLTAGES AND EQUIVALENT CIRCUIT

Electric machine equivalent circuit development can generally be classified as coupled coil based or revolving field based. The former treats the machine wind-

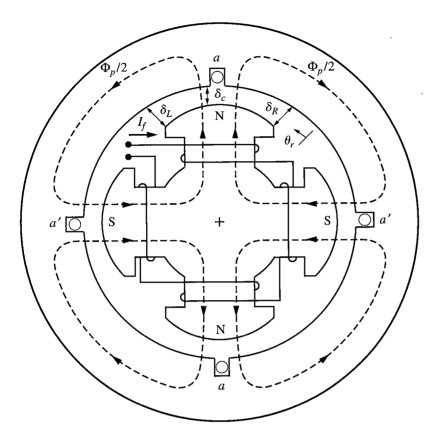

Figure 7.10
Four-pole, salient-pole synchronous machine

ings as mutually coupled coils and draws heavily on circuit theory. The latter approach develops an equivalent circuit while explicitly maintaining the identity of the air gap magnetic field quantities. The approach adopted for the round-rotor synchronous machine is coupled coil based. The revolving field approach will be introduced later in the analysis techniques for salient-pole synchronous machines.

7.3.1 COIL FLUXES AND VOLTAGES

The air gap flux density described by [7.1] in the rotor frame of reference is seen along the phase *a* coil axis in Fig. 7.12 as a traveling wave described by

$$B_{sa} = B_{mrm}\sin\left[\frac{p}{2}(\theta_s - \omega_s t)\right] \qquad \textbf{[7.2]}$$

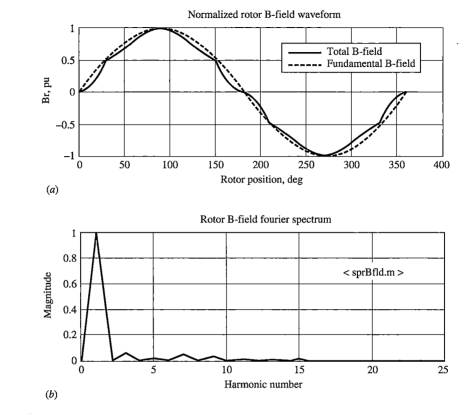

Figure 7.11
Salient-pole rotor-produced field flux density. (a) Actual B-field with fundamental component superimposed. (b) Fourier spectrum of B-field.

Using [7.2], the mutual flux of the rotor field appearing along the axis of the phase a coil is determined by integrating over the mechanical span of a magnetic pole $(2\pi/p)$.

$$\phi_{fm} = \int B_{sa}\,dA = \int_{-\frac{\pi}{p}}^{\frac{\pi}{p}} B_{mrm}\sin\left[\frac{p}{2}(\theta_s - \omega_s t)\right]\left(\frac{D}{2}\ell d\theta_s\right) \qquad \textbf{[7.3]}$$

where D is the stator bore diameter and ℓ is the stator lamination stack depth.

After integrating [7.3], applying trigonometric simplification, and using $\omega_s = (2/p)\omega$, the flux along the axis of the phase a coil is given by

$$\phi_{fm} = -\frac{2D\ell}{p}B_{mrm}\sin\,\omega t = -\Phi_m\sin\,\omega t \qquad \textbf{[7.4]}$$

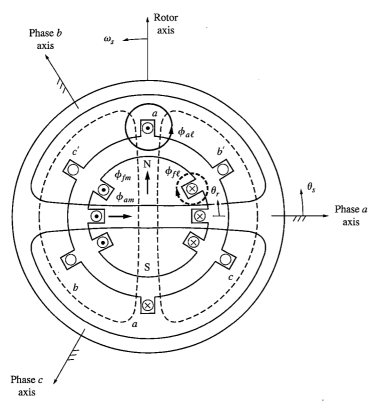

Figure 7.12
Fluxes due to field winding and phase a winding

Faraday's law and Lenz's law can be applied to [7.4] to find the voltage induced behind the N_{eff}-turn phase a coil where the value for N_{eff} is determined by [6.1].

$$e_{aa'} = -N_{eff}\frac{d\phi_{fm}}{dt} = N_{eff}\omega\Phi_m\cos\omega t \qquad \textbf{[7.5]}$$

Based on [4.6], the rms phasor transformation of [7.5] yields

$$\bar{E}_f = \bar{E}_{aa'} = 4.44\,N_{eff}f\Phi_m\angle0° \qquad \textbf{[7.6]}$$

where f is the frequency of $e_{aa'}$ in Hz. Voltage $\bar{E}_f$ is commonly called the *excitation voltage*. Owing to the 120° delay and advance of fluxes ϕ_b and ϕ_c, respectively, the phasor values of the voltage induced behind the coils of phases b and c are

$$E_{bb'} = 4.44N_{eff}f\Phi_m\angle-120° \qquad \textbf{[7.7]}$$

$$E_{cc'} = 4.44N_{eff}f\Phi_m\angle\,120° \qquad \textbf{[7.8]}$$

7.3.2   PER PHASE EQUIVALENT CIRCUIT: ROUND ROTOR CASE

For the balanced, three-phase synchronous machine, all effective stator coils have an identical number of turns (N_{eff}) and identical per phase values of leakage ($L_{s\ell}$), self (L_{ss}), and mutual (M_s) inductances. Consequently, any inductance or reactance need only be determined for one of the phase windings. The common assumption of an infinitely permeable ferromagnetic structure allows superposition addition of fluxes.

Using the leakage flux ($\phi_{a\ell}$) and mutual flux (ϕ_{am}) for phase a winding as indicated in Fig. 7.12, the self-inductance of a phase winding is given by [3.22] as

$$L_{ss} = \frac{N_{eff}(\phi_{a\ell} + \phi_{am})}{i_a} = L_{s\ell} + L_{sm} \qquad \textbf{[7.9]}$$

where $i_b = i_c = i_f = 0$. The positive axes of phases b and c coils are located 120° counterclockwise and 120° clockwise, respectively, from the positive axis of phase a coil. The component of phase a flux along the positive axes of both phase b and phase c coils is given by

$$\phi_m = \phi_{am}\cos(120°) = -\frac{1}{2}\phi_{am} \qquad \textbf{[7.10]}$$

Hence, the mutual inductance between stator phases is determined as

$$M_s = \frac{N_{eff}\phi_m}{i_a} = \frac{N_{eff}\left(-\frac{1}{2}\phi_{am}\right)}{i_a} = -\frac{1}{2}L_{sm} \qquad \textbf{[7.11]}$$

If the three phases are excited by a balanced set of currents with the field winding de-energized while the rotor revolves at synchronous speed, the flux linkages of coil a are given by

$$\lambda_a = L_{ss}i_a + M_si_b + M_si_c \qquad \textbf{[7.12]}$$

Substituting [7.9] and [7.11] into [7.12] and using $i_b + i_c = i_a$ yields

$$\lambda_a = \left(\frac{3}{2}L_{sm} + L_{s\ell}\right)i_a \qquad \textbf{[7.13]}$$

Thus, the effective inductance of a phase, including accounting for mutual coupling of other phases, is

$$L_s = \frac{\lambda_a}{i_a} = \frac{3}{2}L_{sm} + L_{s\ell} \qquad \textbf{[7.14]}$$

This value of inductance is called the *synchronous inductance*.

The induced voltage behind phase a coil for the case of generator action is given by [7.5]. This induced voltage acts to produce a current flowing out of the a terminal of Fig. 7.8. In addition to the inductance of [7.14], it is recognized that the coil winding has a resistance R_s. Thus, the line-to-neutral terminal voltage of phase a for a wye-connected synchronous generator follows as

$$v_{an} = -R_s i_a - L_s \frac{di_a}{dt} + N_{eff}\omega\Phi_m\cos\omega t \qquad \textbf{[7.15]}$$

Equation [7.15] can be transformed to the phasor domain after use of [7.14] to give

$$\overline{V}_{an} = -R_s\overline{I}_a - j\omega\frac{3}{2}L_{sm}\overline{I}_a - j\omega L_{s\ell}\overline{I}_a + \overline{E}_f \qquad \textbf{[7.16]}$$

The second and third coefficients of $\overline{I}_a$ are recognized as reactances, and the following definitions are made:

$$X_\phi = \frac{3}{2}\omega L_{sm} \qquad \textbf{[7.17]}$$

$$X_\ell = \omega L_{s\ell} \qquad \textbf{[7.18]}$$

X_ϕ is known as the *magnetizing reactance* as it is associated with the flux established by the stator windings that traverse the whole of the machine magnetic circuit. X_ℓ is called the *leakage reactance* owing to its association with the leakage flux fields.

Using the reactance definitions of [7.17] and [7.18], [7.16] can be written as
Generator:

$$\overline{V}_{an} = -R_s\overline{I}_a - jX_\phi\overline{I}_a - jX_\ell\overline{I}_a + \overline{E}_f = -R_s\overline{I}_a - jX_s\overline{I}_a + \overline{E}_f \qquad \textbf{[7.19]}$$

where $X_s = X_\phi + X_\ell$ is known as the *synchronous reactance*. An equivalent circuit that satisfies [7.19] for the synchronous generator on a per phase basis is shown by Fig. 7.13a. Since any electric machine is capable of both generator and motor action, the equivalent circuit of Fig. 7.13a is valid for motor action wherein power is absorbed by excitation source $\overline{E}_f$. However, the phase angle

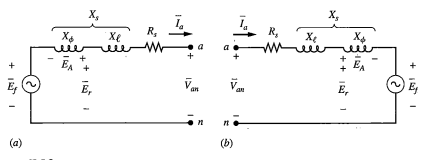

Figure 7.13
Synchronous machine per phase equivalent circuit. (a) Generator. (b) Motor.

between $\overline{V}_{an}$ and $\overline{I}_a$ must necessarily be in the range from $\pi/2$ to $3\pi/2$ for motor action resulting in negative power factor values. In order to avoid the case of negative-valued power factors and the associated confusion surrounding leading and lagging power factor, the assumed direction of current $\overline{I}_a$ will be reversed for the case of motor action giving the equivalent circuit of Fig. 7.13b for the synchronous motor. The line-to-neutral voltage for the synchronous motor is

Motor: $\qquad \overline{V}_{an} = R_s\overline{I}_a + jX_\phi\overline{I}_a + jX_\ell\overline{I}_a + \overline{E}_f = R_s\overline{I}_a + jX_s\overline{I}_a + \overline{E}_f$ **[7.20]**

The armature reaction voltage $\overline{E}_A$ and resultant voltage $\overline{E}_r$ labeled on Fig. 7.13 will be explained in later sections.

7.3.3   MAGNETIC LINEARIZATION

The formulation of [7.15] used

$$\frac{d\lambda_a}{dt} = \frac{d(L_s i_a)}{dt} = L_s\frac{di_a}{dt}$$

where it was assumed that L_s is independent of i_a. Further, the induced voltage $e_{aa'}$ given by [7.5] depends on the flux established by the field winding mmf when the stator coil currents are zero. The synchronous inductance of [7.14] depends on the flux established by the stator coil mmf's when the field current is zero. In order to use flux dependent $e_{aa'}$ and L_s in the describing equation of [7.15] where both field mmf and stator coil mmf's are nonzero, superposition of fluxes must hold. Flux superposition and constant-valued L_s can be a reality only if the ferromagnetic structure of the machine is magnetically linear. Since the actual machine is not magnetically linear, the conclusion must be drawn that use of [7.19] and [7.20] yields analysis of a fictitious magnetically linear machine. In order to avoid potential pitfalls in analysis, it is important to establish correlation between the actual nonlinear machine and the linearized machine implied by [7.19] and [7.20].

Based on [7.6], the OCC vertical axis of Fig. 7.14b can be scaled to give the maximum value of the air gap flux that mutually couples the rotor and stator.

$$\Phi_m = \frac{V_{OC}/\sqrt{3}}{4.44N_{eff}f}$$

The OCC horizontal axis can be converted to mmf values if each point is multiplied by field winding turns per pole N_f. Thus, the OCC is a scaled replica of the flux-mmf characteristic of the actual machine magnetic circuit. A particular value of mmf impressed on the magnetic circuit will produce a corresponding particular value of flux regardless of whether the impressed mmf is established by a single source or multiple sources.

When operating with both the field coil and stator coils conducting current, the three-phase synchronous machine actually has four active mmf sources—one field

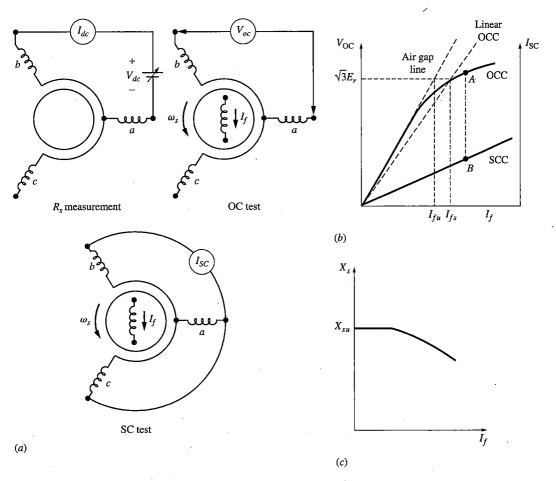

Figure 7.14
Test determination of synchronous machine equivalent circuit values. (a) Test setups. (b) Open-circuit and short-circuit curves. (c) Saturation nature of X_s.

coil and three stator coils. However, it has been established in Sec. 6.3.3 that the three stator coils, when conducting a balanced set of three-phase currents, can be represented by a single mmf source that produces a traveling mmf wave. Hence, the magnetic circuit of the synchronous machine with stator currents present can justifiably be modeled with two series-connected mmf sources as illustrated by Fig. 7.15a. The mmf established by the field winding is sinusoidally distributed in space; however, as the field structure rotates by the phase a stator coil, the mmf seen by the coil varies sinusoidally in time so that a time-phasor representation $(\overline{\mathcal{F}}_f)$ is justified. By the same argument, the traveling mmf wave produced by the three stator coils appears to phase a coil as a mmf varying sinusoidally in time with the same period as $\overline{\mathcal{F}}_f$. This so-called *armature reaction mmf* as seen by phase a coil

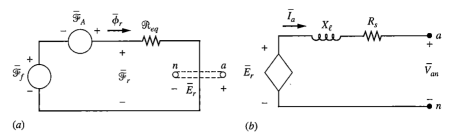

Figure 7.15
Synchronous generator magnetic and electric circuits. (a) Magnetic circuit. (b) Resulting electric circuit.

is labeled $\overline{\mathcal{F}}_A$ in Fig. 7.15a. Since $\overline{\mathcal{F}}_f$ and $\overline{\mathcal{F}}_A$ do not peak at the same point in time, there is a phase angle difference between the two quantities. Phasor addition of $\overline{\mathcal{F}}_f$ and $\overline{\mathcal{F}}_A$ yields the *resultant mmf* $\overline{\mathcal{F}}_r$ that acts on the magnetic circuit of the machine to produce the *resultant flux* $\overline{\phi}_r$ linking the phase a coil superimposed on Fig. 7.15a. It is this resultant flux $\overline{\phi}_r$ passing through the phase a coil that produces the *resultant voltage* $\overline{E}_r = 4.44N_{eff}\overline{\phi}_r f$ of Fig. 7.15b. The resultant flux $\overline{\phi}_r$ is the flux that mutually couples the stator and rotor. Phase a coil will still have a leakage flux that does not couple the rotor leading to leakage reactance X_ℓ. This leakage reactance and the coil resistance R_s result in voltage drops between the resultant voltage $\overline{E}_r$ and the phase a coil terminal voltage $\overline{V}_{an}$.

The resultant voltage $\overline{E}_r$ of Fig. 7.15b must necessarily be the same voltage $\overline{E}_r$ as labeled in the equivalent circuit of Fig. 7.13a. MMF $\overline{\mathcal{F}}_r$ produces flux $\overline{\phi}_r$ that in turn induces voltage $\overline{E}_r$. Comparing Figs. 7.13a and 7.15, it is logical to think that if the magnetic circuit were magnetically linear, then a flux $\overline{\phi}_f$ produced by $\overline{\mathcal{F}}_f$ induces $\overline{E}_f$ and a flux $\overline{\phi}_A$ produced by $\overline{\mathcal{F}}_A$ induces $\overline{E}_A$ where $\overline{\phi}_r$ is the superposition sum of $\overline{\phi}_f$ and $\overline{\phi}_A$. These ideas will be further developed in Sec. 7.7. The objective at hand is to correlate the linear magnetic model with the OCC.

It has been established that the flux actually flowing in the machine magnetic circuit is ϕ_r, which produces the resultant voltage E_r. If the stator coil currents were all zero, $\mathcal{F}_A$ vanishes in Fig. 7.15a and E_A vanishes in Fig. 7.13a. Then, ϕ_r of Fig. 7.15a in such case is due only to $\mathcal{F}_f$, and $E_f = E_r = V_{an}$ in Fig. 7.13a. Since this condition is one valid point of operation, it is concluded that the linearized OCC is a straight line from the origin through $\sqrt{3}E_r$ as indicated on Fig. 7.14b. The $\sqrt{3}$ is necessary since V_{OC} is plotted as a line voltage and E_r is a phase voltage. If the value of E_r changes, then a different linear OCC is implied. For best accuracy in analysis, the value of L_s, thus X_s, should be evaluated using V_{OC} taken from the point of intersection of the actual OCC and the linear OCC. As a practical matter, determination of E_r requires a knowledge of X_ℓ, R_s, I_a, and PF. Then, E_r is determined by $\overline{E}_r = \overline{V}_{an} + \overline{I}_a(R_s + jX_\ell)$. Frequently, X_ℓ is not known. In such case, the accuracy is only compromised by a small percentage if it is assumed that $E_r \cong V_{an}$ for purposes of establishing the linear OCC.

It should be emphasized that the value of E_r labeled on Fig. 7.14b is the value $E_r = E_f$ when stator coil currents are zero. For other than zero stator coil currents, the linear OCC of Fig. 7.14b gives the relationship between E_f of Fig. 7.13 and field current I_f.

7.4 EQUIVALENT CIRCUIT PARAMETERS FROM TEST DATA

If an equivalent circuit is to be useful in performance predictions, it is imperative that practical methods exist to determine the values for the elements. The dc value of stator winding resistance can be measured by impressing a dc voltage (V_{dc}) across a pair of stator winding line terminals and measuring the resulting current (I_{dc}) as indicated in Fig. 7.14a to yield the value[1]

$$R_{sdc} = \frac{1}{2} \frac{V_{dc}}{I_{dc}} \qquad \textbf{[7.21]}$$

The value of R_s for use in the equivalent circuit is determined by appropriate modification of R_{sdc}. The value must be adjusted to the average operating temperature of the stator windings.[2] In addition, cross-slot fluxes (proximity effect) and a slight skin effect in actual machine operation lead to less than ideal uniform current density distribution across the conductor area.[1] Consequently, the value of resistance should be increased 3 to 5 percent to give what is known as the *ac value of resistance* for use in equivalent circuit analysis.

With the value of R_s determined, the open-circuit and short-circuit tests of Fig. 7.14a can be performed and the resulting data from the two tests plotted on a common field current axis as shown by Fig. 7.14b. If a set of values for V_{OC} and I_{SC} are selected at the same value of I_f (for example, points A and B of Fig. 7.14b), then the synchronous impedance is given by

$$Z_s = |R_s + jX_s| = \frac{V_{OCA}/\sqrt{3}}{I_{SCB}} \qquad \textbf{[7.22]}$$

and

$$X_s = \sqrt{Z_s^2 - R_s^2} \qquad \textbf{[7.23]}$$

For machines of rating greater than approximately 10 kVA, $R_s \ll X_s$, thus, $Z_s \cong X_s$.

The open-circuit voltage curve of Fig. 7.14b reflects the effect of magnetic saturation. If the machine did not exhibit saturation, then the OCC would simply follow the *air gap line* labeled on Fig. 7.14b. However, for short-circuit operation, the air gap mmf produced by the stator windings is nearly equal in magnitude and positioned nearly 180° from the field mmf so that the resultant air gap mmf is small; thus, significantly smaller values of flux flow through the magnetic circuit for short-circuit conditions than for the open-circuit case. As a consequence, there is negligible magnetic saturation present, yielding a short-circuit curve that is a near straight line.

If values of synchronous reactance are determined by [7.22] and [7.23] for various values of field current, the values of X_s will decrease as the open-circuit voltage curve moves into magnetic saturation, as indicated by the plot of Fig. 7.14c. Typically, a value of X_s determined for near rated voltage is chosen for analysis work under the assumption that $V_{an} \cong E_r$.

The above procedure for determining the synchronous reactance does not offer any way to separate X_s into its components X_ϕ and X_ℓ. For most synchronous machine analysis problems, knowledge of the component values is of no concern. One noteworthy exception is the regulation problem to be addressed in Sec. 7.5.3. The leakage reactance X_ℓ typically has a value that ranges from 10 to 20 percent of the unsaturated synchronous reactance X_{su} as indicated in Fig. 7.14c. Using the mean of the range for X_ℓ, a good approximation is

$$X_\ell \cong 0.15 X_{su} \qquad\qquad \textbf{[7.24]}$$

Then,
$$X_\phi \cong X_s - X_\ell \qquad\qquad \textbf{[7.25]}$$

If a more accurate determination of X_ℓ is necessary, then the Potier reactance method should be used.[1] The Potier method requires additional test work to determine field current requirement to operate at zero power factor conditions.

Example 7.3

The open-circuit line-to-line voltage and short-circuit current curves for a 13.8-kV, 60 Hz, 3600 rpm, 250 MVA, 0.8 PF lagging, round-rotor alternator are shown on a common field current axis by Fig. 7.16. Determine the value of synchronous reactance (X_s) and approximate values of magnetizing reactance (X_ϕ), and leakage reactance (X_ℓ) appropriate for rated voltage operation. Assume R_s is negligible and use $E_r \cong V_{an}$ for linearization.

For rated line voltage on the OCC, $I_{SC} = 10,000$ A. By [7.22], with $R_s = 0$,

$$X_s = \frac{V_{OC}/\sqrt{3}}{I_{SC}} = \frac{13,800/\sqrt{3}}{10,000} = 0.796\ \Omega$$

The unsaturated synchronous reactance can be determined by using any value of voltage along the air gap line and the corresponding value of short-circuit current. Arbitrarily choose $V_{OC} = 14,000$. The corresponding $I_{SC} = 9,000$ A gives

$$X_{su} = \frac{V_{OC}/\sqrt{3}}{I_{SC}} = \frac{14,000/\sqrt{3}}{9000} = 0.898\ \Omega$$

Using [7.24] and [7.25],

$$X_\ell \cong 0.15 X_{su} = 0.15(0.898) = 0.135\ \Omega$$

$$X_\phi \cong X_s - X_\ell = 0.796 - 0.135 = 0.661\ \Omega$$

The MATLAB program ⟨Xs.m⟩ uses the stored open-circuit voltage and field current array along with a single short-circuit stator current-field current data point to generate a plot of X_s as a function of field current, giving a quantitative measure

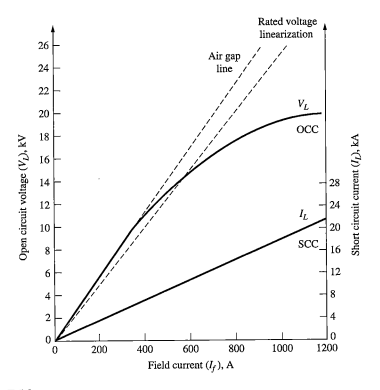

Figure 7.16
250-MVA, 60-Hz, 13.8-kV alternator

of magnetic saturation impact. Fig. 7.17 is such a plot for the data of a 400-MVA, 24-kV, three-phase alternator.

If the OCC is available but the SCC is not available, the value of synchronous reactance can be readily determined using only the OCC provided the value of unsaturated synchronous reactance (X_{su}) is known from either manufacturer-supplied data or using typical values. If the line voltage OCC of Fig. 7.14b were entered at $\sqrt{3}\,E_r$, the value of field current (I_{fu}) for unsaturated condition and the value of field current (I_{fs}) for saturated condition can be directly determined. A *saturation factor* is then formed as

$$k_s = \frac{I_{fu}}{I_{fs}} \le 1 \qquad\qquad \textbf{[7.26]}$$

which has the same characteristic shape as the synchronous reactance plot of Fig. 7.14c. A reasonable value of the synchronous reactance at the point of operation is given by

$$X_s = X_\ell + k_s X_{\phi u} \cong 0.15 X_{su} + 0.85 k_s X_{su} \qquad\qquad \textbf{[7.27]}$$

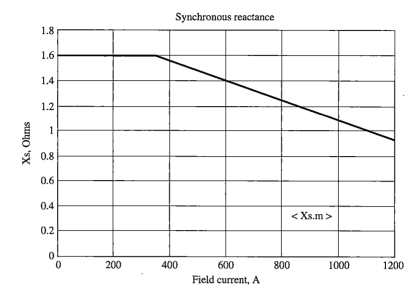

Figure 7.17
Effect of saturation on synchronous reactance X_s

Example 7.4   If the alternator of Example 7.3 is operating at nameplate conditions, determine the value of required field current.

Assume $\bar{V}_{an}$ on the reference,

$$\bar{I}_a = \frac{S}{\sqrt{3}V_L}\angle-\cos^{-1}(PF) = \frac{250 \times 10^6}{\sqrt{3}\,(13,800)}\angle-\cos(0.8) = 10,459\angle-36.87°$$

From Fig. 7.13a,

$$\bar{E}_f = \bar{V}_{an} + jX_s\bar{I}_a = \frac{13,800}{\sqrt{3}}\angle0° + 0.796\angle90°(10,459\angle-36.87°)$$

$$\bar{E}_f = 14,573.6\angle27.19°$$

Enter Fig. 7.16 at $\sqrt{3}E_f = 25,241.4$ on the rated voltage linearization curve to read $I_f \cong$ 1000 A.

7.5 GENERATOR PERFORMANCE

Generator mode applications of the synchronous machine can be subdivided into the two categories of isolated generators and interconnected generators. Each of these applications emphasizes different characteristics of the synchronous machine. Prior to study of the resulting performance exhibited, the phasor diagram representation and the nature of the energy conversion will be introduced.

7.5.1 SYNCHRONOUS GENERATOR PHASOR DIAGRAM

The phase load impedance of the isolated generator, or the average and reactive power flow from an interconnected generator, determines the terminal power factor. Consequently, the power factor may range from lagging to leading, although the former is the common case. The phasor diagram associated with the equivalent circuit of Fig. 7.13a is shown by Fig. 7.18 for the case of a lagging power factor.

The MATLAB program ⟨smphasor.m⟩ uses the polar plotting feature to produce the phasor diagram for a synchronous machine. The program interacts with the user to determine motor or generator mode of operation and the value of the power factor. Per unit or normalized values of voltage, current, and impedance are used; however, the user could modify the code to plot actual values. Figure 7.19 presents plots for a synchronous generator for the case of unity PF and 0.866 PF leading. Review of Figs. 7.18 and 7.19 leads to the conclusion that for constant terminal voltage and current operation, E_f (thus, I_f) is greater for lagging PF than for leading PF.

7.5.2 ELECTROMECHANICAL DEVELOPED TORQUE

The total power converted $(3P_d)$ from mechanical to electrical form for the three-phase synchronous generator is three times the power flowing out of the excitation voltage source (E_f) of Fig. 7.13a. Based on the phasor diagram of Fig. 7.20a,

$$3P_d = 3E_f I_a \cos \beta \qquad \textbf{[7.28]}$$

In order to obtain the result in terms of desired variables, start with the length of construction line aE_r which can be expressed in two ways.

$$aE_r = E_r \sin \delta_r = I_a X_\phi \cos \beta \qquad \textbf{[7.29]}$$

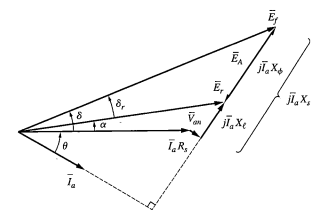

Figure 7.18
Phasor diagram of a synchronous generator for case of lagging PF

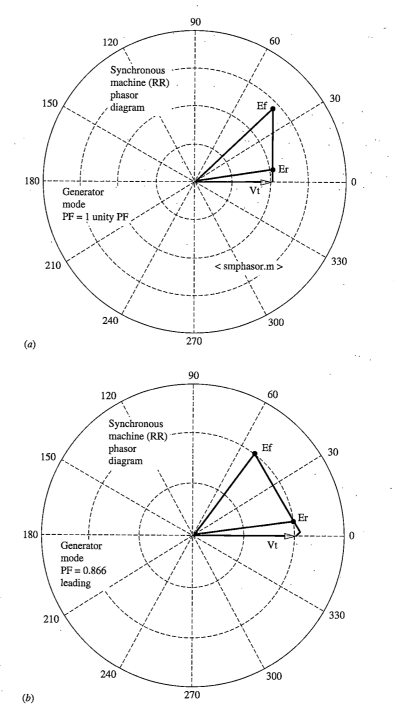

(a)

(b)

Figure 7.19
Phasor diagram of a synchronous generator. (a) Unity PF. (b) Leading PF.

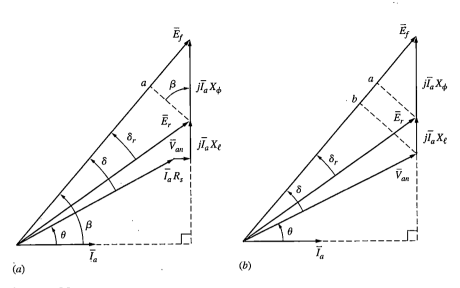

Figure 7.20
Phasor diagram of synchronous generator with $\bar{I}_a$ as reference. (a) $R_s \neq 0$. (b) $R_s = 0$.

Solve [7.29] for cos β and substitute the result into [7.28] to yield the total converted power as

$$3P_d = \frac{3E_f E_r}{X_\phi} \sin \delta_r \qquad\qquad \textbf{[7.30]}$$

Since $3P_d = 3T_d \omega_s$, the total developed torque follows as

$$3T_d = \frac{3E_f E_r}{\omega_s X_\phi} \sin \delta_r \qquad\qquad \textbf{[7.31]}$$

Since the developed torque expression has been derived based on power converted from mechanical form, the torque will be a positive value for generator action.

While [7.31] is valid for $R_s \neq 0$ and should be used in such a case, it does require the determination of E_r, X_ϕ, and δ_r, which frequently leads to additional computation. For larger synchronous machines, $R_s \ll X_s$ and can be neglected. In such case, the phasor diagram degenerates to that of Fig. 7.20b where from the two similar triangles aE_rE_f and $bV_{an}E_f$

$$\frac{E_r \sin \delta_r}{X_\phi} = \frac{V_{an} \sin \delta}{X_s} \qquad\qquad \textbf{[7.32]}$$

Substitution of [7.32] into [7.31] gives the total developed torque if $R_s = 0$ as

$$3T_d = \frac{3E_f V_{an}}{\omega_s X_s} \sin \delta \qquad\qquad \textbf{[7.33]}$$

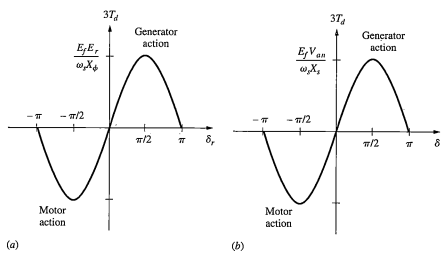

Figure 7.21
Synchronous machine developed torque

Recalling that excitation voltage E_f increases directly with field current, it is apparent that maximum value of developed torque depends directly on the value of I_f. The angles δ_r and δ of [7.31] and [7.33], respectively, are known as the *torque angles* or *power angles* of the synchronous machine. For generator action, the values of δ_r and δ are positive; however, it will later be seen that for motor action, phasor $\bar{E}_f$ lags $\bar{V}_{an}$; thus, δ_r and δ are negative in value, yielding a developed torque of opposite direction for the motor case as would be expected. Both [7.31] and [7.33] give the developed torque derived from power converted from mechanical to electrical form. Consequently, positive values from these expressions represent the torque acting opposite in direction to shaft rotation.

Figure 7.21 gives a graphical description of synchronous machine developed torque. If δ or δ_r were to exceed $\pi/2$ for steady-state generator operation, the machine rotor will begin acceleration to a speed greater than synchronous speed and is said to have lost synchronism. This undesirable occurrence would initiate a transient condition, characterized by high currents and time-varying torque, that results in overcurrent breaker trips and potential mechanical damage. Similar conditions result for a synchronous motor if δ_r or $\delta < -\pi/2$.

Example 7.5 | Determine the value of developed torque for the 2-pole, 60-Hz alternator of Example 7.4 for which $R_s = 0$ and $X_s = 0.796 \, \Omega$ as calculated in Example 7.3.

Since $\bar{V}_{an}$ was taken on the reference, the angle of $\bar{E}_f$ is the torque angle δ. By [7.33],

$$3T_d = \frac{3E_f V_{an}}{\omega_s X_s}\sin\delta = \frac{3(14{,}576.6)\left(13{,}800/\sqrt{3}\right)}{\frac{2}{2}(2\pi 60)(0.796)}\sin(27.19°) = 5.305 \times 10^5 \, \text{N·m}$$

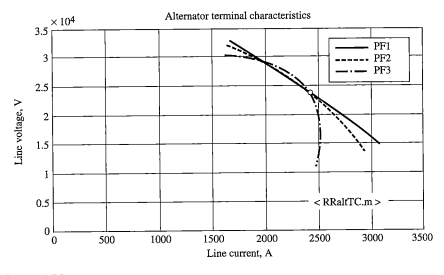

Figure 7.22
Alternator terminal characteristics

7.5.3 ISOLATED SYNCHRONOUS GENERATORS

If an isolated alternator were operated with a fixed field current setting as the electrical load varies, the resulting plot of terminal voltage vs. load current is known as the *terminal characteristics*. The MATLAB program ⟨RRaltTC.m⟩ plots the terminal characteristics with PF as a parameter. The program sets conditions for each PF so that the curves pass through rated voltage and current. Figure 7.22 is the results for a 24-kV, 100-MVA alternator with typical values for X_s and R_s with PF values of $PF1 = 0.8$ lagging, $PF2 = 1.0$, and $PF3 = 0.8$ leading. The curves are plotted for only a small range on either side of rated conditions so that the assumption of a constant value for X_s is a reasonable approximation. Inspection of the results leads to the conclusion that an alternator must operate with a closed-loop voltage regulator to adjust field current as load varies if terminal voltage is to be maintained constant in value.

Voltage Regulation A quantitative assessment of the terminal characteristics is known as *voltage regulation* or percentage change in the magnitude of terminal voltage from full-load (FL) to no-load (NL) conditions with field excitation (I_f) unchanged from the value that produces full-load rated (R) conditions. Expressed in equation form,

$$Reg = \frac{(V_{NL} - V_{FL})\,100\%}{V_{FL}} \Bigg|_{I_f = I_{fR}} \qquad \textbf{[7.34]}$$

where the voltage values are determined as line or phase quantities as convenient. For a synchronous generator with typical magnetic saturation, an accurate value of

Reg cannot be obtained from the linearized circuit model of Fig. 7.13*a*, wherein excitation voltage E_f is a fictitious value for a magnetically linearized model.

Example 7.6

For the synchronous generator of Example 7.3, determine the voltage regulation.

From Example 7.4, $E_f = 14{,}573.6$ V at full-load conditions with an associated field current $I_f \cong 1000$ A. If the machine were unloaded while I_f is unchanged, the no-load line voltage is read from the OCC of Fig. 7.16 as $V_{NL} \cong 19.3$ kV. By [7.32],

$$Reg = \frac{V_{NL} - V_{FL}}{V_{FL}} \times 100 = \frac{19.3 - 13.8}{13.8} \times 100 = 39.85\%$$

The large percentage excursion of terminal voltage determined by the above example supports the earlier conclusion that a voltage regulator to control field current for an isolated synchronous generator is a necessity.

Efficiency Calculation   Figure 7.23 shows the power flow diagram for a synchronous generator where it is assumed that the exciter (impresses voltage V_f on field winding giving field current I_f) is integrally fixed to the alternator shaft so that the mechanical input power ($T_s \, \omega_s$) includes power to both the alternator and the exciter. As a consequence, the friction and windage losses (P_{FW}) are the total for both alternator and exciter. The core losses (P_c) represent both the alternator and exciter core losses plus the exciter copper losses. If an efficiencyless exciter is to be determined, then the exciter losses must be appropriately subtracted from shaft power and ignored in loss accounting. When auxiliary pumps and blowers are used for lubrication and cooling, the associated power (P_{aux}) must be added to the shaft power for total input power and added to the tally of losses. For synchronous generators above 150 kW output power ratings, there are load current dependent losses, such as eddy current losses in metallic end-turn structures, and added tooth ripple core losses in the rotor slot area due to shift in flux paths when a stator mmf is present. These losses are known as *stray load losses,* and they are accounted for by adjustment of stator copper losses using the factor k_{SL}.

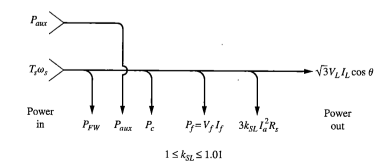

$$1 \leq k_{SL} \leq 1.01$$

Figure 7.23
Power flow diagram for synchronous generator

Once the losses have been determined for a point of operation, the efficiency is given by

$$\eta = \frac{P_{out}}{P_{in}} \times 100\% = \frac{P_{out}}{P_{out} + \text{losses}} \times 100\%$$

$$\eta = \frac{\sqrt{3}\, V_L I_L PF}{\sqrt{3}\, V_L I_L PF + P_{FW} + P_c + P_f + P_{aux} + 3k_{SL}I_a^2 R_s} \times 100\% \quad \textbf{[7.35]}$$

Example 7.7

The alternator of Example 7.3 is driven at open circuit by a calibrated dc motor at 3600 rpm. The alternator field current is set for 620 A with $V_f = 362$ V, giving alternator line voltage of 15.2 kV. At the point of operation, the power supplied to the dc motor is 2.05 MW and its efficiency is 94.1 percent. Power supplied to all blowers and lubricant circulation pumps is known to be 151 kW. The per phase alternator resistance is $R_s = 6\,\text{m}\Omega$. Determine the full-load efficiency of this alternator.

The alternator voltage for the above described test is approximately equal to the expected value of $\sqrt{3}E_r$ under full load; thus, the core losses will not be significantly different from the value exhibited for the no-load test condition. Rotational losses at full load will be unchanged from the no-load test. Hence,

$$P_c + P_{FW} \cong P_{motor} - V_{fNL}I_{fNL} = (2.05 \times 10^6)(0.941) - (362)(620) = 1.705\,\text{MW}$$

Assuming that the field winding temperature does not change significantly and recalling from Example 7.4 that full-load field current is 1000 A,

$$P_f = I_f^2 R_f = (1000)^2 \left(\frac{362}{620}\right) = 0.584\,\text{MW}$$

The full-load stator current was found to be 10,459 A in Example 7.4. By [7.35] using the reasonable value $k_{SL} = 1.01$,

$$\eta = \frac{\sqrt{3}(13.8)(10.459)(0.8) \times 100\%}{\sqrt{3}(13.8)(10.459)(0.8) + 1.705 + 0.584 + 0.151 + 3(1.01)(10.459)^2(0.006)}$$

$$\eta = 98.03\%$$

Qualitative Performance by Phasor Diagram The phasor diagram can serve as a useful analysis tool to predict qualitative performance of a synchronous generator. In such analysis approach, it is expedient to neglect R_s and assume small enough changes in the operating point so that X_s can be treated as constant in value.

Example 7.8

The isolated synchronous generator of Fig. 7.24a is driven by a speed-governed prime mover with reserve power capability. If field current I_f were increased by 10 percent, predict the qualitative changes in E_f, I_a, V_{an}, θ, and δ. Assume the change in I_f is small enough that X_s can be assumed constant in value.

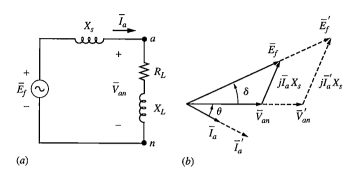

Figure 7.24
Phasor diagram qualitative analysis of isolated synchronous generator. (a) Equivalent circuit with fixed load. (b) Phasor diagram for two different values of I_f.

Terminal current and voltage are given by

$$\bar{I}_a = \frac{\bar{E}_f}{R_L + j(X_s + X_L)} \qquad \textbf{[1]} \qquad \bar{V}_{an} = \frac{R_L + jX_L}{R_L + j(X_s + X_L)}\bar{E}_f \qquad \textbf{[2]}$$

If I_f increases by 10 percent, E_f increases by 10 percent. By [1], I_a increases 10 percent, but the phase angle between $\bar{E}_f$ and $\bar{I}_a$ is unchanged. Thus, $\theta + \delta$ is unchanged. From [2], E_f increases by 10 percent; however, the phase angle between $\bar{V}_{an}$ and $\bar{E}_f$ is unchanged. It is concluded that δ is unchanged; hence, θ is unchanged. The phasor diagram of Fig. 7.24*b* displays the original and new conditions where the phasors with prime marks indicate the new conditions.

7.5.4   INTERCONNECTED SYNCHRONOUS GENERATORS

The electric utility grid is a network of distributed loads and synchronous generators. An alternator operating within a utility grid will typically represent 5 percent or less of the total system generator capacity. As a result, a single generator can only alter voltage values a small amount in a local area. It cannot increase the value of system frequency unless it adds sufficient energy to accelerate the rotating masses of all electric machines (motors and generators) of the network. Consequently, an interconnected synchronous generator must operate under the constraints of near constant voltage and constant frequency, resulting in a different performance nature than is the case for isolated operation.

Synchronization   Prior to placing an alternator in service on the electric utility grid, it must be matched in voltage magnitude, frequency, and phase sequence with the interconnecting grid in a process called *synchronization.* Although the grid is made up of a large number of distributed alternators and loads interconnected through transmission lines, it can be considered a linear network. Thus, Thevenin's theorem says that each phase can be modeled by a voltage source in series with an

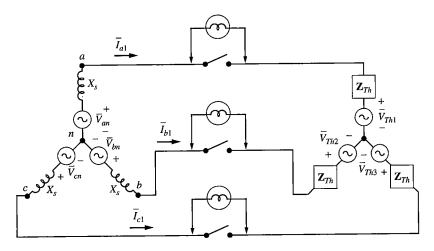

Figure 7.25
Synchronization setup

impedance as shown on the right-hand side of Fig. 7.25. If the network is a balanced three-phase system, the impedance of each equivalent phase will be equal and the phase voltage sources form a balanced three-phase set. Upon closure of the contactors, the three line currents of Fig. 7.25 are

$$\bar{I}_{a1} = \frac{\bar{V}_{an} - \bar{V}_{Th1}}{Z_{Th} + jX_s} \qquad \bar{I}_{b1} = \frac{\bar{V}_{bn} - \bar{V}_{Th2}}{Z_{Th} + jX_s} \qquad \bar{I}_{c1} = \frac{\bar{V}_{cn} - \bar{V}_{Th3}}{Z_{Th} + jX_s}$$

In order for the oncoming alternator to connect to the grid with negligible transient conditions, all three line currents should ideally be zero immediately after contactor closure. The necessary condition for this desired transient-free closure is that

$$\bar{V}_{an} = \bar{V}_{Th1} = 0 \qquad \bar{V}_{bn} - \bar{V}_{Th2} = 0 \qquad \bar{V}_{cn} - \bar{V}_{Th3} = 0$$

A little thought leads to the conclusion that four conditions must be satisfied:

1. All voltages are equal in magnitude ($V_{an} = V_{Th1} = V_{bn} = V_{Th2} = V_{cn} = V_{Th3}$).
2. The frequency of the oncoming alternator and the grid must be identical.
3. The phase sequence of the oncoming generator must match that of the grid.
4. The phase angles of each voltage pair must be identical ($\angle\bar{V}_{an} = \angle\bar{V}_{Th1}$, $\angle\bar{V}_{bn} = \angle\bar{V}_{Th2}$, $\angle\bar{V}_{cn} = \angle\bar{V}_{Th3}$).

The three lamps connected across the open contactors serve as indicators to assure that the above four conditions are met.

1. A lamp will never completely extinguish if the voltage magnitudes do not match.
2. A lamp will blink off and on if the frequency of the oncoming alternator and the electric utility are not equal.

3. For a slight difference in frequencies, the lamps will blink simultaneously for matched phase sequence; the lamps will blink sequentially if the phase sequence is not matched.

4. If a phase difference exists between each voltage pair with matched frequency, voltage magnitudes, and phase sequence, each lamp will be luminated at a constant intensity.

The obvious condition for proper contact closure is that all lamps are extinguished.

The program ⟨synphas.m⟩ uses the polar plot capability of MATLAB to visualize via phasors the conditions that can occur prior to contactor closure as the synchronization process is attempted. The program takes advantage of the fact that phasors of two different frequencies can be shown on a common origin with one phasor stationary provided the second phasor revolves at an angular speed equal to the difference in angular frequency of the two sinusoidal time functions represented by the two phasors. The program interacts with the user to set up small or large frequency differences, equal or unequal voltage magnitudes, and matched or unmatched phase sequence. The normalized voltages across open contactors are the distance between the phasor tips of the associated voltage pair of that phase indicated by the dotted green lines of the graphical displays. Fig. 7.26 shows one frame of the display gen-

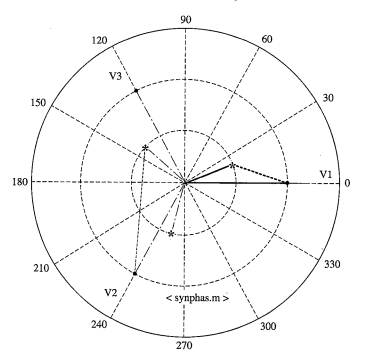

Phasor diagrams for synchronizing alternator

Figure 7.26
Phasor diagrams for synchronization

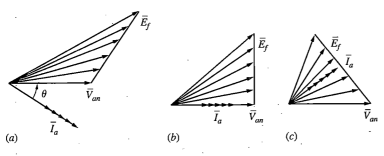

Figure 7.27
Phasor diagram of alternator with varying load. (a) Lagging PF. (b) Unity PF. (c) Leading PF.

erated by ⟨synphas.m⟩ for a small frequency difference, a magnitude difference, and a phase sequence mismatch set up by the screen display that follows:

```
» synphas
Frequency difference small? ( Y or N ) Y
Voltage magnitude different? ( Y or N ) Y
Phase sequence different? ( Y or N ) Y
```

Excitation Requirements The near constant voltage constraint posed by grid connection justifies the three phasor diagrams of Fig. 7.27, where each diagram represents various electrical loads for a fixed value of PF. The approximation of $R_s = 0$ has been introduced. Since excitation voltage E_f is proportional to field current I_f, the following qualitative judgments concerning field current requirements for constant terminal voltage operation can be made.

1. For cases of unity and lagging PF loads, increased load current always requires increased field current.

2. For a leading PF load, increased load results in a decrease in field current requirement for light load. As load is further increased, the field current requirement reaches a minimum value beyond which it begins to increase.

3. For any particular value of load current, the field current progressively decreases as the load PF changes from lagging to leading.

Although the qualitative assessment of field requirements above is an interesting instructional study, quantitative determination of excitation values necessitates an approach that yields more accuracy. As the load point varies with load current and power factor, the point of magnetic linearization shifts along the OCC curve changing the value of X_s and the proportionality between E_f and I_f. The MATLAB program ⟨excreq.m⟩ uses the OCC and the saturation factor (k_s) of [7.26] to determine the field current (I_f) requirement as load current (I_a) varies for any specified terminal voltage and PF. Figure 7.28 shows a plot generated for a 24-kV, 400-MVA alternator with PF of 0.8 lag, unity, and 0.8 lead. This field requirement plot is often called the *field compounding curves*.

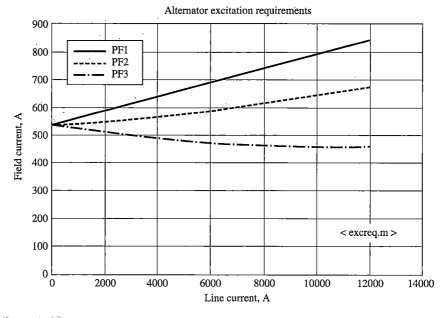

Figure 7.28
Field compounding curves

Reactive Capability Curves A synchronous generator is limited by both ability to cool the field windings and ability to cool the stator windings. A graphical presentation that defines the safe operating area based on the I_f and I_a limits, but scaled to express the results in terms of reactive and average power, is known as the *reactive capability curves*.

The phasor diagram of Fig. 7.29 explicitly shows the condition of rated V_L, S, and PF. If the tip of phasor $\bar{E}_f$ were to move along the arc centered at the tip of $\bar{V}_{an}$, then I_a would be maintained at rated value. If $\bar{E}_f$ were to move along the arc centered at the origin of the phasor diagram, then I_f would be maintained at rated value. Hence, it can be concluded that if the tip of $\bar{E}_f$ were located at any point in the shaded area, then neither I_f nor I_a exceed rated value, or the shaded area indicates a safe area of operation for rated terminal voltage.

If the vertical and horizontal projections of $j\bar{I}_a X_s$ are multiplied by the constant $3V_{an}/X_s$, then the values of average and reactive power supplied by the alternator result.

$$I_a X_s \cos\theta \left(\frac{3V_{an}}{X_s} \right) = 3V_{an} I_a \cos\theta = P_T$$

$$I_a X_s \sin\theta \left(\frac{3V_{an}}{X_s} \right) = 3V_{an} I_a \sin\theta = Q_T$$

Thus, if a P_T-Q_T axis were set up at the tip of $\bar{V}_{an}$ on Fig. 7.29 and the axes were scaled by $3V_{an}/X_s$, the resulting plot of P_T vs. Q_T shows the average and reactive

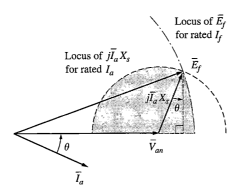

Figure 7.29
Phasor diagram for rated I_a and I_f values

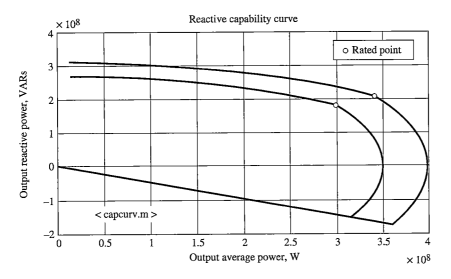

Figure 7.30
Reactive capability curves for 400-MVA alternator for two different hydrogen pressures

power producing capabilities of the alternator. By common practice the axes are interchanged to give a plot of Q_T vs. P_T to present the reactive capability curves. For large hydrogen-cooled alternators, the manufacturer may assign different ratings according to the pressure maintained for the hydrogen gas. In such case, multiple reactive capability curves may be plotted with coolant pressure as a parameter.

The MATLAB program ⟨capcurv.m⟩ plots the reactive capability curves for the apparent power ratings stored in the array S. The first entry in array S should be the largest value associated with the highest hydrogen pressure in order to set adequate axis scaling for the plot. Figure 7.30 displays the reactive capability curves for a 24-kV alternator rated at 400 MVA for the highest allowable hydrogen pressure and a rating of 350 MVA for some reduced hydrogen pressure. Negative values of reactive power correspond to leading PF operation. The reactive capability curves have been truncated at 0.9 PF leading—a typical value, but the actual value is specified by the manufacturer. For machines with ferromagnetic field-winding retainer rings, the reduced field current of leading PF operation can lead to unsaturated conditions in the retainer ring material, allowing pulsating magnetic fields supported by stator winding end turns to produce losses in ring. At sufficiently leading PF, the end-ring heating from these losses can damage field winding insulation. In addition to the potential end-ring heating problem, the reduced field current for leading PF excitation can result in the alternator operating with a reduced maximum developed torque capability, leaving it less likely to survive a transient stability disturbance without loss of synchronism.

Qualitative Performance by Phasor Diagram   Just as for the case of the isolated synchronous generator, the phasor diagram can serve as a useful analysis tool to predict qualitative performance of the synchronous generator interconnected with the utility grid. The constraint of near constant terminal voltage actually results in a simpler problem by removing a degree of freedom that existed in the isolated synchronous generator case.

Example 7.9   An alternator is connected to the electric utility grid where its terminal voltage may be considered constant. The field current is increased by 10 percent while the power supplied by the prime mover is unchanged. Neglect R_s and assume lagging PF. Qualitatively describe the changes in E_f, δ, I_a, θ, and Q_T if for this small change in I_f, X_s can be assumed constant.

   If I_f increases by 10 percent, E_f increases by 10 percent. Since the average power is unchanged, $\bar{E}_f$ must move along a horizontal line while $\bar{I}_a$ moves along a vertical line maintaining $I_a X_s \cos\theta$ and $I_a \cos\theta$ constant, respectively. The phasor diagram of Fig. 7.31

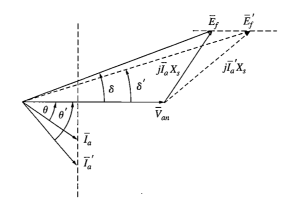

Figure 7.31
Qualitative phasor diagram analysis

displays the original conditions by the solid-line phasor. The conditions after the change are shown by the dashed-line phasors with the new phasors denoted by prime marks. Inspection of the phasor diagram shows that I_a and θ increase while δ decreases. Further, $I_a \sin \theta$ has increased in value; hence, total reactive power $Q_T = 3V_{an}I_a\sin\theta$ is now larger in value.

7.6 MOTOR PERFORMANCE

The round-rotor synchronous machine is capable of bilateral power flow; thus, it can operate in the motor mode modeled by the per phase equivalent circuit of Fig. 7.13b. Synchronous motors in practice are typically salient-pole machines and, as such, should be analyzed by the two-reaction theory of the next section for the best accuracy; however, no loss in qualitative assessment is suffered by their study using round-rotor theory.

Application of KVL to the equivalent circuit of Fig. 7.13b gives the excitation voltage as

$$\overline{E}_f = \overline{V}_{an} - \overline{I}_a(R_s + jX_s) \qquad \textbf{[7.36]}$$

The phasor diagram of Fig. 7.32a represents [7.36], where it has been assumed that the PF is lagging. However, if the field current (I_f) were increased, excitation voltage (E_f) must increase in magnitude. Equation [7.36] remains valid so that with sufficient increase in field current the phasor diagram of Fig. 7.32b results. A characteristic unique to synchronous motors has developed. The motor is now exhibiting a leading PF. While still supplying power to a coupled mechanical load, the machine supplies leading VARs to the connecting bus. This leading PF characteristic allows the synchronous motor to be used for power factor correction. When so applied, the size power factor correction capacitor bank required by an industrial facility to maintain the composite PF at a sufficiently small lagging value is reduced. As a result, an energy-pricing penalty from the electric utility may be avoided.

Examination of the synchronous motor phasor diagrams of Fig. 7.32 shows that the torque angles δ and δ_r are now negative in value. The developed torque is still given as appropriate by [7.31] or [7.33]; however, due to negative values of δ or δ_r, $3T_d$ is negative, indicating that the torque acts in the direction of speed (ω_s) as necessary for motor action.

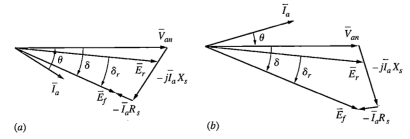

(a)   (b)

Figure 7.32
Phasor diagram of synchronous motor. (a) Lagging PF. (b) Leading PF.

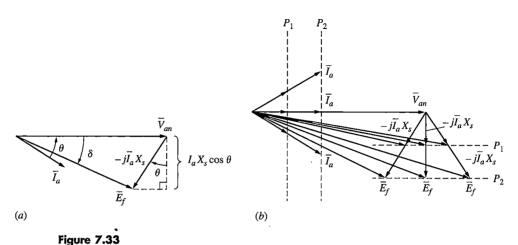

Figure 7.33
Synchronous motor V-curve analysis. (a) Lagging PF with $R_s = 0$. (b) Constant power, variable PF.

A qualitative assessment of synchronous motor performance is graphically presented by a plot of field current (I_f) vs. armature current (I_a) for constant terminal voltage with power as a parameter. The resulting plot has a characteristic V-shape and thus is known as the *V-curve*. As a first approximation to the V-curves, consider the synchronous motor lossless ($P_c = P_{FW} = I_a^2 R_s = 0$) so that output power equals input power. With $R_s = 0$, the phasor diagram of Fig. 7.33a results for lagging PF operation. Under the lossless assumption,

$$P_s = P_T = 3 V_{an} I_a \cos\theta \qquad \textbf{[7.37]}$$

Since V_{an} is constant, it is apparent from [7.37] that the product $I_a \cos\theta$ must be constant. Given a particular value of $P_s \doteq P_T = P_1$, the tip of the $\overline{I}_a$ phasor must lie along a vertical line and the tip of the $\overline{E}_f$ phasor must lie along a horizontal line for all possible points of operation at the chosen value of $P_s = P_T = P_1$. Figure 7.33b shows the composite phasor diagram for the cases of a lagging, a unity, and a leading PF. As the PF moves from lagging to leading, the current I_a passes through a minimum value for unity PF as the excitation voltage $E_f \alpha I_f$ continuously increases. Thus, for a I_f-I_a plot, the trough of the V-shape occurs at the unity PF point. Figure 7.33b also shows the composite phasor diagram for a second power case $P_s = P_T = P_2 > P_1$.

The MATLAB program ⟨vcurves.m⟩ generates an accurate set of V-curves for a synchronous motor using the OCC values. A lossless motor is not assumed. In addition, the program adjusts the value of synchronous reactance (X_s) to account for saturation effects using the procedure of [7.26] and [7.27]. Typical per unit values of X_{su}, X_ℓ, and R_s are assumed; however, more accurate values could be inserted if known. Figure 7.34 presents a plot of the V-curves for a 10-MVA, 2400-V (11,000-hp) synchronous motor. In typical operation at any given power point, the field current would be increased until either I_a or I_f heating limit is

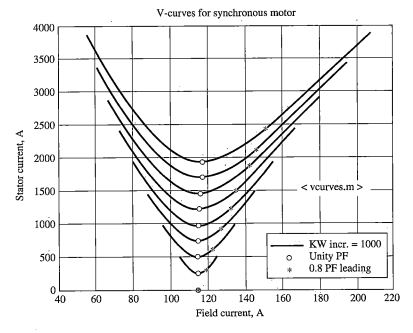

Figure 7.34
V-curves for a synchronous motor

reached, thereby operating at the maximum possible leading PF in order to maximize the leading VARs supplied, thus giving the maximum possible PF correction contribution.

7.7 SALIENT-POLE MACHINE PERFORMANCE

The phasor diagram of Fig. 7.18 describes the voltages $\bar{E}_f$, $\bar{E}_A$, and $\bar{E}_r$ of Fig. 7.13a induced behind the terminals of phase a winding for a round-rotor (RR) synchronous generator. The subscripts refer to field, armature, and resultant, respectively. By Faraday's law, it can be argued that there must be a flux linking the turns of phase a winding to have produced each of these voltages that vary sinusoidally with time. The phasor diagram of Fig. 7.18 has been redrawn in Fig. 7.35, except $\bar{E}_f$ rotated to the reference, to show the fluxes associated with each of these voltages. Each flux leads its associated voltage by 90° on the phasor diagram—a necessary condition that can be verified by Faraday's and Lenz's laws. The three fluxes (ϕ_f, ϕ_A, ϕ_r) linking the phase a coil are seen as time-varying sinusoidal quantities when viewed along the magnetic axis of the coil. Each flux could be represented by time phasors as done in Sec. 7.3.3; however, the fluxes are established by the field and armature reaction air gap traveling mmf waves that move in space around the air gap at a constant speed. This speed is synchronous

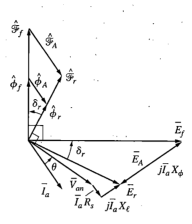

Figure 7.35
MMF's and fluxes of round-rotor synchronous generator

speed ($\omega_s = (2/p)\omega$) determined by the electrical radian frequency of the phase winding voltages. Consequently, when viewed in space, these fluxes can be thought of as traveling waves of speed ω_s. From this concept, the fluxes $\hat{\phi}_f, \hat{\phi}_A$, and $\hat{\phi}_r$ shown in Fig. 7.35 are known as *space phasors*. A diamond-shaped tip is used to designate them as space phasors.

Flux $\hat{\phi}_f$ and $\hat{\phi}_A$ are component fluxes established by the traveling field mmf and armature reaction mmf waves, respectively. Flux $\hat{\phi}_r$ is the resultant flux that physically exists due to the combination of $\hat{\phi}_f$ and $\hat{\phi}_A$ by vector addition. Making the common assumption of highly permeable ferromagnetic material and a uniform air gap width, space phasor mmf's $\hat{\mathscr{F}}_f, \hat{\mathscr{F}}_A$, and $\hat{\mathscr{F}}_r$, as shown in Fig. 7.35, can be associated with each flux using $\mathscr{F} = \phi\mathscr{R}_g$, where $\mathscr{R}_g$ is the air gap reluctance over the phase winding span.

Salient-Pole Machine Space Phasors The space phasor mmf's are unchanged if the round rotor were removed and replaced by a salient-pole (SP) rotor; however, the air gap is not uniform. The field mmf $\hat{\mathscr{F}}_f$ acts along a low reluctance path, commonly called the *direct-axis* or *d-axis* fixed to the rotor as indicated in Fig. 7.36a. On the other hand, the armature reaction mmf $\hat{\mathscr{F}}_A$ acts along a higher-reluctance path due to inclusion of a part of the interpolar region as indicated in Fig. 7.36a. The exact location of the armature reaction path depends on the machine PF since $\hat{\mathscr{F}}_A$ must peak along the axis of phase a coil at the same instant that current $\bar{I}_a$ peaks in value through coil a. As a consequence of this nonuniform air gap, there is no simple value of air gap reluctance $\mathscr{R}_g$ by which the space phasor fluxes can be determined using $\phi = \mathscr{F}/\mathscr{R}_g$ as in the case of the round-rotor machine.

In order to relate the space phasor fluxes to mmf's, the armature reaction mmf is decomposed into a component $\hat{\mathscr{F}}_d$ along the *d*-axis and a component $\hat{\mathscr{F}}_q$ acting

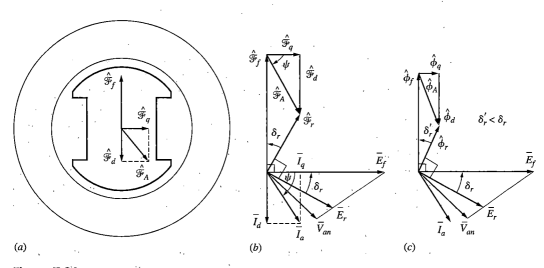

Figure 7.36
MMF's and fluxes of a salient-pole synchronous generator. (a) MMF orientations. (b) Decomposition of $\mathcal{F}_A$. (c) Resulting flux space phasors.

along a *quadrature axis* or *q-axis* that, as the name implies, is 90° (electrical) from the *d*-axis as indicated in Fig. 7.36*a*. The vector sum

$$\hat{\mathcal{F}}_A = \hat{\mathcal{F}}_d + \hat{\mathcal{F}}_q \qquad \textbf{[7.38]}$$

must hold. Current $\bar{I}_a$ can be decomposed into components associated with the *d*- and *q*-axis mmf's, giving

$$\bar{I}_a = \bar{I}_d + \bar{I}_q \qquad \textbf{[7.39]}$$

where
$$I_d = I_a \sin \psi \qquad I_q = I_a \cos \psi \qquad \textbf{[7.40]}$$

The reluctance values $\mathcal{R}_d$ and $\mathcal{R}_q$ along the *d*-axis and *q*-axis, respectively, are fixed values so that

$$\hat{\phi}_d = \frac{\hat{\mathcal{F}}_d}{\mathcal{R}_d} \qquad \hat{\phi}_q = \frac{\hat{\mathcal{F}}_q}{\mathcal{R}_q} \qquad \textbf{[7.41]}$$

regardless of the spatial location of $\hat{\mathcal{F}}_A$. Further,

$$\hat{\phi}_A = \hat{\phi}_d + \hat{\phi}_q \qquad \textbf{[7.42]}$$

The flux space phasors are shown by Fig. 7.36*c*. Since $\mathcal{R}_d < \mathcal{R}_q$, $\hat{\phi}_A$ would not align with $\hat{\mathcal{F}}_A$ if the phasor diagrams of Fig. 7.36*b* and *c* were placed on a common origin. More important, $\hat{\phi}_r$ and $\hat{\mathcal{F}}_r$ do not align. As a direct consequence, a single reactance (X_ϕ) cannot be determined to model armature as done in the case of the RR machine.

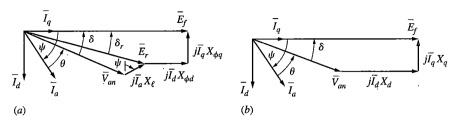

Figure 7.37
Voltages of salient-pole synchronous generator. (a) With d and q armature reactances defined. (b) With d and q synchronous reactances defined.

Salient-Pole Machine Voltages   A voltage may be associated with each of these armature reaction fluxes (ϕ_d, ϕ_q) and a corresponding reactance of armature reaction defined.

$$X_{\phi d} = \frac{\omega N_{eff}\phi_d}{I_d} \qquad X_{\phi q} = \frac{\omega N_{eff}\phi_q}{I_q} \qquad \textbf{[7.43]}$$

Using the reactances of [7.43], the phasor diagram of Fig. 7.37a can be constructed. Analogous to X_ϕ in the RR machine case, reactances $X_{\phi d}$ and $X_{\phi q}$ are associated with the mutual flux that links both the rotor and stator as a result of $\hat{\mathscr{F}}_A$. An accounting of the voltage drop associated with the stator coil leakage flux must also be introduced. The phasor $j\overline{I}_a X_\ell$ may be decomposed into $j\overline{I}_d X_\ell + j\overline{I}_q X_\ell$ and *d-axis* and *q-axis synchronous reactances* defined as

$$X_d = X_\ell + X_{\phi d} \qquad X_q = X_\ell + X_{\phi q} \qquad \textbf{[7.44]}$$

With this current decomposition and the reactances of [7.44], the phasor diagram of Fig. 7.37a can be redrawn as Fig. 7.37b. This analysis approach based on X_d and X_q is known as the *two-reaction theory* of synchronous machines.

  Although the phasor diagram of Fig. 7.37b could be used as a basis to draw an equivalent circuit containing current-controlled voltage sources for the salient-pole synchronous machine, no such action will be taken in this text. Rather, all analysis will be based solely on the phasor diagram model. For convenience, R_s has been neglected in the presentation; however, the $\overline{I}_a R_s$ or $R_s(\overline{I}_d + \overline{I}_q)$ voltage drop could be simply added to Fig. 7.37b if R_s were not negligible, giving the following equation for the SP synchronous generator for $R_s \neq 0$:

Generator: $$\overline{E}_f = \overline{V}_{an} + R_s\overline{I}_d + R_s\overline{I}_q + jX_d\overline{I}_d + jX_q\overline{I}_q \qquad \textbf{[7.45]}$$

  The phasor diagram of the salient-pole synchronous motor is formed by reversing the direction of current $\overline{I}_a$ of Fig. 7.37b, thereby rotating the phasor so that it is within the range of $\pm90°$ of $\overline{V}_{an}$, reflecting the appropriate input power factor of the motor. Figure 7.38 displays the phasor diagram for a salient-pole synchronous motor with lagging input PF. The describing equation for the SP synchronous motor with $R_s \neq 0$ is

Motor: $$\overline{E}_f = \overline{V}_{an} - R_s\overline{I}_d - R_s\overline{I}_q - jX_d\overline{I}_d - jX_q\overline{I}_q \qquad \textbf{[7.46]}$$

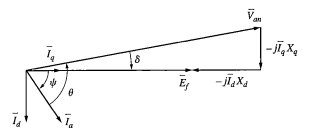

Figure 7.38
Phasor diagram for salient-pole synchronous motor

A 10-kVA, three-phase, 480 V, 60 Hz salient-pole synchronous machine has parameters R_s = 0.1Ω, X_d = 8Ω, and X_q = 4Ω. The machine is operating as a generator while supplying rated apparent power at rated voltage and frequency to a 0.8 PF lagging load. Determine the torque angle δ, as labeled in Fig. 7.37b, for the point of operation.

Example 7.10

Determination of the torque angle δ is frequently the starting point in SP synchronous machine analysis. The PF, thus θ, is typically known. If δ can be determined, then $\psi = -(\delta + \theta)$, allowing decomposition of $\bar{I}_a$ into $\bar{I}_d$ and $\bar{I}_q$. From the phasor diagram of Fig. 7.37b with $R_s \neq 0$,

$$\bar{E}_f = \bar{V}_{an} + R_s\bar{I}_a + jX_d\bar{I}_d + jX_q\bar{I}_q \qquad [1]$$

Equation [7.39] can be solved for $\bar{I}_q$. Substituting the result into [1] and rearranging yields

$$\bar{E}_f = \bar{V}_{an} + R_s\bar{I}_a + jX_q\bar{I}_a + j(X_d - X_q)\bar{I}_d \qquad [2]$$

A phasor diagram of [2] is sketched as Fig. 7.39a, where a new phasor $\bar{E}_f'$ that is collinear with $\bar{E}_f$ can be defined as

Generator: $\qquad\qquad \bar{E}_f' = \bar{V}_{an} + \bar{I}_a(R_s + jX_q) \qquad [3]$

Once $\bar{E}_f'$ is known,

$$\delta = \angle\bar{E}_f' - \angle\bar{V}_{an} \qquad [4]$$

For the generator under consideration,

$$I_a = \frac{S}{\sqrt{3}\,V_L} = \frac{10{,}000}{\sqrt{3}(480)} = 12.028 \text{ A}$$

$$\theta = \cos^{-1}(PF) = \cos^{-1}(0.8) = 36.87°$$

Assume $\bar{V}_{an}$ on the reference and apply [3] to find

$$\bar{E}_f' = \bar{V}_{an} + \bar{I}_a(R_s + jX_q)$$

$$= \frac{480}{\sqrt{3}}\angle 0° + 12.028\angle -36.87°(0.1 + j4) = 309.25\angle 7.02°$$

By [4],

$$\delta = \angle\bar{E}_f' - \angle\bar{V}_{an} = 7.02°$$

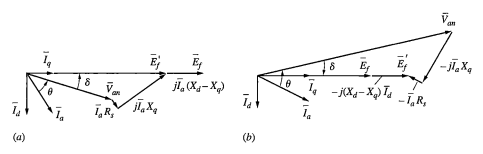

Figure 7.39
Collinear phasor $\bar{E}_f'$ defined. (a) Generator. (b) Motor.

An analogous approach for the case of a SP motor determines the collinear phasor $\bar{E}_f'$ as

Motor:
$$\bar{E}_f' = \bar{V}_{an} - \bar{I}_a(R_s + jX_q) \tag{5}$$

The phasor diagram of Fig. 7.39b results where the torque angle δ is found by [4]. Development of [5] is left as a problem at the end of this chapter.

Determination of X_d **and** X_q Data for determination of the direct- and quadrature-axis synchronous reactances of a SP synchronous machine can be measured from the *slip test* setup shown by Fig. 7.40. With the field winding unexcited, the machine rotor is driven by a mechanical source at near rated synchronous speed while the stator windings are excited by balanced three-phase source of rated frequency (ω) with voltage magnitude significantly less than rated to avoid magnetic saturation and to yield unsaturated values for X_d and X_q. As the rotor direct axis slips past the traveling mmf wave produced by the stator winding, variation in I_a magnitude can be observed as the stator mmf wave is alternately in and out of alignment with the d-axis of the rotor. If the rotor speed differs from synchronous speed by ε, and ε is small enough that the analog ammeter reading I_a can fully respond to the current variation, then the unsaturated d- and q-axis synchronous reactances are found by

$$X_d \cong \frac{V_L/\sqrt{3}}{I_{a\min}} \qquad X_q \cong \frac{V_L/\sqrt{3}}{I_{a\max}} \tag{7.47}$$

For typical machines, X_q/X_d ranges from 0.6 to 0.8. Since X_q is determined for the high-reluctance interpolar magnetic path, it experiences negligible saturation effects. On the other hand, X_d is associated with the lower-reluctance magnetic path centered about the small air gap portion of the rotor poles. It experiences the same saturation impact as the synchronous reactance (X_s) of the RR machine. Accounting for saturation effects can be handled with reasonable accuracy using the same procedures introduced in Sec. 7.4 for the RR machine.

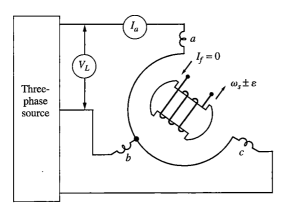

Figure 7.40
X_d and X_q determination by test

Electromechanical Developed Torque The developed torque formula will be derived for a lossless SP synchronous generator for which the power converted from mechanical to electrical form must be equal to the output power; hence,

$$3\,P_d = 3\,V_{an}I_a\cos\theta \qquad\qquad \textbf{[7.48]}$$

From the phasor diagram of Fig. 7.37b,

$$I_qX_q = V_{an}\sin\delta \qquad\text{or}\qquad I_q = \frac{V_{an}\sin\delta}{X_q} \qquad\qquad \textbf{[7.49]}$$

$$I_dX_d = E_f - V_{an}\cos\delta \qquad\text{or}\qquad I_d = \frac{E_f - V_{an}\cos\delta}{X_d} \qquad\qquad \textbf{[7.50]}$$

$$I_a\cos\theta = I_d\sin\delta + I_q\cos\delta \qquad\qquad \textbf{[7.51]}$$

Substitute [7.49] and [7.50] into [7.51], use $\cos\delta\sin\delta = \tfrac{1}{2}\sin(2\delta)$, and rearrange the results to obtain

$$I_a\cos\theta = \frac{E_f\sin\delta}{X_d} + \frac{V_{an}(X_d - X_q)\sin(2\delta)}{2X_dX_q} \qquad\qquad \textbf{[7.52]}$$

If [7.52] is used in [7.48] and the expression is divided by speed ω_s, the total developed torque is

$$3T_d = \frac{3P_d}{\omega_s} = \frac{3V_{an}E_f}{\omega_sX_d}\sin\delta + \frac{3V_{an}^2(X_d - X_q)}{2\omega_sX_dX_q}\sin(2\delta) \qquad \textbf{[7.53]}$$

The first term on the right-hand side of [7.53] is the same as given by [7.33] for the RR machine. If $I_f = 0$, E_f vanishes, leading to the conclusion that this component of developed torque is produced by interaction of the stator and field mmf's. The second term of [7.53] is independent of field current. Further,

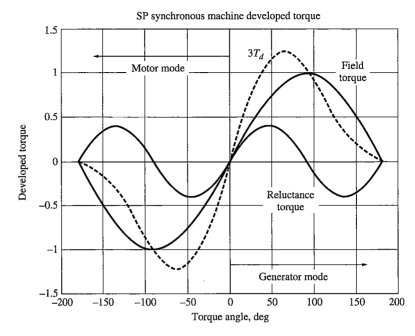

Figure 7.41
Developed torque for salient-pole synchronous machine

if the rotor periphery were uniform, $X_d = X_q(\mathcal{R}_d = \mathcal{R}_q)$ and this component of total developed torque vanishes. The term is fittingly known as a *reluctance torque*.

As the machine moves from generator to motor mode, the torque angle δ takes on negative values as noted from the phasor diagram of Fig. 7.38. A sketch of the components and composite developed torques for a SP synchronous machine is displayed by Fig. 7.41, from which it is apparent that the reluctance torque acts to increase the maximum torque available without loss of synchronism. By nature of the contribution of the reluctance torque, SP machines operate with a smaller torque angle δ than RR machines. Further, $d3T_d/d\delta_r$ is greater in value for the SP machine. Consequently, the oscillatory movement of the rotor during sudden load changes is less for the SP machine.

7.8 SELF-SYNCHRONOUS MOTORS

In order to maintain brevity of presentation, the scope of study for self-synchronous machines will be limited to motoring mode operation, although the physical machines are capable of bilateral energy conversion. A self-synchronous motor is actually an integrated system consisting of a synchronous machine, a shaft position

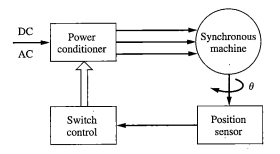

Figure 7.42
Block diagram of self-synchronous motor

sensor, a switch logic controller, and a power conditioner. This latter element is an array of power-level switches to direct current flow through the synchronous machine windings so as to control the angular position of the stator field based upon the position of the rotor. The synchronous machine may have an active field rotor or be a reluctance principle machine. Figure 7.42 is a functional block diagram clarifying the self-synchronous motor concept.

7.8.1   BRUSHLESS DC MOTORS

The brushless dc motor (BDCM) is a self-synchronous motor with performance characteristics similar to a separately excited commutator dc motor, hence the fitting name. Specifically, speed control is implemented by increase or decrease of armature impressed voltage. The synchronous machine can be either wound field or permanent magnet (PM) excited. Within the permanent magnet machines, there are two basic types—air gap magnets and interior magnets. The former produces a trapezoidal shape generated voltage, while the latter generates a sinusoidal shape voltage. The present scope of study will be limited to the air gap magnet synchronous machine.

For simplicity in discussion, a two-pole PM synchronous machine is illustrated by Fig. 7.43; however, four-, six-, and eight-pole machines are more common in the BDCM application. With the particular synchronous machine identified, the functional block diagram can be expanded to show the machine and power conditioner details as displayed by Fig. 7.44. When drawing the schematic of the PM synchronous machine, it has been recognized that each phase winding has a resistance, an inductance, and an induced voltage. Due to the ideally constant flux produced by the PMs, the magnitude of the induced voltages varies directly with speed. The position sensor sends a signal to the base drive logic each 60° of shaft movement. This signal is decoded to produce a repetitive six-step set of conduction logic for the inverters as indicated by Table 7.1 on page 467.

Each 60°, one transistor is turned ON and one transistor is turned OFF. Any one transistor conducts for 120° of the 360° step sequence. With this switching

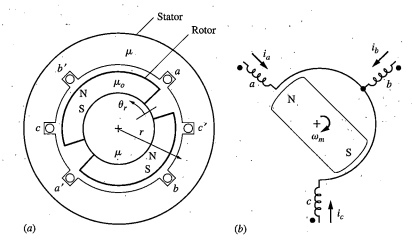

Figure 7.43
Permanent magnet synchronous machine. (a) Physical. (b) Schematic.

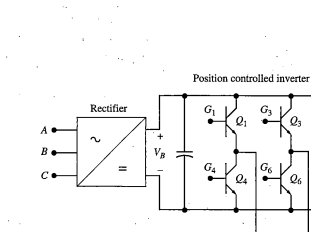

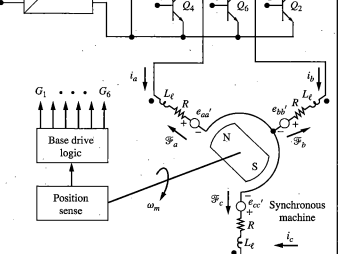

Figure 7.44
BDCM schematic

Step	I	II	III	IV	V	VI	I
Switch	Q_1			Q_4			Q_1
Active		Q_2			Q_5		
	Q_6			Q_3		Q_6	

Table 7.1
Inverter conduction logic

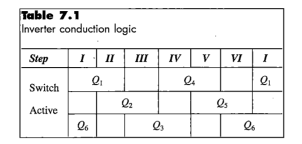

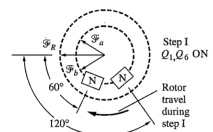

Step I
Q_1, Q_6 ON

Rotor travel during step I

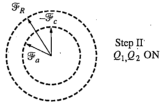

Step II
Q_1, Q_2 ON

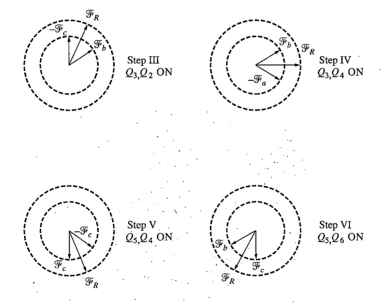

Step III
Q_3, Q_2 ON

Step IV
Q_3, Q_4 ON

Step V
Q_5, Q_4 ON

Step VI
Q_5, Q_6 ON

Figure 7.45
Discrete steps of BDCM stator mmf wave

pattern, ideally two of the motor phase windings are conducting current at any instant in time. The stator produced mmf wave ($\mathscr{F}_R$) travels in discrete steps with location in space as illustrated by Fig. 7.45. As the rotor magnets move toward alignment, the stator field continues to advance, producing continuous motion of the rotor.

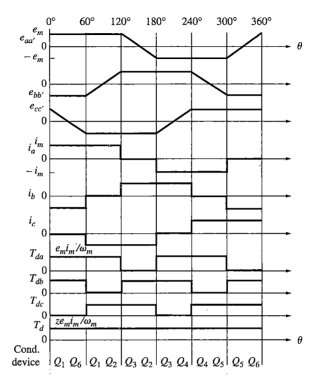

Figure 7.46
Timing diagram for BDCM

The control logic functions to assure the phase relationships between the coil induced trapezoidal voltages and currents shown by the timing diagram of Fig. 7.46. The total power converted from electrical to mechanical form is given by

$$3P_d = \sum_{a,b,c} e_k i_k = e_{aa'} i_a + e_{bb'} i_b + e_{cc'} i_c \qquad \textbf{[7.54]}$$

The total electromagnetic developed torque at speed ω_m follows as

$$3T_d = \frac{3P_d}{\omega_m} = \frac{\displaystyle\sum_{a,b,c} e_k i_k}{\omega_m} \qquad \textbf{[7.55]}$$

The torque contributions of each phase (T_{da}, T_{db}, T_{dc}) are easily determined with the aid of the timing diagram of Fig. 7.46. The individual phase contributions to developed torque are added to give total developed torque as

$$3T_d = \frac{2e_m i_m}{\omega_m} \qquad \textbf{[7.56]}$$

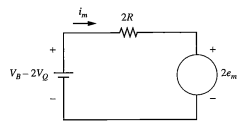

Figure 7.47
Equivalent circuit of BDCM

As apparent from Fig. 7.43a, flux paths of the PM synchronous machine include a radial path across the PM material which has a permeability that approaches that of free space. Consequently, the reluctance of coil coupling flux paths is large; thus, coil inductance values are small. For operating speeds that result in a frequency (f) of coil induced voltages such that $T = 1/f \gg L_\ell/R$, the coil currents are near the ideal waveforms of Fig. 7.46. The dc output voltage (V_B) of the rectifier of Fig. 7.44 at any point in time is impressed across two phase windings that present a total induced voltage of magnitude $2e_m$. The simple equivalent circuit of Fig. 7.47 then models the BDCM where the impressed voltage appearing at the motor terminals has been reduced by $2V_Q$—the ON-state voltage drop across the two conducting transistors.

A BDCM is operating with a dc link voltage $V_B = 325$ V. The motor phase winding resistance is 2.3 Ω. At a speed of 1000 rpm with the motor open circuit, the magnitude $(2e_m)$ of the trapezoidal line-to-line voltage was determined to be 123 V. If the BDCM is operating at 2500 rpm, find the magnitude of phase currents and total developed torque. Assume that the forward voltage drop of a conducting transistor is 0.8 V. | **Example 7.11**

The induced voltage of the two series phase coils is determined by direct ratio of speed.

$$2\,e_m = \frac{2500}{1000}(123) = 307.5\,\text{V}$$

Based on the equivalent circuit of Fig. 7.47,

$$i_m = \frac{V_B - 2V_Q - 2e_m}{2R} = \frac{325 - 2(0.8) - 307.5}{2(2.3)} = 3.456\,\text{A}$$

By [7.56],

$$3T_d = \frac{2\,e_m i_m}{\omega_m} = \frac{(307.5)(3.456)}{2500\,(\pi/30)} = 4.059\,\text{N·m}$$

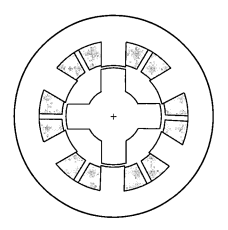

Figure 7.48
6/4 SRM

7.8.2   SWITCHED RELUCTANCE MOTORS

The switched reluctance motor (SRM) is a doubly salient reluctance machine with a set of concentrated excitation coils wound on the poles of one member, usually the stator, as illustrated by Fig. 7.48. This figure shows a set of rotor poles aligned with the stator poles of phase *a* coil—a minimum reluctance (or maximum inductance) position for phase *a*. If phase *a* coil were de-energized at this point and phase *b* coil energized, the pair of rotor poles on the vertical axis would attempt to align with phase *b* stator poles. At the point at which phase *b* coil was energized, the coil of phase *b* saw a large reluctance (or small inductance). Phase *c* coil would be energized and phase *b* coil de-energized when a pair of rotor poles aligned with phase *b*. From this description for counterclockwise rotation of the SRM of Fig. 7.48, it is concluded that it is the coil currents that are switched and, further, that this position-sensitive switching can be specified based on the reluctance characteristics of the coils. With this background, the reasoning behind the name *switched reluctance* is understood.

In Example 3.11, it was found that $L = N^2/\mathcal{R}$, or inductance varies inversely with reluctance. Consequently, the SRM is frequently characterized in terms of the coil inductances. One could argue that the machine should be rightfully referred to as a switched inductance motor; however, such a nomenclature has never been widely adopted. The idealized inductance-position profile (L_a) for the phase *a* coil of the SRM of Fig. 7.48 is given by Fig. 7.49 for the first half revolution from the shown position. In order for this motion to have transpired, the phase coils would have each been excited for 30° of rotation according to the sequence *b·c·a·b·c·a*. The inductance profiles for the coils of phases *b* and *c* are identical in shape but shifted in angular position by 30° and 60°, respectively. Figure 7.50 places the inductance profiles and coil excitation currents on a common position axis.

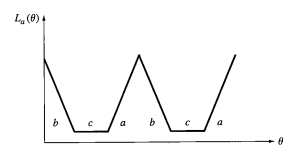

Figure 7.49
Idealized inductance profile

The electromagnetic developed torque produced by each coil is described by [3.71]. For the idealized inductance and current profiles of Fig. 7.50, it is seen that over the conduction interval of a particular coil, i is a constant value and $dL/d\theta$ likewise is a constant. Thus, by [3.71] the torque developed by an energized phase coil is

$$T_d = \frac{1}{2}i^2\frac{dL}{d\theta} = \frac{1}{2}I_m^2\frac{L_{\max} - L_{\min}}{\pi/6} \qquad \textbf{[7.57]}$$

The individual phase torques (T_a, T_b, T_c) and their sum (T_T), which is a constant value, are added to Fig. 7.50. In order to produce the phase current profiles of Fig. 7.50 from a common dc source, a circuit (*converter*) topology such as shown by Fig. 7.51 is necessary. The converter also provides a path for continuity of inductor currents through the shunting diodes when excitation is removed from a coil.

The above idealized presentation has conveyed the principle of SRM operation. Actual inductance profiles will tend to have some rounding in the transition areas due to flux fringing and to reflect the dependency on the values of magnetic saturation when the flux density increases in the ferromagnetic material as pole alignment occurs. The coil currents will experience delay rounding on the leading edges and trailing edges. Consequently, the total developed electromagnetic torque will have a resulting position-dependent ripple. Although the accuracy of the idealized analysis may be questionable, the concept of the operation remains valid.

The SRM torque production can also be explained in terms of [3.70]. The flux linkage–current (λ-i) plot of Fig. 7.52 displays the flux linkages of one of the SRM phase coils for the two cases of poles aligned and poles unaligned, or the maximum inductance and minimum inductance positions. If the coil is energized with a current $i = I_1$ in the unaligned position, the magnetic coenergy is given by the area enclosed by *oab*. As the rotor moves to the position of full alignment, the

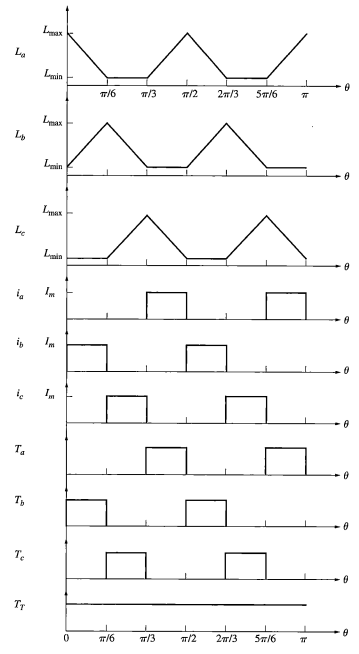

Figure 7.50
Idealized inductance, current, and torque profiles for SRM

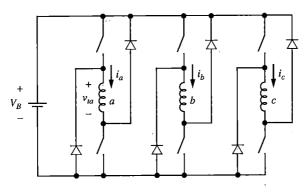

Figure 7.51
Converter to supply the current profiles of SRM

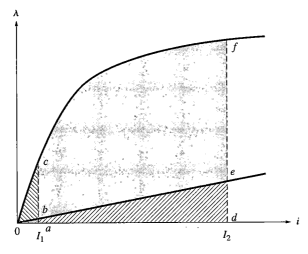

Figure 7.52
Flux linkage—current plot

magnetic coenergy is given by the enclosed area *oac*. Based on [3.70], the torque average produced during the rotor movement is

$$T_d \cong \frac{\Delta W_f'}{\Delta \theta} = \frac{(\text{area } oac) - (\text{area } obc)}{\Delta \theta}$$

[7.58]

Clearly, the average torque is proportional to the area *obc* between the unaligned and aligned flux linkage curves. The SRM is only capable of producing high torque density ($N \cdot m/m^3$ or $ft \cdot lb/in^3$) if the area between the two flux linkage plots is large. This can be accomplished by driving the machine into saturation for the aligned condition with a coil excitation current such as $i = I_2$ of Fig. 7.52. In such case, the average torque is given by

$$T_d \cong \frac{\Delta W_f'}{\Delta \theta} = \frac{(\text{area } odf) - (\text{area } ode)}{\Delta \theta} = \frac{\text{area } oef}{\Delta \theta} \qquad \textbf{[7.59]}$$

Dynamic operation of the SRM of Fig. 7.51 will not result in the rectangular current waveforms of Fig. 7.50 unless the speed of the motor is near zero in value. For motor action, the switches for each phase coil of Fig. 7.51 are turned on when the poles of that phase are in the unaligned orientation where the inductance is small in value; however, the circuit is an *R-L* circuit and the current rise time is finite. A simple mathematical expression to describe the current trajectory is not possible since this is a *R-L* circuit in which the inductance is increasing in value as the poles move toward alignment. The switches of each phase are turned off when the poles are aligned. At this point, there is significant energy stored in the magnetic field of the coil that must be transferred back to the battery V_B of Fig. 7.51 as continuity of the coil current is sustained through the pair of diodes in each phase circuit until the current is reduced to zero.

The λ-i plots of Fig. 7.52 are valid for the unaligned and aligned cases. Similar plots can be generated for intermediate positions to give a family of plots as illustrated by Fig. 7.53. For the 6/4 SRM of Fig. 7.48, these plots would represent the flux linkages at 5° increments of movement. Consider the phase *a* coil of Fig. 7.51 with a resistance R and a flux linkage $\lambda_a (i_a, \theta)$ as described by Fig. 7.52. The coil terminal voltage is given by

$$v_{ta} = R i_a + \frac{d\lambda_a}{dt} = R i_a + \frac{\partial \lambda_a}{\partial i}\frac{di}{dt} + \frac{\partial \lambda_a}{\partial \theta}\omega_m \qquad \textbf{[7.60]}$$

where the rotational speed $\omega_m = d\theta/dt$. The coefficient of di/dt in [7.60] is the incremental inductance of the phase coil evaluated at the instantaneous values of i_a and θ. The coefficient of the speed term in [7.60] can be thought of as a cemf coefficient with units of V·s/rad that depends upon the rate of change in coil flux linkage with position at the instantaneous values of i_a and θ. Thus, KVL allows [7.60] to be written as

$$kV_B = R i_a + L(i_a, \theta)\frac{di}{dt} + K(i_a, \theta)\omega_m \qquad \textbf{[7.61]}$$

where $k = 1$ if the switches are on and $k = -1$ if the switches are off and $i_a > 0$. With a set of curves such as Fig. 7.53, the incremental inductance and cemf coefficient are approximated by

$$L \cong \frac{\Delta \lambda_a}{\Delta i_a}\bigg|_{i_a, \theta} \qquad K \cong \frac{\Delta \lambda_a}{\Delta \theta}\bigg|_{i_a, \theta} \qquad \textbf{[7.62]}$$

Based on the above work, an equivalent circuit for phase *a* of the SRM can be drawn as illustrated by Fig. 7.54. Similar equivalent circuits can be constructed for

SRM flux linkage

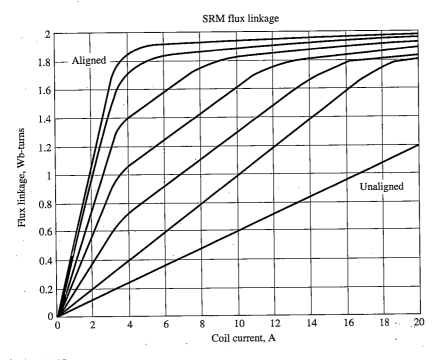

Figure 7.53
Flux linkage–current plots for SRM at numerous positions of pole alignment

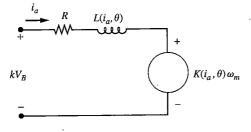

Figure 7.54
Per phase equivalent circuit of SRM

phases b and c. The instantaneous power converted from electrical to mechanical form by phase a of the SRM is the power flowing into the controlled voltage source.

$$p_a(t) = \frac{\partial \lambda_a}{\partial \theta} \omega_m i_a = K(i_a, \theta) \omega_m i_a \qquad \textbf{[7.63]}$$

The instantaneous developed power of phase a follows as

$$\tau_{da}(t) = \frac{p_a(t)}{\omega_m} = \frac{\partial \lambda_a}{\partial \theta} i_a = K(i_a, \theta) i_a \qquad \textbf{[7.64]}$$

The average value of developed torque is simply the average value of the instantaneous developed torque (τ_d) of all three phases.

$$T_d = \frac{1}{T}\int_0^T (\tau_{da} + \tau_{db} + \tau_{dc})dt = \frac{1}{T}\int_0^T [K(i_a,\theta)i_a + K(i_b,\theta)i_b + K(i_c,\theta)i_c]dt$$

[7.65]

Example 7.12 | If the λ-i plots of Fig. 7.53 model the 6/4 SRM of Fig. 7.48 at 5° increments and the motor at a particular instant is operating at a speed of 100 rad/s with the poles of phase a aligned and $i_a = 10$ A, determine the values of L, K, and τ_d for phases a and b.

Enter Fig. 7.48 at $i_a = 10$ A and aligned position to determine that for phase a,

$$L(i_a,\theta) \cong \frac{\Delta \lambda_a}{\Delta i_a} = \frac{0.02}{4} = 0.005 \text{ H}$$

$$K(i_a,\theta) \cong \frac{\Delta \lambda_a}{\Delta \theta} = \frac{0.03}{5(\pi/180)} = 3.39 \text{ V·s/rad}$$

$$\tau_{da} \cong K(i_a,\theta)i_a = 3.39(10) = 33.9 \text{ N·m}$$

At the described point, phase b will just be switching on so that $i_b = 0$.

$$L(i_b,\theta) \cong \frac{\Delta \lambda_b}{\Delta i_b} = \frac{0.1}{1.9} = 0.053 \text{ H}$$

$$K(i_b,\theta) \cong \frac{\Delta \lambda_b}{\Delta \theta} = \frac{0}{5(\pi/180)} = 0$$

$$\tau_{db} = K(i_b,\theta)i_b = 0$$

7.9 SYNCHRONOUS MACHINE DESIGN

The chosen object of design study is the large, high-speed synchronous generator known as the turboalternator for electric utility power plant application. The prime mover for these machines is either a steam turbine or a combustion gas turbine. The round rotor is of either a two-pole or a four-pole design. The scope of the presentation is limited to only electrical and magnetic performance considerations. References are cited that give the reader additional details of derivation and guidance into the all-important aspects of cooling and mechanical design that are not adequately addressed by the limited scope.[3-5]

7.9.1 STANDARDS AND CLASSIFICATIONS

Turboalternators are typically purchased through a contractual arrangement between the manufacturer and the electric utility. There may well be some degree of package customization for each application, although the central cross section of the machine has commonality within a design series produced by the manufacturer. The American National Standards Institute publications ANSI C50.10, ANSI C50.13, ANSI C50.14, and ANSI C50.30 are the principal documents that serve to establish expected performance of turboalternators. The topics covered by these documents can be loosely organized into rating, cooling, and abnormal service categories.

Rating The output rating of a turboalternator is specified in apparent power along with voltage, frequency, and power factor. These ratings are understood to be applicable only if the associated specified cooling conditions are maintained.

Cooling Turboalternators are classified as either directly cooled or indirectly cooled according to the mechanism of heat transfer to the cooling medium—air, hydrogen gas, or water. Directly cooled machines are designed so that the cooling medium contacts the conductors directly while for indirectly cooled machines, heat must flow across the conductor ground insulation prior to reaching the cooling medium. The type of cooling for the rotor and stator does not necessarily have to be the same. Three common cooling schemes for turboalternators are:

1. Air-cooled. Forced ventilation of air gives indirect cooling.
2. Hydrogen-cooled. Self-ventilation of hydrogen gas under pressure produces indirect cooling.
3. Water-cooled. Ionized water is pumped through hollow stator conductors to give direct cooling of the stator. Typically, the rotor is indirectly cooled with self-ventilation of pressurized hydrogen gas.

Abnormal Service Some guidelines or specifications are usually provided regarding operation for overcurrent (field or armature time limits), overvoltage, nonrated frequency, and variance in ambient temperature and coolant pressures.

7.9.2 VOLUME AND BORE SIZING

A sectional view of a turboalternator is shown in Fig. 7.55, where principal sizing dimensions are indicated. The *specific electric loading,* or equivalent surface current density along the stator bore diameter (D), of a turboalternator is given by

$$J_s = \frac{N_{eff}I_a}{\pi D} \qquad \textbf{[7.66]}$$

and

$$I_a = \frac{\pi D J_s}{N_{eff}} \qquad \textbf{[7.67]}$$

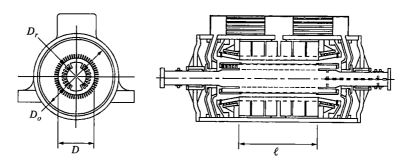

Figure 7.55
Section view of turboalternator

The *specific magnetic loading,* or average flux density over a pole span, can be computed in terms of the peak value of the assumed fundamental flux per pole.

$$B_{av} = \frac{(2/\pi)\Phi_m}{\pi D\ell/p} \qquad \textbf{[7.68]}$$

whence

$$\Phi_m = \frac{\pi^2 D\ell B_{av}}{2p} \qquad \textbf{[7.69]}$$

Based on [7.6] with neglect of leakage reactance and coil resistance, the per phase terminal voltage is

$$V_{an} = 4.44N_{eff}f\Phi_m \qquad \textbf{[7.70]}$$

The total apparent power of a three-phase machine is formed as

$$S = 3V_{an}I_a \qquad \textbf{[7.71]}$$

Substitute [7.69] into [7.70]. Then use the result along with [7.67] in [7.71] to find

$$S = 6(4.44)\,\pi^2 f J_s B_{av}\frac{(\pi/4)D^2\ell}{p} \qquad \textbf{[7.72]}$$

For any particular type of cooling, there is a value of J_s beyond which the ohmic losses of the stator coils cannot be removed without overheating. Likewise, the ferromagnetic material establishes an upper limit on B_{av}. If J_s and B_{av} are maintained at the practical limits and f is fixed, then from [7.72] it is seen that the apparent power rating of a turboalternator varies directly on the air gap volume and inversely with the number of poles. With the principle for sizing established, by [7.72] the following equation that reflects successful design experience can be used for initial selection of D and ℓ proportions for turboalternator design work.

$$\frac{D^2\ell}{Sp} = C_o \qquad \textbf{[7.73]}$$

where S is the apparent power in MVA, D and ℓ are in units of inches, and the constant C_o depends on the method of cooling.

$$C_o = 1400 \text{ (air-cooled)} \qquad\qquad \textbf{[7.74]}$$

$$C_o = 700 \text{ (hydrogen-cooled)}$$

$$C_o = 375 \text{ (water-cooled)}$$

Example 7.13

Size the stator bore diameter (D) and active rotor length (ℓ) for a 100-MVA, 24-kV, 3600 rpm water-cooled turboalternator if the rotor peripheral speed at rated conditions is 35,000 ft/min.

The peripheral speed sets the rotor diameter as

$$D_r = \frac{v_r}{\pi n_s} = \frac{35,000(12)}{\pi(3600)} = 37.14 \text{ in}$$

Assume that a 1.5-in air gap length is reasonable; then,

$$D = D_r + 2\delta = 37.14 + 2(1.5) = 40.14 \text{ in}$$

From [7.73] and [7.74],

$$\ell = \frac{C_o p S}{D^2} = \frac{350(2)(100)}{(40.14)^2} = 43.44 \text{ in}$$

7.9.3 STATOR DESIGN

Since the turboalternator stator is identical in function and concept to that of the induction motor, the logic flowchart of Fig. 6.45 is applicable to the design process. If attainment of an acceptable design is possible otherwise, the stator bore and length dimensions are decided solely by the above procedure and maintained outside the iterative process of stator design.

Number of Stator Slots The standard practice of a double-layer winding inherently results in the number of conductors per slot (C_s) being an even integer. The scope of study is limited to the case of the slots per pole being an integer multiple of 3. As a result, the number of stator slots (S_1) is limited to one of the selections from Table 7.2.

Table 7.2
Stator slot selection

Pole (p)	No. Stator Slots (S₁)											
24	6	12	18	24	30	36	42	48	54	60	66	72
4		12		24		36		48		60		72

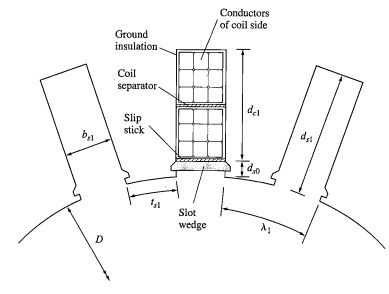

Figure 7.56
Stator slot section view

All the choices of Table 7.2 will not necessarily result in a good design. The stator slot pitch of Fig. 7.56 is given by

$$\lambda_1 = \frac{\pi D}{S_1} \qquad \textbf{[7.75]}$$

Good design practice is to select the number of stator slots so that

$$0.04D \leq \lambda_1 \leq 0.07D \qquad \textbf{[7.76]}$$

Slot Design   The stator tooth width should be chosen so that

$$0.4\lambda_1 \leq t_{s1} \leq 0.5\lambda_1 \qquad \textbf{[7.77]}$$

Hence, the slot width is

$$b_{s1} = \lambda_1 - t_{s1} \qquad \textbf{[7.78]}$$

For reasons of short-circuit transient protection, it is desirable that turboalternators have a leakage reactance that is 10 to 15 percent of the synchronous reactance. This goal is usually achieved if the slot is proportioned so that

$$3b_{s1} \leq d_{s1} \leq 7b_{s1} \qquad \textbf{[7.79]}$$

In addition to containment of the two coil sides, the slots provide adequate insulation between coil sides (coil separator) and have a means of retaining the coils (slot wedge) as illustrated by Fig. 7.56. Coil ground insulation for turboalternators typically encapsulates the individual coil sides. The following material thicknesses are suggested:

1. Use 65 V/mil ground insulation.
2. Use 0.375-in slot wedge.
3. Use 0.125-in coil separator and top stick.

Conductors per Slot Let C_s be the conductors per slot and $N_s = S_1/3p$ be the slots per pole per phase; then the effective series turns per phase is

$$N_{eff} = \frac{k_w N_s C_s p}{2a} \qquad \text{[7.80]}$$

Use this result in [7.6] to yield

$$E_{aa'} = \frac{2.22 k_w N_s C_s pf \Phi_m}{a} \qquad \text{[7.81]}$$

Allowing a typical value of 10 percent voltage drop across the leakage reactance and assuming a wye-connected stator winding,

$$E_{aa'} \cong 1.1 V_{anR} \qquad \text{[7.82]}$$

where V_{anR} is the rated line-to-neutral voltage. Substitute [7.81] into [7.80] to determine the peak value of the flux per pole.

$$\Phi_m = \frac{1.1 a V_{anR}}{2.22 k_w N_s C_s pf} \qquad \text{[7.83]}$$

Solve [7.68] for Φ_m, equate the result to [7.83], and rearrange to find an expression for the conductors per slot.

$$C_s = \frac{0.1 a V_{anR}}{k_w N_s D \ell f B_{av}} \qquad \text{[7.84]}$$

A value of $B_{av} = 0.6\,\text{T}$ is typical for the specific loading of turboalternators. The value of C_s determined by use of [7.84] must be rounded to an even integer.

Conductor sizing For indirectly cooled machines, the stator conductor is solid copper as illustrated by the turboalternator coils of Fig. 7.57. If the machine is indirectly cooled, the hollow water passageway should be approximately 25 percent of the conductor cross-section area (s_a). Allow an additional 15 percent of both height and width of the conductors for tolerance and for Roebel transformation of the conductors within the slot.

The conductor current density is

$$\Delta_1 = \frac{I_{aR}}{a s_a} \qquad \text{[7.85]}$$

Figure 7.57
Stator coils for turboalternator. (*Courtesy of National Electric Coil, Columbus, Ohio. Photo by Chuck Drake.*)

For proper cooling, good design practice is

Air-cooled: $\Delta_1 \leq 2500$ A/in^2 **[7.86]**

Hydrogen-cooled: $\Delta_1 \leq 4000$ A/in^2

Water-cooled: $\Delta_1 \leq 7000$ A/in^2

Owing to the large size of turboalternator conductors, standard wire sizes do not provide a match. The conductors will be typically a special size of drawn copper. In order to minimize the magnitudes of eddy current resulting from cross-slot flux, the conductor should be formed from multiple strands where no dimension exceeds 0.25 in. Individual conductor strands should be served with no less than 2 mil high-dielectric-strength film insulation.

Core Diameter The outside diameter of the stator laminations must be chosen sufficiently large so that the field winding mmf requirement to drive flux over the lengthy yoke path does not create an undue burden on the rotor design. Further, the high-efficiency design goal for turboalternators demands that the flux density of the voluminous yoke ferromagnetic material be such that core losses are reasonably small. An acceptable design can usually be obtained if the stator core OD is sized according to the guidelines below:

Two-pole: $D_o \cong 2.1 D$ **[7.87]**

Four-pole: $D_o \cong 1.7 D$

B-check The average flux density of the stator tooth at one-third slot depth from the stator bore should be checked prior to proceeding to the rotor design phase. All turboalternators have long lamination stack lengths and must be provided with radial air ducts to allow radial flow of air or hydrogen from the rotor. The following design guideline is suggested to determine the net length (ℓ_n) of active ferromagnetic material in the stator lamination stack.

Air-cooled: $\qquad\qquad \ell_n \cong 0.8\,\ell$ $\qquad\qquad\qquad$ **[7.88]**

Hydrogen-cooled: $\qquad\quad \ell_n \cong 0.9\,\ell$

Water-cooled: $\qquad\qquad \ell_n \cong 0.95\,\ell$

The tooth width at one-third slot depth is

$$t_{s13} = \frac{\pi(D + 2/3d_{s1})}{S_1} - b_{s1} \qquad\qquad \textbf{[7.89]}$$

A lamination stacking factor $SF = 0.95$ is typical for 29-gage ESS. The average value of flux density at one-third slot depth is found as

$$B_{t13} = \frac{p(2/\pi)\Phi_m}{S_1 t_{s13}\ell_n SF} \qquad\qquad \textbf{[7.90]}$$

Acceptable magnetic circuit performance usually results if B_{t13} does not exceed 1.2 T (68 kilolines/in^2).

7.9.4 AIR GAP SIZING

An empirical formula for initial sizing of the air gap in units of inches is

$$\delta = 4.5 \times 10^{-5}\frac{k_w C_s N_s I_{aR}}{a} \qquad\qquad \textbf{[7.91]}$$

Since the air gap length, rotor diameter, stator bore diameter, and the number of stator conductors are interrelated, it is difficult to avoid at least one complete iteration in the design process.

7.9.5 ROTOR DESIGN

As turboalternators increase in power level, the rotor becomes a mechanical design challenge. For a fixed mechanical speed, the machine power rating depends on $D^2\ell$. As the machine length increases, the rotor is prone to exhibit critical frequencies at lower speeds that can result in shaft flexure to the point that the rotor strikes the stator bore. Even with best design, a turboalternator typically has a rated speed somewhere between the first and second critical frequencies. The rotor is manufactured as a forging of steel alloy with high yield strength. In order to keep the length short as possible, the rotor diameter is sized to operate at near the stress limit of the steel alloy. Consequently, the maximum allowable peripheral speed of the rotor is a central consideration in turboalternator design. With present-day steel alloys, rotor peripheral speeds of 50,000 ft/min represent the design limit.

Example 7.14 | A two-pole, 60-Hz alternator has a synchronous speed of 3600 rpm. If the machine must be capable of sustaining a 20 percent overspeed and the maximum allowable rotor peripheral speed $v_r = 50,000$ ft/min, determine the maximum allowable rotor diameter (D_r).

$$D_{r\max} = \frac{v_r(12)}{1.2n_s\pi} = \frac{50,000(12)}{1.2(3600)\pi} = 44.21 \text{ in}$$

The concept of the turboalternator rotor has been introduced by Fig. 7.8. A section view of a two-pole rotor is depicted by Fig. 7.58, wherein dimensions are indicated for use in the design process. The scope of study is limited to the case of total rotor slots $S_2 = kp$, where k is an even integer. The particular values of $k = 6,8,10,12$ are likely to produce a practical design.

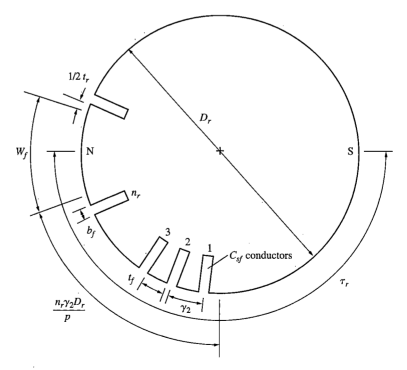

Figure 7.58
Section view of turboalternator rotor

Rotor Slot Selection The center of each pole has a span devoid of slots of width $W_f + t_f$, where

$$0.2\tau_r \leq W_f \leq 0.3\tau_r \qquad \textbf{[7.92]}$$

Rotor pole pitch:

$$\tau_r = \frac{\pi D_r}{p} \qquad \textbf{[7.93]}$$

Slots per pole half:

$$n_r = \frac{S_2}{2p} \qquad \textbf{[7.94]}$$

Angular slot pitch:

$$\gamma_2 = \frac{\pi D_r - pW_f}{2D_r n_r} \text{ elec. rad} \qquad \textbf{[7.95]}$$

Rotor tooth width:

$$t_f = \frac{D_r}{2}\gamma_2 - b_f \qquad \textbf{[7.96]}$$

A decision on the number of rotor slots can be made by trial-and-error execution of ⟨rdrotmmf.m⟩ to determine a value of S_2 that produces an acceptably small spectrum of harmonics.

Conductors per Slot Each field coil, and thus rotor slot, is assumed to contain C_{sf} conductors each carrying current I_f. If successive partial, yet symmetric, excitation of the field windings could be accomplished as illustrated in Fig. 7.59a for the case of the four slots labeled by 1, and if Ampere's circuital law given by [3.11] were applied for each case of partial excitation, the coil set mmf's of Fig. 7.59b result. Summing of the n_r in-phase fundamental components from a Fourier series of the mmf sets of Fig. 7.59b determines the peak value of the fundamental component of field mmf.

$$\mathscr{F}_{f1} = \frac{4}{\pi}C_{sf}I_f\sum_{i=1}^{n_r}\cos\left[(2i-1)\gamma_2/2\right] \qquad \textbf{[7.97]}$$

The field mmf must be large enough to overcome the armature mmf and to provide sufficient additional mmf to drive the air gap flux around the machine magnetic circuit. The final value of field mmf to operate at the rated condition point is not known until the complete winding and magnetic circuit is specified. However, the field winding design for turboalternators is usually adequate if the field mmf magnitude of [7.97] produced at the design maximum value of field current is 170 percent of the armature mmf magnitude given by [6.10] for the value

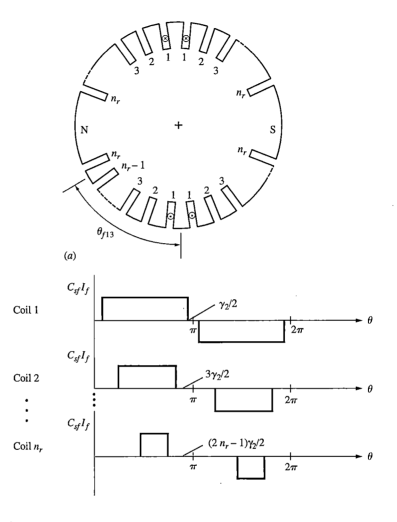

Figure 7.59
Partial symmetric field excitation. (a) 1-coils excited. (b) Coil set mmf's.

of rated stator winding current. Write the magnitude of [6.10] in terms of rms rated current and conductors per pole per phase. Then,

$$\mathscr{F}_{r1} = 1.7\mathscr{F}_{g1}$$

$$\frac{4}{\pi}C_{sf}I_{fR}\sum_{i=1}^{n_r}\cos[(2i-1)\gamma_2/2] = 1.75\frac{3\sqrt{2}}{\pi}k_wC_sN_s\frac{I_{aR}}{a}$$

or

$$I_{fR} \cong \frac{1.8k_wN_sC_sI_{aR}}{aC_{sf}\sum_{i=1}^{n_r}\cos[(2i-1)\gamma_2/2]} \qquad \textbf{[7.98]}$$

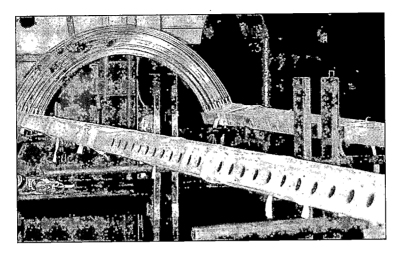

Figure 7.60
Indirectly cooled field coils for turboalternator. (*Courtesy of National Electric Coil, Columbus, Ohio. Photo by Jane Hutt.*)

Field winding conductors are cooled by one of two methods—indirect cooling and direct cooling. For the case of indirect cooling, heat generated within the field conductor must flow across the ground insulation to reach the cooling medium, which may be, for example, a channel beneath the rotor slot carrying an axial flow of air. The direct cooling methods are schemes that allow the cooling medium to directly contact the field conductor by passing through holes or slots that pierce the wide side as seen in Fig. 7.60. The usual cooling medium is hydrogen gas. The field winding should not exceed an average temperature of 125°C. This desired temperature can be maintained with the following current densities:

1. Indirect cooling $\Delta_f = 2000 \text{ A/in}^2$ **[7.99]**

2. Direct cooling $\Delta_f = 3500 \text{ A/in}^2$

The field winding of each pole is an arrangement of n_r concentric coils. Each coil has C_{sf} conductors having the general shape of a rectangle with rounded corners. A sufficiently accurate expression for the mean length turn of the ith field coil is given by

$$MLT_i = 2\left(W_f + \ell + i\frac{\pi D_r \gamma_2}{p} \right)$$

The total mean length of the field winding follows as

$$MLT_f = pn_r C_{sf} \sum_{i=1}^{n_r} MLT_i = 2n_r C_{sf}\left[p(W_f + \ell) + \pi D_r \gamma_2 \sum_{i=1}^{n_r} i \right] \quad \textbf{[7.100]}$$

The field winding is to be supplied by a voltage source with maximum value $V_{f\max}$. If 30 percent is to be held in reserve for field forcing, then at rated conditions

$$0.7V_{f\max} = I_{fR}R_f = I_{fR}\frac{\rho MLT_f}{s_f} = \Delta_f \rho MLT_f \qquad \textbf{[7.101]}$$

where s_f is the cross-section area of the field conductor, Δ_f is the allowable current density from [7.99], and $\rho = 1.014 \times 10^{-6}\,\Omega\cdot\text{in}$ is the resistivity of copper at 125°C. Substitution of [7.100] into [7.101] and solution for C_{sf} gives the following expression for the number of field conductors per slot:

$$C_{sf} = \frac{0.7V_{f\max}}{2n_r \rho \Delta_f \left[p(W_f + \ell) + \pi D_r \gamma_2 \displaystyle\sum_{i=1}^{n_r} i \right]} \qquad \textbf{[7.102]}$$

The calculated result for C_{sf} must be rounded to an integer.

Slot Sizing   Using a value of field conductor current density from [7.99], the cross-section area of the field conductor can be calculated.

$$s_f = \frac{I_{fR}}{\Delta_f} = \frac{1.8k_w N_s C_s I_{aR}}{a\Delta_f C_{sf} \displaystyle\sum_{i=1}^{n_r} \cos[(2i-1)\gamma_2/2]} \qquad \textbf{[7.103]}$$

A section view of the rotor slot that defines basic dimensions is shown by Fig. 7.61. The following values are suggested for good design:

Slot liner: 0.015 in for $V_{f\max} = 125\ V$
                  0.030 in for $V_{f\max} = 250\ V$
                  0.075 in for $V_{f\max} = 600\ V$

Turn insulation: 0.015 in

Creepage block: 0.063 in

The slot wedge is nonmagnetic stainless steel sized in thickness to withstand, without yielding, the bending and shear stress resulting from the slot conductor centrifugal force at 120 percent rated speed. Suggested initial values for rotor slot dimensions are given.

$$\frac{0.4D_r\gamma_2}{p} \le b_f \le \frac{0.5D_r\gamma_2}{p} \qquad \textbf{[7.104]}$$

$$2b_f \le d_{f1} \le 4b_f \qquad \textbf{[7.105]}$$

$$0.75 \le d_{f2} \le 1.00\ \text{in} \qquad \textbf{[7.106]}$$

$$d_{f3} = 0.25\ \text{in} \qquad \textbf{[7.107]}$$

$$0.50 \le t_w \le 1.00\ \text{in} \qquad \textbf{[7.108]}$$

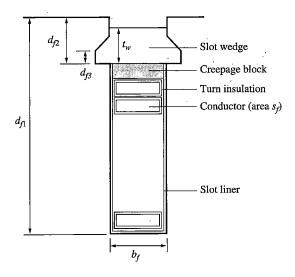

Figure 7.61
Rotor slot section view

Flux Density Checks The rotor of the turboalternator, like the dc machine, has teeth with the smallest cross-section area adjacent to the slot bottom. However, unlike the dc machine, the rotor flux density is sinusoidally distributed over the pole span rather than uniformly distributed. Consequently, the tooth root between slot n_r and slot $n_r - 1$ of Fig. 7.59a experiences the highest value of flux density. This area is singled out for examination. The tooth root will be saturated in this area so that significant flux will travel parallel to the tooth through the slot area.

The slot pitch measured at one-third tooth depth from the slot bottom is

$$\lambda_{f13} = \frac{(D_r - 4/3d_{f1})\gamma_2}{2} \qquad [7.109]$$

The electrical angular distance θ_{f13} from the position of zero flux density to the center of the tooth of concern is indicated on Fig. 7.59a.

$$\theta_{r13} = (n_r - 1)\gamma_2 \qquad [7.110]$$

The flux that enters the stator bore over the rotor slot pitch projected to the stator bore must be the apparent flux that crosses the rotor tooth area $\ell\lambda_{f13}$; thus, apparent maximum flux density of the critical tooth-slot area is

$$B_{f13} = \frac{\Phi_m}{\pi D\ell/p}\sin(\theta_{r13})\frac{\dfrac{D}{2}\gamma_2\dfrac{2}{p}}{\lambda_{f13}} = \frac{\Phi_m\gamma_2}{\pi\lambda_{f13}\ell}\sin(n_r - 1)\gamma_2 \qquad [7.111]$$

Good design practice is to assure that $B_{f13} \leq 65$ kilolines/in². However, on the larger, two-pole turboalternators, it may not be possible to attain this measure of

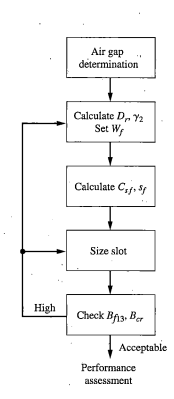

Figure 7.62
Logic flowchart for rotor design

goodness without significant increase in the lamination stack length. If a compromise situation arises, the lamination stack length should be increased to the point that no more than one-third of the mmf per pole is required to support rated flux through the ferromagnetic material paths.

The rotor core flux density will typically not be a problem area of design. However, it should be verified that

$$B_{cf} = \frac{\Phi_m}{(D_r - 2d_{f1})(\ell + 2D_r\gamma_2/p)} \leq 80 \text{ kilolines/in}^2 \qquad \textbf{[7.112]}$$

where the axial length of the rotor core flux path includes the path under the rotor coil overhang.

The logic flowchart of Fig. 7.62 organizes the design process for the turboalternator rotor. At this stage of the design, it is advisable to calculate an OCC for the alternator that gives insight into the headroom available insofar as magnetic saturation is concerned. If any portion of the magnetic circuit is borderline on saturation, the evaluation at 10 to 20 percent above rated flux will identify such areas for consideration. The later design example will introduce a MATLAB program for determination of the OCC.

7.9.6   EQUIVALENT CIRCUIT PARAMETERS

Analysis of the turboalternator under balanced conditions to assess the performance with regard to design specifications requires determination of the equivalent circuit parameters of Fig. 7.13.

Stator Coil Resistance R_s The stator coil resistance per phase for a temperature of 125°C for the copper conductors with a 5 percent increase for ac values is given by

$$R_s = \frac{1.014 \times 10^{-6} C_\phi MLT_a}{a s_a} \qquad \textbf{[7.113]}$$

where MLT_a is the mean length turn of a stator coil (in) and s_a is the cross-section area (in²) of a coil conductor. C_ϕ is the total series conductors per phase.

$$C_\phi = \frac{S_1 C_s}{3a} \qquad \textbf{[7.114]}$$

Reactance of Armature Reaction X_ϕ An unsaturated value for X_ϕ with dimensions in units of inches is

$$X_\phi = \frac{19.15 \times 10^{-8} f (k_w C_\phi)^2 D \ell_n SF}{p^2 \delta_e} \qquad \textbf{[7.115]}$$

where ℓ_n is the net stator lamination stack length determined by subtraction of the widths of all radial cooling ducts from the overall length ℓ. The effective air gap length is calculated by

$$\delta_e = \frac{\lambda_1 (5\delta + b_{s1})\delta}{\lambda_1 (5\delta + b_{s1}) - b_{s1}^2} \qquad \textbf{[7.116]}$$

Stator Leakage Reactance X_ℓ The stator leakage reactance is made up of two components—slot leakage reactance (X_{ss}) and coil end-turn reactance (X_{es}). The slot leakage and end-turn reactances are determined by the same procedure followed for the induction motor using [6.157] and [6.165]. With X_{ss} and X_{es} evaluated,

$$X_\ell = X_{ss} + X_{es} \qquad \textbf{[7.117]}$$

7.9.7   DESIGN REFINEMENT

With the equivalent circuit parameters known, performance predictions for the turboalternator can be made and compared against specifications. The MATLAB program ⟨TAperf.m⟩ can alleviate much of the tedious work of analysis and design refinement that might be necessary. This program reads a data file ⟨TAdata.m⟩ that contains general descriptive information on the turboalternator as well as dimensional data. ⟨TAperf.m⟩ extends the design procedure presented in that it calculates rotational losses and core losses to allow realistic efficiency evaluation. Further, it determines the OCC of the turboalternator, permitting assessment of magnetic saturation.

7.9.8 SAMPLE DESIGN

The turboalternator to be designed is specified as follows:

> 24 kV   500 MVA   60 Hz   0.8 PF   3600 rpm
>
> Directly cooled stator (H$_2$O)   Directly cooled rotor (H$_2$)
>
> Minimum full-load efficiency 98.5%

This large power rating machine should have as large a bore diameter as possible so that the length is as short as possible. The rotor diameter of Example 7.14 represents the limit of technology; thus, use $D_r = 44.21$ in. Based on experience, it is anticipated that an air gap of 5 in is near proper value. Thus,

$$D = D_r + 2\delta = 44.21 + 2(5) = 54.21 \cong 54 \text{ in}$$

Based on [7.73] and [7.74],

$$\ell = \frac{C_o Sp}{D^2} = \frac{375(500)(2)}{(54)^2} = 128.6 \cong 129 \text{ in}$$

The rated current is

$$I_{aR} = \frac{S_R}{\sqrt{3}\,V_{LR}} = \frac{500 \times 10^6}{\sqrt{3}(24 \times 10^3)} = 12{,}028 \text{ A}$$

In order to avoid excessive conductor size, use $a = 2$. Hence,

$$I_c = \frac{I_{aR}}{2} = \frac{12{,}028}{2} = 6014 \text{ A per conductor}$$

The number of stator slots is chosen as $S_1 = 60$ to place the slot pitch λ_1 near the midrange suggested by [7.76]. By [7.75],

$$\lambda_1 = \frac{\pi D}{S_1} = \frac{\pi(54)}{60} = 2.827 \text{ in}$$

The winding factor can now be determined by [A.15].

$$\frac{S_1}{p} = \frac{60}{2} = 30 \text{ slots per pole}$$

$$\gamma = \frac{180°}{30} = 6° \text{ per slot}$$

$$N_s = \frac{S_1}{mp} = \frac{60}{3(2)} = 10 \text{ slots per pole per phase}$$

A 5/6 coil pitch is chosen since the pitch is known to be a good choice for low fifth and seventh harmonics.

$$\rho = \frac{5}{6}(180°) = 150°$$

$$k_w = \frac{\sin(N_s\gamma/2)\sin(\rho/2)}{N_s\sin(\gamma/2)} = \frac{\sin(10 \times 6/2)\sin(150/2)}{10\sin(6/2)} = 0.9228$$

The conductors per slot can be estimated by [7.84].

$$C_s = \frac{0.1aV_{anR}}{k_wN_sD\ell fB_{av}} = \frac{0.1(2)(24{,}000/\sqrt{3})(39.37)^2}{0.9228(10)(54)(129)(60)(0.6)} = 1.86$$

Since C_s must be an even integer, use $C_s = 2$ conductors per slot.

The air gap length choice can now be checked using [7.91] to see if it lies within the appropriate range.

$$\delta = 4.5 \times 10^{-5}\frac{k_wC_sN_sI_{aR}}{a} = 4.5 \times 10^{-5}\frac{(0.9228)(2)(10)(12{,}028)}{2} = 4.99 \text{ in}$$

The actual air gap is

$$\delta = \frac{D - D_r}{2} = \frac{54 - 44.21}{2} = 4.895 \text{ in}$$

The actual value is sufficiently close to the suggested value so that no adjustment in the design is necessary.

The slot can now be sized based on the guidelines of [7.77] to [7.79]. Select $b_{s1} = 1.250$ in; then,

$$t_{s1} = \lambda_1 - b_{s1} = 2.827 - 1.250 = 1.577 \text{ in}$$

A coil design that keeps the conductor strands less than 0.25 in to reduce eddy current losses, satisfies the suggested ground insulation, and allows for 15 percent looseness for tolerance and conductor transposition is shown by Fig. 7.63a. Using this coil design with a 0.375-in slot wedge gives $d_{s1} = 6.00$ in.

$$\frac{d_{s1}}{b_{s1}} = \frac{6.00}{1.250} = 4.8$$

Since the slot depth-to-width ratio is less than the minimum value of 5 suggested by [7.79], the leakage reactance will probably be less than 10 percent of the synchronous reactance. If there were a specification on the leakage reactance, the height of the slot above the slot wedge could be increased with little adjustment of the design other than a slight increase in the stator lamination outside diameter.

The area of each of the 24 strands of the conductor is 0.03636 in², giving a full-load current density of

$$\Delta_1 = \frac{I_{aR}}{as_a} = \frac{12{,}028}{2(24 \times 0.03636)} = 6892 \text{ A/in}^2$$

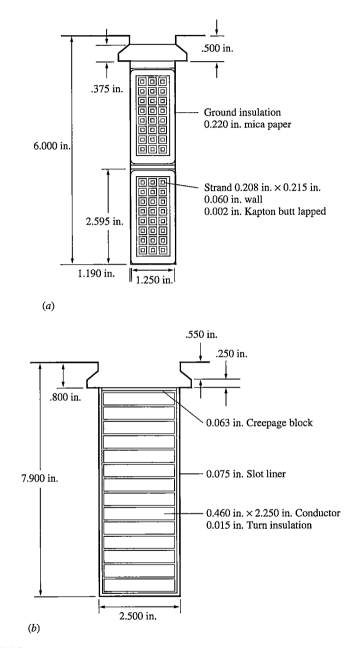

(a)

(b)

Figure 7.63
Sample design slot detail. (a) Stator slot. (b) Rotor slot.

This value satisfies [7.86] for direct water cooling of the stator coils.

The peak value of flux per pole is given by [7.83] when calculated in lines.

$$\Phi_m = \frac{1.1 \times 10^8 a V_{anR}}{2.2 k_w N_s C_s pf} = \frac{1.1 \times 10^8 (2)(24{,}000/\sqrt{3})}{2.22(0.9228)(10)(2)(2)(60)} = 6.2 \times 10^8 \text{ lines}$$

By use of [7.89] and [7.90],

$$t_{s13} = \frac{\pi\left(D + \frac{2}{3}d_{s1}\right)}{S_1} - b_{s1} = \frac{\pi\left(54 + \frac{2}{3}6.00\right)}{60} - 1.25 = 1.787 \text{ in}$$

$$\ell_n = 0.95\ell = 0.95(129) = 122.55$$

$$B_{t13} = \frac{\rho\left(\frac{2}{\pi}\right)\Phi_m}{S_1 t_{s13}\ell_n SF} = \frac{2\left(\frac{2}{\pi}\right)(6.2 \times 10^8)}{60(1.787)(122.55)(0.95)} = 63.2 \text{ kilolines/in}^2$$

This value of tooth flux density is within the acceptable range.

Size the stator core outside diameter according to [7.87].

$$D_o = 2.1D = 2.1(54) = 113.4 \text{ in}$$

Set $D_o = 114$ in to give a simple dimensional value.

In Example 7.1, $\langle$rdrotmmf.m$\rangle$ was executed with $S_2 = 20$ rotor slots and $W_f = 0.25\tau_r$. The resulting Fourier spectrum of Fig. 7.9b shows favorably small values of fifth and seventh harmonics. Hence, this value of rotor slots will be used in the sample design.

$$n_r = \frac{S_2}{2p} = \frac{20}{2(2)} = 5 \text{ slots per pole half}$$

$$W_f = 0.25\tau_r = 0.25\frac{\pi D_r}{p} = 0.25\frac{\pi(44.21)}{2} = 17.36 \text{ in}$$

The angular rotor slot pitch in radians is

$$\gamma_2 = \frac{\pi D_r - p W_f}{2D_r n_r} = \frac{\pi(44.21) - 2(17.36)}{2(44.21)(5)} = 0.2356 \text{ rad}$$

It is decided to use a 600-V exciter. The specification calls for direct hydrogen cooling of the rotor; thus, by [7.99] the field conductor current density can be up to 3500 A/in². From [7.102],

$$C_{sf} = \frac{0.7 V_{f\max}}{2n_r\rho\,\Delta_f\left[p(W_f + \ell) + \pi D_r\gamma_2 \displaystyle\sum_{i=1}^{n_r} i\right]}$$

$$C_{sf} = \frac{0.7(600)}{2(5)(1.014 \times 10^{-6})(3500)[2(17.36 + 129) + \pi(44.21)(0.2356)(15)]}$$

$$= 15.1$$

It is decided to use $C_{sf} = 14$ conductors per slot to give a slight improvement of slot fill factor for this large machine where rotor tooth flux density can easily be a problem.

The field conductor cross-section area is determined by [7.103], where γ_2 is in electrical degrees.

$$s_f = \frac{1.8 k_w N_s C_s I_{aR}}{a \Delta_f C_{sf} \sum\limits_{i=1}^{n_r} \cos[(2i - 1)\gamma_2/2]} = \frac{1.8(0.9228)(10)(2)(12{,}028)}{2(3500)(14)(3.93)} = 1.037 \text{ in}^2$$

A rotor slot layout using a 0.460 in $\times$ 2.250 in field conductor is shown by Fig. 7.63b. This conductor satisfies the calculated value of s_f to yield $\Delta_f = 3500$ A/in^2 at full-load field current. The resulting values for b_f, d_{f1}, d_{f2}, d_{f3}, and t_w are indicated on the figure also.

The rotor slot pitch at one-third slot depth is determined from [7.109].

$$\lambda_{f13} = \frac{\left(D_r - \dfrac{4}{3}d_{f1}\right)\gamma_2}{2} \cdot = \frac{\left(44.21 - \dfrac{4}{3} \times 7.250\right)0.2356}{2} = 4.069$$

By [7.111],

$$B_{f13} = \frac{\Phi_m \gamma_2}{\pi \lambda_{f13}\ell} \sin(n_r - 1)\gamma_2 = \frac{6.2 \times 10^8 (0.2356)}{\pi(4.069)(129)} \sin(4 \times 13.5°)$$

$$B_{f13} = 71.66 \text{ kilolines/in}^2$$

Since this value is greater than the suggested allowable value of 65 kilolines/in^2, the impact on total ferromagnetic path mmf requirement will have to be assessed when the magnetization curve of the machine is determined.

The data for this sample design were entered in ⟨TAdata.m⟩ and ⟨TAperf.m⟩ was executed. The resulting magnetization curve indicated that in excess of 40 percent of the mmf per pole was required by the ferromagnetic material for the rated flux excitation. Since the rotor diameter D_r cannot be increased to reduce saturation, the lamination stack length was increased to 132 in, a modest 2.3 percent increase, to reduce the saturation until approximately one-third of the total mmf per pole is required by the ferromagnetic path. During the saturation study, it was noted that the yoke flux density could be increased without significant negative impact of the overall ferromagnetic path mmf requirement. Consequently, D_o was reduced from 114 to 109 in. This savings in material more than offsets the material added to lengthen stack length ℓ.

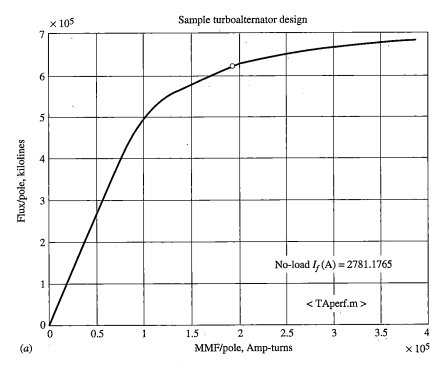

No-load I_f(A) = 2781.1765

< TAperf.m >

(a) MMF/pole, Amp-turns $\times 10^5$

Figure 7.64
Sample turboalternator design. (a) Flux vs. mmf. (b) OCC. (c) Efficiency. (d) Field current requirement.

The final performance calculations for the 500-MVA turboalternator after the adjustment of ℓ and D are displayed by Fig. 7.64a to d. The full-load efficiency has a value of 98.86 percent, which satisfies the design specification. In addition, ⟨TAperf.m⟩ displays to the screen calculated values for machine reactances, values of winding resistance, winding current densities at full load, and overall lengths of the stator and rotor measured to include the end-turn winding projections. The screen display information is as follows:

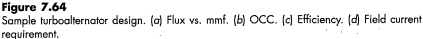

```
    X1pu      Xphipu    Xspu
    0.0602    0.9022    0.9625

    Rapu      Rf (ohm)
    0.0025    0.1080

    Deltas(Apsi)    Deltaf(Apsi)
    1.0e+003  *
```

```
    6.8805    3.1966
    ─────
LosW (in)   LofW (in)
199.5262    185.8363
```

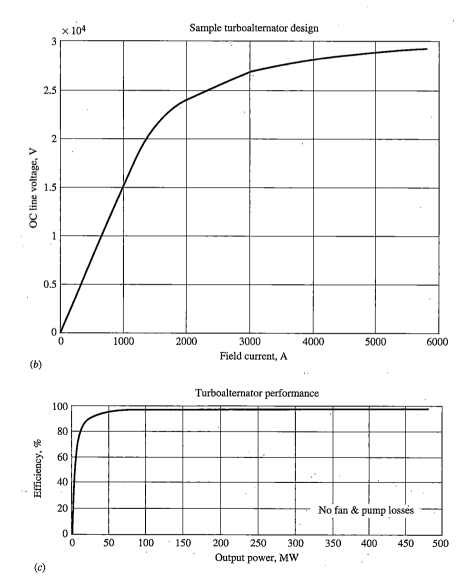

(b)

(c)

Figure 7.64 (Continued)

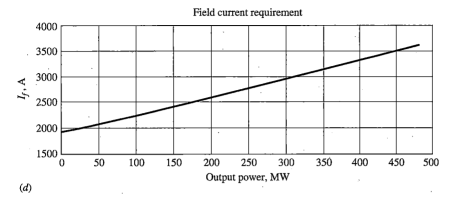

(d)

Figure 7.64 (Continued)

7.10 COMPUTER ANALYSIS CODE

```
%%%%%%%%%%%%%%%%%%%%%%%%%%%%%%%%%%%%%%%%%%%%%%%%%%%%%%%%%%%%%%%%%%%%%%%%%%%%
%
% rdrotmmf.m - Program plots the resulting normalized B-field
%              mmf wave over two pole pitches for a round
%              rotor synchronous machine and gives the Fourier
%              spectrum for assessment of harmonic content.
%              Effect of rotor slot width is included. Total
%              harmonic distortion (THD)is calculated and
%              displayed to screen.
%
%%%%%%%%%%%%%%%%%%%%%%%%%%%%%%%%%%%%%%%%%%%%%%%%%%%%%%%%%%%%%%%%%%%%%%%%%%%%
clear; clf;
p=2;        % No. poles
S2=20;      % No. rotor slots
bf=2;       % Rotor slot width (in.)
Dr=41;      % Rotor diameter (in.)
taur=pi*Dr/p; Wf=0.25*taur; nr=S2/2/p;
gam2=(pi*Dr-p*Wf)/2/Dr/nr; lamr=Dr/p*gam2;
nlamr=fix(gam2/2/pi*1024);nrmp=fix(bf/pi/Dr*1024);
if nrmp/2~=fix(nrmp/2); nrmp=nrmp-1; end; % Even integer
mag=0; k=1;
for i=1:256 % Build first quarter of B-field wave
   if (i==(2*k-1)/2*nlamr)&(k<nr+1); k=k+1; mag=mag+1/nr;
   else; end
   Ff(i)=mag;
end
k=1; % Add slot ramps
for i=1:256
   if (i==(2*k-1)/2*nlamr)&(k<nr+1); k=k+1;
```

```
      for j=i-nrmp/2+1:i+nrmp/2-1
        Ff(j)=Ff(i-nrmp/2)+1/nr/nrmp*(j-i+nrmp/2);
      end
   else; end
end
Ff=[Ff fliplr(Ff)]; Ff=[Ff -Ff]; % Form balance of waveform
theta=linspace(0,360,1024);
x=abs(fft(Ff));       % Determine Fourier spectrum
y=x(2)/512*sin(theta*pi/180);     % Form fundamental of B-field
subplot(2,1,1); plot(theta,Ff,theta,y,'--'); grid;
title('Normalized rotor B-field waveform');
xlabel('Rotor position, deg'); ylabel('Br, pu');
legend('Total B-field','Fundamental B-field',1);
subplot(2,1,2); plot([0:25],x(1:26)/512); grid;
title('Rotor B-field Fourier spectrum');
xlabel('Harmonic number'); ylabel('Magnitude');

% Total harmonic distortion
THD=0; for i=3:50; THD=THD+(x(i)/x(2))^2; end
THD=sqrt(THD)*100; disp(' ');
disp(['TOTAL HARMONIC DISTORTION(%) = ',num2str(THD)]);


%%%%%%%%%%%%%%%%%%%%%%%%%%%%%%%%%%%%%%%%%%%%%%%%%%%%%%%%%%%%%%%%%%%%%%%%%%
%
% sprBfld.m - Program plots the normalized B-field wave
%             form for a SP synchronous machine and gives
%             the Fourier spectrum for assessment of
%             harmonic content. Total harmonic distortion
%             (THD) is displayed to screen.
%             The pole arc design is nonconcentric with
%             rotor center.
%
%%%%%%%%%%%%%%%%%%%%%%%%%%%%%%%%%%%%%%%%%%%%%%%%%%%%%%%%%%%%%%%%%%%%%%%%%%
clear, clf;
Dr=15;                % Rotor dia (in.)
delc=0.125;           % Air gap length @ pole center (in.)
k=0.95;               % pu dia of nonconcentric pole arc
psi=0.6667;           % Pole arc/pole pitch
b=k*Dr/2;             % Pole arc radius
a=Dr/2-b-delc;        % Offset for pole arc center
thetap=psi*pi/2;      % Span of half pole arc

% Build permeance over pole arc
theta=linspace(0,pi/2,256); npts=length(theta);
for i=1:npts;
   if theta(i) >= thetap; nedge=i-1; break; end
end
for i=1:nedge
   x=a*cos(theta(i))+sqrt(a^2/2*(1-cos(2*theta(i)))+b^2);
```

```
    del=Dr/2-x;
    P(i)=1/del;
end
for i=nedge+1:npts % Parabolic decay beyond pole shoe edge
    c=P(nedge)/(theta(nedge)-pi/2)^2;
    P(i)=c*(theta(i)-pi/2)^2;
end
P=[ fliplr(P) P]; P=[ P -fliplr(P) ]; % Balance of waveform
Bf=abs(fft(P)); Bf=Bf/max(Bf); % Form normalized FFT
phi=linspace(0,360,1024);
y=Bf(2)*sin(phi*pi/180); % Form fundamental of Bf
subplot(2,1,1); plot(phi,P/max(P),phi,y,'--'); grid;
title('Normalized rotor B-field waveform');
xlabel('Rotor position, deg'); ylabel('Br, pu');
legend('Total B-field','Fundamental B-field',1);
subplot(2,1,2); plot([0:25],Bf(1:26)); grid;
title('Rotor B-Field Fourier spectrum');
xlabel('Harmonic number'); ylabel('Magnitude');

% Total harmonic distortion
THD=0; for i=3:50; THD=THD+(Bf(i)/Bf(2))^2; end
THD=sqrt(THD)*100; disp(' ');
disp(['TOTAL HARMONIC DISTORTION(%) = ',num2str(THD)]);
disp(' ');


%%%%%%%%%%%%%%%%%%%%%%%%%%%%%%%%%%%%%%%%%%%%%%%%%%%%%%%%%%%%%%%%%%%%%%%%%%%%%
%
% Xs.m - Using the OCC & SCC arrays stored, this program
%       plots the synchronous reactance vs. field current
%       of a round-rotor synchronous machine.
%
%%%%%%%%%%%%%%%%%%%%%%%%%%%%%%%%%%%%%%%%%%%%%%%%%%%%%%%%%%%%%%%%%%%%%%%%%%%%%
clear;
% OC Line Voltage Data
Voc=[0 17.39 19.3 21.12 22.78 24.39 25.84 27.13 28.45 29.66 ...
30.78 31.65 32.52 33.66 34.08 34.43 34.78 34.95]*1000;
% Field Current
Ifoc=[0 350 400 450 500 550 600 650 700 750 800 850 900 ...
1000 1050 1100 1150 1200];
% SC Data, Straight line assumed
Isc=8960; Ifsc=500; Isc=Isc/Ifsc*Ifoc;

Q=input(' Plot OC/SC curve?( ''Y'' or ''N'' ) = ');
if Q=='Y'
% Plot OC & SC with air gap line
for i=1:length(Voc) % Determine air gap line
    Vag=Voc(2)/Ifoc(2)*Ifoc(i);
    if Vag>max(Voc); break; end
end
```

```
plot(Ifoc,Voc,Ifoc,Isc,'-.',[0 Ifoc(i)],[0 Vag],'--' );grid;
title('OC & SC characteristics');
xlabel('Field current, A'); ylabel('Line Voc, V, Isc, A');
legend('Voc', 'Isc', 'Air gap line', 2);
else; end
Xs=[];
for i=2:length(Ifoc) % Calculate Xs
  Xs=[Xs;Voc(i)/sqrt(3)/Isc(i)];
end
Xs=[Xs(1);Xs]; figure(2); legend;
plot(0,0,Ifoc,Xs); grid; title('Synchronous reactance');
xlabel('Field current, A'); ylabel('Xs, Ohms');


%%%%%%%%%%%%%%%%%%%%%%%%%%%%%%%%%%%%%%%%%%%%%%%%%%%%%%%%%%%%%%%%%%%%%%%%%%%
%
% smphasor.m - Draws phasor diagram for a round rotor
%              synchronous machine. Per unit values used.
%
%%%%%%%%%%%%%%%%%%%%%%%%%%%%%%%%%%%%%%%%%%%%%%%%%%%%%%%%%%%%%%%%%%%%%%%%%%%
clear;
mode=input('Operating mode ( gen or mot ) ','s');
PF=input('Power Factor ( numerical value ) ');
if PF < 0 | PF > 1; disp('WARNING - Improper PF value'); pause;
else; end
if PF ~= 1.0
PFS=input('PF Sense ( leading or lagging ) ','s');
else; PFS='unityPF'; end
VL=1.0; Ia=1.0; Vt=VL;
Ra=0.05; Xs=0.9; Xl=0.15*Xs; thet=acos(PF);
if PFS == 'lagging'
  Ia=Ia*exp(-j*thet);
elseif PFS == 'leading'
  Ia=Ia*exp(j*thet);
elseif PFS == 'unityPF'
  Ia=Ia+j*0;
else; end
if mode == 'mot'
  p1=0; p2=Vt; p3=Vt-Ra*Ia;
  p4=Vt-(Ra+j*Xs)*Ia; p5=Vt-(Ra+j*Xl)*Ia;
elseif mode == 'gen'
  p1=0; p2=Vt; p3=Vt+Ia*Ra;
  p4=Vt+(Ra+j*Xs)*Ia; p5=Vt+(Ra+j*Xl)*Ia;
else; end
P=[p1 p2 p3 p4 p1 p5];
R=abs(P);Rm=1.1*max(R);
A=angle(P); figure(1);
polar(0,Rm); hold; % Size plot scale
polar(A,R); polar(A(2),0.98*R(2),'>'); % Vt arrow head
polar(A(4),R(4),'.');      % Dot at tip of Ef
```

```
polar(A(6),R(6),'.'); hold off;   % Dot at tip of Er
if mode == 'gen'; text(0.8*R(2),-0.1*R(2),'Vt');
else; text(0.8*R(2),0.1*R(2),'Vt'); end
text(1.05*real(p4),1.1*imag(p4),'Ef');
text(1.05*real(p5),1.1*imag(p5),'Er');
text(-0.9*Rm,0.4*Rm,'Synchronous');
text(-0.9*Rm,0.3*Rm,'machine (RR)');
text(-0.9*Rm,0.2*Rm,'phasor');
text(-0.9*Rm,0.1*Rm,'diagram');
if mode == 'mot'
   text(-0.9*Rm,-0.1*Rm,'Motor mode');
else
   text(-0.9*Rm,-0.1*Rm,'Generator mode');
end
text(-0.9*Rm,-0.2*Rm,['PF = ',num2str(PF),' ',PFS]);
text(-0.9*Rm,-0.3*Rm,date);



%%%%%%%%%%%%%%%%%%%%%%%%%%%%%%%%%%%%%%%%%%%%%%%%%%%%%%%%%%%%%%%%%%%%%%%%%
%
% RRaltTC.m - plots the terminal characteristics ( VL vs. IL )
%             of an isolated, RR, 3-phase alternator with constant
%             field excitation maintained at the value for rated
%             voltage & current with load PF as a parameter.
%             Constant Xs is assumed to avoid iterative solution.
%
%%%%%%%%%%%%%%%%%%%%%%%%%%%%%%%%%%%%%%%%%%%%%%%%%%%%%%%%%%%%%%%%%%%%%%%%%
clear; clf;
VLR=24000; S=100e6;      % Rated line voltage, apparent power
Xs=6.94; Rs=0.15;        % Syn. reactance, phase resistance
% Any 3 arbitrary PF angles(radians), +'ve for lagging PF
theta=[acos(0.8) acos(1) -acos(0.8) ];
IR=S/sqrt(3)/VLR;        % Rated current
VL=[]; IL=[];
for n=1:length(theta);  % PF loop
Ef=abs( VLR/sqrt(3)+IR*exp(-j*theta(n))*(Rs+j*Xs));
ZLR=VLR/sqrt(3)/IR*exp(j*theta(n));      % Load @ rated condition
ZL=linspace(0.5*ZLR,2*ZLR,1000); % Range of phase load
   for i=1:length(ZL)      % Voltage loop
   Ia(i)=Ef/abs(Rs+j*Xs+ZL(i));
   Va(i)=Ia(i)*abs(ZL(i));
   end
VL=[VL;sqrt(3)*Va]; IL=[IL;Ia];
end
n=length(ZL);
plot(IL(1,1:n),VL(1,1:n),IL(2,1:n),VL(2,1:n),'--',IL(3,1:n), ...
VL(3,1:n),'-.', IR,VLR,'o',0,0); grid
title('Alternator terminal characteristics');
xlabel('Line current, A'); ylabel('Line voltage, V');
legend('PF1','PF2','PF3',1)
```

```
%%%%%%%%%%%%%%%%%%%%%%%%%%%%%%%%%%%%%%%%%%%%%%%%%%%%%%%%%%%%%%%%%%%%%%%%%%%%%
%
% synphas.m - plots phasors for a synchronous generator and
%             equivalent grid prior to synchronization to
%             assess voltage across contactors to be closed.
%
%%%%%%%%%%%%%%%%%%%%%%%%%%%%%%%%%%%%%%%%%%%%%%%%%%%%%%%%%%%%%%%%%%%%%%%%%%%%%
clear;
fdiff=input(' Frequency difference small?( Y or N ) ','s');
mdiff=input(' Voltage magnitude different?( Y or N ) ','s');
pdiff=input(' Phase sequence different?( Y or N ) ','s');
if fdiff == 'Y'; n=100; else; n=15; end;     % Set angle increment
if mdiff == 'Y'; m=0.5; else; m=1; end       % Set magnitudes
if pdiff == 'Y'; k=-1; else; k=1; end        % Set phase sequence
% 123. - Grid, abc* - oncoming alternator
theta=linspace(0,2*pi,n);A1=0; A2=-2*pi/3; A3=2*pi/3;
for i=1:n; pause(0.0001)
Aa=theta(i); Ab=Aa-k*2*pi/3; Ac=Aa+k*2*pi/3;
polar(0,1.25); hold on

% Configurable phasors of the alternator - asterisks at the tip
polar([0 Aa],[0 m],'r');   polar(Aa,m,'*'); % Red
polar([0 Ab],[0 m],'y-.'); polar(Ab,m,'*'); % Yellow
polar([0 Ac],[0 m],'b--'); polar(Ac,m,'*'); % Blue

% Fixed phasors of the grid - dots at the tip
polar([0 A1],[0 1],'r');   polar(A1,1,'.'); % Red
polar([0 A2],[0 1],'y-.'); polar(A2,1,'.'); % Yellow
polar([0 A3],[0 1],'b--'); polar(A3,1,'.'); % Blue

% Contactor voltages indicated by dotted green lines
polar([A1 Aa],[1 m],'g:'); polar([A2 Ab],[1 m],'g:');
polar([A3 Ac],[1 m],'g:'); hold off
end
text(1.1,0.1,'V1'); text(-0.7,-1,'V2'); text(-0.7,1,'V3');
name=['Phasor diagrams for synchronizing alternator';
  '                                                ';
  '                                                ']; title(name);




%%%%%%%%%%%%%%%%%%%%%%%%%%%%%%%%%%%%%%%%%%%%%%%%%%%%%%%%%%%%%%%%%%%%%%%%%%%%%
%
% excreq.m - plots required field current for rated terminal
%            voltage vs. load current with fixed PF for a RR
%            3-phase alternator
%
%%%%%%%%%%%%%%%%%%%%%%%%%%%%%%%%%%%%%%%%%%%%%%%%%%%%%%%%%%%%%%%%%%%%%%%%%%%%%
clear; clf;
VLR=24000; S=400e6;      % Rated line voltage, apparent power
X1=0.24; Ra=0.08;        % Leakage reactance, phase resistance
```

```
Xsu=1.6;                       % Unsaturated synchronous reactance
% Any 3 arbitrary PF angles(radians), +'ve for lagging PF
theta=[acos(0.8) acos(1) -acos(0.8) ];
IR=S/sqrt(3)/VLR;    % Rated current
% OC Line Voltage Data
Voc=[0 17.39 19.3 21.12 22.78 24.39 25.84 27.13 28.45 29.66 ...
30.78 31.65 32.52 33.66 34.08 34.43 34.78 34.95]*1000;
% Field Current
Ifoc=[0 350 400 450 500 550 600 650 700 750 800 850 900 ...
1000 1050 1100 1150 1200];
Ia=linspace(0,1.25*IR,100);         % Current range
m=length(Ia);
for n=1:3                       % PF loop
   for i=1:m                    % Current loop
   Er=abs(VLR/sqrt(3)+Ia(i)*exp(-j*theta(n))*(Ra+j*Xl));
   Ifs=interp1(Voc/sqrt(3),Ifoc,Er); Ifg=Ifoc(2)/Voc(2)/sqrt(3)*Er;
   Xs=(Xsu-Xl)*Ifg/Ifs+Xl;
   Ef=abs(VLR/sqrt(3)+Ia(i)*exp(-j*theta(n))*(Ra+j*Xs));
   If(n,i)=interp1(Voc/sqrt(3),Ifoc,Er)*Ef/Er;
   end
end
plot(0,0,Ia,If(1,1:m),Ia,If(2,1:m),'--',Ia,If(3,1:m),'-.'); grid
title('Alternator excitation requirements');
xlabel('Line current, A'); ylabel('Field current, A');
legend('PF1','PF2','PF3',2)



%%%%%%%%%%%%%%%%%%%%%%%%%%%%%%%%%%%%%%%%%%%%%%%%%%%%%%%%%%%%%%%%%%%%%%%%%%%%
%
% capcurv.m - plots the reactive capability curves for a
%             RR, 3-phase alternator at two rating values
%             dependent on H2 pressure. Constant Xs assumed.
%             0.9 PF leading assumed limit for stability
%             margin.
%
%%%%%%%%%%%%%%%%%%%%%%%%%%%%%%%%%%%%%%%%%%%%%%%%%%%%%%%%%%%%%%%%%%%%%%%%%%%%
clear; clf;
VLR=24000; S=400e6;       % Rated line voltage
S=[400e6 350e6];          % Rated apparent power - [HiPr LoPr]
Xs=1.6; Ra=0.08;          % Syn. reactance, phase resistance
PFR=0.85;                 % Rated PF
for k=1:length(S)
   IR=S(k)/sqrt(3)/VLR;      % Rated current
   Ef=VLR/sqrt(3)+IR*exp(-j*acos(PFR))*(Ra+j*Xs); % Rated condition
   % Ia-limit region
   theta=linspace(-acos(0.9),acos(PFR),200);
   m=length(theta); P=[0]; Q=[0];
   for i=1:m
     P=[P sqrt(3)*VLR*IR*cos(theta(i))];
     Q=[Q sqrt(3)*VLR*IR*sin(theta(i))];
```

```
  end
  % If-limit region
  delta=linspace(angle(Ef),0,200); n=length(delta);
  for i=2:n
    Ia=(abs(Ef)*exp(j*delta(i))-VLR/sqrt(3))/(Ra+j*Xs);I(i)=abs(Ia);
    PC=sqrt(3)*VLR*conj(Ia);
    P=[P real(PC)]; Q=[Q imag(PC)];
  end
plot(P(m),Q(m),'o',P,Q); legend('Rated point'); hold on
end
title('Reactive capability curve'); grid
xlabel('Output average power, W');
ylabel('Output reactive power, VARs');



%%%%%%%%%%%%%%%%%%%%%%%%%%%%%%%%%%%%%%%%%%%%%%%%%%%%%%%%%%%%%%%%%%%%%%%%%%%%%%
%
% vcurves.m - calculates data & plots V-curves for synchronous
%             machine in motor mode.
%
%%%%%%%%%%%%%%%%%%%%%%%%%%%%%%%%%%%%%%%%%%%%%%%%%%%%%%%%%%%%%%%%%%%%%%%%%%%%%%
clear;
VL=2400; KVA=10000; PF=0.8; PR=KVA*PF;   % Rated values
% Assume Xsu=1pu, Xl=0.1pu, 1% Cu losses
Xsu=VL^2/(KVA*1000); Xl=0.1*Xsu; Ra=1e-5*VL^2/KVA;

% OC sat curve data using line voltage
Voc=[0 1700 1950 2200 2300 2400 2500 2600 2700 2800 2900 ...
  3000 3100 3300]';
Ifoc=[ 0 70 83 98 105 115 126 140 160 180 205 235 280 400]';

% Set plot for rated Ia & If
IaR=1000*KVA/VL/sqrt(3)*(PF+j*sin(acos(PF)));
Er=abs(VL/sqrt(3)-j*IaR*Xl-IaR*Ra);
Ifs=interp1(Voc/sqrt(3),Ifoc,Er); Ifg=Ifoc(2)/(Voc(2)/sqrt(3))*Er;
Xss=(Xsu-Xl)*Ifg/Ifs+Xl;    % Saturated Xs
Ef=abs(VL/sqrt(3)-j*IaR*Xss-IaR*Ra);
IfR=interp1(Voc/sqrt(3),Ifoc,Er)*Ef/Er;
axis([0, IfR, 0, abs(IaR)]);

% Set family of output power values
ncurv=9; Po=linspace(0, PR, ncurv)*1000;

% Set PF angle range
ang=linspace(-60, 60, 50)*pi/180; n=length(ang);
for i=1:ncurv
   for k=1:n
   Ia(k)=Po(i)/sqrt(3)/VL/cos(ang(k));
   I=Ia(k)*cos(ang(k))+j*Ia(k)*sin(ang(k));
   Er=abs(VL/sqrt(3)-j*I*Xl-I*Ra);
```

```
      Ifs=interp1(Voc/sqrt(3),Ifoc,Er); Ifg=Ifoc(2)/Voc(2)/sqrt(3)*Er;
      Xss=(Xsu-Xl)*Ifg/Ifs+Xl;
      Ef=abs(VL/sqrt(3)-j*I*Xss-I*Ra);
      If(k)=interp1(Voc/sqrt(3),Ifoc,Er)*Ef/Er;
   end
Iau=Po(i)/sqrt(3)/VL; % Unity PF point
Er=abs(VL/sqrt(3)-j*Iau*Xl-Iau*Ra);
Ifs=interp1(Voc/sqrt(3),Ifoc,Er); Ifg=Ifoc(2)/Voc(2)/sqrt(3)*Er;
Xss=(Xsu-Xl)*Ifg/Ifs+Xl;
Ef=abs(VL/sqrt(3)-j*Iau*Xss-Iau*Ra);
Ifu=interp1(Voc/sqrt(3),Ifoc,Er)*Ef/Er;
Ial=Po(i)/sqrt(3)/VL/0.8; % 0.8 lead PF point
I=Ial*(0.8+j*0.6);
Er=abs(VL/sqrt(3)-j*I*Xl-I*Ra);
Ifs=interp1(Voc/sqrt(3),Ifoc,Er); Ifg=Ifoc(2)/Voc(2)/sqrt(3)*Er;
Xss=(Xsu-Xl)*Ifg/Ifs+Xl;
Ef=abs(VL/sqrt(3)-j*I*Xss-I*Ra);
Ifl=interp1(Voc/sqrt(3),Ifoc,Er)*Ef/Er;
plot(If,Ia,Ifu,Iau,'o',Ifl,Ial,'*'); grid; hold on;
end
title('V-curves for synchronous motor');
xlabel('Field current, A'); ylabel('Stator current, A');
legend(['KW incr. = ',num2str((Po(2)-Po(1))/1000)], ...
   'Unity PF', '0.8 PF leading',4);
hold off;



%%%%%%%%%%%%%%%%%%%%%%%%%%%%%%%%%%%%%%%%%%%%%%%%%%%%%%%%%%%%%%%%%%%%%%%%%%%
%
% TAperf.m - Calculates magnetization & open circuit sat-
%            uration curves for turboalternator. Then cal-
%            culates load performance.
%            Calls TAdata.m for input data.
%
%%%%%%%%%%%%%%%%%%%%%%%%%%%%%%%%%%%%%%%%%%%%%%%%%%%%%%%%%%%%%%%%%%%%%%%%%%%%
clear; clf;
TAdata,
npts=50; Fp=zeros(1,npts+1);      % npts=points for sat curve
Ns=S1/p/m; Cphi=Ns*Cs*p/a; lam1=pi*D/S1; del=(D-Dr)/2;
tau=pi*D/p; taur=pi*Dr/p; Wf=0.25*taur; nr=S2/2/p;
gam2=(pi*Dr-p*Wf)/2/Dr/nr; lamr=Dr/p*gam2;
% Carter coefficients
qty=lam1*(5*del+bs); ks=qty/(qty-bs^2);
qty=lamr*(5*del+bf); kr=(qty/(qty-bf^2)-1)*(1-Wf/taur)+1;
% Stator winding factor
gam=p*pi/S1; kd=sin(Ns*gam/2)/Ns/sin(gam/2);
rho=pitch*gam; kp=cos(pi/2-rho/2); kw=kd*kp;

% Magnetization curve calculations
PHI=1.10e08*VL/sqrt(3)/2.22/kw/Cphi/f;
```

```
phi=linspace(0,1.1*PHI,npts); phim=[phi';PHI];
for i=1:npts+1;       % Calculation loop
Fg=0.09972*p*phim(i)*ks*kr*del/D/ln/SF;   % Air gap mmf
t13=pi*(D+2*d1s/3)/S1-bs;                  % Stator teeth mmf
Bts=p*phim(i)/S1/t13/ln/SF;Fts=HM690(Bts)*d1s;
t13=(Dr-4*d1f/3)*gam2/p-bf;                % Rotor teeth mmf
Ar13=pi*(Dr-4*d1f/3)/p-2*nr*bf;
Btr=phim(i)/Ar13/l; Ftr=H1010(Btr)*d1f;
ls=pi*(Do+D+2*d1s)/12/p;                   % Stator yoke mmf
Bys=0.5*phim(i)/(Do-D-2*d1s)/ln/SF;
Fys=(HM610(Bys)+HM610(sqrt(3)*Bys)+HM610(2*Bys))*ls;
lr=pi*(Dr-2*d1f)/12/p;                     % Rotor core mmf
Brc=0.5*phim(i)/(Dr-2*d1f-2*hc)/(l+2*(1+nr*lamr));
Frc=(H1010(Brc)+H1010(sqrt(3)*Brc)+H1010(2*Brc))*lr;
Fp(1,i)=Fg+Fts+Ftr+Fys+Frc;
end

% Plot magnetization & open circuit saturation curves
flux=phim'/1000; Ifnl=Fp(npts+1)/nr/Csf;
plot(Fp(1,1:npts),flux(1,1:npts),Fp(1,npts+1),flux(1,npts+1),'o');
title(TTL); grid;     % 'o' indicates rated point
xlabel('MMF/pole, Amp-turns'); ylabel('Flux/pole, kilolines');
text(0.7*max(Fp),0.15*max(flux),['No-load I_f(A) = ',num2str(Ifnl)]);
Voc(1:npts)=flux(1:npts)*sqrt(3)*2.22e-5*kw*Cphi*f;
Ifoc(1:npts)=Fp(1:npts)/nr/Csf;
figure(2); plot(Ifoc,Voc); title(TTL); grid;
xlabel('Field current, A'); ylabel('OC line voltage, V');

% Resistance & reactance calculations
Zb=VL^2*1e-6/MVA;
alfa=asin((bs+0.5)/lam1); lalf=rho*tau/(2*pi*cos(alfa));
Ra=1.065e-6*(1+2+2*(lalf+d1s))*Cphi/a/sa;
sum=0; for i=1:nr; sum=sum+2*i; end
Rf=1.014e-6*nr*Csf*(2*p*(Wf+1)+pi*Dr*gam2*sum)/sf;
pupitch=p*pitch/S1;        % Phase factor
if pupitch <= 0.5; Ks=pupitch;
elseif pupitch <= 2/3; Ks=-0.25+1.5*pupitch;
else; Ks=0.25+0.75*pupitch; end
if S2<S1; ks1=1; else; ks1=S1/S2; end
a1=(pi*D/S1-bs-bf)/2; if a1<0; a1=0; end
Xphi=19.15e-8*f*(kw*Cphi)^2*D*ln*SF/p^2/(ks*kr*del);
Xslt=2.0055e-7*m*f*Cphi^2*l*Ks/S1*(d1c/3/bs+d0s/bs);
Xe=0.3192e-7*f*(kw*Cphi)^2*(0.5+lalf)*Ks/p*(1+ ...
   log(3.54*D/S1/sqrt(bs*d1s)));
Xl=Xslt+Xe; Xsu=Xphi+Xl;
Xphipu=Xphi/Zb; Xlpu=Xl/Zb; Xspu=Xphipu+Xlpu;  % pu reactances

% Performance calculations
ang=-acos(PF);        % Lagging PF assumed
Pw=3e-11*pi*Dr*(l+2*(1+nr*lamr))*(2*pi*f*Dr/p)^3;        % F&W losses
```

```
Pf=2*4.62e-6*db^2*lb*(120*f/p)^2; PFW=Pf+Pw;
Wtt=(pi/4*((D+2*d1s)^2-D^2)-S1*bs*d1s)*ln*SF*0.283;    % Teeth weight
Wys=pi/4*(Do^2-(D+2*d1s)^2)*ln*SF*0.283;               % Yoke weight
npts=120; Po=linspace(0, 1.2*MVA*PF, npts)*1e06;       % Set output power
for i=1:npts
   Ia(i)=Po(i)/sqrt(3)/VL/cos(ang);
   I=Ia(i)*cos(ang)+j*Ia(i)*sin(ang);
   Er=abs(VL/sqrt(3)+j*I*Xl+I*Ra);
   Ifs=interp1(Voc/sqrt(3),Ifoc,Er);
   Ifu=Ifoc(2)/(Voc(2)/sqrt(3))*Er;
   Xss=(Xsu-Xl)*Ifu/Ifs+Xl; Xs(i)=Xss;
   Ef=VL/sqrt(3)+j*I*Xss+I*Ra;
   If(i)=abs(Ef)/Er*Ifs;
   PFWT(i)=PFW; PCuT(i)=3*Ia(i)^2*Ra; PfT(i)=If(i)^2*Rf;
   phil=Er/(2.22e-8*kw*f*Cphi);
   t13=pi*(D+2*d1s/3)/S1-bs; Bts=p*phil/S1/t13/ln/SF;
   Bys=0.866*phil/(Do-D-2*d1s)/ln/SF;
   PcT(i)=(PM690(Bts)*Wtt+PM610(Bys)*Wys)*2.5;
   Pin=3*real(Ef*conj(I))+PFW+PcT(i)+PfT(i);
   Pout=3*real(VL/sqrt(3)*conj(I));
   eff(i)=Pout/Pin*100;
end

figure(3); subplot(211); plot(Po/1e6,eff); grid
title('Turboalternator performance');
xlabel('Output power, MW'); ylabel('Efficiency, %');
text(0.7*max(Po/1e6),0.15*max(eff),'No fan & pump losses');
subplot(212); plot(Po/1e6,If); grid
title('Field current requirement');
xlabel('Output power, MW'); ylabel('I_f, A');

figure(4); subplot(211); plot(Po/1e6,Ia); grid
title('Turboalternator performance');
xlabel('Output power, MW'); ylabel('Current, A');
figure(5); subplot(111);
plot(Po/1e6,PFWT,Po/1e6,PcT,'--',Po/1e6,PCuT,'-.', ...
   Po/1e6,PfT,':'); grid
title('Core, copper, field, and F&W losses');
xlabel('Output power, MW'); ylabel('Losses, W');
figure(6); plot(0,0,Po/1e6,Xs,'-.'); grid
title('Synchronous reactance');
xlabel('Output power, MW'); ylabel('Xs, Ohms');

% Display equivalent circuit parameters
disp(' '); disp('  ')
disp(['   Xlpu   ' ' Xphipu  ' ' Xspu  ']);
disp([ Xlpu Xphipu Xspu ]);
disp(' '); disp('  ')
disp(['   Rapu ' '    Rf (ohm) ']);
disp([ Ra/Zb Rf ]);
```

```
% Display rated point current densities
Deltas=Ia(100)/a/sa; Deltaf=If(100)/sf;
disp(' '); disp(' ')
disp([' Deltas(Apsi) ' 'Deltaf(Apsi)     ']);
disp([ Deltas Deltaf]);

% Display length over windings for stator & rotor
LosW=l+1+2*(d1s+la1f*sin(alfa)); LofW=l+2*(1+nr*lamr);
disp(' '); disp(' ')
disp(['  LosW (in) ' 'LofW (in)      ']);
disp([ LosW LofW ]);


%%%%%%%%%%%%%%%%%%%%%%%%%%%%%%%%%%%%%%%%%%%%%%%%%%%%%%%%%%%%%%%%%%%%%%%%%%%%%%
%
% TAdata.m - Provides input data for TAperf.m
%
%              ( All dimensions in inches & areas in sq.in.)
%
%%%%%%%%%%%%%%%%%%%%%%%%%%%%%%%%%%%%%%%%%%%%%%%%%%%%%%%%%%%%%%%%%%%%%%%%%%%%%%
TTL = 'Sample turboalternator design';
%Winding & rating data
VL=24000; MVA=500; f=60; PF=0.8;
p=2; m=3; a=2;
S1=60; S2=20; Cs=2; pitch=25;
Csf=14;

% Conductor data
sa=0.8726; sf=1.035;
d1c=6.000-0.625;

% General dimensions
D=56.00; Dr=44.00; Do=109;
l=132; ln=0.95*l; SF=0.95;

% Stator slot dimensions
bs=1.250; d1s=6.000; d0s=0.625;

% Rotor slot dimensions
d1f=7.900; d2f=0.800; d3f=0.250;
bf=2.500; tw=0.800; hc=0;

% Bearing dimensions
db=21.25; lb=21.25;

% NOTE: 1) Blank lines & blank spaces not important.
%       2) Semicolons are important.
%       3) Anything to right of % is a comment—not important.
%       4) MATLAB is case sensitive, that is B & b are different.
%       5) Watchout for 1 (one) & l (lower case L).
```

SUMMARY

- Synchronous machines in practice range in size from tens of watts PC disk drive motors to hundreds of megawatts power plant turboalternators.

- Rotor structures of synchronous machines are one of two designs—round rotor or salient pole.

- A precise relationship exists between shaft speed and stator coil currents given by

$$\omega_s = \frac{2}{p}\omega \qquad \text{or} \qquad n_s = \frac{120f}{p}.$$

- The basic concept of torque produced by a synchronous machine can be explained by the rotor produced magnetic field attempting to align the air gap produced stator field, which is a traveling wave with speed ω_s.

- A simple, single-loop per phase equivalent circuit, consisting of a dependent voltage source in series with an inductance and resistance, describes the round-rotor synchronous machine in steady-state operation; however, the impact of magnetic saturation on both the voltage source and the inductance must be accounted for to yield accurate performance predictions.

- For accurate performance predictions, the salient-pole machine must be analyzed using two-reaction theory.

- The phasor diagram is a valuable tool for use in qualitative performance analysis of synchronous machines.

- The synchronous motor with increased field excitation is unique among electric machines in that it can exhibit a leading PF.

- The brushless dc motor is a member of a class of machines known as self-synchronous machines. Its stator coil excitation is based upon sensed rotor position to assure that synchronism always exists regardless of operating speed.

- The SR machine is a self-synchronous machine that develops a torque based on the principle of reluctance variation. It exhibits a power density competitive with other electric machine types when operated at conditions of high magnetic saturation.

- Torque produced by a turboalternator depends on the air gap volume. Maximum allowable rotor peripheral speed limits the permissible diameter of turboalternators, resulting in a design for high power levels that has a ratio of length to diameter greater than for other electrical machine types.

PROBLEMS

7.1 Plot X_s vs. I_f for the alternator of Fig. 7.16.

7.2 Using the value of X_{su} determined in Example 7.3, predict the value of X_s for the alternator by [7.27] and explain any discrepancy.

7.3   Rework Example 7.3 if a dc resistance measurement taken between two line terminals gives a value of $R_{sdc} = 0.03 \; \Omega$ at 20°C. Assume that rated average stator winding temperature is 150°C and that the value of $R_{sac}/R_{sdc} = 1.04$

7.4   Use ⟨cratio.m⟩ and ⟨csum.m⟩ to determine the excitation voltage $\overline{E}_f$ for the alternator of Example 7.3 Use X_s as determined in Prob. 7.2. Print a copy of the screen session.

7.5   Determine the developed torque for the alternator of Example 7.4 using [7.31] and compare the result with Example 7.5 solution.

7.6   A RR, three-phase, 5-kVA, 208-V, 60-Hz, 1800 rpm synchronous generator has $R_s \cong 0$, $X_\ell = 1 \; \Omega$, and $X_s = 8 \; \Omega$. The machine is connected to a 208-V infinite bus. It is supplying rated apparent power at 0.866 PF lagging. Determine the value of (a) excitation voltage E_f, (b) torque angle δ, and (c) total developed torque $3T_d$.

7.7   The synchronous generator of Prob. 7.6 is operating with the same value of supplied shaft power P_s. However, the field current has been increased so that the excitation voltage E_f is 20 percent greater in value. Find the values of (a) I_a, (b) PF, and (c) Q_T.

7.8   The synchronous generator of Prob. 7.6 has $R_s = 0.15 \, \Omega$ and all else is unchanged. Calculate (a) the value of excitation voltage E_f and (b) the value of total developed torque $3T_d$.

7.9   For the synchronous generator of Prob. 7.6, the torque supplied by the prime mover is increased so that the machine is operating at the point that synchronism would be lost. X_s can be assumed constant in value. The field current I_f has not been changed. At this new point of operation, determine the values of (a) average output power P_T, (b) line current I_a, and (c) output reactive power Q_T.

7.10 An 1800 rpm, 2400-V, 60-Hz, RR synchronous generator has $R_s = 0$ and $X_s = 4 \; \Omega$. The machine is operating at rated voltage and frequency. The field current is adjusted so that $E_f = 2300$ V and $d = 25°$. (a) Find the value of average power P_T output. (b) Determine the value of line current I_a. (c) Calculate the terminal PF.

7.11 A three-phase synchronous generator is operating at a lagging power factor condition on an infinite bus. Treat the machine as lossless. If the prime mover power supplied to the generator is increased, but the field current is adjusted so that the output reactive power is unchanged, qualitatively describe the changes in I_a, E_f, θ, and δ.

7.12 Figure 7.65 shows the a phases of two RR synchronous generators operating in parallel. The total three-phase load is described by 480 V (line voltage), 10 kW, 0.8 PF lagging. Machine 1 is supplying 50 percent of the average load power and 60 percent of the reactive load power. Determine Thevenin equivalent $\overline{E}_{feq}$ and X_{seq} to replace the two machines with an equivalent machine. Use $\overline{V}_{an} = 480/\sqrt{3} \angle 0°$.

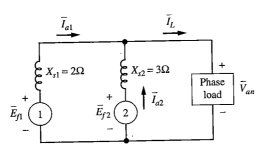

Figure 7.65

7.13 Rework Example 7.6 assuming that the no-load voltage can be predicted directly from the equivalent circuit of Fig. 7.13a as E_f. Compare the result with the solution of Example 7.6 by calculating the error in the regulation prediction if saturation is not taken into account.

7.14 Calculate the half-load efficiency for the alternator of Example 7.7. It is known that for the full-load condition, $P_c = P_{FW}$. Make the reasonable assumption that core losses vary as E_r^2.

7.15 The isolated synchronous generator of Fig. 7.24a is driven by a speed-governed prime mover. If the load impedance increases 10 percent in magnitude with no change in impedance angle and the field current of the generator is unchanged, use a phasor diagram to predict the qualitative changes in E_f, I_a, V_{an}, θ, and δ.

7.16 Show that the reactive power supplied by a three-phase alternator operating at zero PF lagging is given by $Q_T = 3(E_f V_{an} - V_{an}^2)/X_s$. If E_f is the value of excitation voltage for rated conditions, present convincing argument that this value of Q_T is the value of the vertical axis intercept for the reactive capability curve of Fig. 7.30.

7.17 An alternator with negligible R_s is operating with a lagging PF on the electric utility grid when the power supplied by the prime mover is increased by 10 percent while field current is unchanged. By means of a phasor diagram, qualitatively describe changes in I_a, θ, δ, and Q_T.

7.18 An alternator has just been synchronized on the electric utility grid with $\overline{V}_{an} \cong \overline{E}_f$. If the power supplied by the prime mover is increased without adjustment of the field current, use a phasor diagram to determine for the new increased power condition whether the alternator is operating at leading or lagging PF.

7.19 A 3-phase, 6-pole, 150-hp, 60-Hz, 480-V, RR synchronous motor has $R_s = 0.06\ \Omega$, $X_\ell = 0.15\ \Omega$, and $X_\phi = 1.5\ \Omega$. The motor is fed by rated voltage and frequency while driving a 150-hp coupled mechanical load. The motor field current has been adjusted so that the input power factor is 0.85 leading. It is known that for the point of operation, P_{FW} is 5 percent of the total input power. Core losses can be neglected. For the point of operation, determine the values of (a) excitation voltage E_f, (b) torque angle δ, and (c) efficiency η.

7.20 A 3-phase, 20-kVA, 220-V, 60-Hz, 6-pole synchronous motor can be considered a round-rotor machine with $X_s = 3\ \Omega$. The motor is initially operating at rated voltage, current, and speed with a unity PF. Rotational losses can be neglected. The coupled mechanical load is slowly increased until synchronism is lost. Determine the value of load torque and I_a at the point of "pull-out."

7.21 Given a synchronous motor with $R_s = 0.25\ \Omega$, $X_s = 3.8\ \Omega$, $\bar{E}_f = 457\angle -8°$, $\bar{V}_{an} = 480\angle 0°$, and total rotational losses are 700 W. It is also known that field voltage and current are 150 V and 3 A, respectively. Calculate P_T, Q_T, $3P_d$, P_s (shaft output power), and efficiency of the motor at this point of operation.

7.22 An industrial plant presently has a total three-phase load of 500 kVA at 0.65 PF lagging. Consequently, the electric utility charges them a power factor penalty on energy usage. A 200-hp synchronous motor with 93 percent efficiency is to be added to drive a new air compressor being installed. The motor is to be operated with its field overexcited to correct the plant PF to 0.85 lagging. Determine (a) the kVA of the plant with the motor added and (b) the kVA rating and PF of the synchronous motor.

7.23 Justify that the voltages $\omega N_{eff}\phi_d$ and $\omega N_{eff}\phi_q$ in the numerators of [7.43] are correct.

7.24 Show that the decomposition of $j\bar{I}_a X_\ell$ of Fig. 7.37a into $j\bar{I}_d X_\ell + j\bar{I}_q X_\ell$ is valid.

7.25 Draw the phasor diagram for a salient-pole synchronous generator with a leading PF.

7.26 Draw the phasor diagram for a salient-pole synchronous motor with a leading PF.

7.27 Show that equation [5] of Example 7.10 is valid and, thus, the salient-pole motor case phasor diagram of Fig. 7.39b is justified.

7.28 Determine the value of E_f for the generator of Example 7.10.

7.29 Given a 10-kVA, 4-pole, 480-V, 60-Hz, 3-phase salient-pole synchronous machine for which $R_s = 0.1\ \Omega$, $X_d = 8\ \Omega$, and $X_q = 4\ \Omega$. The machine is operating as a motor with rated voltage, current, and frequency. The field current has been adjusted so that the input PF = 0.707 leading. The rotational losses are 225 W. Determine (a) the mechanical power supplied to the coupled mechanical load and (b) the total developed torque.

7.30 For the synchronous motor of Prob. 7.29, assume $R_s = 0$ and all else is unchanged. (a) Use [7.53] to calculate the total developed torque. (b) Determine the value of power supplied to the coupled mechanical load.

7.31 A 20-MVA, 13.8-kV, 0.866 PF lagging, 360 rpm, waterwheel driven SP synchronous generator has $X_d = 9\ \Omega$, $X_q = 5\ \Omega$, and $R_s \cong 0$. If the generator is operating at nameplate rated conditions, determine the values of (a) excitation voltage (E_f) and (b) total power converted from mechanical to electrical form ($3P_d$).

7.32 The BDCM of Example 7.11 is an eight-pole, three-phase machine wound with full pitch, concentrated coils forming a two-layer winding. Each coil has 10 turns. The stator bore diameter is 4 in and the stator lamination stack length is 3 in. Determine the air gap flux density under the span of the PM poles.

7.33 The SRM of Fig. 7.48 has the inductance profile of Fig. 7.50. The rotor pole arc spans 25°. The stator bore diameter is 6 in, the pole axial length is 4 in, and the air gap is 0.020 in. Each coil has 30 turns. For the condition of analysis, the ferromagnetic material can be considered infinitely permeable. Determine the value of L_{max}.

7.34 Design a turboalternator to satisfy the following specifications:

    13.8 kV   25 MVA   60 Hz   0.8 PF   1800 rpm

    Indirectly air-cooled stator and rotor

    Minimum full-load efficiency—98 percent

REFERENCES

1. *IEEE Standard* 118, IEEE Test Code for Resistance Measurement.
2. *IEEE Standard* 115, IEEE Guide: Test Procedures for Synchronous Machines.
3. J. H. Kuhlmann, *Design of Electrical Apparatus,* 3d ed., John Wiley & Sons, New York, 1950.
4. A. K. Shawhney, *Electric Machine Design,* 4th ed., Dhanpat Rai & Sons, Delhi, 1977.
5. A. Still and C. S. Siskind, *Elements of Electrical Machine Design,* 3d ed., McGraw-Hill, New York, 1954.

APPENDIX

A

WINDING FACTORS

A.1 DISTRIBUTION FACTOR

If all conductors per pole (C'_s) belonging to one phase of an ac machine (*phase group*) were concentrated in a single slot, the slot (and coil) would become excessively large with negative impact on lamination design and end-turn size. A practical solution is to divide the coil into multiple coils with C_s conductors per side and distribute the coil sides over several slots as illustrated in Fig. A.1

If a dc current i_c were to flow through the concentrated coil of Fig. A.1a, the resulting total and fundamental components of the air gap mmf of Fig. A.2 would exist. By Fourier series analysis, the fundamental component of the concentrated coil mmf is given by

$$\mathcal{F}_{1c} = \frac{2}{\pi} C'_s i_c \sin\theta = \frac{6}{\pi} C_s i_c \sin\theta \qquad \textbf{[A.1]}$$

The same dc current i_c flowing through the center single coil of the distributed winding of Fig. A.1b will produce the total and fundamental mmf waveforms of Fig. A.2b. The fundamental component is described by

$$\mathcal{F}_{1d} = \frac{2}{\pi} C_s i_c \sin\theta \qquad \textbf{[A.2]}$$

The fundamental components of the mmf for the coils on either side of this center coil are given by [A.2] if θ is replaced by $\theta - \gamma$ or $\theta + \gamma$ as appropriate.

Since both $\mathcal{F}_{1c}$ and $\mathcal{F}_{1d}$ are sinusoidal quantities with the same period, space phasor representation is justified. Using a sine reference, the fundamental mmf for the concentrated coil and the total fundamental mmf for the distributed coil phase group are represented, respectively, in the space phasor domain as

$$\hat{\mathcal{F}}_{1c} = \frac{6}{\pi} C_s i_c \angle 0° \qquad \textbf{[A.3]}$$

$$\hat{\mathcal{F}}_{1dT} = \frac{2}{\pi} C_s i_c \angle -\gamma + \frac{2}{\pi} C_s i_c \angle 0° + \frac{2}{\pi} C_s i_c \angle \gamma \qquad \textbf{[A.4]}$$

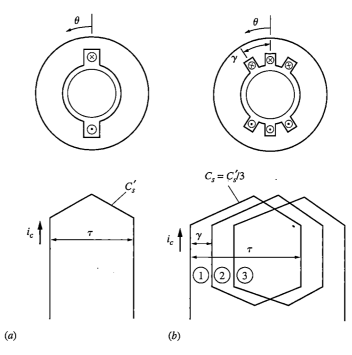

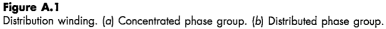

Figure A.1
Distribution winding. (a) Concentrated phase group. (b) Distributed phase group.

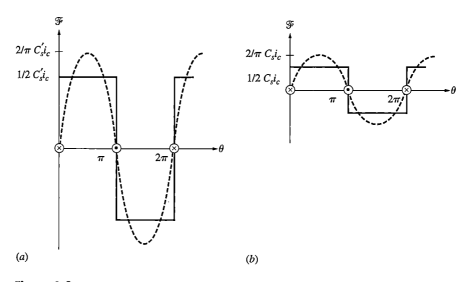

Figure A.2
Air gap mmf. (a) Concentrated coil. (b) Single distributed coil.

The coil *distribution factor* (k_d) is defined as a quantity that can be multiplied by the concentrated coil mmf to yield the mmf of the distributed phase group of coils. Mathematically,

$$k_d \triangleq \frac{\text{distributed mmf}}{\text{concentrated mmf}} \qquad \text{[A.5]}$$

For the particular case of Fig. A.1,

$$k_d = \frac{\mathscr{F}_{1dT}}{\mathscr{F}_{1c}} = \frac{\frac{2}{\pi} C_s i_c \left|(1 \angle -\gamma) + (1 \angle 0°) + (1 \angle \gamma)\right|}{\frac{6}{\pi} C_s i_c} = \frac{1 + 2\cos\gamma}{3}$$

Although the space phasors could be added as above to obtain a value of k_d provided all phase groups are identical, the process can be simplified by derivation of an algebraic formula for the distribution factor. The number of coils in a phase is

$$N_s = \frac{\text{slots/pole}}{\text{phase}} \qquad \text{[A.6]}$$

There are then N_s coil fundamental component mmf's to sum by space phasor addition to determine $\hat{\mathscr{F}}_{1dT}$ as illustrated by Fig. A.3. The right-hand diagram shows the individual coil space phasor mmf's added by graphical vector addition to form $\hat{\mathscr{F}}_{1dT}$. The perpendicular bisectors of each coil mmf phasor intersect at the point 0. From the expanded view of angles around point B, it can be seen that

$$\left(\frac{\pi}{2} - \frac{\theta}{2}\right) + \left(\frac{\pi}{2} - \frac{\theta}{2}\right) + \gamma = \pi \qquad \text{or} \qquad \theta = \gamma$$

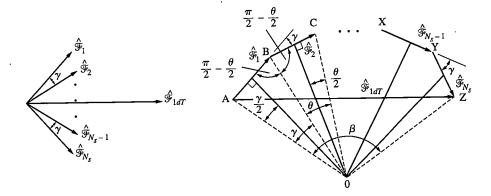

Figure A.3
General case distributed coil mmf's

Hence, $\beta = N_s\gamma$ Now,

$$OA = \frac{\frac{1}{2}AB}{\sin(\gamma/2)} = \frac{\mathscr{F}_1}{2\sin(\gamma/2)} \qquad \text{[A.7]}$$

Also,

$$OA = \frac{\frac{1}{2}AZ}{\sin(\beta/2)} = \frac{\mathscr{F}_{1dT}}{2\sin(N_s\gamma/2)} \qquad \text{[A.8]}$$

Equate [A.7] and [A.8], then rearrange to find

$$\mathscr{F}_{1dT} = \frac{\mathscr{F}_1 \sin(N_s\gamma/2)}{\sin(\gamma/2)} \qquad \text{[A.9]}$$

However, if the N_s coils were concentrated in a single slot,

$$\mathscr{F}_{1c} = N_s\mathscr{F}_1 \qquad \text{[A.10]}$$

Substitution of [A.9] and [A.10] into [A.5] results in the desired general algebraic formula for the distribution factor k_d.

$$k_d = \frac{\sin(N_s\gamma/2)}{N_s\sin(\gamma/2)} \qquad \text{[A.11]}$$

The distribution factor for the nth harmonic of the mmf waveform is given by [A.11] if γ is replaced by $n\gamma$.

A.2 PITCH FACTOR

When placing coils in the slots of an ac machine, the sides of a coil are frequently spaced less than a pole pitch (τ) apart as an effective method of reducing the low-order harmonic components of the mmf waveform. Figure A.4 depicts the total and fundamental components of the mmf for a single coil with a fractional pitch of $\rho < \pi$. By Fourier series, the fundamental component of the coil mmf is

$$\mathscr{F}_{1p} = \frac{2}{\pi} C_s i_c \cos\left(\frac{\pi - \rho}{2}\right)\sin\theta \qquad \text{[A.12]}$$

The coil *pitch factor* (k_p) is defined as a quantity that if multiplied by the full-pitch coil mmf yields the fractional-pitch coil mmf. Expressed mathematically,

$$k_p = \frac{\text{fractional-pitch mmf}}{\text{full-pitch mmf}} = \frac{\mathscr{F}_{1d}}{\mathscr{F}_{1p}} \qquad \text{[A.13]}$$

Substitution of [A.2] and [A.12] into [A.13] gives the coil pitch factor as

$$k_p = \cos\left(\frac{\pi - \rho}{2}\right) = \sin(\rho/2) \qquad \text{[A.14]}$$

The nth harmonic pitch factor is found by replacing ρ by $n\rho$ in [A.14]

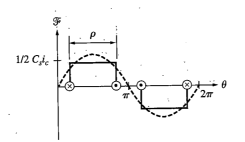

Figure A.4
MMF waveform for a reduced-pitch coil

A.3 WINDING FACTOR

The total effect on mmf from both distribution of the phase group coils and fractional pitch of each individual coil is known as the coil *winding factor* (k_w) determined as the product

$$k_w = k_d k_p = \frac{\sin(N_s\gamma/2)\sin(\rho/2)}{N_s\sin(\gamma/2)} \qquad \textbf{[A.15]}$$

Consequently, if an ac machine has N_ϕ series turns per phase, the actual winding would be replaced by a concentrated winding with N_{eff} turns where

$$N_{eff} = k_w N_\phi \qquad \textbf{[A.16]}$$

insofar as the fundamental component analysis is concerned.

A three-phase, four-pole motor with 36 stator slots (S_1) has a coil pitch of 1 and 9 (8 slots). Determine (*a*) the fundamental component winding factor and (*b*) the seventh-harmonic winding factor. | **Example A.1**

(*a*) $$N_s = \frac{36/4}{3} = 3 \text{ slots/pole/phase}$$

$$\gamma = \frac{p\pi}{S_1} = \frac{4\pi}{36} = 0.349 \text{ rad/slot} = 20°/\text{slot}$$

$$\rho = \frac{8}{9}\pi = 2.793 \text{ rad} = 160° \text{ coil pitch}$$

By [A.15],

$$k_w = \frac{\sin(3 \times 20°/2)\sin(160°/2)}{\sin(20°/2)} = 0.946$$

The fundamental component of the winding mmf has been reduced by 5.4 percent from the value it would have if the coils were full-pitch and all coil sides in a phase group were concentrated in a single slot.

(*b*) For the seventh harmonic,

$$k_{w7} = \frac{\sin(N_s n\gamma/2)\sin(n\rho/2)}{N_s \sin(n\gamma/2)} = \frac{\sin(3 \times 7 \times 20°/2)\sin(7 \times 160°/2)}{\sin(7 \times 20°/2)} = 0.0607$$

The seventh harmonic has been significantly reduced to a value of 6.07 percent of the value that it would have for the case of a concentrated, full-pitch winding.

B

CONVERSION FACTORS

Entity	SI Unit	Equivalent
Magnetic flux	1 weber (Wb)	10^8 maxwells or lines 10^5 kilolines
Flux density	1 tesla (T)	1 Wb/m^2 10^4 gauss 64.52 kiloline/in^2
Magnetomotive force (mmf)	1 ampere-turn (A-t)	1.257 gilberts
Magnetic field intensity	1 ampere/meter (A/m)	2.54×10^{-2} ampere/in 1.257×10^{-2} oersted
Resistivity	1 ohm·meter ($\Omega \cdot$ m)	$10^2 \ \Omega \cdot$ cm 39.37 $\Omega \cdot$ in
Back emf constant	1 volt · second/radian	104.7 V/k rpm
Velocity	1 radian/second (rad/s)	$30/\pi$ rpm = 9.549 rpm
Length	1 meter (m)	39.37 in
Area	1 meter^2 (m^2)	1550 in^2 10^4 cm^2 10.764 ft^2 1.974×10^9 circular mil
Volume	1 meter^3 (m^3)	$6.1024 = 10^4$ in^3 10^6 cm^3 35.315 ft^3
Mass	1 kilogram (kg)	2.205 lb 35.27 oz
Mass density	1 kilogram/meter^3 (kg/m^3)	6.243×10^{-2} lb/ft^3 3.613×10^{-5} lb/in^3 5.780×10^{-4} oz/in^3
Force	1 newton (N)	1 m · kg/s^2 0.2248 pound (lb$_f$) 3.597 ounces (oz$_f$) 10^5 dynes
Torque	1 newton·meter (N·m)	141.61 oz · in 8.85 lb · in 0.738 lb · ft 1.02×10^4 g · cm

Entity	SI Unit	Equivalent
Energy	1 joule (J)	1 W · s 9.478×10^{-4} Btu
Power	1 watt (W)	1 J/s 1/746 hp
Current density	1 ampere/meter^2 (A/m^2)	10^{-4} A/cm^2 6.452×10^{-4} A/in^2 5.066×10^{-10} A/circular mil

APPENDIX

C

MAGNETIC WIRE TABLES

C.1 ROUND WIRE WITH FILM INSULATION

AWG	Bare Diameter (in)	Total Diameter (in)	Weight (lb/1000 ft)	Resistance (20°C) (ohms/1000 ft)
4	0.2053	0.2098	127.20	0.2485
5	0.1828	0.1872	100.84	0.3134
6	0.1628	0.1671	80.00	0.3952
7	0.1450	0.1491	63.51	0.4981
8	0.1292	0.1332	50.39	0.6281
9	0.1150	0.1189	39.98	0.7925
10	0.1024	0.1061	31.74	0.9988
11	0.0912	0.0948	25.16	1.26
12	0.0812	0.0847	20.03	1.59
13	0.0724	0.0757	15.89	2.00
14	0.0644	0.0682	12.60	2.52
15	0.0574	0.0609	10.04	3.18
16	0.0511	0.0545	7.95	4.02
17	0.0455	0.0488	6.33	5.05
18	0.0405	0.0437	5.03	6.39
19	0.0361	0.0391	3.99	8.05
20	0.0322	0.0351	3.18	10.1
21	0.0286	0.0314	2.53	12.8
22	0.0254	0.0281	2.00	16.2
23	0.0227	0.0253	1.60	20.3
24	0.0202	0.0227	1.26	25.7
25	0.0180	0.0203	1.00	32.4
26	0.0160	0.0182	0.794	41.0
27	0.0143	0.0164	0.634	51.4
28	0.0127	0.0147	0.502	65.3
29	0.0114	0.0133	0.405	81.2
30	0.0101	0.0119	0.318	104.0
31	0.0090	0.0108	0.253	131.0
32	0.0081	0.0098	0.205	162.0
33	0.0072	0.0088	0.162	206.0

C.2   SQUARE WIRE WITH FILM INSULATION

AWG	Bare Width (in)	Total Width (in)	Area (in²)	Weight (lb/1000 ft)	Resistance (20°C) (ohms/1000 ft)
0	0.3279	0.3329	0.10420	403.1	0.07815
1	0.2922	0.2972	0.08232	318.4	0.09895
2	0.2602	0.2652	0.06498	251.4	0.1253
3	0.2317	0.2367	0.05125	198.4	0.1589
4	0.2063	0.2113	0.04037	156.4	0.2018
5	0.1837	0.1887	0.03171	122.9	0.2568
6	0.1636	0.1686	0.02536	98.39	0.3211
7	0.1457	0.1507	0.01994	77.42	0.4085
8	0.1298	0.1348	0.01563	60.74	0.5212
9	0.1155	0.1205	0.01251	48.66	0.6514
10	0.1029	0.1079	0.00980	38.19	0.8310
11	0.0917	0.0967	0.00788	30.74	1.033
12	0.0818	0.0868	0.00619	24.16	1.316
13	0.0730	0.0780	0.00496	19.41	1.641
14	0.0651	0.0701	0.00389	15.24	2.095

index

A

Actuator, 1
Admittance, 8
Air gap, equivalent or effective, 60, 285, 390, 491
Air gap field
    induction motor, 367–368
    synchronous machine, 426–428
Air gap fringing, 60
Air gap line, 437
Air gap sizing
    dc machine, 285, 294
    induction motor, 384
    synchronous machine, 483
All-day efficiency, 169
Alternator, 421
Ampere's circuit law, 53
Angular coil pitch, 323
Angular slot pitch, 323
Apparent inductance, 77
Apparent power, 28, 29
Armature, 424
Armature design, dc motor, 278
Armature reaction, 248–250
Armature reaction mmf, 435
Armature resistance control, dc motor, 273
Armature voltage, 234–239
Armature voltage control, dc motor, 274
Autotransformer, 175–177
Autotransformer starting, 357
Auxiliary starting winding, 373–375
Average power, 28, 29
Average volts per bar, 28, 29

B

Balanced conditions, 3–phase set, 16–21
Base current, 153
Base impedance, 153
Base power, 153
Base voltage, 153
Biot-Savart law, 50
Blocked-rotor test, 336

$B\lambda v$ rule, 52
Brushless dc motor (BDCM), 465–469
Bus, 15

C

Capacitor-start motor, 374
Coefficient of coupling, 78
Coenergy, 98
Coercive force, 83
Coil distribution factor, 325, 517–520
Coil pitch, 322
Coil pitch factor, 325, 520
Coil resistance, 196, 391, 491
Coil winding factor, 521–522
Commutation, 247
Commutation interval, 239
Commutation pole, 247
Commutator flashover, 247
Compensating winding, 250
Complex power, 28, 30
Concentric-layered winding, transformer, 133
Conductor, 238
Consequent pole, 360
Constant-torque pole changing connection, 361
Converter, 471
Converter, phase-controlled, 274
Core loss resistance, 193
Core-type transformer, 132, 189
Counter emf, 238
Cumulative compound dc generator, 256
Cumulative compound dc motor, 266
Curie temperature, 57
Current
    base, 153
    delta connection, 19–21
    inrush, 172
    line, 15
    phase, 15
    pickup, 105
Current transformer, 187–188
Curves
    demagnetization, 85

field compounding, 451
reactive capability, 452
V-, 456
Cylindrical rotor, 426

D

d-axis synchronous reactance, 460
DC generator, 251–258
DC machines, 229–311
    magnetic fields in, 243–249
    physical construction of, 229–232
    voltage and torque principles of, 232–241
DC motor, 258–270
    brushless, 465–469
    design of, 275–297
    speed control of, 271–275
    starting control of, 270–271
DC motor nameplate, 275
Delta connection, 15, 19–21, 179
Delta-wye transformation, 21
Demagnetization curves, 85
Developed power, 239, 340
Developed torque, 239, 331, 341–351, 441–444
Differential compound dc generator, 258
Differential compound dc motor, 269
Direct (d) axis, 458
Distribution factor, coil, 325, 517–520
Diverter resistor, 272
Double-layer winding, 235, 319
Double-revolving field theory, 367
Double-window magnetic circuit, 69

E

Eddy current, 147
Efficiency, 166–171
    all-day, 169
Electric steel sheet (ESS), 55
    grain-oriented, 56
    nonoriented, 55
Electromagnetic field, 49

Electromagnetic torque, 239–241, 331
Electromechanical energy conversion,
    90–103
    coenergy in, 98–101
    doubly-excited systems in, 102
    force and energy in, 92–98
    model formulation of, 90
EMF constant, 238
Energy density, 80
Equivalent circuit
    approximate (transformer), 163
    dc machine, 242
    induction motor, 333–335, 389
    single-phase motor, 368–370
    synchronous machine,
        432–440, 491
    transformer, 141–143, 150–155,
        192–197
Euler equation, 5
Excitation field, 422
Excitation voltage, 431

F

Faraday's law, 52
Ferromagnetic core properties,
    transformer, 143–152
Ferromagnetic materials, 54
Ferromagnetism, 54–57
Field compounding curves, 451
Field control, dc motor, 272
Field pole design, dc motor, 285
Field resistance line, 256
Field weakening, 264
Field winding design, dc motor, 288
Field winding
    dc machine, 241
    synchronous machine, 426–428
Fixed tap adjustment, transformer, 184
Flat compounding, 258
Flux
    leakage, 61
    resultant, 436
Flux density, residual, 83
Flux linkage, 76
Form-wound coils, 318
Fringing (air gap), 60

G

Generator, 1
    cumulative compound, 256
    dc, 251–258
    differential compound, 258

separately excited, 253
    series excited, 256
    shunt, 254
    synchronous, 440–455
Grain-oriented electric steel
    sheet, 56

H

Hysteresis, 57
Hysteresis loop, 83, 144

I

Ideal transformer, 133–141, 176
Impedance
    base, 153
    in phasor domain, 7
    principles of, 7–10
Incremental inductance, 77
Inductance, 76–80
    apparent, 77
    incremental, 77
    magnetizing, 143
    mutual, 79
    self-, 78
    synchronous, 432
Induction, 317
Induction motor, 317–417
    classifications of, 317–319
    design of, 375–403
    equivalent circuit for, 333–335
    performance calculations of,
        340–355
    physical construction of, 317–319
    reduced voltage starting of,
        355–358
    rotor action in, 328–333
    single-phase, 366–375
    speed control of, 358–366
    stator windings in, 319–328
Inrush current, 172
Instantaneous power, 28
Instrument transformer, 187–188
Interpole, 247
Inverter
    six-step, 363
    sinusoidal pulse-width modulation
        (SPWM), 363

K

Knee, in magnetic saturation, 54

L

Lap winding, 238
Leakage flux, 61
Leakage reactance, 194, 393, 433, 491
Lenz's law, 52
Line current, 15
Line voltage, 15
Linear region, magnetic materials, 54
Load line, 87
Load line method, 66
Lorentz force equation, 49
Lossless-core transformer, 141–143

M

Magnetic circuit
    analysis methods of, 58–60
    double window, 69
    doubly-excited, 102–103
    ferromagnetism in, 54–57
    in dc machine, 243, 287
    laws and rules in, 49–54
    parallel, 69–75
    permanent magnet in, 83–90
    series, 61–69
    sinusoidal excitation in, 81–83
    solenoid design in, 103–112
Magnetic field, dc machine, 232,
    243–249
Magnetic saturation, 54–57
Magnetizing inductance, 143
Magnetizing reactance, 193, 390, 433
Magnetomotive force (mmf), 53
    armature reaction, 435
    resultant, 436
Magnetostriction, 57
Mean length turn, 152
Modulation index, 363
Motor, 1
    brushless dc, 465–469
    capacitor-start, 374
    cumulative compound, 266
    dc, 258–270
    differential compound, 269
    induction, 317–417
    self-synchronous, 464–476
    series excited, 263
    shunt excited, 260, 272
    single-phase induction, 366–375
    split-phase, 374
    switched reluctance, 470–476
    synchronous, 455–457
Motor control, dc, 270–275
Mutual inductance, 79

N

Neutral (conductor), 15, 23–28
No-load magnetic field, 232
No-load test (induction motor), 337
Nonlinear analysis (energy
    conversion), 101

O

Open-circuit characteristic (OCC), 244
Open-circuit test, 158
Open-delta circuit, 47

P

Parallel magnetic circuit, 69–75
Parallel path, 238
Per phase friction and windage
    loss, 340
Per phase output shaft power, 340
Per phase output shaft torque, 341
Permanent magnets (PM), 83–90
Permeability, 51
Phase, 15
Phase current, 15
Phase group, 517
Phase sequence, 16
Phase voltage, 15
Phase-controlled converter, 274
Phasor
    definition of, 6
    in sinusoidal steady state, 5–10
    space, 458
Phasor diagram, 6
Pitch factor, coil, 325, 520
Point of operation, 87
Pole changing, 360
Pole pitch, 322
Pole, consequent, 360
Potential transformer, 187
Power
    apparent, 28, 29
    average, 28, 29
    base, 153
    complex, 28, 30
    developed, 239, 340
    instantaneous, 28
    reactive, 28, 29
Power angle, 444
Power factor, 30
Power factor correction, 35–38
Power flow
    in dc generator, 252

    in dc motor, 259–260
    in induction motor, 340–341
    in sinusoidal steady-state circuit,
        28–38
    in transformer, 178
    measurement of, 34–35
    power factor correction of, 35–38
Power measurement, 34–35
Power triangle, 29
Primary winding, transformer, 133
Primary, induction motor, 317

Q

q-axis synchronous reactance, 460
quadrature (q) axis, 459

R

Random-wound coils, 318
Reactance
    d-axis synchronous, 460
    leakage, 194, 393, 433, 491
    magnetizing, 193, 390, 433
    q-axis synchronous, 460
    synchronous, 433, 460
Reactive capability curves, 452
Reactive factor, 30
Reactive power, 28, 29
Recoil characteristic, 84
Reduced voltage starting, 355–358
Relative permeability, 51
Reluctance, 58
Reluctance torque, 464
Residential distribution transformer,
    173–175
Residual flux density, 83
Resistance
    ac value of, 437
    coil, 196, 391, 491
    core loss, 193
    rotor, 392
Resistor, diverter, 272
Resultant flux, 436
Resultant mmf, 436
Resultant voltage, 436
Right-hand rule, 52
Rotor
    cylindrical, 426
    induction motor, 318, 328–333, 384
    round, 426
    salient-pole, 427
    squirrel-cage, 328

    synchronous motor, 426–428
    turboalternator, 483–490
    wound, 328
Rotor lamination stack, 318
Rotor resistance, 392
Rotor winding, 426–428
Round rotor (RR), 426

S

Salient-pole (SP) rotor, 427
Salient-pole synchronous machine,
    457–464
Saturation factor, 439
Saturation in magnetic circuits,
    54–57
Saturation region, 54
Secondary (induction motor), 318
Secondary winding, transformer, 133
Self-inductance, 78
Self-synchronous motor, 464–476
Separately-excited dc generator, 253
Series field winding, 241
Series magnetic circuit, 61–69
Series-excited dc generator, 256
Series-excited dc motor, 263
Shell-type transformer, 133, 189
Short-circuit test, 157
Shunt dc generator, 254
Shunt field winding, 241
Shunt-excited dc motor, 260, 272
Single-phase induction motor,
    366–375
Single-phase sinusoidal steady state,
    10–14
Sinusoidal excitation, 81–83
Sinusoidal steady state, 81
Sinusoidal steady-state circuits, 5–44
    impedance in, 5–10
    multiple frequency, 38–39
    phasors in, 5–10
    power flow in, 28–38
    single-phase, 10–14, 28–32
    three-phase, 14–28, 32–34
Slip, 329
Slip test, 462
Slot skew, 322
Solenoid, 103
    design of, 103–112
    pickup current in, 105
Solid-state starter, 358
Space phasors, 458
Specific electric loading, 477
Specific magnetic loading, 478

Speed control
   dc motor, 271
   induction motor, 358–366
Split-phase motor, 374
Squirrel-cage rotor, 328
Squirrel-cage rotor winding, 318
Stabilized shunt motor, 272
Starter, solid-state, 358
Starting, autotransformer, 357
Starting, reduced voltage, 355
Starting control, dc motor, 270
Stator
   induction motor, 317, 379
   turboalternator, 479–483
Stator lamination stack, 317
Stator winding, 319–328, 424–425
   double-layer, 319
Step-down autotransformer, 177
Step-up autotransformer, 176–177
Stray load losses, 446
Switched reluctance (SRM) motor,
   470–476
Synchronization, 448
Synchronous inductance, 432, 433
Synchronous machines, 421–511
   design of, 476–499
   equivalent circuit of, 432–440, 491
   generator, 440–455
   motor, 455–457
   physical construction of, 422–428
   salient-pole, 457–464
   self-, 464–476
Synchronous reactance, 460
Synchronous speed, 327

T

Tap change under load, transformer,
   184, 186
Terminal characteristics, 445
Three-phase network
   analysis of, 23–28
   balanced conditions in, 16–21
   connection arrangements in, 15–16
   connection transformations in,
      21–23
Three-phase set, 16

Three-phase transformer, 179–184
Torque
   developed, 239, 331, 341–351,
      441–444
   electromagnetic, 239, 331
   reluctance, 464
Torque angle, 444
Torque constant, 238, 280
Transformer, 131–222
   auto-, 175–178
   core-type, 132, 189
   current, 187–188
   ideal, 133–141, 176
   instrument, 187–188
   lossless-core, 141–143
   potential, 187
   residential distribution, 173–175
   shell-type, 133, 189
   step-down, 177
   step-up, 176–177
   three-phase, 179–184
Transformer coil polarity, 155
Transformer concentric-layered
   winding, 133
Transformer design, 188–202
Transformer efficiency, 166
Transformer inrush current, 172
Transformer nameplate, 155
Transformer open-circuit test, 158
Transformer parameters, 156–161
Transformer performance assessment,
   161–173, 182
Transformer primary winding, 133
Transformer secondary winding, 133
Transformer short-circuit test, 157
Transformer tap change, 184, 186
Transformer vertical (disk)-layered
   winding, 133
Transformer voltage regulation, 171
Transformer winding taps, 184–186
Traveling wave, air gap
   induction motor, 326–327,
      331–333, 367–368
   synchronous machine, 429, 457
Triplen harmonics, 324
Two-reaction theory (synchronous
   mach.), 460
Two-wattmeter method, 35

V
V-curve, 456
Vertical (disk) layered-winding,
   transformer, 133
Voltage
   armature, 234–239
   base, 153
   excitation, 431
   line, 15
   phase, 15
   resultant, 436
   wye connection, 17–19
Voltage controller, 360
Voltage regulation, 171–172, 445–446

W
Ward-Leonard system, 274
Wave winding, 238
Winding
   auxiliary starting, 373–375
   compensating, 250
   concentric-layered, 133
   double-layer, 235, 319
   lap, 238
   primary, 133
   rotor, 426–428
   secondary, 133
   series field, 241
   shunt field, 241
   squirrel-cage, 318
   stator, 319–328, 424–425
   vertical (disk)-layered, 133
   wave, 238
   wound-rotor, 318
Winding distribution, 322
Winding factor, 521–522
Winding mmf, 319
Winding taps, transformer, 184–186
Wound field, 423
Wound rotor, 328
Wound-rotor winding, 318
Wye connection, 15, 17–19, 179
Wye-delta transformation, 22